Partitional Clustering Algorithms

M. Emre Celebi
Editor

Partitional Clustering Algorithms

Springer

Editor
M. Emre Celebi
Department of Computer Science
Louisiana State University Shreveport
Shreveport, LA, USA

ISBN 978-3-319-09258-4 ISBN 978-3-319-09259-1 (eBook)
DOI 10.1007/978-3-319-09259-1
Springer Cham Heidelberg New York Dordrecht London

Library of Congress Control Number: 2014946973

Printed on acid-free paper

Springer is part of Springer Science+Business Media (www.springer.com)

Preface

Clustering, the unsupervised classification of patterns into groups, is one of the most important tasks in exploratory data analysis. Primary goals of clustering include gaining insight into, classifying, and compressing data. Clustering has a long and rich history that spans a variety of scientific disciplines including anthropology, biology, medicine, psychology, statistics, mathematics, engineering, and computer science. As a result, numerous clustering algorithms have been proposed since the early 1950s. Among these algorithms, partitional (nonhierarchical) ones have found many applications, especially in engineering and computer science.

The goal of this volume is to summarize the state of the art in partitional clustering. The intended audience includes researchers and practitioners, who are increasingly using partitional clustering algorithms to analyze their data.

The volume opens with a chapter on model-based clustering entitled "Recent Developments in Model-Based Clustering with Applications." In this chapter, Melnykov et al. review the latest developments in this field including semi-supervised clustering, nonparametric mixture modeling, initialization strategies, merging mixture components, and handling spurious solutions. In "Accelerating Lloyd's Algorithm for k-Means Clustering," Hamerly and Drake present a survey of triangle inequality-based acceleration techniques for the celebrated k-means clustering algorithm. Based on extensive experiments, the authors conclude that a suitable application of the triangle inequality can provide dramatic speedups of up to 40x over a naive implementation of the standard Lloyd's algorithm. In another k-means related chapter entitled "Linear, Deterministic, and Order-Invariant Initialization Methods for the k-Means Clustering Algorithm," Celebi and Kingravi investigate the empirical performance of six linear, deterministic, and order-invariant k-means initialization methods on a large and diverse collection of data sets from the UCI Machine Learning Repository. Their results demonstrate that two relatively unknown hierarchical initialization methods outperform the remaining four methods with respect to two objective effectiveness criteria. They also show that one of the most recent initialization methods performs surprisingly poorly. In "Nonsmooth

Optimization Based Algorithms in Cluster Analysis," Bagirov and Mohebi approach the problem of partitional clustering using nonsmooth and nonconvex optimization. Based on this formulation, they design an efficient incremental algorithm similar to k-means that can handle ℓ_1 and ℓ_∞ norms besides the commonly used ℓ_2 norm.

In "Fuzzy Clustering Algorithms and Validity Indices for Distributed Data," Vendramin et al. present a framework to generalize several fuzzy clustering algorithms to handle distributed data without resorting to approximations. This framework also allows the exact calculation of a variety of relative validity indices to evaluate the quality of fuzzy partitions. The authors also describe a procedure based on this framework for the estimation of the number of clusters in parallel and distributed settings.

In "Density Based Clustering: Alternatives to DBSCAN," Braune et al. propose two algorithms similar to the celebrated DBSCAN algorithm. Unlike DBSCAN, both algorithms require only one input parameter. One of these algorithms gives similar results to DBSCAN while being able to assign multiple cluster labels to a data point, whereas the other one is significantly faster than DBSCAN.

In "Nonnegative Matrix Factorization for Interactive Topic Modeling and Document Clustering," Kuang et al. propose a novel formulation of the nonnegative matrix factorization (NMF) problem based on the block coordinate descent algorithm. The authors present a clustering algorithm based on this formulation and prove its convergence. In addition to extending this framework to sparse and weakly supervised clustering, the authors describe a method to determine the number of clusters based on random sampling and consensus clustering. Experiments on various real-world document data sets demonstrate the advantage of the proposed NMF clustering algorithm in terms of clustering quality, convergence behavior, sparseness, and consistency.

In "Overview of Overlapping Partitional Clustering Methods," Ben N'Cir et al. review the fundamental concepts of partitional overlapping clustering and present a survey of widely known partitional overlapping clustering algorithms as well as techniques to evaluate the quality of non-disjoint partitions. Furthermore, the authors investigate the ability of various clustering algorithms to generate overlapping partitions from multi-labeled real-world data sets.

In "On Semi-Supervised Clustering," Bongini et al. present a survey of semi-supervised clustering (SSC). The authors first give a conceptual overview of the field and then discuss some of the most important algorithms for SSC including COP-COBWEB, COP-kMeans, HMRF k-means, seeded k-means, constrained k-means, and active fuzzy constrained clustering. The authors conclude with a discussion of future directions for this relatively new field.

In "Consensus of Clusterings Based on High-Order Dissimilarities," Aidos and Fred describe a novel dynamic clustering algorithm based on a recently proposed dissimilarity measure called "dissimilarity increments." Starting from an initial partition, this algorithm incorporates a merging strategy based on either a likelihood-ratio test or a test based on the minimum description length principle. The authors address the initialization dependence of their algorithm using a consensus function-based combination strategy. Finally, the best partition is selected using a criterion based on dissimilarity increments. Experimental results demonstrate the effectiveness of the proposed algorithm on a variety of synthetic as well as real-world data sets.

In "Hubness-Based Clustering of High-Dimensional Data," Tomašev et al. investigate the hubness phenomenon observed in k-nearest neighbor graphs of high-dimensional data. The authors demonstrate that hubness complicates the cluster discovery process by reducing the separability of clusters. The authors then review some recent clustering algorithms that exploit the hubness phenomenon and then introduce a kernel-based clustering algorithm that does not restrict the shape of the clusters to hyperspheres.

A chapter entitled "Clustering for Monitoring Distributed Data Streams" by Barouti et al. completes the volume. The authors consider an application of clustering to monitoring data streams in a distributed system. Unlike conventional clustering algorithms that group similar data points into clusters, monitoring requires that clusters with dissimilar points cancel each other as much as possible. The authors devise a novel clustering algorithm to tackle this problem and demonstrate that it yields a reduction in communication load.

We hope that this volume focused on partitional clustering will demonstrate the significant progress that has occurred in this field in recent years. We also hope that the developments reported in this volume will motivate further research in this exciting field.

Shreveport, LA, USA M. Emre Celebi

Contents

Chapter 1
Recent Developments in Model-Based Clustering with Applications

Volodymyr Melnykov, Semhar Michael, and Igor Melnykov

Abstract Model-based clustering is a popular technique relying on the notion of finite mixture models that proved to be efficient in modeling heterogeneity in data. The underlying idea is to model each data group by a particular mixture component. This relationship between mixed distributions and clusters forms an attractive interpretation of groups: each cluster is assumed to be a sample from the corresponding distribution. In practice, however, there are many issues that have to be accounted for by the researcher. The area of model-based clustering is very dynamic and rapidly developing, with many questions yet to be answered. In this paper, we review and discuss the latest developments in model-based clustering including semi-supervised clustering, non-parametric mixture modeling, choice of initialization strategies, merging mixture components for clustering, handling spurious solutions, and assessing variability of obtained partitions. We also demonstrate the utility of model-based clustering by considering several challenging applications to real-life problems.

Keywords Model-based clustering • Finite mixture model • Initialization strategy • Merging mixture components • Semi-supervised clustering • Variable selection • Spurious solutions

1.1 Introduction

Finite mixture models have been widely used in cluster analysis over the last several decades, but their first applications date back to more than 120 years ago [131, 139]. The main appeal of these models lies in their ability to associate each potential cluster with its own component of the mixture. As motivational examples that illustrate the connection between finite mixture models and cluster analysis, we

V. Melnykov (✉) • S. Michael
The University of Alabama, Tuscaloosa, AL 35487, USA
e-mail: vmelnykov@ua.edu; skmichael@ua.edu

I. Melnykov
Colorado State University–Pueblo, Pueblo, CO 81001, USA
e-mail: igor.melnykov@colostate-pueblo.edu

M.E. Celebi (ed.), *Partitional Clustering Algorithms*,
DOI 10.1007/978-3-319-09259-1_1

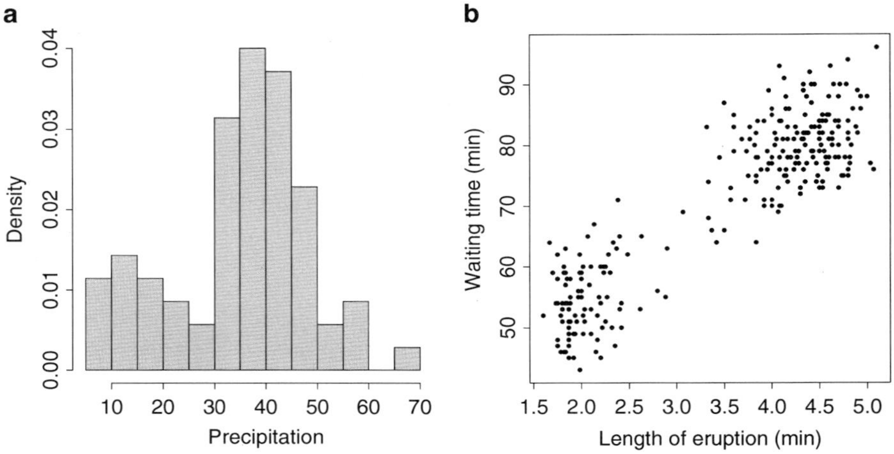

Fig. 1.1 (**a**) Histogram of the rainfall precipitation data and (**b**) scatterplot of the eruption and waiting times for the Old Faithful geyser

consider two popular datasets, both available from the R library *datasets*. The first one is a univariate dataset called *precip* [112]. It provides information on the amount of rainfall precipitation given in inches for 70 cities in the United States and Puerto Rico. The histogram illustrating the data is provided in Fig. 1.1a. The other example is based on the bivariate dataset called *faithful* [4]. The dataset contains records on the duration of eruptions and waiting time between eruptions in minutes for the Old Faithful geyser from the Yellowstone National Park in the United States. The scatterplot of the 272 observations available is provided in Fig. 1.1b. From the both displays, we can conclude that the data are not homogeneous and should be clustered into several groups. In plot (a), we can observe evidence of the presence of at least two modes. In plot (b), we can observe two groups of relatively dense data points. Unfortunately, no standard probability density function allows straightforward modeling of these datasets. In the meantime, it is considerably easier to find distributions that can be helpful in describing patterns in each data group individually. This logic immediately suggests considering a mixture of several distributions with each of them being responsible for modeling a particular data group. These simple examples explain our motivation to study finite mixture models as a powerful tool to tackle many cluster analysis problems.

To simplify further reading of the paper, the most important or frequently used notation is summarized in Table 1.1. In some rare places, the meaning of the notation can differ from the declared one in order to keep the notation used in our paper consistent with that established in the existing literature. In such cases, we specifically declare the particular interpretation in the corresponding parts of the paper.

Table 1.1 List of notation used in the paper

Expression	Description
Y_i and y_i	Random variable and observed value associated with the ith data point
$\mathbf{Y}_i$ and $\mathbf{y}_i$	Random and observed vectors associated with the ith data point
n	Sample size
p	Data dimensionality
K	Number of mixture components
τ_k	kth mixing proportion
$\boldsymbol{\tau}$	Vector of mixing proportions
$\boldsymbol{\vartheta}_k$	Parameter vector of the kth mixture component
$f_k(\cdot\|\boldsymbol{\vartheta}_k)$	kth mixture component distribution
$\boldsymbol{\Psi}$	Overall parameter vector
$\hat{\boldsymbol{\Psi}}$	Maximum likelihood estimate of the parameter vector
$f(\boldsymbol{\Psi})$	Finite mixture distribution
$\ell(\boldsymbol{\Psi})$	Log likelihood function
$\ell_c(\boldsymbol{\Psi})$	Complete-data log likelihood function
$Q(\boldsymbol{\Psi})$	Conditional expectation of the complete-data log likelihood function
$\ell_{c,P}(\boldsymbol{\Psi})$	Complete-data penalized log likelihood function
$h_\lambda(\boldsymbol{\Psi})$	Penalty function
Z_i and z_i	Random variable and observed value of the ith data point's membership label
$\hat{z}_i$	Estimated classification of the ith data point
(b)	Iteration number
π_{ik}	Posterior probability of the membership for the ith data point
$\boldsymbol{\pi}$	Matrix of posterior probabilities
$\nabla\pi_{ik}$	Gradient vector of the posterior probability function
$\boldsymbol{\mu}_k$	kth mean vector
$\boldsymbol{\Sigma}_k$	kth covariance matrix
$\boldsymbol{\beta}$	Vector of coefficients associated with explanatory variables
$\mathbf{X}_i$	ith matrix of predictor variables
λ_j	jth eigenvalue of covariance matrix
$\boldsymbol{\Lambda}$	Diagonal matrix of eigenvalues
$\boldsymbol{\Gamma}$	Matrix of eigenvectors
$\mathbf{Y}^{(1)}$	Set of variables included in the model
$\mathbf{Y}^{(2)}$	Set of variables under consideration for inclusion in the model
$\mathbf{Y}^{(3)}$	Set of variables to be studied
$\mathbf{I}^{-1}(\hat{\boldsymbol{\Psi}})$	Inverse of the observed information matrix
$\mathbf{P}$	Transition probability matrix

Consider a random sample $\mathbf{Y}_1, \ldots, \mathbf{Y}_n$ from a p-variate probability distribution

$$f(\mathbf{y}|\boldsymbol{\tau}) = \sum_{k=1}^{K} \tau_k f_k(\mathbf{y}), \tag{1.1}$$

where mixing proportions $\boldsymbol{\tau} = (\tau_1, \ldots, \tau_K)'$ comply to the restrictions $0 \leq \tau_k < 1$ and $\sum_k \tau_k = 1$ and each τ_k represents the probability that an observation originates from the kth component of the mixture f_k. Here, K is the number of components in the mixture. In the majority of applications, the functional form of f_k is assumed to be known. Such an assumption allows rewriting the model (1.1) in a parametric form

$$f(\mathbf{y}|\boldsymbol{\Psi}) = \sum_{k=1}^{K} \tau_k f_k(\mathbf{y}|\boldsymbol{\vartheta}_k), \tag{1.2}$$

with $\boldsymbol{\Psi} = (\tau_1, \ldots, \tau_{K-1}, \boldsymbol{\vartheta}'_1, \ldots, \boldsymbol{\vartheta}'_K)'$ denoting the parameter vector of this model and $\boldsymbol{\vartheta}_k$ representing the parameter vector of a specific mixture component f_k. By far, the most common kind of mixture investigated in the literature is the multivariate Gaussian mixture model [23, 51, 102, 106, 110, 121, 169]. Other models considered by researchers include those with Poisson [86, 160], skew-normal [10, 88, 89], Kent [141], t-distribution [85, 140, 166], and other types of components. Although the functional form of f_k is usually assumed known, the parameter vector $\boldsymbol{\Psi}$ is typically unknown and its estimation can pose significant challenges. Some early works on this topic considered the method of moments estimation [139], but the maximum likelihood approach is being used almost exclusively nowadays due to its computational advantages and appealing asymptotic properties of the estimators [29, 107, 155].

The corresponding log likelihood function for the mixture (1.2) is given by

$$\ell(\boldsymbol{\Psi}) = \sum_{i=1}^{n} \log \sum_{k=1}^{K} \tau_k f_k(\mathbf{y}_i|\boldsymbol{\vartheta}_k) \tag{1.3}$$

and usually is not easily handled by direct optimization due to its complicated form. The most common way of optimizing (1.3) consists in implementing the expectation-maximization (EM) algorithm [38, 108]. This approach relies on the consideration of the complete-data log likelihood function ℓ_c that utilizes unknown labels $z_1, \ldots, z_n$ identifying the origins of all observations. Thus,

$$\ell_c(\boldsymbol{\Psi}) = \sum_{i=1}^{n} \sum_{k=1}^{K} I(z_i = k) \left\{\log \tau_k + \log f_k(\mathbf{y}_i|\boldsymbol{\vartheta}_k)\right\}, \tag{1.4}$$

where $I(z_i = k) = 1$ only if $\mathbf{y}_i$ originated from the component k of the mixture; otherwise, $I(z_i = k) = 0$. Unlike (1.3), the complete-data log likelihood function (1.4) has a more tractable form. The EM algorithm iteratively performs two steps: expectation (E-step) and maximization (M-step). During the E-step, one finds the conditional expectation of (1.4) given observed data, commonly referred to as the Q-function

$$Q(\boldsymbol{\Psi}|\boldsymbol{\Psi}^{(b-1)}) = \sum_{i=1}^{n} \sum_{k=1}^{K} \pi_{ik}^{(b)} \left\{\log \tau_k + \log f_k(\mathbf{y}_i|\boldsymbol{\vartheta}_k)\right\}, \tag{1.5}$$

where $\pi_{ik}^{(b)}$ is the current estimate of the posterior probability that the ith data point belongs to the kth component. Posterior probabilities at iteration b are obtained according to the following expression:

$$\pi_{ik}^{(b)} = \frac{\tau_k^{(b-1)} f_k(\mathbf{y}_i|\boldsymbol{\vartheta}_k^{(b-1)})}{\sum_{k'=1}^K \tau_{k'}^{(b-1)} f_{k'}(\mathbf{y}_i|\boldsymbol{\vartheta}_{k'}^{(b-1)})}.$$

During the M-step, (1.5) is maximized with respect to the parameter vector $\boldsymbol{\Psi}$ either analytically or numerically if a closed form solution does not exist. The two-step cycle of the EM algorithm is repeated until a specified stopping criterion is satisfied. Upon convergence, the EM algorithm yields the maximum likelihood estimate $\hat{\boldsymbol{\Psi}}$ and estimated posterior probabilities $\hat{\pi}_{ik}$. The criteria that can be used for stopping include the relative change in estimated parameters or relative change in log likelihood values. These approaches can result in slow convergence of the EM algorithm. Hence, Böhning et al. [21] proposed using the Aitken acceleration-based stopping criterion relying on

$$\ell_\infty(\boldsymbol{\Psi}^{(b+1)}) \equiv \ell(\boldsymbol{\Psi}^{(b)}) + \frac{1}{1-c(\boldsymbol{\Psi}^{(b)})}(\ell(\boldsymbol{\Psi}^{(b+1)}) - \ell(\boldsymbol{\Psi}^{(b)})),$$

where $c(\boldsymbol{\Psi}^{(b)}) = (\ell(\boldsymbol{\Psi}^{(b+1)}) - \ell(\boldsymbol{\Psi}^{(b)}))/(\ell(\boldsymbol{\Psi}^{(b)}) - \ell(\boldsymbol{\Psi}^{(b-1)}))$. The EM algorithm is stopped when $|\ell_\infty(\boldsymbol{\Psi}^{(b+1)}) - \ell_\infty(\boldsymbol{\Psi}^{(b)})|$ is less than some prespecified tolerance level. For more details, we refer the reader to [21, 109]. Some additional methods that have been proposed to overcome slow convergence of the EM algorithm include creative specification of missing data in the complete-data likelihood function [91, 92, 125], modification of the algorithm itself such as the Incremental EM (IEM) algorithm [130], Expectation Conditional Maximization (ECM) algorithm [126], and the Monte Carlo EM (MCEM) algorithm [167], geometric considerations improving the trajectory of the algorithm [18], use of extrapolation [149] or more sophisticated access to data points [128]. In many problems, the number of components K is also unknown and should be estimated along with the parameter vector $\boldsymbol{\Psi}$. In this case, an information criterion such as AIC [1] or BIC [152] is usually employed to select a better model. Alternatively, one can apply bootstrapping of the likelihood ratio test statistic [47, 106].

It should be noted that the EM algorithm is a climbing procedure, i.e., the values of the log likelihood function (1.3) form a non-decreasing sequence over the course of the algorithm's iterations. This property ensures that the procedure will converge to a local maximum in the vast majority of cases with exceptions when the algorithm is trapped in a saddle point or is on its way to infinity for unbounded likelihood functions. The latter might pose a serious concern in optimizing the likelihood function, subject to a specific mixture model form. In particular, this is the case when the mixture model (1.2) has Gaussian components with unrestricted covariance matrices $\boldsymbol{\Sigma}_k$. Then, $f_k(\mathbf{y}|\boldsymbol{\vartheta}_k) \equiv \phi_p(\mathbf{y}|\boldsymbol{\mu}_k, \boldsymbol{\Sigma}_k) = [(2\pi)^p|\boldsymbol{\Sigma}_k|]^{-\frac{1}{2}} \exp\{-\frac{1}{2}(\mathbf{y}-\boldsymbol{\mu}_k)'\boldsymbol{\Sigma}_k^{-1}(\mathbf{y}-\boldsymbol{\mu}_k)\}$, where $\phi_p(\mathbf{y}|\boldsymbol{\mu}_k, \boldsymbol{\Sigma}_k)$ stands for the p-variate Gaussian density function with the mean vector $\boldsymbol{\mu}_k$ and covariance matrix $\boldsymbol{\Sigma}_k$. The resulting likelihood function is unbounded due to a potential of obtaining singular covariance matrices in the process of their estimation. When a global maximizer of the likelihood function does not exist, it is desirable to obtain a satisfactory local maximizer as it has been shown that a consistent and asymptotically efficient local maximizer exists in the interior of the parameter space [77]. An alternative but less popular approach is to restrict the parameter space with the intention to obtain a global maximizer within the restricted space [64, 110].

In spite of the difficulties that arise in the implementation of the EM algorithm, the flexibility of the mixture model with normal components makes it very popular. In the case of a p-variate Gaussian mixture model with unequal covariance matrices, the Q-function shown in (1.5) assumes the following form:

$$Q(\boldsymbol{\Psi}) = -\frac{pn}{2}\log 2\pi - \frac{1}{2}\sum_{i=1}^{n}\sum_{k=1}^{K}\pi_{ik}\{-2\log\tau_k + \log|\boldsymbol{\Sigma}_k| + (\mathbf{x}_i - \boldsymbol{\mu}_k)'\boldsymbol{\Sigma}_k^{-1}(\mathbf{x}_i - \boldsymbol{\mu}_k)\}, \tag{1.6}$$

where $\boldsymbol{\Psi} = \left(\tau_1, \ldots, \tau_{K-1}, \boldsymbol{\mu}_1', \cdots, \boldsymbol{\mu}_K', \text{vech}\{\boldsymbol{\Sigma}_1\}', \cdots, \text{vech}\{\boldsymbol{\Sigma}_K\}'\right)'$ and vech operator stacks subcolumns of a symmetric matrix so that the produced vector includes only unique elements. The maximization of (1.6) with respect to $\boldsymbol{\Psi}$ yields closed-form expressions for the parameter estimates given by

$$\tau_k^{(b)} = \frac{1}{n}\sum_{i=1}^{n}\pi_{ik}^{(b)}, \quad \boldsymbol{\mu}_k^{(b)} = \frac{\sum_{i=1}^{n}\pi_{ik}^{(b)}\mathbf{y}_i}{\sum_{i=1}^{n}\pi_{ik}^{(b)}}, \quad \boldsymbol{\Sigma}_k^{(b)} = \frac{\sum_{i=1}^{n}\pi_{ik}^{(b)}(\mathbf{y}_i - \boldsymbol{\mu}_k^{(b)})(\mathbf{y}_i - \boldsymbol{\mu}_k^{(b)})'}{\sum_{i=1}^{n}\pi_{ik}^{(b)}}.$$

Final posterior probabilities returned by the EM algorithm are used to form the model-based clustering partition. According to the Bayes decision rule $\hat{z}_i = \arg\max_k \hat{\pi}_{ik}$ with $\hat{z}_i$ being the estimated classification of $\mathbf{y}_i$, each observation is assigned to a component associated with the largest posterior probability. This defines the connection between mixture modeling and model-based clustering frameworks. For the most comprehensive analysis of finite mixture models, we refer the reader to the work by McLachlan and Peel [109].

Often, more than one component is required for modeling a specific data group. In this case the attractive one-to-one relationship between components and clusters is ruined. Some remedies in this case include merging mixture components and nonparametric mode-based inference. Generally, the log likelihood function (1.3) has multiple modes, making the search for the global maximizer or the best local maximizer quite difficult. The existence of multiple local maxima elevates the importance of initialization of the EM algorithm. Starting the algorithm with a promising set of initial parameter values can lead to its convergence to a better solution as well as shorter convergence time. Another feature of mixture models is the potential for leading to so-called spurious solutions. These solutions can occur due to minor random patterns in data when a certain group of points almost falls into a lower-dimensional subspace. Such a solution manifests itself in the form of a relatively large local maximum that can be reached for some specific initial parameter values. Spurious solutions rarely have practical significance and are usually removed from consideration. This paper recaptures these and other recent trends and developments in the theory and applications of model-based clustering and related aspects of finite mixture modeling.

1.2 Methodological Developments

1.2.1 Initialization

The likelihood function corresponding to finite mixture models is nearly always multimodal. Upon convergence, the EM algorithm finds a local maximizer with no guarantee that the obtained solution is the best one. To enhance the chances of identifying the global maximizer or the best local one when the global maximizer does not exist, the practitioner has to employ a proper strategy for choosing the set of initial parameter values $\boldsymbol{\Psi}^{(0)}$ or posterior probabilities $\boldsymbol{\pi}^{(0)}$. Thus, the results of mixture modeling and therefore model-based clustering are entirely governed by the point in the parameter space from which the EM algorithm is started.

Although the choice of a proper initialization strategy is absolutely crucial even for a perfectly specified model, the current literature provides very little insight into the problem. We illustrate the importance of the discussed matter on the popular classification dataset called *Crabs* [27] fitted with a mixture of multivariate Gaussian densities with unrestricted covariance matrices. The dataset includes 200 5-variate observations measuring frontal lobe size, rear width, carapace length, carapace width, and body depth of blue and orange crabs divided into equal male and female groups. As a result, there are 50 subjects in each of the following four groups: *Blue Male*, *Blue Female*, *Orange Male*, and *Orange Female*. As discussed by Hennig [66], the popular R package MCLUST [51] suggests a two-component solution based on the BIC value of 3064.47 ($\ell = -1423.62$) with only 28% of observations classified correctly. If the correct number of groups ($K = 4$) is assumed to be known, MCLUST finds a solution with BIC of 3179.51 ($\ell = -1369.87$) and 45.5% of correctly classified observations. In the paper by Hennig [66], the author explains that the choice of the model employed, perhaps, is not entirely valid for the dataset. At the same time, Bouveyron and Brunet [22] refer to this dataset making an argument that the clustering complexity can be relaxed by means of the dimensionality reduction. While high dimensionality and model invalidity can contribute to the difficulty of grouping, Melnykov [117] remarks that the issue is entirely related to the choice and implementation of an initialization strategy. The author reports the detected model with BIC value of 2887.15 ($\ell = -1223.67$) and 92.5% of correctly classified observations. As a matter of fact, the same result can be obtained if the EM algorithm is initialized by model parameters estimated based on known class memberships of data points. This example highlights the importance of an effective initialization choice in the mixture modeling framework.

One of the most traditional approaches to the initialization of the EM algorithm is to apply another clustering technique first. The obtained classification vector can then be used to obtain starting parameter estimates. The k-means algorithm [48] is, perhaps, the most commonly used for the purpose of initialization. Despite the popularity of this strategy, such an approach can hardly be recommended in general since the provided starting point in the parameter space reflects the characteristics of the particular clustering procedure employed rather than those of the considered model. For example, the k-means algorithm is known to be equivalent to model-based clustering by means of the CEM algorithm based on a mixture of Gaussian components with equal spherical covariance matrices and equal mixing proportions, where CEM represents the EM algorithm with an additional classification (C) step. As a result, the solution provided by the k-means algorithm can be far away from the best solution in the parameter space, for instance, in the case of elongated clusters with unequal representations. In addition, the initialization of k-means is a standalone problem that requires attention [28, 122]. Similar challenges can be observed for other clustering procedures used as an initialization strategy for the EM algorithm.

An algorithm called *emEM* [20] starts by randomly selecting K points $\mathbf{y}_1^*, \ldots, \mathbf{y}_K^*$ from a dataset without replacement. Then, the remainder of the dataset is split into K groups according to the shortest Euclidean distance from every observation to one of the K seeds, i.e., $\hat{z}_i = \arg\min_k\{\|\mathbf{y}_i - \mathbf{y}_k^*\|\}$. For the obtained partition, model parameters are estimated and the EM algorithm, called *short EM*, runs for a small number of iterations or until some rough convergence criterion is met. The entire process is repeated multiple times to find the best candidate point in terms of the likelihood value. Starting from this point, the main EM algorithm, also known as *long EM*, operates until the final convergence is reached.

It is possible to restrict every *short EM* run with a single iteration to increase the number of candidate points; such a modification is called *Rnd-EM* and was proposed by Maitra [96]. A simulation study conducted by Melnykov and Melnykov [121] for $K = 10$ and $K = 20$ showed that the *emEM* algorithm outperformed *Rnd-EM* in all considered settings. However, as remarked by Melnykov [115], such a modification might be beneficial in cases with well-separated clusters but should not be preferred over *emEM* in the presence of groups with more considerable overlaps. When clusters overlap substantially, it is more beneficial to run the EM algorithm for more iterations. On the contrary, for well-separated clusters, additional iterations of the EM algorithm do not improve the likelihood value much as the convergence is usually fast. In the latter case, it is more important to have every data group represented by exactly one seeded point. As a result, the performance of the *emEM* and *Rnd-EM* algorithms degrades with the increase in the number of groups. It is important to mention that both *emEM* and *Rnd-EM* are stochastic initialization strategies that often lead to good results due to trying multiple points in the parameter space. In the meantime, in many situations, it is highly unlikely or even impossible that these procedures will find a candidate point that leads to the best solution possible. This limitation can be explained by the use of Euclidean distances to form the original partition. Therefore, one might examine numerous initial points but the benefit of restarting can be marginal if all of them are far away from the ideal solution in the parameter space by default.

As remarked by Melnykov and Melnykov [121], a particular mixture model form should be taken into consideration as early as at the initialization stage. The authors developed a new strategy for Gaussian mixture models, called Σ*-EM*, that starts with a small number of observations that are likely to belong to the same cluster. Then, the covariance matrix of the underlying mixture component is estimated iteratively based on a truncated Gaussian distribution. At each iteration, more points fall into an extending confidence hyperellipsoid, thus allowing to improve the precision of parameter estimates. When no new points can be added to the cluster, it is declared detected and the initial parameter estimates of the related mixture component are calculated. Then, the corresponding points get excluded from the dataset to avoid redundancy and the entire process starts over in search of another cluster. The estimates of mixture model parameters obtained this way are often so close to the best possible solution that the convergence can be reached within very few iterations. This approach can also handle high numbers of clusters. In their paper, Melnykov and Melnykov [121] provided an illustrative example with $K = 50$ in which they identified the best available solution while all alternative initialization procedures failed. This attractive procedure, however, has its own limitations. It shows the best performance in low dimensions since it becomes more challenging to estimate covariance matrices based on few initial data points in higher dimensions. Another related issue occurs when clusters overlap considerably and foreign points that belong to other groups are mistakenly included into the current cluster. This can cause the undesired elongation of the confidence hyperellipsoid accompanied by capturing several clusters. The authors manage this situation by checking every cluster produced for the presence of subclusters by means of BIC.

Some other initialization strategies include the deterministic multi-stage approach based on finding the best local modes [96] and model-based hierarchical clustering [49, 50]. As reported by Melnykov [115], the former shows mixed performance but usually loses the competition to the *emEM* and Σ*-EM* algorithms. The latter procedure relies on the notion of classification likelihood and its approximate maximization. The main disadvantage of both approaches is their deterministic nature, i.e., the set of initial parameter values for a specific dataset cannot be changed and the EM algorithm cannot be run from other points in the parameter space. Therefore, if the best solution is not obtained starting from these values, it will not be found at all.

The development of flexible and practical initialization strategies is an open problem that requires more attention. Overall, it can be recommended to employ multiple initialization procedures and choose the solution producing the highest likelihood value. Besides initialization issues, there are many other factors that affect the complexity of clustering. A comprehensive simulation study by Michael and Melnykov [127] showed that the number of clusters along with overlap between components are more influential factors than the dimensionality and sample size. In their study, the authors avoided issues related to the choice of the initialization strategy by starting the EM algorithm from the true parameter values.

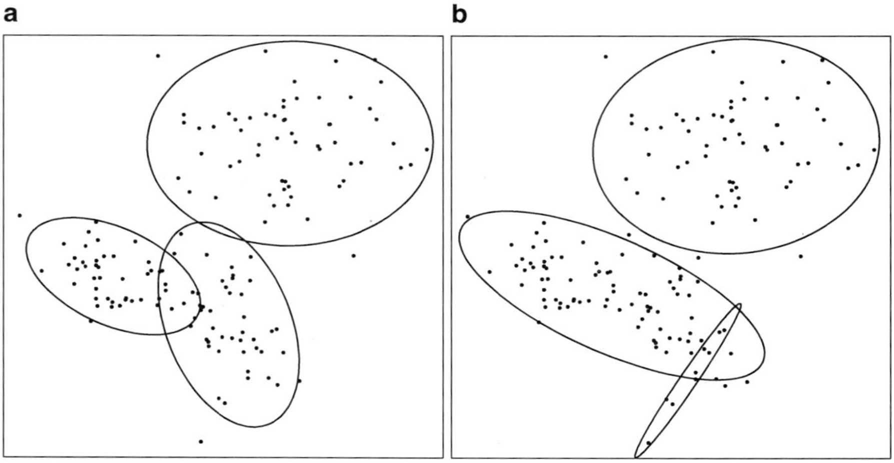

Fig. 1.2 Sample data with two 3-component Gaussian mixture model solutions: (**a**) legitimate solution and (**b**) spurious solution based on a local random pattern captured by the elongated component

1.2.2 Spurious Solutions

There is a problem closely related to the process of solution selection and the issue of unbounded likelihood. Quite often, practitioners can observe mixture components constructed on very few observations lying close to a lower-dimensional parameter subspace. In the case of bivariate data, for example, these data points will be located along a straight line. Such nearly degenerated components are known as spurious and are quite common in mixture modeling. Figure 1.2 illustrates the difference between a legitimate (display (a)) and spurious (display (b)) solutions detected in a mixture of three Gaussian distributions. Usually, likelihood values associated with fake solutions are somewhat higher than those representing other local maxima. As a result, there is a high chance that a spurious maximizer will be preferred over competing solutions even though it typically has unclear or no practical meaning. A general recommendation is to ignore such solutions as ones emphasizing a local random pattern in data rather than a systematic cluster structure. As mentioned by McLachlan and Peel [109], spurious components can be easily identified by low data representation and small generalized variance relative to those of other mixture components. One issue is that the terms "low" and "small" have to be appropriately quantified to separate reasonable solutions from fake ones. Another problem related to spurious solutions is the necessity of rerunning the EM algorithm every time when a spurious solution is detected.

Some work related to spurious solutions and unbounded likelihood employs the penalized likelihood function [33, 34]. These papers recommend adding a penalty term and maximizing the newly formed likelihood over the modified parameter space. Such an adjustment helps prevent or at least relax the singularity-related problem. Mainly, current research stream is devoted to reducing the chances of selecting a spurious solution.

In the paper by Garcia-Escudero et al. [55], the authors considered mixtures of multivariate Gaussian distributions and employed the trimmed log likelihood function [54, 132] defined by

$$\ell_t(\boldsymbol{\Psi}) = \sum_{i=1}^{n} t_i \log \sum_{k=1}^{K} \tau_k \phi_p(y_i | \boldsymbol{\mu}_k, \boldsymbol{\Sigma}_k), \tag{1.7}$$

where $t_i = 0$ for trimmed observations and $t_i = 1$ otherwise. The original idea behind the use of $\ell_t(\boldsymbol{\Psi})$ was the robust parameter estimation in the presence of outliers. The authors incorporated an eigenvalue-ratio constraint $M/m \leq c$, where $M = \max_{k=1,\dots,K} \max_{j=1,\dots,p} \lambda_j(\boldsymbol{\Sigma}_k)$ and $m = \min_{k=1,\dots,K} \min_{j=1,\dots,p} \lambda_j(\boldsymbol{\Sigma}_k)$ with $\lambda_j(\boldsymbol{\Sigma}_k)$ denoting the jth eigenvalue of the covariance matrix associated with the kth Gaussian component, and c is some fixed value. The values of c close to the unity restrict covariance matrices to be more spherical and similar to each other. Larger values of c allow higher flexibility in forms of covariance matrices. By controlling c, the authors show that the risk of observing spurious solutions can be considerably reduced. The choice of c, however, is the major issue here since low values of c can be too restrictive while high c-values are unlikely to relieve the problem with spurious solutions.

Another approach that yields excellent results and is not very restrictive at the same time was proposed by Seo and Kim [153]. The authors considered the modified log likelihood function given by $\ell_{-s}(\boldsymbol{\Psi}) = \ell(\boldsymbol{\Psi}) - \sum_{i\in\{i_1,\dots,i_s\}} \ell_i$, where $\ell_i = \log f(\mathbf{y}_i; \boldsymbol{\Psi})$ is the log likelihood contribution of the ith data point and s is the number of observations whose contributions are excluded. Then, the corresponding modified maximum likelihood estimator is defined by

$$\hat{\boldsymbol{\Psi}}_{-s} = \arg\max \ell_{-s}(\boldsymbol{\Psi}).$$

The authors investigate two ways of deciding which contributions have to be eliminated: one is based on the log likelihood magnitude while the other relies on the score function of a local maximizer at each data point. The latter method, called by the authors *the gradient-based k-deleted MLE*, demonstrates somewhat better performance and should be generally preferred over the former one. The continuation of this work can be found in [78] where the authors prove the consistency of the gradient-based k-deleted MLE and apply it to investigate the performance of several likelihood-based model selection criteria.

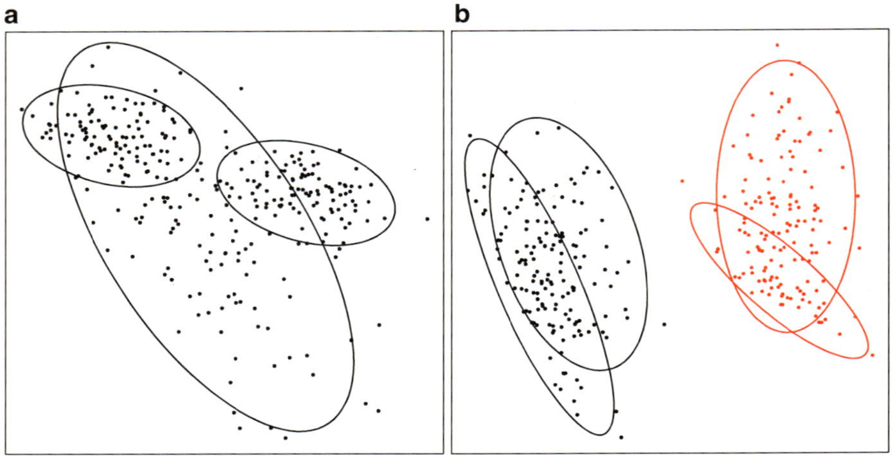

Fig. 1.3 Illustrative examples for two situations when merging mixture components for clustering can be helpful: (**a**) complex structure of a cluster requiring several components, (**b**) misspecified model—data simulated from a mixture of two skew-normal distributions are fitted with a 4-component Gaussian mixture model

1.2.3 Merging Mixture Components for Clustering

The connection between finite mixture modeling and model-based clustering is established through the Bayes decision rule, i.e., each data point gets assigned to a mixture component with the highest posterior probability of being the origin of the considered observation. Assuming that each data group can be well-modeled by means of a single component distribution, there is a one-to-one relationship between distributions and clusters. Several papers have been recently devoted to tackling a challenging situation when this one-to-one correspondence is not adequate. This can happen due to various reasons such as complex multimodal cluster structure (Fig. 1.3a), model misspecification (Fig. 1.3b), and presence of spurious components (Fig. 1.2b). In such cases, more than one distribution may be needed to model each group of data and the attractive model-based clustering interpretation breaks. Fortunately, good clustering results can often be obtained by merging mixture components. A general idea is to measure the level of interaction between mixture components. Those distributions that overlap considerably are likely (although not guaranteed) to represent the same data cluster and should be merged into a common group of mixture components. As a result, each cluster is modeled by a set of components rather than a single one. The best fitting mixture model has to be found first and then substantially overlapping components are merged to establish the correspondence with clusters. All existing procedures focus on one-by-one agglomerative hierarchical merging.

Recently, Baudry et al. [13] proposed a criterion based on the change in the entropy of fuzzy classification associated with merging two groups of mixture components, $\mathscr{G}_r$ and $\mathscr{G}_v$. The criterion is defined by

$$\kappa^{\mathscr{G}_r\mathscr{G}_v} = \sum_{i=1}^{n} ((\pi_{i\mathscr{G}_r} + \tau_{i\mathscr{G}_v}) \log (\pi_{i\mathscr{G}_r} + \pi_{i\mathscr{G}_v}) - \pi_{i\mathscr{G}_r} \log \pi_{i\mathscr{G}_r} - \pi_{i\mathscr{G}_v} \log \pi_{i\mathscr{G}_v}), \tag{1.8}$$

where $\pi_{i\mathscr{G}_r} = \sum_{k\in\mathscr{G}_r} \pi_{ik}$ and $\pi_{i\mathscr{G}_v}$ is defined similarly. When $\kappa^{\mathscr{G}_r\mathscr{G}_v}$ is close to the lower bound of 0, attainable only when $\pi_{i\mathscr{G}_r}\pi_{i\mathscr{G}_v} = 0$ for all values of i, the entropy change is not substantial, meaning that groups are well-separated and merging should not be recommended. Large values of $\kappa^{\mathscr{G}_r\mathscr{G}_v}$ are observed when merging component groups is reasonable. The authors recommended using their criterion hierarchically after all mixture components are detected by BIC. Unfortunately, there is no immediate recipe for deciding when the merging process has to be stopped. The authors considered two possible approaches for detecting the optimal number of clusters. One is based on the integrated completed likelihood criterion (ICL) developed by Biernacki et al. [19]. ICL represents a version of BIC penalized by a fuzzy classification entropy term: $ICL = BIC - 2\sum_{i=1}^{n}\sum_{k=1}^{K} \pi_{ik} \log \pi_{ik}$. By construction, ICL is more conservative than BIC as instead of finding the optimal number of components, it aims at detecting the number of clusters. If all components are well-separated, the entropy term does not contribute to ICL and both criteria are equivalent. In the paper by Melnykov [118], the author examined the performance of ICL through simulation studies and concluded that the results produced by the criterion depend on the sample size. Specifically, ICL tends to find a higher number of clusters in cases when groups are heavily populated as opposed to the cases with lower cluster sizes. The other approach proposed by Biernacki et al. [19] is based on a piecewise linear regression of the entropy change versus the number of clusters. The authors claim that the inflection point will correspond to the correct number of clusters. Unfortunately, this empirical rule may be misleading when there are several inflection points or even inapplicable when the number of mixture components is low (*e.g.*, $K = 2$ or 3).

An approximate distribution of the criterion (1.8) was derived by Melnykov [117] by means of the multivariate Delta method and matrix differential calculus for multivariate Gaussian mixture models. The author remarked that posterior probabilities can be seen as functions of model parameters and showed that $\kappa^{\mathscr{G}_r\mathscr{G}_v}(\hat{\boldsymbol{\Psi}})$ is asymptotically normally distributed with mean $\kappa^{\mathscr{G}_r\mathscr{G}_v}(\boldsymbol{\Psi})$ and variance that can be estimated by $\nabla\kappa^{\mathscr{G}_r\mathscr{G}_v}(\hat{\boldsymbol{\Psi}})'\mathbf{I}^{-1}(\hat{\boldsymbol{\Psi}})\nabla\kappa^{\mathscr{G}_r\mathscr{G}_v}(\hat{\boldsymbol{\Psi}})$, where $\mathbf{I}^{-1}(\boldsymbol{\Psi})$ is the inverse of the observed information matrix and $\nabla\kappa^{\mathscr{G}_r\mathscr{G}_v}(\boldsymbol{\Psi}) = \sum_{i=1}^{n} \nabla\kappa_i^{\mathscr{G}_r\mathscr{G}_v}(\boldsymbol{\Psi})$ is the gradient vector with

$$\nabla\kappa_i^{\mathscr{G}_r\mathscr{G}_v}(\boldsymbol{\Psi}) = \left(\left(\frac{\partial\kappa_i^{\mathscr{G}_r\mathscr{G}_v}(\boldsymbol{\Psi})}{\partial\tau_s}\right)'_{s=1,\dots,K-1}, \left(\frac{\partial\kappa_i^{\mathscr{G}_r\mathscr{G}_v}(\boldsymbol{\Psi})}{\partial\boldsymbol{\mu}_s}\right)'_{s=1,\dots,K}, \left(\frac{\partial\kappa_i^{\mathscr{G}_r\mathscr{G}_v}(\boldsymbol{\Psi})}{\partial\mathrm{vech}\{\boldsymbol{\Sigma}_s\}}\right)'_{s=1,\dots,K}\right)'$$

evaluated at $\hat{\boldsymbol{\Psi}}$. For more details on vech operator and corresponding partial derivatives, we refer the reader to the original paper by Melnykov [117]. The author proposed using the derived distribution to assess closeness of the observed value $\kappa^{\mathscr{G}_r\mathscr{G}_v}$ to zero.

Ray and Lindsay [145] studied the topography of multivariate Gaussian mixtures. In their discussion, they suggested that the ridgeline elevation plot can be used to assess the number of modes in pairs of Gaussian components. For any K-component mixture of p-dimensional Gaussian distributions, they showed that there is a $(K-1)$-dimensional manifold that contains all critical points. For $K = 2$, this manifold, also known as a ridgeline function, is given by

$$\mathbf{y}^*(a) = [(1-a)\boldsymbol{\Sigma}_1^{-1} + a\boldsymbol{\Sigma}_2^{-1}]^{-1}[(1-a)\boldsymbol{\Sigma}_1^{-1}\boldsymbol{\mu}_1 + a\boldsymbol{\Sigma}_2^{-1}\boldsymbol{\mu}_2]$$

for $a \in [0, 1]$. According to the authors, if there is a single mode, the considered components should be merged to represent one cluster. The details of this approach with some drawbacks are discussed by Hennig [66].

Another approach to measuring the degree of overlap between components was proposed by Hennig [66]. His procedure relies on directly estimated misclassification probabilities (DEMP) that can be computed from the set of posterior probabilities returned by the EM algorithm. In this sense, the DEMP method is similar to the criterion (1.8) which also relies on posterior probabilities entirely. The probability that an observation that belongs to the vth component is incorrectly classified to the rth component is given by

$$\Pr(\tilde{Z} = r|Z = v) = \frac{\Pr(\tilde{Z} = r, Z = v)}{\tau_v} \tag{1.9}$$

with Z and $\tilde{Z}$ being random variables associated with the genuine and assigned class labels, respectively. The author recommends estimating the joint probability $\Pr(\tilde{Z} = r, Z = v)$ by

$$\hat{\Pr}(\tilde{Z} = r, Z = v) = \frac{1}{n}\sum_{i=1}^{n} I(\hat{Z}_i = r)\hat{\pi}_{iv}, \tag{1.10}$$

where $I(\cdot)$ represents the indicator function such that $I(\mathscr{A}) = 1$ if $\mathscr{A}$ contains the true statement and $I(\mathscr{A}) = 0$ if $\mathscr{A}$ is false. $\hat{Z}_i$ stands for the classification of the ith data point obtained for the observed data. Combining (1.9)–(1.10) and generalizing the result from the case with two distributions to the framework with groups of mixture components $\mathscr{G}_r$ and $\mathscr{G}_v$, the following estimator of the misclassification probability can be derived:

$$\hat{\Pr}(\tilde{Z} \in \mathscr{G}_r|Z \in \mathscr{G}_v) = \frac{\sum_{i=1}^{n} I(\hat{Z}_i \in \mathscr{G}_r)\hat{\pi}_{i\mathscr{G}_v}}{n\sum_{k\in\mathscr{G}_v}\hat{\tau}_k}. \tag{1.11}$$

The author suggests estimating all pairwise misclassification probabilities and merging those groups of components that yield the highest misclassification probability. Repeating this process hierarchically, a traditional dendrogram can be constructed. Then, one can decide on the number of clusters according to a pre-specified misclassification level: observed pairwise misclassification probabilities higher than this value indicate that there is a considerable overlap and corresponding groups have to be combined. On the contrary, lower misclassification probabilities imply that the degree of interaction between groups of components is not sufficient for merging. As per discussion in [118], the estimator (1.11) can often work well but can also yield misleading results when mixture components overlap considerably. He also remarked that both pairwise misclassification probabilities should be taken into consideration rather than just the largest one. The sum of these probabilities is already defined and known as pairwise overlap [97] which can be calculated by means of the R package *MixSim* [124] and c package *CARP* [120] for Gaussian mixture models.

Another approach for estimating pairwise misclassification probabilities for component groups $\mathscr{G}_r$ and $\mathscr{G}_v$ was proposed by Melnykov [118]. The author suggests employing the Bayes decision rule and Monte Carlo simulations. First, a large sample $\mathbf{Y}_1, \ldots, \mathbf{Y}_N$, where N represents the sample size, is simulated from the mixture $\sum_{k \in \mathscr{G}_v} \tau_k^* f_k(\mathbf{y}_i; \boldsymbol{\Psi}_k)$. Here, τ_k^* represents the standardized probability calculated by $\tau_k^* = \tau_k / \sum_{k' \in \mathscr{G}_v} \tau_{k'}$ for all $k \in \mathscr{G}_v$. Then, the misclassification probability can be estimated by the following expression:

$$\hat{\Pr}(\tilde{Z} \in \mathscr{G}_r | Z \in \mathscr{G}_v) = \frac{1}{N} \sum_{i=1}^{N} I\left(\sum_{k \in \mathscr{G}_v} \tau_k f_k(\mathbf{y}_i; \boldsymbol{\Psi}_k) < \sum_{h \in \mathscr{G}_r} \tau_h f_h(\mathbf{y}_i; \boldsymbol{\Psi}_h) \right),$$

where $I(\cdot)$ is again the indicator function. The author shows through simulation studies that this estimator provides good results in different settings when N is sufficiently large.

As a final remark on merging mixture components, we can note that such an approach does not guarantee yielding good partitions in all circumstances. On the other hand, it is often an effective remedy in the case when a one-to-one correspondence between mixture components and clusters is not realistic. Perhaps, even more importantly, merging mixture components for clustering considerably relieves the problem of model misspecification, which is almost always a concern in statistical data modeling in general.

1.2.4 Nonparametric Clustering

One popular alternative to merging mixture components is known as nonparametric clustering. In this setting, it is traditionally assumed that clusters are associated with density bumps or modes instead of mixture components. This idea is rather intuitive and automatically resolves the problem discussed in the previous section as the one-to-one relationship is established between density bumps and data groups. Observations are assigned to clusters in accordance with the attraction zones of detected modes. Despite immediate advantages of this method, there are also certain challenges such as the need of extensive computing resources for finding modes and attraction zones and the lack of meaning for some identified groups. While some mode-based methods assume specific functional forms of mixture components [84], the others use nonparametric density estimation, usually by means of the kernel method.

Although mode-based analysis is not new in literature [63, 168], it has been given more attention recently. The focus of [156] was on estimating the hierarchical modal structure, also known as a cluster tree of a density, employing the connection between the minimal spanning tree and nearest neighbor density estimation. A graph-based method capable of approximating a cluster tree corresponding to a density estimate was proposed by Stuetzle and Nugent [157]. The authors used excess mass to measure the size of graph branches in order to address the problem of spurious modes and branches associated with local random patterns in data. An EM-like nonparametric algorithm, called the Modal EM, associating observations with local modes of the density function was proposed by Li et al. [87]. The authors also developed its hierarchical extension based on the Gaussian kernel density estimation with increasing bandwidths. The authors developed an algorithm to find the ridgeline separating two density bumps. The functionality of the proposed approach is realized in the R package *Modalclust* [144].

An R package devoted to clustering via nonparametric density estimation by the kernel method is called *pdfCluster* [5]. It is based on the idea presented in [6]. Clusters are constructed in association with connected components whose estimated density exceeds the pre-specified threshold value.

Another approach to nonparametric mixture modeling is proposed by Benaglia et al. [14, 16]. It is based on a modification of the EM algorithm and employs the kernel-based estimation of mixture component distributions. The authors consider the model

$$f(\mathbf{y}|\boldsymbol{\tau}) = \sum_{k=1}^{K} \tau_k \prod_{j=1}^{p} f_{kj}(y_j)$$

assuming the conditional independence of coordinates in $\mathbf{Y} = (Y_1, \ldots, Y_p)'$. While this assumption is somewhat restrictive in a general case, it is realistic for the framework with repeated measurements. The authors develop an EM-like algorithm with the E-step given by

$$\pi_{ik}^{(b)} = \frac{\tau_k^{(b-1)} \prod_{j=1}^{p} f_{kj}^{(b-1)}(y_{ij})}{\sum_{k'=1}^{K} \tau_{k'}^{(b-1)} \prod_{j=1}^{p} f_{k'j}^{(b-1)}(y_{ij})},$$

the M-step provided by

$$\tau_k^{(b)} = \frac{\sum_{i=1}^{n} \pi_{ik}^{(b)}}{n},$$

and an additional nonparametric density estimation step:

$$f_{kj}^{(b)}(u) = \frac{\frac{1}{h}\sum_{i=1}^{n} \pi_{ik}^{(b)} \mathscr{K}\left(\frac{u-y_{ij}}{h}\right)}{\sum_{i=1}^{n} \pi_{ik}^{(b)}},$$

where $\mathscr{K}(\cdot)$ specifies the kernel density function and h is the corresponding bandwidth. The authors consider the Gaussian kernel with the bandwidth chosen according to $h = 0.9\min\{s, \frac{IQR}{1.34}\}/\sqrt[5]{np}$, where s and IQR are sample standard deviation and interquartile range for all np data points, respectively. The functionality of the developed methodology is available through the R package *mixtools* [15].

1.2.5 Variable Selection for Clustering

One of the most active areas of model-based clustering research is dimensionality reduction and variable selection in high-dimensional data. Excellent interpretability of cluster analysis relying on mixture models can be severely affected by the presence of multiple variables. The main problem is model's dramatic overparameterization in high dimensions. In the meantime, important clustering information almost always lies in some parameter subspace, thus making the focus on the entire space unnecessary at least for the purpose of cluster analysis. Moreover, different groups of observations may belong to different subspaces and this fact can be taken into consideration.

One possible way to avoid the issue of overparameterization is to use more restrictive models. For example, in the mixture of Gaussian distributions, unrestricted covariance matrices can be forced to be equal, diagonal, or even spherical. The papers by Banfield and Raftery [8] and Celeux and Govaert [29] consider 14 different models obtained by incorporating various constraints on the parameters of the covariance matrix eigenvalue decomposition: $\boldsymbol{\Sigma}_k = \rho_k \boldsymbol{\Gamma}_k \boldsymbol{\Lambda}_k \boldsymbol{\Gamma}'_k$, where ρ_k is the volume parameter, $\boldsymbol{\Gamma}_k$ is the matrix of eigenvectors, and $\boldsymbol{\Lambda}_k$ is the diagonal matrix of eigenvalues. The R package *mclust* [51] is capable of fitting all models from this family.

An alternative way of fighting high dimensionality is to reduce the number of variables and focus only on those variables that carry discriminating information for clusters. It is important to mention that the task of variable selection cannot be considered aside from the problem of clustering itself. Figure 1.4a shows that the variable y_2 used alone discriminates two clusters, y_1 distinguishes three clusters, and both variables are needed for detecting all four groups. Hence, the set of variables selected depends on the result the researcher looks for. Traditional dimensionality reduction approaches such as the principal component analysis cannot be used in this setting as the variable that shows the lowest variability in data can be the most important for discriminating clusters. This idea is illustrated in Fig. 1.4b. The variable y_2 has low variation compared to that of y_1 but carries important information about groupings.

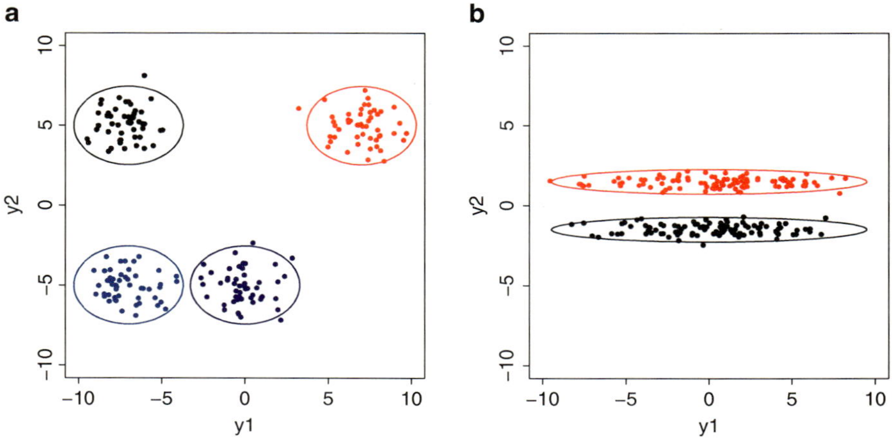

Fig. 1.4 Illustrative examples: (**a**) variables y_1 and y_2 are both important for detecting the 4-cluster structure, (**b**) variable y_2 shows low variation in data compared with y_1 but is crucial for the discrimination of two clusters

In the paper by Pan and Shen [135], the authors focused on the penalized Q-function constructed as the conditional expectation of the complete-data penalized log likelihood function given by

$$\ell_{c,P}(\boldsymbol{\Psi}) = \sum_{i=1}^{n} \sum_{k=1}^{K} I(z_i = k) \{\log \tau_k + \log f_k(\mathbf{y}_i | \boldsymbol{\vartheta}_k)\} - h_\lambda(\boldsymbol{\Psi}),$$

where $h_\lambda(\boldsymbol{\Psi})$ is the penalty function with the penalty parameter λ. The authors consider Gaussian mixtures and assume common covariance matrices of the diagonal form $\boldsymbol{\Sigma}_k = \{\sigma_1^2, \sigma_2^2, \ldots, \sigma_p^2\}$ and mean coordinates $\mu_{kj} = \mu_j + \delta_{kj}$ for $j = 1, 2 \ldots, p$ and all K components. The penalty function allows shrinking variables toward the global mean. The variables j for which $\delta_{kj} = 0$ for all $k = 1, 2, \ldots, K$ are considered irrelevant and can be excluded from the consideration. Various penalty functions have been proposed in literature. The L_1 penalty $h_\lambda(\boldsymbol{\Psi}) = \lambda \sum_{k=1}^{K} \sum_{j=1}^{p} |\mu_{kj}|$ was considered by Pan and Shen [135]. Another penalty $h_\lambda(\boldsymbol{\Psi}) = \lambda \sum_{j=1}^{p} \max_k\{|\mu_{1j}|, |\mu_{2j}|, \ldots, |\mu_{Kj}|\}$ was proposed by Wang and Zhu [165] who claimed that it is a better choice as it takes into consideration the fact that μ_{kj}s for $k = 1, \ldots, K$ correspond to the same variable j. Pairwise variable selection with the penalty $h_\lambda(\boldsymbol{\Psi}) = \lambda \sum_{j=1}^{p} \sum_{1 \le k \le k' \le K} |\mu_{kj} - \mu_{k'j}|$ was studied by Guo et al. [58]. To overcome limitations of the restrictive assumption that covariance matrices are diagonal, [175] extended the penalized model-based clustering idea to the case with common unconstrained covariance matrices. As it was remarked by Xie et al. [170], unconstrained matrices help model relationships between variables but involve too many parameters. Thus, the authors considered another approach to variable selection through penalized mixtures of factor analyzers.

In the paper by Raftery and Dean [142], the authors took a different approach dividing all variables into three sets: variables included in the model ($\mathbf{Y}^{(1)}$), variables currently under consideration ($\mathbf{Y}^{(2)}$), and variables to be studied ($\mathbf{Y}^{(3)}$). The authors proposed considering two models $\mathscr{M}_1$ and $\mathscr{M}_2$ specified as follows:

$$\mathscr{M}_1 : \Pr(\mathbf{Y}^{(1)}, \mathbf{Y}^{(2)}, \mathbf{Y}^{(3)}|\mathbf{Z}) = \Pr(\mathbf{Y}^{(3)}|\mathbf{Y}^{(1)}, \mathbf{Y}^{(2)}) \Pr(\mathbf{Y}^{(2)}|\mathbf{Y}^{(1)}) \Pr(\mathbf{Y}^{(1)}|\mathbf{Z})$$

$$\mathscr{M}_2 : \Pr(\mathbf{Y}^{(1)}, \mathbf{Y}^{(2)}, \mathbf{Y}^{(3)}|\mathbf{Z}) = \Pr(\mathbf{Y}^{(3)}|\mathbf{Y}^{(1)}, \mathbf{Y}^{(2)}) \Pr(\mathbf{Y}^{(1)}, \mathbf{Y}^{(2)}|\mathbf{Z})$$

with $\mathbf{Z}$ being the vector of class memberships. In the formulation of $\mathscr{M}_1$, it is assumed that the inclusion of the variables $\mathbf{Y}^{(2)}$ does not contribute to the model improvement as the clustering information is already contained in the set of already included variables $\mathbf{Y}^{(1)}$. The models can be conveniently compared through the Bayes factor $B_{12} = \frac{\Pr(\mathbf{Y}^{(2)}|\mathbf{Y}^{(1)},\mathscr{M}_1)\Pr(\mathbf{Y}^{(1)}|\mathscr{M}_1)}{\Pr(\mathbf{Y}^{(1)},\mathbf{Y}^{(2)}|\mathscr{M}_2)}$ that can be estimated by means of BIC. The functionality of the proposed methodology is available through the R package *clustvarsel* [36]. The papers by Maugis et al. [102, 103] extended the proposed idea noting that irrelevant and informative variables can be independent of each other and organizing variables in unsplittable blocks. This modification might improve the variable selection performance dramatically [30]. Finally, [104] extended the approach of [103] by adding capabilities for handling missing values.

A variable selection idea related to the Modal EM algorithm discussed in Sect. 1.2.3 was proposed by Lee and Li [84]. Instead of the kernel density estimation used in the original paper of [87], the authors focused on the mixture of Gaussian distributions. The Modal EM algorithm is run to detect all local modes. If several components produce one mode, they are responsible for modeling one cluster. In the paper by Lee and Li [84], the authors define a ridgeline separability measure and apply a forward variable selection procedure in order to maximize the so-called aggregated cluster separability.

The effective treatment of heterogeneous high-dimensional data is extremely important and attracts much attention of researchers and practitioners. For a more comprehensive review on the current state of the discipline in handling high-dimensional data by means of mixture models, we refer the reader to an excellent review provided by Bouveyron and Brunet [22].

1.2.6 Semi-Supervised Clustering

Semi-supervised machine learning is a popular direction of cluster analysis [11, 12, 70, 176] that assumes that some additional information about classes is available in the form of constraints or restrictions on class memberships. According to [12], semi-supervised clustering methods can be divided into two large groups: metric-based and constraint-based. The former uses the available labeled data to make adjustments to the metric utilized by the algorithm [9, 79, 171], while the latter rely on some modifications made to the clustering algorithm to accommodate the set of existing labels and constraints [11, 37, 163].

From the perspective of the nature of constraints, there are also different possibilities depending on an application in question. In particular, it may be the case that the memberships of certain observations are known [109]. For example, a set of medical data may consist of observations that describe two kinds of tumors: cancerous and benign, and have a subset of observations for which the classification has been performed by a human expert. Obtaining the data labeled in such a way can be expensive or require substantial resources, therefore, its portion in the overall amount of data is usually relatively small, but it may still provide considerable improvements to the classification [12]. Another possible framework in regards to restrictions affecting the clustering solution is the scenario when certain points must belong to the same cluster in the proposed classification (*positive equivalence constraint*) or, on the contrary, belong to different clusters (*negative equivalence constraint*) [154]. This setup presents more challenges than the one where some of the class labels are known due to the potential complexity of pairwise relations between points determined by positive and negative constraints.

In the framework of model-based clustering, the necessary constraints can be introduced into the model by making appropriate modifications to the EM algorithm. Finite mixture models are often considered in the literature on semi-supervised clustering, with the Gaussian mixtures receiving the most attention [42, 83, 154]. The applications of this framework are widespread and deal with image recognition and image segmentation [100, 154], speech and audio processing [42, 70], text classification [133], and many others.

The researchers emphasize the considerable difference that exists between positive and negative equivalence constraints [60, 154]. Thus, positive constraints are transitive, i.e., if points $\mathbf{y}_i$ and $\mathbf{y}_j$ must be in the same class and points $\mathbf{y}_j$ and $\mathbf{y}_k$ also have to be in the same class, it follows that $\mathbf{y}_i$ and $\mathbf{y}_k$ will be in the same class as well. Taking all such constraints into account, one can obtain *blocks* of observations that have to belong to the same cluster. Observations that are not involved in any constraints can still be viewed as trivial blocks consisting of singletons. On the contrary, negative constraints do not share the transitive property. Thus, if $\mathbf{y}_i$ and $\mathbf{y}_j$ have to be in different classes and $\mathbf{y}_j$ and $\mathbf{y}_k$ also must belong to different classes, it does not follow that $\mathbf{y}_i$ and $\mathbf{y}_k$ are necessarily placed in different classes. This difference in the nature of the two kinds of restrictions makes positive equivalence constraints easier to model. It is argued by Hammer et al. [60] that negative equivalence constraints, when generated at random, do not provide substantial gain in the performance of classification procedures compared to those with only positive constraints in place. However, in practice, the constraints are usually determined by the context of the application rather than randomly. Although the implementation of negative constraints is harder computationally, their necessity might be dictated by a classification task at hand.

A thorough treatment of semi-supervised clustering in the case of Gaussian mixture models with both positive and negative equivalence constraints involved is given by Shental et al. [154]. In the case of positive constraints, the authors make a distinction between two sampling situations that lead to block formation. The first possibility is for the blocks to be sampled according to the weights of their corresponding sources. This scenario occurs when blocks are obtained due to the nature of sampling, for instance, from sequential data in a Markovian process. The second possibility arises when data points are sampled from the mixture distribution and the formation of blocks occurs afterwards. For example, this situation can occur when positive equivalence constraints are defined during distributed learning. The modifications of the EM algorithm that would occur in both sampling situations are shown by Shental et al. [154]. To accommodate negative equivalence constraints, the authors consider a Markov network and propose using Pearl's junction tree algorithm [138] in the computation of posterior probabilities during the E-step of the EM algorithm.

It should be mentioned that in addition to the models considering positive and negative equivalence constraints, there are also approaches to describing models that are not as strict in regards to constraints used. Such models influence the classification of points via penalties or weights that may encourage or discourage the inclusion of certain points in the same cluster, but do not determine such an inclusion unequivocally [83, 94, 136].

1.2.7 Diagnostics and Assessment of Partition Variability

Another area of potentially high importance that researchers paid little attention to is diagnostics in model-based clustering including the detection of outliers, scatter, and influential observations. The problem of outliers has been addressed in several different ways. One popular approach is to introduce an additional uniform component defined on a hypercube constructed over the data domain [8]. Presumably, scatter data points that are not fitted well by the rest of components, will be assigned to this uniform distribution, thus collecting noise observations and automatically cleaning the dataset. Despite good performance in many practical scenarios, this approach is not robust to breakdowns as shown by Hennig [65]. Also, the asymptotic properties of the obtained estimator are not satisfactory as a proper maximum likelihood estimator is not defined in this setting. In the paper by Hennig and Coretto [67], the authors recommend employing an improper uniform distribution which is not dependent on data and claim that this method is more robust than the original one.

Another possible approach of handling outliers is based on a mixture of distributions with heavy tails such as t and skew t distributions [85, 109, 140]. This method does not automatically detect outliers but provides a better fit than traditional Gaussian mixtures. If the goal of the analysis is to find outliers, it is possible to consider $(1-\alpha)$-confidence regions for all estimated components and declare observations beyond these regions outliers. In Sect. 1.2.2, there were mentioned methods based on the trimmed log likelihood function. Although they are primarily devoted to the robust parameter estimation in presence of outliers, they can also be used for detecting unusual observations.

While the problem of outliers and scatter in mixture models is considered in literature, the detection of influential observations is nearly fully omitted. As per definition given by Jolliffe et al. [74], influential observations in cluster analysis are data points whose exclusion from the data would lead to different group assignments of other points. As remarked by Melnykov [115], influential observations can be different in nature. Some can influence results as important observations that affect cluster shape and structure. In particular, spurious components discussed in Sect. 1.2.2 are associated with random patterns in data and often involve influential observations. Others can affect the produced partition due to their location between components. Such observations often have high classification uncertainty. Hence, the problem of assessing uncertainty in the estimated classification vector is closely related to the subject discussed.

Recently, Melnykov [117] showed that in the case of Gaussian mixture models, the posterior probability $\pi_{ik}(\hat{\boldsymbol{\Psi}})$ is asymptotically normally distributed with mean $\pi_{ik}(\boldsymbol{\Psi})$ and variance that can be approximated by $\nabla\pi_{ik}(\hat{\boldsymbol{\Psi}})' \, \mathbf{I}^{-1}(\hat{\boldsymbol{\Psi}})\nabla\pi_{ik}(\hat{\boldsymbol{\Psi}})$ with the gradient vector $\nabla\pi_{ik}(\boldsymbol{\Psi}) = \left(\left(\frac{\partial\pi_{ik}(\boldsymbol{\Psi})}{\partial\tau_s}\right)'_{s=1,\ldots,K-1}, \left(\frac{\partial\pi_{ik}(\boldsymbol{\Psi})}{\partial\boldsymbol{\mu}_s}\right)'_{s=1,\ldots,K}, \left(\frac{\partial\pi_{ik}(\boldsymbol{\Psi})}{\partial \mathrm{vech}\{\boldsymbol{\Sigma}_s\}}\right)'_{s=1,\ldots,K}\right)'$ and partial derivatives specified by the following expressions:

$$\frac{\partial\pi_{ik}(\boldsymbol{\Psi})}{\partial\tau_k} = \frac{\pi_{ik}\pi_{iK}}{\tau_K} + \frac{\pi_{ik}(1-\pi_{ik})}{\tau_k}, \qquad \frac{\partial\pi_{ik}(\boldsymbol{\Psi})}{\partial\tau_s} = \frac{\pi_{ik}\pi_{iK}}{\tau_K} - \frac{\pi_{ik}\pi_{is}}{\tau_s}, \quad s \neq k,$$

$$\frac{\partial\pi_{iK}(\boldsymbol{\Psi})}{\partial\tau_s} = -\frac{\pi_{iK}(1-\pi_{iK})}{\tau_K} - \frac{\pi_{iK}\pi_{is}}{\tau_s},$$

$$\frac{\partial\pi_{ik}(\boldsymbol{\Psi})}{\partial\boldsymbol{\mu}_k} = \pi_{ik}(1-\pi_{ik})\boldsymbol{\Sigma}_k^{-1}(\mathbf{y}_i-\boldsymbol{\mu}_k), \qquad \frac{\partial\pi_{ik}(\boldsymbol{\Psi})}{\partial\boldsymbol{\mu}_s} = -\pi_{ik}\pi_{is}\boldsymbol{\Sigma}_s^{-1}(\mathbf{y}_i-\boldsymbol{\mu}_s), \quad s \neq k,$$

$$\frac{\partial\pi_{ik}(\boldsymbol{\Psi})}{\partial \mathrm{vech}\{\boldsymbol{\Sigma}_k\}} = \frac{1}{2}\pi_{ik}(1-\pi_{ik})\mathbf{G}'\mathrm{vec}\left\{\boldsymbol{\Sigma}_k^{-1}((\mathbf{y}_i-\boldsymbol{\mu}_k)(\mathbf{y}_i-\boldsymbol{\mu}_k)'\boldsymbol{\Sigma}_k^{-1} - \mathbf{I}_p)\right\},$$

$$\frac{\partial\pi_{ik}(\boldsymbol{\Psi})}{\partial \mathrm{vech}\{\boldsymbol{\Sigma}_s\}} = -\frac{1}{2}\pi_{ik}\pi_{is}\mathbf{G}'\mathrm{vec}\left\{\boldsymbol{\Sigma}_s^{-1}((\mathbf{y}_i-\boldsymbol{\mu}_s)(\mathbf{y}_i-\boldsymbol{\mu}_s)'\boldsymbol{\Sigma}_s^{-1} - \mathbf{I}_p)\right\}, \quad s \neq k.$$

The vec operator stacks matrix columns on top of each other, vech operator stacks columns of the lower triangular part of a symmetric matrix, and $\mathbf{G}$ is the unique matrix of ones and zeros such that $\mathrm{vec}(\boldsymbol{\Sigma}) = \mathbf{G}\mathrm{vech}(\boldsymbol{\Sigma})$. Since posterior probabilities can often be close to the boundary of the [0, 1]-interval, the author recommends employing the well-known transformation $\gamma_{ik}(\boldsymbol{\Psi}) = \arcsin(\sqrt{\pi_{ik}(\boldsymbol{\Psi})})$ and the corresponding backward transformation $\pi_{ik}(\boldsymbol{\Psi}) = \sin^2(\gamma_{ik}(\boldsymbol{\Psi}))$. An approach to identify observations with uncertain class assignments for several popular classification datasets was considered by Melnykov [117].

1.2.8 Miscellaneous Topics

In this section, we briefly mention some other important topics related to mixture modeling and model-based clustering.

The Bayesian approach to modeling provides a way to incorporate available prior information about parameters in a coherent way. In this setup, inference is made based on the posterior distribution of parameters, which can be formulated according to Bayes' theorem as

$$p(\boldsymbol{\Psi}|\mathbf{y}_1, \ldots, \mathbf{y}_n) \propto p(\mathbf{y}_1, \ldots, \mathbf{y}_n|\boldsymbol{\Psi})p(\boldsymbol{\Psi}),$$

where $p(\mathbf{y}_1, \ldots, \mathbf{y}_n|\boldsymbol{\Psi})$ is essentially the likelihood function and $p(\boldsymbol{\Psi})$ is the prior distribution of the parameters. The posterior distribution is used then to provide inference about estimated parameters. Unfortunately, except for the simplest models, the analytical form of $p(\boldsymbol{\Psi}|\mathbf{y}_1, \ldots, \mathbf{y}_n)$ is unavailable or difficult to find. As a result, researchers rely on numerical methods such as Markov Chain Monte Carlo (MCMC) techniques to approximate the posterior distribution and find its summaries. One of the first papers considering the MCMC approach in the course of the mixture modeling estimation with a known number of components is [41]. Since then, there were substantial developments in mixture modeling under the Bayesian framework [45, 53, 95, 129]. It is easy to see that Bayesian estimation based on finding the mode of a posterior distribution becomes equivalent to the likelihood-based inference when the prior is uniform. Thus, Bayesian inference provides a somewhat more general setting, in which additional information about model parameters can be taken into account. For example, the inclusion of a prior on the number of components was investigated by Richardson and Green [146]. This additional information can help relax issues with overfitting the order of mixture models. Some of the main drawbacks of the Bayesian approach to mixture modeling include cumbersome calculations and the so-called label switching problem which complicates posterior simulations and is closely related to the lack of identifiability of $\boldsymbol{\Psi}$.

Identifiability is the ability to recover the probability distribution given its parameters. It is a necessary model characteristic for providing consistent inference. The family of distributions is called identifiable if $f(\mathbf{y}|\boldsymbol{\Psi}) = f(\mathbf{y}|\boldsymbol{\Psi}^*)$ holds if and only if $\boldsymbol{\Psi} = \boldsymbol{\Psi}^*$. In finite mixture models, the problem of identifiability occurs due to possible label switching and is usually solved by imposing restrictions on mixing proportions or other model parameters. There is an extensive literature dealing with the identifiability issue for various mixture models [31, 69, 76, 159, 172]. In general, the problem of identifiability can be successfully overcome in the mixture modeling context. Some recent papers dealing with label switching and relabeling algorithms include [52, 72, 137, 148].

Variational approximations are another topic that arises frequently in the Bayesian framework. The idea behind it is to find an approximate distribution such that it is close to the target posterior distribution which has a complicated form and is difficult to obtain. The Kullback-Leibler divergence is traditionally used to measure the closeness of the target and approximating distributions. In general, variational methods aim at maximizing a lower bound for the log of the marginal distribution of data $p(\mathbf{y}_1, \ldots, \mathbf{y}_n)$. This bound is specified by the inequality given by

$$\log p(\mathbf{y}_1, \ldots, \mathbf{y}_n) \geq \int \sum_{\mathbf{z}} q(\boldsymbol{\Psi}, \mathbf{z}|\mathbf{y}_1, \ldots, \mathbf{y}_n) \log \frac{p(\mathbf{y}_1, \ldots, \mathbf{y}_n, \mathbf{z}, \boldsymbol{\Psi})}{q(\boldsymbol{\Psi}, \mathbf{z}|\mathbf{y}_1, \ldots, \mathbf{y}_n)} d\boldsymbol{\Psi},$$

where $p(\mathbf{y}_1, \ldots, \mathbf{y}_n, \mathbf{z}, \boldsymbol{\Psi})$ is the joint distribution and $q(\boldsymbol{\Psi}, \mathbf{z}|\mathbf{y}_1, \ldots, \mathbf{y}_n)$ is commonly known as the variational approximation of the posterior. Thus, the right-hand side of the above inequality is maximized over the choice of $q(\boldsymbol{\Psi}, \mathbf{z}|\mathbf{y}_1, \ldots, \mathbf{y}_n)$. A variational approximation in the context of Gaussian mixtures was provided by McGrory and Titterington [105]. Some important papers on similar topics include [3, 35]. Consistency and other theoretical aspects of estimators obtained by variational approximations are discussed in [59, 164].

An interesting approach to modeling Gaussian mixtures that is worth mentioning lies in the use of Gaussian graphical models. Such a model puts an observation $\mathbf{Y}$ from a Gaussian density $\phi_p(\mathbf{y}; \boldsymbol{\mu}, \boldsymbol{\Sigma})$ in correspondence with an undirected graph $G = (V, E)$, where V is the set of p vertices, matching the dimensionality of $\mathbf{Y}$, and E is the set of edges designed in such a manner that the edge between vertices i and j is present only if the matching (i, j)th entry of $\boldsymbol{\Sigma}^{-1}$ is non-zero [174]. At the same time, the absence of an edge between vertices i and j reflects the conditional independence of the ith and jth coordinates of $\mathbf{Y}$ given the rest of the coordinates of this vector [82]. [93] consider a mixture where each cluster is represented by an individual Gaussian graphical model. They focus on sparse cluster distributions and implement a penalized EM algorithm [38] with L_1 penalty that stimulates the sparsity of precision matrices $\boldsymbol{\Sigma}_k^{-1}$, i.e., leads to a large number of zero off-diagonal entries in these matrices. This approach helps with problems associated with the estimation of a large number of parameters that arise in the analysis of high-dimensional data.

1.3 Modern Applications of Model-Based Clustering

In this section, we present some recent applications of model-based clustering. The list of considered topics is by no means comprehensive, however, we attempt to show the diversity of applications.

1.3.1 Clustering Tree Ring Sequences

Dendrochronology is an area of science that studies tree ring data and has numerous applications in other fields. The width of tree rings, also known as growth rings, represents the amount of volume gained over a period of time. This growth can be affected by climate and environmental conditions. In climatology, tree rings are used to reconstruct past temperature and climate trends [46]. In archaeology, they are used to identify the origin of wooden artifacts [24]. In this process, researchers often rely on subjective assessments based on the resemblance of the tree ring sequence of an artifact to that of known sequences from different localities. Clustering and group identification approaches are often useful here; however, previous attempts of employing traditional clustering approaches in dendrochronology research have encountered substantial difficulties [62]. In light of this, [123] applied model-based clustering of Gaussian regression time series to identify groups of trees with similar annual tree ring behavior over 281 years from 1700 to 1980. The analyzed dataset included 160 trees from 9 locations in western regions of the United States. The goal of the study was to learn about climate patterns in this area.

The authors let $\mathbf{y}_i = (y_{i1}, \ldots, y_{iT})'$ denote a T-variate observed time series vector from a sample of n observations representing tree ring widths. $\mathbf{X}_i$ denotes a $T \times b$ matrix constructed on b explanatory variables. The first column of $\mathbf{X}_i$ consists of ones and is responsible for modeling the intercept, while the second column represents the current tree age. Then, the regression model is given by $\mathbf{y}_i = \mathbf{X}_i \boldsymbol{\beta} + \boldsymbol{\nu}_i$, where $\boldsymbol{\beta}$ is the vector of coefficients and $\boldsymbol{\nu}_i$ is the error term following an autoregressive moving average process of orders p and q, respectively. The kth mixture component has the form $\phi_T(\mathbf{y}; \mathbf{X}\boldsymbol{\beta}_k, \boldsymbol{\Sigma}_k)$, where $\mathbf{X}\boldsymbol{\beta}_k$ is the mean vector and $\boldsymbol{\Sigma}_k$ is the covariance matrix of the T-variate Gaussian density. The process of parameter estimation via the EM algorithm is thoroughly described by Melnykov and Michael [123]. To avoid operations with high-dimensional covariance matrices, the authors expressed the likelihood function via the Kalman filter [75]. The best model chosen based on BIC included nine components. The authors used hues of the red, blue, and green colors to represent the solution and to show that their solution made good geographical sense as the clusters were formed along the north-south and west-east directions.

Other works related to clustering Gaussian regression time series and considering other interesting applications can be found in papers of [32] and [114].

1.3.2 Identification of Differentially Expressed Genes

Gene expression is the transformation of information contained in a DNA sequence into a protein. There are two steps involved in this process: first, the DNA sequence is transcribed to messenger RNA and, second, the messenger RNA is translated into a protein. Microarray analysis has made it possible to have an insight into thousands of gene expressions simultaneously. In microarray analysis of genes, one of the objectives is to identify genes that are differentially expressed between two or more mutually exclusive groups. For example, the interest could lie in finding the genes that have increased or decreased expression between normal and abnormal cells or control and treatment groups. To distinguish the genes that are differentially expressed from those that are not, a multiple comparison of means that describe the groups can be conducted. The tests commonly used to compare means are t-test (two groups) and ANOVA (more than two groups). These procedures require the control of the error rate at a pre-specified level. One approach to tackle this problem is to use a Bonferroni-type adjustment which controls the familywise error rate (FWER). However, as the number of comparisons can be extremely large, depending on the number of genes to be evaluated, this approach can result in the reduced power of the procedure making it impractical. Another approach is to control the expected value of the proportion of genes incorrectly identified as differentially expressed, also known as the false discovery rate (FDR) [17]. The paper by Efron et al. [44] considers a mixture model approach to determine differentially expressed genes using the FDR as described below.

Let $g_1, g_2, \ldots, g_n$ be a set of genes from $x_1, x_2, \ldots, x_s$ samples. Let G_1 be the group of genes that are not differentially expressed and G_2 the group of genes that are differentially expressed. Let τ_1 be the prior probability of gene membership in G_1 and hence $1 - \tau_1 = \tau_2$ is the prior probability of its membership in G_2. After conducting a t-test for each gene, let the test statistic for the ith gene g_i be denoted by Y_i. Then, the mixture density for g_i can be written as

$$f(y_i|\Psi) = \tau_1 f_1(y_i|\boldsymbol{\vartheta}_1) + \tau_2 f_2(y_i|\boldsymbol{\vartheta}_2), \tag{1.12}$$

where f_1 denotes the density of the test statistics of non-differentially expressed genes (G_1) and f_2 stands for the density of differentially expressed genes (G_2). The posterior probabilities, π_{i1} and π_{i2}, represent the probabilities of the ith gene coming from G_1 and G_2, respectively. Hence, π_{i1} can be interpreted as the FDR for the ith gene.

The work by McLachlan et al. [111] considers a two-component mixture model by transforming the p-value p_i obtained from the test statistic y_i into a z-score. The z-scores are calculated as $z_i = \Phi^{-1}(1 - p_i)$, where Φ is the cdf of the standard normal distribution. Then, (1.12) is modified by replacing y_i with z_i and normal distributions are used as mixture components. The authors proposed the first component to be a standard normal, corresponding to the null distribution of z-scores, and second component to be a normal distribution with mean μ_2 and variance σ_2^2 to be estimated. If the null distribution does not appear to be standard normal, then it will be replaced by a normal distribution with mean μ_1 and variance σ_1^2. The estimation of parameters is carried out using the EM algorithm. In their article, [111] applied the proposed model to three real-life datasets and showed that the two-component mixture model fits the empirical distribution of z-scores very well. Since this work, there have been several others that considered the general framework of two-component mixture models with different density functions in the roles of f_1 and f_2 and p_i instead of y_i in (1.12). For example, Robin et al. [147] used a pre-specified parametric distribution for f_1 such as the uniform distribution and a weighted kernel function to estimate f_2. The paper by Markitsis and Lai [99] proposed a censored beta distribution to account for the p-values that are close to zero. Another work by Jiao and Zhang [73] used t-distributions instead of Gaussian densities for both f_1 and f_2 and implemented the control of FWER rather than FDR.

1.3.3 Analysis of Customer Navigation Patterns

With the growth of web-usage and web-related commerce, it has become important to develop techniques capable of extracting and analyzing information from massive datasets containing information about customers. One direction of research that can help in identifying customer trends and preferences is the analysis of web-routes taken by users. A sequence of sites or web-pages visited is traditionally called clickstream. One challenge related to the analysis of clickstream information is its categorical nature. Also, this type of data is not static and thus can be best modeled by approaches reflecting its dynamic characteristics. Among the papers devoted to clustering clickstreams, there are works by Inbarani and Thangavel [71], Liu [90], and Ypma and Heskes [173].

Model-based clustering of clickstream data using a mixture of first-order Markov models was first proposed by Cadez et al. [26]. Let $\mathbf{y}_i = (y_{i1}, \ldots, y_{iT_i})'$ be the observed clickstream sequence for the ith user with T_i representing the length of the ith sequence. Given J unique sites (states) in the system, one can introduce a transition probability matrix given by

$$\mathbf{P} = \begin{Vmatrix} p_{11} & p_{12} & \cdots & p_{1J} \\ p_{21} & p_{22} & \cdots & p_{2J} \\ \vdots & \vdots & \ddots & \vdots \\ p_{J1} & p_{J2} & \cdots & p_{JJ} \end{Vmatrix},$$

where $p_{jj'} = \Pr(Y_{it} = j'|Y_{i(t-1)} = j)$ for $i = 1, \ldots, n$. Denoting $\alpha_j = \Pr(Y_{i1} = j)$, the proposed mixture model is given by

$$f(y_{i1}, \mathbf{x}_i|\Psi) = \sum_{k=1}^{K} \tau_k \prod_{j=1}^{J} \alpha_{kj}^{I(y_{i1}=j)} \prod_{j=1}^{J} \prod_{j'=1}^{J} p_{kjj'}^{x_{ijj'}},$$

where $\mathbf{x}_i$ is a $J \times J$ matrix with elements $x_{ijj'}$ representing the number of transitions from state j to state j' by the ith user, and $p_{kjj'}$ is the transition probability from j to j' for the kth component. Estimation of the parameter vector Ψ is outlined in [26]. The developed methodology was used to cluster web-users of the msnbc.com site for a data collected over one day. A training set of 100,023 clickstreams was used first to build the model which was applied to cluster 98,687 clickstream sequences producing around 60 clusters.

The paper by Melnykov [119] provided an extension to the model of [26] developing an algorithm with a capability to group similar web-pages and customers at the same time. This setting, in which objects are clustered along with data features, is called biclustering. Another biclustering approach for grouping customers and products by means of finite mixture models was proposed by Vicari and Alfó [161]. Their model had a hierarchical structure based on customer- and product-specific segments. The authors claimed that their methodology can be used for any dataset when one is interested in grouping both observations and their features.

1.3.4 Data Clustering on Unit-Hypersphere

Multivariate data are called directional if the direction of a vector observation is relevant as opposed to its magnitude. Some examples of applications in which directional data occur frequently include the analysis of gene expression profiles [7, 43], text documents [2, 7], and wind information [101]. Directional data are usually obtained by a transformation of some original data. For example, text documents are processed as follows. First, all words in analyzed documents are listed. Then, the frequency of all words gets recorded for each document. After cleaning the data from noisy words with too high or too low frequencies, each document-specific vector is normalized. As a result, all observations lie on a unit hypersphere. The most popular choice for modeling such data is the von Mises-Fisher (vMF) distribution. In the context of model-based clustering, finite mixtures of vMF distributions can be used. A p-variate vMF distribution is defined for a p-dimensional random unit vector $\mathbf{y}$, subject to the restriction $\|\mathbf{y}\| = 1$. Therefore, $\mathbf{y} \in S^{p-1}$, where S^{p-1} denotes a $(p-1)$-dimensional hypersphere. If $p = 2$, S^{p-1} reduces to a unit circle. The kth vMF mixture component is given by

$$f_k(\mathbf{y}|\boldsymbol{\vartheta}_k) = C_p(\eta_k) \exp\left\{\eta_k \boldsymbol{\mu}_k^T \mathbf{y}\right\},$$

where $\boldsymbol{\vartheta}_k = (\eta_k, \boldsymbol{\mu}_k')'$ is the parameter vector built on the concentration parameter η_k and mean $\boldsymbol{\mu}_k'$, $C_p(\eta_k) = \eta_k^{p/2-1}\{(2\pi)^{p/2}\mathscr{I}_{p/2-1}(\eta_k)\}^{-1}$ and $\mathscr{I}_r$ is the order r modified Bessel function of first kind.

The work by Banerjee et al. [7] discusses the estimation of model parameters based on the EM algorithm. The corresponding Q-function has the form

$$Q(\boldsymbol{\Psi}|\boldsymbol{\Psi}^{(b-1)}) = \sum_{i=1}^{n}\sum_{k=1}^{K} \pi_{ik}^{(b)} \{\log \tau_k + \log C_p(\eta_k) + \eta_k \boldsymbol{\mu}_k^T \mathbf{y}_i\} - \sum_{k=1}^{K} \lambda_k(\boldsymbol{\mu}_k^T \boldsymbol{\mu}_k - 1),$$

where λ_ks are Lagrange multipliers accounting for the restriction $\boldsymbol{\mu}_k^T \boldsymbol{\mu}_k = 1$ for $k = 1, \ldots, K$. A closed-form expression can be obtained for parameters τ_k and $\boldsymbol{\mu}_k$:

$$\boldsymbol{\mu}_k^{(b)} = \frac{\sum_{i=1}^{n} \pi_{ik}^{(b)} \mathbf{y}_i}{\|\sum_{i=1}^{n} \pi_{ik}^{(b)} \mathbf{y}_i\|} \quad \text{and} \quad \tau_k^{(b)} = \frac{\sum_{i=1}^{n} \pi_{ik}^{(b)}}{n}.$$

Taking the derivative of the Q-function with respect to η_k and setting it equal to zero results in

$$\frac{C_p'(\eta_k)}{C_p(\eta_k)} = -\frac{\|\sum_{i=1}^{n} \pi_{ik} \mathbf{y}_i\|}{\sum_{i=1}^{n} \pi_{ik}}.$$

The analytical expression for η_k is not readily available as it involves finding the derivative of the Bessel function. In their paper, Banerjee et al. [7] used an approximation for η_k given by $\eta_k \approx (p\bar{r} - \bar{r}^3)(1 - \bar{r}^2)^{-1}$, where

$\bar{r} = \| \sum_{i=1}^{n} \pi_{ik} \mathbf{y}_i \| / n$. Other works provide alternative ways to approximate η_k [40, 98, 158]. Applications of the developed model-based clustering approach to the analysis of gene expressions and text documents are considered in [7].

More recently, mixtures of vMF distributions were considered in the Bayesian framework [56]. A Dirichlet process mixture model of vMF distribution for text document clustering was proposed by Anh et al. [2]. Another interesting piece of work was conducted by Dortet-Bernadet and Wicker [43] who used the inverse stereographic projection of $(p - 1)$-variate Gaussian mixtures to cluster gene expression data. The authors remarked that they prefer this model as mixtures of vMF distributions tend to produce spherical clusters.

1.3.5 Analysis of Mass Spectrometry Data

Mass spectrometry is an analytical technique used to determine the composition of compounds and chemical structure of molecules. Secondary ion mass spectrometry (SIMS) is one among many existing mass spectrometry methods. In SIMS, a sample is bombarded by a primary ion which generates charged particles (secondary ions) from the sample surface. The secondary ions are then captured, their mass-to-charge ratios are measured and the corresponding spectrum is constructed. Analyzing a substance sample in a liquid matrix, SIMS has certain advantages over other methods. One is that it has higher ion yields and lower fragmentation or noise [39]. Moreover, the liquid matrix can be used to study the stages of chemical reactions as pointed out by Melnikov et al. [113]. The use of the liquid matrix, however, leads to producing a mixed mass spectrum that includes peaks originating from both the liquid matrix and sample. In the meantime, it is easy to observe a mass spectrum of the liquid alone. Thus, it is essential to separate the secondary ions coming from the studied substance by making a comparison between the mixed and liquid matrix spectra. The problem can be seen as a classical task of extracting signal that is observed in the presence of noise.

In the paper by Melnykov [116], the author proposed a model-based clustering approach to extract the informative part of the mixed spectrum. Two observed samples of mass-to-charge ratio measurements representing the liquid matrix $\mathbf{x} = (x_1, \ldots, x_{n_x})'$ and studied substance $\mathbf{y} = (y_1, \ldots, y_{n_y})'$ are provided. Here, n_x and n_y stand for the total number of particles associated with each sample. Since mass-to-ratio values can be measured with specific accuracy, one observes multiple particles at the same locations. This allows for more convenient notational form given by $\mathbf{x} = \{(x_{(1)}, n_{x_{(1)}}), \ldots, (x_{(r_x)}, n_{r_{(x)}})\}$ and $\mathbf{y} = \{(y_{(1)}, n_{y_{(1)}}), \ldots, (y_{(r_y)}, n_{r_{(y)}})\}$, where $n_{x_{(i)}}$ and $n_{y_{(j)}}$ denote observed frequencies at peak locations $x_{(i)}$ and $y_{(j)}$, respectively, and $r_{(x)}$ and $r_{(y)}$ represent the number of distinct peaks in each sample. The author used the discrete mixture component distribution

$$f_k(x|\boldsymbol{\vartheta}_k) = \Phi\left(\frac{x + 0.5 - \mu_k}{\sigma_k}\right) - \Phi\left(\frac{x - 0.5 - \mu_k}{\sigma_k}\right),$$

where Φ is the cdf of the standard normal distribution and $\boldsymbol{\vartheta}_k = (\mu_k, \sigma_k)'$. The liquid matrix can be modeled by a mixture of K components given by

$$f_x(x|\boldsymbol{\Psi}_x) = \sum_{k=1}^{K} \tau_k f_k(x|\boldsymbol{\vartheta}_k)$$

and the mixed spectrum can be modeled by the mixture of $K + M$ components given by

$$f_y(y|\boldsymbol{\Psi}_y) = c\sum_{k=1}^{K} \tau_k f_k(y|\boldsymbol{\vartheta}_k) + \sum_{k=K+1}^{K+M} \tau_k f_k(y|\boldsymbol{\vartheta}_k).$$

Here, the first K components represent the liquid and c is a factor assuring that all mixing proportions sum up to the unity. Then, the corresponding Q-function has the form

$$Q(\boldsymbol{\Psi}|\mathbf{x}, \mathbf{y}) = \sum_{i=1}^{r_x} n_{x_{(i)}} \sum_{k=1}^{K} \pi_{x_{(i)}k}\{\log \tau_k + \log f_k(x_{(i)}|\mu_k, \sigma_k)\} + \sum_{i=1}^{r_y} n_{y_{(i)}} \sum_{k=K+1}^{K+M} \pi_{y_{(i)}k}$$
$$\times \{\log \tau_k + \log f_k(y_{(i)}|\mu_k, \sigma_k) + I(k \leq K) \log c\} - \lambda_1 \left(\sum_{k=1}^{K} \tau_k - 1\right)$$
$$- \lambda_2 \left(c + \sum_{k=K+1}^{K+M} \tau_k - 1\right),$$

where $\boldsymbol{\Psi}$ denotes the entire parameter vector and λ_1 and λ_2 are Lagrange multipliers associated with the two restrictions on mixing proportions.

Using this model, Melnykov [116] successfully extracted mass spectra for different concentrations of organic dyes in a glycerol liquid matrix. The developed methodology proved to be promising as the obtained peaks were sensible from a biochemical standpoint.

1.3.6 Network Clustering

A network can be defined as a collection of units (also called actors) interacting with each other. For example, networks are used to represent social relationships such as friendship within a group of people. One of the points of interest in the analysis of network data is grouping actors.

A network model usually has the following construction. Let n be the number of actors and let y_{ij} represent the relationship between the ith and jth actors for $i, j = 1, \ldots, n$ such that $i \neq j$. Then, the dataset $\mathbf{Y} = \left(y_{ij}\right)_{n \times n}$ is a matrix of ties among actors. This setup is usually represented by a graph, where the actors are shown as nodes and the edges indicate ties among the actors. In its simplest form, y_{ij} is a binary variable taking values 0 or 1 showing an absence or presence of a tie between actors y_i and y_j. Undirected ties are represented by $y_{ij} = y_{ji}$ for all i and j and directed ones are accounted for in the case when y_{ij} is not necessarily equal to y_{ji}. Depending on whether the tie is directed or not, there will be $n(n-1)$ or $n(n-1)/2$ edges, hence the data could be very high-dimensional.

The paper by Handcock et al. [61] proposed a two-stage model that incorporates the latent position method of [68] and model-based clustering via mixture models. The latent position of actors represents their social standing in an unobserved low-dimensional space. During the first stage, a conditional independence modeling approach is implemented. Let $\mathbf{x}_{ij}$ contain available covariate information for y_{ij}. Assuming that the presence or absence of a tie y_{ij} is independent of other ties given the latent social position of individuals, the conditional independence model is

$$\Pr\{\mathbf{Y}|\mathbf{Z}, \mathbf{X}, \boldsymbol{\varphi}\} = \prod_{i=1}^{n} \prod_{\substack{j=1 \\ j \neq i}}^{n} \Pr\{y_{ij}|\mathbf{z}_i, \mathbf{z}_j, \mathbf{x}_{ij}, \boldsymbol{\varphi}\},$$

where $\boldsymbol{\varphi}$ is the parameter vector of the model, $\mathbf{X}$ is the set of all $\mathbf{x}_{ij}$ vectors, and $\mathbf{Z} = \{\mathbf{z}_1, \mathbf{z}_2, \ldots, \mathbf{z}_n\}$ represents the latent social positions of the actors in a low-dimensional space. $\Pr\{y_{ij}|\mathbf{z}_i, \mathbf{z}_j, \mathbf{x}_{ij}, \boldsymbol{\varphi}\}$ that shows the probability of a tie between y_i and y_j given their latent positions $\mathbf{z}_i$ and $\mathbf{z}_j$ as well as explanatory variables $\mathbf{x}_{ij}$ is modeled using logistic regression as follows:

$$\log\left(\frac{\Pr(y_{ij} = 1|\mathbf{z}_i, \mathbf{z}_j, \mathbf{x}_{ij}, \boldsymbol{\varphi})}{\Pr(y_{ij} = 0|\mathbf{z}_i, \mathbf{z}_j, \mathbf{x}_{ij}, \boldsymbol{\varphi})}\right) = \mathbf{x}_{ij}'\boldsymbol{\beta} - \alpha|\mathbf{z}_i - \mathbf{z}_j|,$$

where $\boldsymbol{\varphi} = (\alpha, \boldsymbol{\beta}')'$ and α is the coefficient of the Euclidean distance between the actors y_i and y_j in the latent space $\mathbf{Z}$. $\boldsymbol{\beta}$ denotes the coefficient vector linked to the covariates specific to y_{ij}. At stage two, $\mathbf{Z}$ is assumed to originate from a mixture of K multivariate Gaussian distributions with means $\boldsymbol{\mu}_k$ and spherical covariance matrices $\sigma_k^2\mathbf{I}$ given by

$$f(\mathbf{z}_i|\boldsymbol{\Psi}) = \sum_{k=1}^{K} \tau_k \phi_p(\mathbf{z}_i|\boldsymbol{\mu}_k, \sigma_k^2\mathbf{I}),$$

where $\mathbf{I}$ stands for an identity matrix.

Two procedures to estimate the parameters of this model are proposed by Handcock et al. [61]. The first one uses maximum likelihood estimation via the EM algorithm. The second approach relies on Bayesian inference. Figure 1.5a provides a network graph created by the R package *network* [25] that represents the classical dataset *monks* [151]. The solution of the latent position clustering model applied to this dataset using the R package *latentnet* [80,81] is provided in Fig. 1.5b. Fast parameter estimation procedures for this model are proposed by Raftery et al.

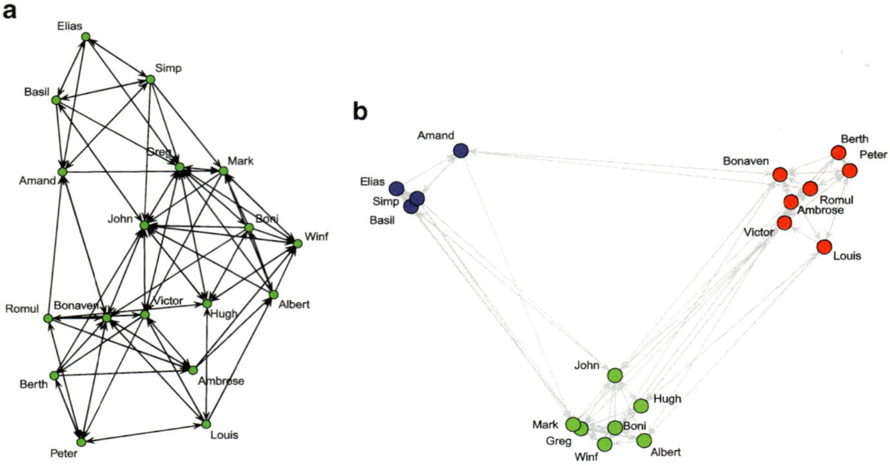

Fig. 1.5 Illustration for the monk data network showing friendship relationships for 18 monks in a monastery: (**a**) network graph and (**b**) groups of individuals identified by latent position clustering model for $K = 3$ and $p = 2$

[143] and Salter-Townshend and Murphy [150]. The latent position clustering model was extended by Gormley and Murphy [57] to account for the covariates in identification of clusters in addition to finding probabilities of relationships between two actors. In more recent work, Vu et al. [162] use the idea of minorization-maximization algorithm of [134] to implement model-based clustering of large network data. Their method was able to analyze more than 131,000 nodes and 17 billion edge variables.

References

1. Akaike H (1973) Information theory and an extension of the maximum likelihood principle. In: Second international symposium on information theory, pp 267–281
2. Anh NK, Tam NT, Van Linh N (2013) Document clustering using Dirichlet process mixture model of von Mises-Fisher distributions. In: Proceedings of the fourth symposium on information and communication technology, New York, pp 131–138
3. Attias H (1999) Inferring parameters and structure of latent variable models by variational Bayes. In: Proceedings of the fifteenth conference on uncertainty in artificial intelligence
4. Azzalini A, Bowman AW (1990) A look at some data on the old faithful geyser. J R Stat Soc C 39:357–365
5. Azzalini A, Menardi G (2013) Package pdfCluster: cluster analysis via nonparametric density estimation. http://cran.r-project.org/web/packages/pdfCluster
6. Azzalini A, Torelli N (2007) Clustering via nonparametric density estimation. Stat Comput 17:71–80
7. Banerjee A, Dhillon IS, Ghosh J, Sra S (2005) Clustering on the unit hypersphere using von Mises-Fisher distributions. J Mach Learn Res 6:1345–1382

8. Banfield JD, Raftery AE (1993) Model-based Gaussian and non-Gaussian clustering. Biometrics 49:803–821
9. Bar-Hillel A, Hertz T, Shental N, Weinshall D (2003) Learning distance functions using equivalence relations. In: Proceedings of the twentieth international conference on machine learning, pp 11–18
10. Basso R, Lachos V, Cabral C, Ghosh P (2010) Robust mixture modeling based on scale mixtures of skew-normal distributions. Comput Stat Data Anal 54(12):2926–2941
11. Basu S, Banerjee A, Mooney R (2002) Semi-supervised clustering by seeding. In: Proceedings of the 19th international conference on machine learning, pp 19–26
12. Basu S, Bilenko M, Mooney RJ (2004) A probabilistic framework for semi-supervised clustering. In: Proceedings of the tenth ACM SIGKDD international conference on knowledge discovery and data mining, pp 59–68
13. Baudry JP, Raftery A, Celeux G, Lo K, Gottardo R (2010) Combining mixture components for clustering. J Comput Graph Stat 19(2):332–353
14. Benaglia T, Chauveau D, Hunter DR (2009) An EM-like algorithm for semi- and nonparametric estimation in multivariate mixtures. J Comput Graph Stat 18(2):505–526
15. Benaglia T, Chauveau D, Hunter DR, S YD (2009) mixtools: an R package for analyzing mixture models. J Stat Softw 32(6):1–29
16. Benaglia T, Chauveau D, Hunter DR (2011) Bandwidth selection in an EM-like algorithm for nonparametric multivariate mixtures. In: Hunter D, Richards DSP, Rosenberger J (eds) Nonparametric statistics and mixture models, A Festschrift in honor of Thomas P Hettmansperger. World Scientific, Singapore, pp 15–27
17. Benjamini Y, Hochberg Y (1995) Controlling the false discovery rate: a practical and powerful approach to multiple testing. J R Stat Soc 57:289–300
18. Berlinet AF, Roland C (2012) Acceleration of the em algorithm: P-em versus epsilon algorithm. Comput Stat Data Anal 56(12):4122–4137
19. Biernacki C, Celeux G, Gold EM (2000) Assessing a mixture model for clustering with the integrated completed likelihood. IEEE Trans Pattern Anal Mach Intell 22:719–725
20. Biernacki C, Celeux G, Govaert G (2003) Choosing starting values for the EM algorithm for getting the highest likelihood in multivariate Gaussian mixture models. Comput Stat Data Anal 413:561–575
21. Böhning D, Dietz E, Schaub R, Schlattmann P, Lindsay B (1994) The distribution of the likelihood ratio for mixtures of densities from the one-parameter exponential family. Ann Inst Stat Math 46(2):373–388
22. Bouveyron C, Brunet C (2014) Model-based clustering of high-dimensional data: a review. Comput Stat Data Anal 71:52–78
23. Bouveyron C, Girard S, Schmid C (2007) High-dimensional data clustering. Comput Stat Data Anal 52(1):502–519. http://lear.inrialpes.fr/pubs/2007/BGS07a
24. Bridge M (2012) Locating the origins of wood resources: a review of dendroprovenancing. J Archaeol Sci 39(8):2828–2834
25. Butts CT, Handcock MS, Hunter DR (2014) Network: classes for relational data. Irvine. R package version 1.9.0, http://statnet.org/
26. Cadez I, Heckerman D, Meek C, Smyth P, White S (2003) Model-based clustering and visualization of navigation patterns on a web site. Data Min Knowl Discov 7:399–424
27. Campbell NA, Mahon RJ (1974) A multivariate study of variation in two species of rock crab of Genus *Leptograsus*. Aust J Zool 22:417–25
28. Celebi ME, Kingravi H, Vela PA (2013) A comparative study of efficient initialization methods for the K-means clustering algorithm. Expert Syst Appl 40(1):200–210
29. Celeux G, Govaert (1995) Gaussian parsimonious clustering models. Comput Stat Data Anal 28:781–93
30. Celeux C, Martin-Magniette ML, Maugis C, Raftery A (2011) Letter to the editor. J Am Stat Assoc 106:383
31. Chandra S (1977) On the mixtures of probability distributions. Scand J Stat 4:105–112

32. Chen WC, Maitra R (2011) Model-based clustering of regression time series data via APECM – an AECM algorithm sung to an even faster beat. Stat Anal Data Min 4:567–578
33. Chen J, Tan X, Zhang R (2008) Consistency of penalized MLE for normal mixtures in mean and variance. Stat Sin 18:443–465
34. Ciuperca G, Ridolfi A, Idier J (2003) Penalized maximum likelihood estimator for normal mixtures. Scand J Stat 30(1):45–59
35. Corduneanu A, Bishop CM (2001) Variational Bayesian model selection for mixture distributions. In: Proceedings eighth international conference on artificial intelligence and statistics, pp 27–34
36. Dean N, Raftery A, Scrucca L (2013) Package clustvarsel: variable selection for model-based clustering. http://cran.r-project.org/web/packages/clustvarsel
37. Demiriz A, Bennett K, Embrechts MJ (1999) Semi-supervised clustering using genetic algorithms. In: Artificial neural networks in engineering (ANNIE-99). ASME Press, New York, pp 809–814
38. Dempster AP, Laird NM, Rubin DB (1977) Maximum likelihood for incomplete data via the EM algorithm (with discussion). J R Stat Soc Ser B 39:1–38
39. Dertinger JJ, Walker AV (2013) Ionic liquid matrix-enhanced secondary ion mass spectrometry: the role of proton transfer. J Am Soc Mass Spectrom 24:348–355
40. Dhillon IS, Modha DS (2001) Concept decompositions for large sparse text data using clustering. Mach Learn 42:143–175
41. Diebolt J, Robert C (1994) Estimation of finite mixture distributions by Bayesian sampling. J R Stat Soc Ser B 56:363–375
42. Digalakis VV, Rtischev D, Neumeyer LG (1995) Speaker adaptation using constrained estimation of Gaussian mixtures. IEEE Trans Speech Audio Process 3(5):357–366
43. Dortet-Bernadet J, Wicker N (2008) Model-based clustering on the unit sphere with an illustration using gene expression profiles. Biostatistics 9(1):66–80
44. Efron B, Tibshirani R, d Storey J, Tusher V (2001) Empirical Bayes analysis of a microarray experiment. J Am Stat Assoc 96:1151–1160
45. Escobar MD, West M (1995) Bayesian density estimation and inference using mixtures. J Am Stat Assoc 90:577–588
46. Esper J, Cook E, Schweingruber F (2002) Low-frequency signals in long tree-ring chronologies for reconstructing past temperature variability. Science 295(5563):2250–2253
47. Feng Z, McCulloch C (1996) Using bootstrap likelihood ratio in finite mixture models. J R Stat Soc B 58:609–617
48. Forgy E (1965) Cluster analysis of multivariate data: efficiency vs. interpretability of classifications. Biometrics 21:768–780
49. Fraley C (1998) Algorithms for model-based Gaussian hierarchical clustering. SIAM J Sci Comput 20:270–281
50. Fraley C, Raftery AE (2002) Model-based clustering, discriminant analysis, and density estimation. J Am Stat Assoc 97:611–631
51. Fraley C, Raftery AE (2006) MCLUST version 3 for R: normal mixture modeling and model-based clustering. Technical Report 504, Department of Statistics, University of Washington, Seattle
52. Frühwirth-Schnatter S (2001) Markov Chain Monte Carlo estimation of classical and dynamic switching and mixture models. J Am Stat Assoc 96:194–209
53. Frühwirth-Schnatter S, Pyne S (2010) Bayesian inference for finite mixtures of univariate and multivariate skew-normal and skew-*t* distributions. Biostatistics 11:317–336
54. Gallegos MT, Ritter G (2009) Trimmed ML estimation of contaminated mixtures. Sankhya Ser A 71:164–220
55. Garcia-Escudero L, Gordaliza A, Mayo-Iscar A (2013) A constrained robust proposal for mixture modeling avoiding spurious solutions. Adv Data Anal Classif 1–17. doi:10.1007/s11634-013-0153-3
56. Gopal S, Yang Y (2014) von Mises-Fisher clustering models. J Mach Learn Res 32:154–162

57. Gormley IC, Murphy TB (2010) A mixture of experts latent position cluster model for social network data. Stat Methodol 7:385–405
58. Guo J, Levina E, Michailidis G, Zhu J (2010) Pairwise variable selection for high-dimensional model-based clustering. Biometrics 66:793–804
59. Hall P, Ormerod JT, Wand MP (2011) Theory of Gaussian variational approximation for a Poisson mixed model. Stat Sin 21:369–389
60. Hammer R, Hertz T, Hochstein S, Weinshall D (2007) Classification with positive and negative equivalence constraints: theory, computation and human experiments. In: Proceedings of the 2nd international conference on advances in brain, vision and artificial intelligence, Springer-Verlag Berlin, pp 264–276
61. Handcock MS, Raftery AE, Tantrum JM (2007) Model-based clustering for social networks. J R Stat Soc Ser A 170:301–354
62. Haneca K, Wazny T, Van Acker J, Beeckman H (2005) Provenancing Baltic timber from art historical objects: success and limitations. J Archaeol Sci 32(2):261–271
63. Hartigan JA (1981) Consistency of single linkage for high-density clusters. J Am Stat Assoc 76:388–394
64. Hathaway RJ (1985) A constrained formulation of maximum-likelihood estimation for normal mixture distributions. Stat Probab Lett 4:53–56
65. Hennig C (2004) Breakdown points for maximum likelihood-estimators of location-scale mixtures. Ann Stat 32:1313–1340
66. Hennig C (2010) Methods for merging Gaussian mixture components. Adv Data Anal Classif 4:3–34
67. Hennig C, Coretto P (2008) The noise component in model-based cluster analysis. In: Preisach C, Burkhardt H, Schmidt-Thieme L, Decker R (eds) Data analysis, machine learning and applications, studies in classification, data analysis, and knowledge organization. Springer, Berlin, Heidelberg, pp 127–138
68. Hoff PD, Raftery AE, Handcock MS (2002) Latent space approaches to social network analysis. J Am Stat Assoc 97:460:1090–1098
69. Holzmann H, Munk A, Gneiting T (2006) Identifiability of finite mixtures of elliptical distributions. Scand J Stat 33:753–763
70. Huang JT, Hasegawa-Johnson M (2009) On semi-supervised learning of Gaussian mixture models for phonetic classification. In: NAACL HLT workshop on semi-supervised learning
71. Inbarani HH, Thangavel K (2009) Mining and analysis of clickstream patterns. In: Abraham A, Hassanien AE, Leon F de Carvalho A, Snášel V (eds) Foundations of computational, intelligence, vol 6. Studies in computational intelligence, vol 206. Springer, Berlin, Heidelberg, pp 3–27
72. Jasra A, Holmes CC, Stephens DA (2005) Markov chain Monte Carlo methods and the label switching problem in Bayesian mixture modeling. Stat Sci 20:50–67
73. Jiao S, Zhang S (2008) The t-mixture model approach for detecting differentially expressed genes in microarrays. Funct Integr Genomics 8:181–186
74. Jolliffe IT, Jones B, Morgan BJT (1995) Identifying influential observations in hierarchical cluster analysis. J Appl Stat 22(1):61–80
75. Kalman RE (1960) A new approach to linear filtering and prediction problems. J Basic Eng 82:35–45
76. Kent J (1983) Identifiability of finite mixtures for directional data. Ann Stat 11(3):984–988
77. Kiefer NM (1978) Discrete parameter variation: efficient estimation of a switching regression model. Econometrica 46:427–434
78. Kim D, Seo B (2014) Assessment of the number of components in Gaussian mixture models in the presence of multiple local maximizers. J Multivar Anal 125:100–120
79. Klein D, Kamvar SD, Manning C (2002) From instance-level constraints to space-level constraints: making the most of prior knowledge in data clustering. In: Proceedings of the nineteenth international conference on machine learning (ICML-2002), pp 307–314
80. Krivitsky PN, Handcock MS (2008) Fitting position latent cluster models for social networks with latentnet. J Stat Softw 24(5). http://statnetproject.org

81. Krivitsky PN, Handcock MS (2009) latentnet: Latent position and cluster models for statistical networks. R package version 2.2-2. http://statnetproject.org
82. Lauritzen SL (1996) Graphical models. Clarendon Press, Oxford
83. Law MHC, Topchy A, Jain AK (2005) Model-based clustering with probabilistic constraints. In: 2005 SIAM international conference on data mining, pp 641–645
84. Lee H, Li J (2012) Variable selection for clustering by separability based on ridgelines. J Comput Graph Stat 21:315–337
85. Lee S, McLachlan G (2013) On mixtures of skew normal and skew t-distributions. Adv Data Anal Classif 7:241–266
86. Li J, Zha H (2006) Two-way Poisson mixture models for simultaneous document classification and word clustering. Comput Stat Data Anal 50(1):163–180
87. Li J, Ray S, Lindsay B (2007) A nonparametric statistical approach to clustering via mode identification. J Mach Learn Res 8:1687–1723
88. Lin TI (2009) Maximum likelihood estimation for multivariate skew normal mixture models. J Multivar Anal 100:257–265
89. Lin TI, Lee JC, Yen SY (2007) Finite mixture modelling using the skew normal distribution. Stat Sin 17:909–927
90. Liu B (2011) Web data mining: exploring hyperlinks, contents, and usage data, 2nd edn. Springer, New York
91. Liu C, Rubin DB (1994) The ECME algorithm: a simple extension of EM and ECM with faster monotone convergence. Biometrika 81:633–648
92. Liu C, Rubin DB, Wu YN (1998) Parameter expansion to accelerate em: the PX-EM algorithm. Biometrika 85:755–770
93. Lotsi A, Wit E (2013) High dimensional sparse Gaussian graphical mixture model. arXiv:13083381v3
94. Lu Z, Leen TK (2007) Penalized probabilistic clustering. Neural Comput 19:1528–1567
95. MacEachern SN, Muller P (1998) Estimating mixtures of Dirichlet process models. J Comput Graph Stat 7:223–238
96. Maitra R (2009) Initializing partition-optimization algorithms. IEEE/ACM Trans Comput Biol Bioinform 6:144–157. http://doi.ieeecomputersociety.org/10.1109/TCBB.2007.70244
97. Maitra R, Melnykov V (2010) Simulating data to study performance of finite mixture modeling and clustering algorithms. J Comput Graph Stat 19(2):354–376. doi:10.1198/jcgs.2009.08054
98. Mardia KV, Jupp PE (2000) Directional statistics. Wiley, New York
99. Markitsis A, Lai Y (2010) The t-mixture model approach for detecting differentially expressed genes in microarrays. Bioinformatics 26:640–646
100. Martinez-Uso A, Pla F, Sotoca J (2010) A semi-supervised Gaussian mixture model for image segmentation. In: International conference on pattern recognition, pp 2941–2944
101. Masseran N, Razali A, Ibrahim K, Latif M (2013) Fitting a mixture of von Mises-distributions in order to model data on wind direction in Peninsular Malaysia. Energy Convers Manag 72:94–102
102. Maugis C, Celeux G, Martin-Magniette ML (2009) Variable selection for clustering with Gaussian mixture models. Biometrics 65(3):701–709
103. Maugis C, Celeux G, Martin-Magniette ML (2009) Variable selection in model-based clustering: a general variable role modeling. Comput Stat Data Anal 53(11):3872–3882
104. Maugis-Rabusseau C, Martin-Magniette ML, Pelletier S (2012) Selvarclustmv: variable selection approach in model-based clustering allowing for missing values. J Soc Fr Stat 153(2):21–36
105. McGrory C, Titterington D (2007) Variational approximations in Bayesian model selection for finite mixture distributions. Comput Stat Data Anal 51(11):5352–5367. doi:10.1016/j.csda.2006.07.020, http://www.sciencedirect.com/science/article/B6V8V-4KMYRPW-1/2/428635340ac2d823187a0c04164508c5. Advances in Mixture Models
106. McLachlan G (1987) On bootstrapping the likelihood ratio test statistic for the number of components in a normal mixture. Appl Stat 36:318–324

107. McLachlan GJ, Basford KE (1988) Mixture models: inference and applications to clustering. Marcel Dekker, New York
108. McLachlan G, Krishnan T (2008) The EM algorithm and extensions, 2nd edn. Wiley, New York
109. McLachlan G, Peel D (2000) Finite mixture models. Wiley, New York
110. McLachlan G, Peel G, Basford K, Adams P (1999) Fitting of mixtures of normal and t-components. J Stat Softw 4:2
111. McLachlan G, Been R, Jones LT (2006) A simple implementation of a normal mixture approach to differential gene expression in multiclass microarrays. Bioinformatics 22:1608–1615
112. McNeil DR (1977) Interactive data analysis. Wiley, New York
113. Melnikov V, Litvinov V, Koppe V, Bobkov V (2008) Sims study of the processes in buffer solutions of bioorganic systems. Bull Russ Acad Sci Phys 72:929–933
114. Melnykov V (2012) Efficient estimation in model-based clustering of Gaussian regression time series. Stat Anal Data Min 5:95–99
115. Melnykov V (2013) Challenges in model-based clustering. Wiley Interdiscip Rev Comput Stat 5:135–148
116. Melnykov V (2013) Finite mixture modelling in mass spectrometry analysis. J R Stat Soc Ser C 62:573–592
117. Melnykov V (2013) On the distribution of posterior probabilities in finite mixture models with application in clustering. J Multivar Anal 122:175–189
118. Melnykov V (2014) Merging mixture components for clustering through pairwise overlap. J Comput Graph Stat (tentatively accepted)
119. Melnykov V (2014) Model-based biclustering of clickstream data. Comput Stat Data Anal (under minor revision)
120. Melnykov V, Maitra R (2011) CARP: software for fishing out good clustering algorithms. J Mach Learn Res 12:69–73
121. Melnykov V, Melnykov I (2012) Initializing the EM algorithm in Gaussian mixture models with an unknown number of components. Comput Stat Data Anal 56:1381–1395
122. Melnykov I, Melnykov V (2014) On k-means algorithm with the use of Mahalanobis distances. Stat Probab Lett 84:88–95
123. Melnykov V, Michael S (2014) Finite mixture modeling of Gaussian regression time series with application to dendrochronology. J Classif (under review)
124. Melnykov V, Chen WC, Maitra R (2012) MixSim: an R package for simulating data to study performance of clustering algorithms. J Stat Softw 51:1–25
125. Meng XL, van Dyk D (1997) The EM algorithm - an old folk song sung to a fast new tune (with discussion). J R Stat Soc Ser B 59:511–567
126. Meng XL, Rubin DB (1993) Maximum likelihood estimation via the ECM algorithm: a general framework. Biometrika 80(2):267–278
127. Michael S, Melnykov V (2014) Studying complexity of model-based clustering. Commun Stat Simul Comput (accepted)
128. Moore A (1998) Very fast EM-based mixture model clustering using multiresolution kd-trees. In: In advances in neural information processing systems 11. MIT Press, Cambridge, pp 543–549
129. Neal R (2000) Markov chain sampling methods for Dirichlet process mixture models. J Comput Graph Stat 9:249–265
130. Neal RM, Hinton GE (1993) A new view of the EM algorithm that justifies incremental and other variants. In: Learning in graphical models. Kluwer, Dordrecht, pp 355–368
131. Newcomb S (1886) A generalized theory of the combination of observations so as to obtain the best result. Am J Math 8:343–366
132. Neykov N, Filzmoser P, Dimova R, Neytchev P (2007) Robust fitting of mixtures using the trimmed likelihood estimator. Comput Stat Data Anal 17:299–308
133. Nigam K, McCallum AK, Thrun S, Mitchell T (2000) Text classification from labeled and unlabeled documents using EM. Mach Learn 39:103–134

134. Ortega JM, Rheinboldt WC (1970) Iterative solutions of nonlinear equations in several variables. Academic, Princeton
135. Pan W, Shen X (2007) Penalized model-based clustering with application to variable selection. J Mach Learn Res 8:1145–1164
136. Pan W, Shen X, Jiang A, Hebbel R (2006) Semisupervised learning via penalized mixture model with application to microarray sample classification. Bioinformatics 22(19):2388–2395
137. Papastamoulis P, Iliopoulos G (2010) An artificial allocations based solution to the label switching problem in Bayesian analysis of mixtures of distributions. J Comput Graph Stat 19:313–331
138. Pearl J (1988) Probabilistic reasoning in intelligent systems: networks of plausible inference. Morgan Kaufmann, Los Altos
139. Pearson K (1894) Contribution to the mathematical theory of evolution. Philos Trans R Soc 185:71–110
140. Peel D, McLachlan G (2000) Robust mixture modeling using the t distribution. Stat Comput 10:339–348
141. Peel D, Whiten W, McLachlan G (2001) Fitting mixtures of Kent distributions to aid in joint set identifications. J Am Stat Assoc 96:56–63
142. Raftery AE, Dean N (2006) Variable selection for model-based clustering. J Am Stat Assoc 101:168–178
143. Raftery AE, Niu X, Hoff PD, Yeung KY (2012) Fast inference for the latent space network model using a case-control approximate likelihood. J Comput Graph Stat 21(4):901–919
144. Ray S, Cheng Y (2014) Package Modalclust: hierarchical modal clustering. http://cran.r-project.org/web/packages/Modalclust
145. Ray S, Lindsay B (2005) The topography of multivariate normal mixtures. Ann Stat 33(5):2042–2065
146. Richardson S, Green PJ (1997) On Bayesian analysis of mixtures with an unknown number of components (with discussion). J R Stat Soc Ser B 59:731–792
147. Robin S, Bar-Hen A, Daudin JJ, Pierre L (2007) A semi-parametric approach for mixture models: application to local false discovery rate estimation. Comput Stat Data Anal 51:5483–5493
148. Rodriguez CE, Walker SG (2014) Label switching in Bayesian mixture models: deterministic relabeling strategies. J Comput Graph Stat 23(1):25–45
149. Saídaoui F (2010) Acceleration of the em algorithm via extrapolation methods: review, comparison and new methods. Comput Stat Data Anal 54(3):750–766
150. Salter-Townshend M, Murphy TB (2013) Variational Bayesian inference for the latent position cluster model for network data. Comput Stat Data Anal 57:661–671
151. Sampson SF (1969) Crisis in a cloister. Ph.D. thesis, Department of Sociology, Cornell University, Ithaca
152. Schwarz G (1978) Estimating the dimensions of a model. Ann Stat 6:461–464
153. Seo B, Kim D (2012) Root selection in normal mixture models. Comput Stat Data Anal 56:2454–2470
154. Shental N, Bar-Hillel A, Hertz T, Weinshall D (2003) Computing Gaussian mixture models with EM using equivalence constraints. In: Advances in NIPS, A Bradford Book, vol 15
155. Steiner P, Hudec M (2007) Classification of large data sets with mixture models via sufficient em. Comput Stat Data Anal 51:5416–5428
156. Stuetzle W (2003) Estimating the cluster tree of a density by analyzing the minimal spanning tree of a sample. J Classif 20:25–47
157. Stuetzle W, Nugent R (2010) A generalized single linkage method for estimating the cluster tree of a density. J Comput Graph Stat 19:397–418
158. Tanabe A, Fukumizu K, Oba S, Takenouchi T, Ishii S (2007) Parameter estimation for von Mises-Fisher distributions. Comput Stat 22:145–157
159. Teicher H (1963) Identifiability of finite mixtures. Ann Math Stat 34:1265–1269
160. Vardi Y, Shepp LA, Kaufman LA (1985) A statistical model for positron emission tomography. J Am Stat Assoc 80:8–37

161. Vicari D, Alfó M (2014) Model based clustering of customer choice data. Comput Stat Data Anal 71:3–13
162. Vu DQ, Hunter DR, Schweinberger M (2013) Model-based clustering of large networks. Ann Appl Stat 7:1010–1039
163. Wagstaff K, Cardie C, Rogers S, Schroedl S (2001) Constrained K-means clustering with background knowledge. In: Proceedings of the eighteenth international conference on machine learning (ICML-2001), pp 577–584
164. Wang B, Titterington D (2006) Convergence properties of a general algorithm for calculating variational Bayesian estimates for a normal mixture model. Bayesian Anal 1(3):625–650
165. Wang S, Zhu J (2008) Variable selection for model-based high-dimensional clustering and its application to microarray data. Biometrics 64:440–448
166. Wang H, Zhang Q, Luo B, Wei S (2004) Robust mixture modelling using multivariate t-distribution with missing information. Pattern Recognit Lett 25:701–710
167. Wei GCG, Tanner MA (1990) A Monte Carlo implementation of the EM algorithm and the Poor Man's data augmentation algorithms. J Am Stat Assoc 85(411):699–704
168. Wishart D (1969) Mode analysis: a generalization of nearest neighbor which reduces chaining effect. In: Cole AJ (ed) Numerical taxonomy. Academic, London, pp 282–311
169. Wolfe JH (1967) NORMIX: computational methods for estimating the parameters of multivariate normal mixture distributions. Technical bulletin USNPRA SRM 6
170. Xie B, Pan W, Shen X (2010) Penalized mixtures of factor analyzers with application to clustering high-dimensional microarray data. Bioinformatics 26:501–508
171. Xing EP, Ng AY, Jordan MI, Russell S (2003) Distance metric learning with application to clustering with side-information. In: Thrun S, Becker S, Obermayer K (eds) Advances in neural information processing systems, vol 15. MIT Press, Cambridge, pp 505–512
172. Yakowitz SJ, Spragins JD (1968) On the identifiability of finite mixtures. Ann Math Stat 39(1):209–214
173. Ypma A, Heskes T (2002) Categorization of web pages and user clustering with mixtures of hidden Markov models. In: Proceedings of the international workshop on web knowledge discovery and data mining WEBKDD'02, Edmonton, pp 31–43
174. Yuan M, Lin Y (2007) Model selection and estimation in the Gaussian graphical model. Biometrika 94:19–35
175. Zhou H, Pan W, X S (2009) Penalized model-based clustering with unconstrained covariance matrices. Electron J Stat 3:1473–1496
176. Zhu X (2005) Semi-supervised learning literature survey. Technical Report 1530, Computer Sciences, University of Wisconsin-Madison

Chapter 2
Accelerating Lloyd's Algorithm for k-Means Clustering

Greg Hamerly and Jonathan Drake

Abstract The k-means clustering algorithm, a staple of data mining and unsupervised learning, is popular because it is simple to implement, fast, easily parallelized, and offers intuitive results. Lloyd's algorithm is the standard batch, hill-climbing approach for minimizing the k-means optimization criterion. It spends a vast majority of its time computing distances between each of the k cluster centers and the n data points. It turns out that much of this work is unnecessary, because points usually stay in the same clusters after the first few iterations. In the last decade researchers have developed a number of optimizations to speed up Lloyd's algorithm for both low- and high-dimensional data.

In this chapter we survey some of these optimizations and present new ones. In particular we focus on those which avoid distance calculations by the triangle inequality. By caching known distances and updating them efficiently with the triangle inequality, these algorithms can provably avoid many unnecessary distance calculations. All the optimizations examined produce the same results as Lloyd's algorithm given the same input and initialization, so are suitable as drop-in replacements. These new algorithms can run many times faster and compute far fewer distances than the standard unoptimized implementation. In our experiments, it is common to see speedups of over 30–50x compared to Lloyd's algorithm. We examine the trade-offs for using these methods with respect to the number of examples n, dimensions d, clusters k, and structure of the data.

Keywords k-Means • Triangle inequality • Caching • Accelerate • Lloyd's algorithm • Clustering • Unsupervised learning

---

G. Hamerly (✉)
Baylor University, 105 Baylor Ave., Waco, TX 76798, USA
e-mail: hamerly@cs.baylor.edu

J. Drake
Hewlett-Packard Company, 14231 Tandem Blvd, Austin, TX 78728, USA
e-mail: jonathan.drake@hp.com

M.E. Celebi (ed.), *Partitional Clustering Algorithms*,
DOI 10.1007/978-3-319-09259-1_2

2.1 Introduction

The k-means clustering algorithm is a very popular tool for data analysis and learning. At its heart is an easily-understood optimization problem: given a set of data points (in some vector space), try to position k other points (called 'centers') at locations that minimize the (squared) distance between each point and its closest center. While it is popular and easy to implement, the naive implementation for solving the problem is inefficient, wasting a lot of processing time on unnecessary and redundant computations.

This chapter discusses simple geometric methods, based on the triangle inequality and keeping cached bounds on computed distances, to reduce wasted computation and build much more efficient algorithms that give exactly the same output. We point out that all the accelerated algorithms we investigate in this chapter give exactly the same answer as Lloyd's standard batch algorithm, given the same initialization. In our experiments, it is common to see speedups of over 30–50x compared to Lloyd's algorithm.

2.1.1 Popularity of the k-Means Algorithm

The k-means algorithm is very widely used. In [40], it is chosen as one of the top ten data mining algorithms. It is implemented in many commercial and open-source statistical data analysis software packages, including MATLAB, SAS, Stata, SPSS, R, and Weka to name a few. A simple search for 'k means clustering' on Google Scholar yields over 2.1 million results, which is greater than the number of results for 'neural network', 'support vector machine', 'nearest neighbor', or 'logistic regression'.

Many applications benefit from using k-means. Just a few are:

- clustering the pixels of an image for image color quantization [7, 19],
- post-processing to decide the memberships in spectral clustering [29],
- selecting the codewords for vector quantization, enabling lossy compression of audio or image data [23],
- image segmentation [14, 19],
- unsupervised feature learning in single-layer neural networks [9],
- identifying self-similar behaviors in dynamic program execution for structured sampling of the behaviors [36],
- finding a good initialization for a more costly learning method [6], and
- finding good locations for basis functions in a radial basis function network [39].

Such a widely-used algorithm deserves to be well-studied and efficiently implemented.

2.1.2 The Standard k-Means Algorithm does a Lot of Unnecessary Work

The k-means algorithm is popular due to its clarity, simplicity, and intuitive optimization function. As a result, it has been implemented, albeit inefficiently, many times over. We investigated the source code of the k-means implementations for the software packages ELKI, graphlab, Mahout, MATLAB, MLPACK, Octave, OpenCV, R, SciPy, Weka, and Yael, and found that none of them use the triangle inequality bound acceleration techniques that we discuss in this chapter, though they would benefit from doing so. Of those, the ones which provide scalable implementations primarily depend on some form of parallel processing, which are compatible with the acceleration methods presented here.

The standard methods for solving the k-means optimization problem are Lloyd's algorithm [24] (a batch algorithm, also known as Lloyd-Forgy [13]), as well as MacQueen's algorithm [26]. Each algorithm spends the vast majority of its time computing distances between the clustered points and the current cluster centers. However, much of the time, these distance calculations are unnecessary, wasted computation. In this study we focus on improving Lloyd's algorithm, which is widely used.

The basic reason why the standard batch methods for optimizing k-means are inefficient is because in each iteration they must identify the closest center for each clustered point. To do this, the methods naively compute all nk distances between each of the n clustered points and each of the k centers. After each iteration, the centers move and these distances may all change, requiring recomputation. But typically the centers don't move much, especially after the first few iterations. Most of the time the closest center in the previous iteration remains the closest center. Thus, keeping track of the closest center for each clustered point is made much more efficient with some caching. When the closest center doesn't change, ideally we shouldn't need to compute the distance between that point and *any* cluster center. And even when the closest center for a point changes, it might be possible to avoid computing the distance from that point to all centers, instead looking only at a few centers that are guaranteed to be closer to that point than all other centers.

2.1.3 Previous Work on k-Means Acceleration

There is a healthy line of research on accelerating learning algorithms, which goes hand in hand with accelerating data retrieval algorithms such as k-nearest neighbor search. The primary methods of acceleration are: algorithmic improvements, parallelization (including threading, multiprocessing, and distributed computation), and approximation. In this chapter, we focus on algorithmic improvements which give exact answers.

2.1.3.1 Algorithmic Improvements

Pelleg and Moore [31] incorporated the popular k-d tree data in the task of accelerating the k-means method (note that the k in k-d tree is a naming clash with the k in k-means). Kanungo et al. [19] developed a similar algorithm. Both algorithms use the k-d tree to structure the data to be clustered. By restructuring the search for each point's closest center, many distance calculations can be avoided. While these approaches are excellent in low dimension, k-d trees perform poorly in dimensions much greater than 8.

Moore [28] developed the anchors hierarchy, a new type of spatial data structure based on metric trees. Applying the triangle inequality, this structure organizes the data by enclosing it in hierarchically organized anchors with associated radii. The anchors hierarchy allows the k-means algorithm to avoid many provably unnecessary distance calculations, even in high dimension. We discuss this structure more in Sect. 2.3.2.

Elkan [12] started a line of research which pairs the triangle inequality directly with cached distance bounds to avoid unnecessary point-center distance calculations. Avoiding complicated hierarchical data structures and preprocessing, his algorithm simply caches $O(nk)$ distance bounds to prune distance calculations. It uses the triangle inequality to efficiently updates the bounds each time centers move. Hamerly [15] reduced this overhead to $O(n)$ bounds, which makes it much faster in practice for low and medium-dimension datasets. Drake [11] bridged the gap between these two approaches by using an adaptive number of distance bounds, $O(nb)$ where $b < k$ and can be learned online.

Other minor algorithmic improvements have also helped accelerate the standard k-means algorithm. Because k-means repeatedly seeks the minimum distance between a point and all k centers, using partial distortion search (PDS, also called partial distance search) [5] and some loop unrolling permits some distance calculations to be cut short without looking at all dimensions. Mean-distance-ordered partial search (MPS) [33] draws a connection between the squared distances and the squared difference of vector sums to help eliminate candidate k-means centers. Pan et al. [30] eliminates unlikely centers using the first and second moments of a vector and portions of the vector.

DHSS (dynamic hyperplanes shrinking search) [37] eliminates unlikely candidate centers by transforming the input space (e.g. using principal components analysis) and then using the new canonical dimensions to bound the closest centers to a point. It is unclear if this method will work well in higher-dimension spaces.

Several researchers, beginning with Kaukoranta et al., identified that some clusters found by Lloyd's algorithm are 'static' over some iterations. In other words, no points join or leave the cluster from one iteration to the next. These static clusters are easily identified, as their centers do not move. Clusters which do change are called active. This information can be used to reduce the number of candidate centers for some points [20–22]. It's worth noting that the triangle inequality methods which are the focus of this chapter implicitly exploit the same information and enjoy similar benefits.

2.1.3.2 Parallelization

Several machine learning packages have been constructed with the intent of improving the scalability and speed of the learning algorithms, making them applicable to large real-world problems. Packages like graphlab [25], Yael [41], and Mahout [2] provide scalable implementations of many machine learning algorithms, including k-means. They use different approaches: Yael focuses on low-level multithreaded optimized implementations, Mahout provides machine learning on a map-reduce infrastructure, and graphlab focuses on graph-structured algorithms and multi-core processing.

2.1.3.3 Alternative Heuristic Methods

Lloyd's algorithm is a gradient descent heuristic algorithm for minimizing k-means distortion criterion. While it is popular and is the focus of this chapter, there are alternative methods which also aim to reduce the k-means distortion using different heuristics.

Agarwal et al. [1] use subsamples of the whole dataset, called core-sets, and optimize a solution based on the sample. This algorithm can be much faster in practice in low-dimensional data, but can be slow and find poor solutions in higher dimensions.

Hartigan and Wong [17] suggested an algorithm for optimizing the k-means distortion which considers the clustered points one by one. Each time a point changes membership, the algorithm updates the affected center locations.

Sculley [35] developed a method that is a hybrid of stochastic gradient descent (as proposed by Bottou and Bengio [6]) and batch k-means. It operates on small samples rather than individual examples. The resulting algorithm is less susceptible to noise caused by individual examples, yet still quite fast.

Each of these alternative algorithms can be quite fast, much faster than Lloyd's batch algorithm. The tradeoff is that they tend to produce different results. We view these as suitable alternatives to Lloyd's algorithm, but focus on accelerating Lloyd's algorithm due to its popularity. Some of these algorithms, especially those which repeatedly make nearest-center queries for the same points, may be compatible with the acceleration methods we discuss here.

2.2 Cluster Distortion and Lloyd's Algorithm

When clustering with k-means, we are trying to minimize the distortion, or sum of squared errors, between the points and their assigned centers. We can improve the distortion in two ways: by changing the points' cluster assignments and moving the cluster centers. Specifically, given a fixed set of points X, we are attempting to minimize the distortion function

Table 2.1 Terms that are used frequently in this chapter

Name and type	Description
$d \in \mathbb{N}$	Dimension of points to cluster and cluster centers.
$n \in \mathbb{N}$	Number of points to cluster.
$k \in \mathbb{N}$	Number of cluster centers.
$X \subset \mathbb{R}^{n \times d}$	Set of points to cluster, indexed as $x(i)$ for $1 \leq i \leq n$.
$C \subset \mathbb{R}^{n \times d}$	Set of centers, indexed as $c(j)$ for $1 \leq j \leq k$.
$n(j) \in \mathbb{R}$	The number of points that are currently assigned to cluster j.
$N = \{i \in \mathbb{N} \mid 1 \leq i \leq n\}$	Indexes of the points in X.
$K = \{j \in \mathbb{N} \mid 1 \leq j \leq k\}$	Indexes of the points in C.
$a : N \rightarrow K$	Index of assigned (closest) center for each point.
$u : N \rightarrow \mathbb{R}$	Upper bound on the distance between each point and its assigned center.
$\ell : N \times K \rightarrow \mathbb{R}$	Lower bound on the distance between each point and each center.
$s : K \rightarrow \mathbb{R}$	Half the distance between a center and its current closest other center.

$$J(X, C) = \sum_{i \in N} \|x(i) - c(a(i))\|^2 \tag{2.1}$$

by choosing the best set of clusters C (Table 2.1).

Lloyd's batch algorithm for minimizing the distortion has three basic steps, which are stated here and in slightly more detail in Algorithm 1.

1. Initialize the centers.
2. Until the algorithm converges:
   a. Assign each point to its currently closest cluster center.
   b. Move each center to the mean of its currently-assigned centers.

Step 1 occurs only once, while steps 2(a) and 2(b) alternate until the algorithm converges. Convergence is guaranteed due to the fact that steps 2(a) and 2(b) both reduce $J(X, C)$, and there is a finite number of ways to partition the n points among k clusters [6].

Much has been written on initializing the centers for k-means [8]. Researchers have used the first several examples [6], chosen k points from X at random [16], and used the furthest-first method [18]. The most effective current method, theoretically and in practice, is the k-means++ initialization [3], which randomly selects a good initialization with high probability, using something akin to furthest-first. In all our experiments, we use k-means++ for initialization.

The remainder of this paper is primarily concerned with optimizing step 2(a) of the algorithm (finding the closest center for each point). The naive implementation of Lloyd's algorithm spends the majority of its time here, and much of the computation done here is unnecessary; the information needed for this step can be derived using some caching and geometry.

Algorithm 1 Lloyd's k-means algorithm—the standard algorithm for minimizing $J(X, C)$. Like other algorithms presented in this chapter, this algorithm's pseudocode is presented simply, without details on efficiency optimizations used in real implementations

```
procedure LLOYD(X, C)
  while not converged do
    for all i ∈ N do {Find the closest center to each x(i).}
      a(i) ← 1
      for all j ∈ K do
        if ‖x(i) − c(j)‖ < ‖x(i) − c(a(i))‖ then
          a(i) ← j
    for all j ∈ K do {Move the centers}
      move c(j) to the mean of {x(i)|a(i) = j}
```

Step 2(b) can be optimized easily by caching sufficient statistics for each cluster: the vector sum of the points assigned to the cluster, and the number of points assigned to the cluster. Keeping these is inexpensive and avoids a sum over all points for each iteration. Each time a point changes cluster membership, the relevant sufficient statistics are updated. After the first few iterations, most points remain in the same cluster for many iterations. Thus these sufficient statistics updates become much cheaper than a sum over all points. All the algorithms we implement for this study use this technique. Thus, it is not explicitly included in algorithmic pseudocode, for clarity.

2.2.1 *Analysis of Lloyd's Algorithm*

The running time of Lloyd's algorithm for k-means is $O(wnkd)$ for w iterations, k centers, and n points in d dimensions. For a fixed dataset and k, the number of iterations w will vary depending on the initialization. In fact, w may be superlinear with respect to n, even exponential in the worst case [38]. However, when considering less extreme cases, Lloyd's algorithm has polynomial smoothed complexity [4]. That is, w is polynomial in n and $1/\sigma$ (where σ is the amount of perturbation allowed on a weakened adversary's challenge dataset).

When viewed as a gradient descent algorithm, Bottou and Bengio [6] showed that (from a given initialization) the distortion converges in Lloyd's batch algorithm at superlinear speed. This is because the variations of the second derivative of the cost function are bounded. Thus, the algorithm is minimizing the distortion in a way that is equivalent to Newton's method.

In this chapter we look at ways to accelerate Lloyd's algorithm. Given that Lloyd's algorithm is so widely used, the goal is to accelerate the exact algorithm (without any approximation), so that the resulting accelerated algorithm can be used anywhere Lloyd's algorithm is used. The accelerations we look at primarily work by avoiding many of the nk interactions between n clustered points and k cluster centers.

2.2.2 MacQueen's Algorithm

MacQueen [26] described a method similar to Lloyd's algorithm, but which updates the location of each affected cluster center whenever a point changes cluster membership. While Lloyd's algorithm moves the centers once per pass over the entire dataset, MacQueen's algorithm will move the centers (by smaller amounts) far more often. Thus, Lloyd's algorithm could be considered a 'batch' algorithm whereas MacQueen's is more 'online'. As Lloyd's algorithm is more popular in practice, and easier to accelerate since the centers move less frequently, we focus primarily on it in this chapter.

2.3 Tree Structured Approaches

Tree structures are effective methods for indexing spatial data in retrieval and learning algorithms. Two approaches in particular, k-d trees and the anchors hierarchy, have shown success in accelerating the k-means algorithm. We review these methods here.

2.3.1 Blacklisting and Filtering Centers with k-d Trees

Pelleg and Moore [31] and Kanungo et al. [19] proposed similar methods for accelerating k-means by constructing and using a k-d tree.[1] A k-d tree is a binary tree that recursively partitions the space of the data it's constructed on using separating hyperplanes (typically axis-aligned). In this discussion, it is applied to the data to be clustered. After construction, the tree's structure plus sufficient statistics kept at each tree node can be used to eliminate point-center calculations. Pelleg and Moore call this approach 'blacklisting' the centers, while Kanungo et al. call it 'filtering' the centers.

Pelleg and Moore's blacklisting algorithm proceeds as shown in Algorithm 2. Initially it constructs a k-d tree on the data that is to be clustered. For each iteration

[1]Note that the k in k-d trees and the k in k-means are two different (clashing) variable names. In k-means, the k refers to the number of centers/clusters sought; in k-d trees k refers to the dimension of the data the structure is built on.

Algorithm 2 Pelleg and Moore's Blacklisting k-means algorithm

```
procedure BLACKLISTING(X, C)
   construct k-d tree T on X
   while not converged do
      Update(T.root, C)
      move centers to the mean of their assigned points
procedure UPDATE(h, C)
   if h is a leaf then
      for each data point x in h do
         find the closest center to x and update that center's counters
   else
      compute the distance between h and each center in C
      remove from C any dominated centers
      if only one center c′ remains in C then
         update the counters for c′ using the data in h
      else
         call Update(h.left, C)
         call Update(h.right, C)
```

of k-means, the algorithm performs a traversal of the tree, searching for regions of the tree that are 'owned' by a single center.

The traversal starts with all k centers at the root of the k-d tree. Then it recursively descends the tree, and at each node attempts to prove that there is one center that 'dominates' (is closer to) one or more of the other centers, with respect to the hyperrectangle enclosing the k-d tree node. If so, it eliminates the dominated center(s). If only the dominating center remains, then the recursion stops and all points below that node in the k-d tree are assigned to that center. If multiple centers remain, it recursively continues its search on both child nodes. If it reaches a leaf, it performs a search of the data at the leaf to assign them to the remaining centers.

By using sufficient statistics kept at each internal node of the tree, identifying a dominating center early in the recursion (close to the root) allows the algorithm to skip not only many distance calculations but also many vector additions. For k-means, the sufficient statistics for a node are the number of points at that node or below, and the vector sum of those points.

While k-d trees can work very well in practice, the following limitations make their use less desirable, especially in high dimension:

- They are efficient only in low dimension—the running time has an exponential dependence on the dimension of the data [27]. If the number of points in the structure are significantly (exponentially) larger than the dimension of the data, k-d trees tend to be efficient relative to linear lookups, but for even moderate dimensions in practical applications they become too slow. Empirically they become too slow (compared to simple linear search) somewhere between 8 and 10 dimensions [28].
- They require extra memory for the tree structure and sufficient statistics. The extra memory required is on the order of the original dataset.

- When clustering, we must construct the k-d tree in advance, which means that we cannot start clustering until this is done.
- A k-d tree is not designed for efficient updates (such as adding new points or removing old points). If many updates are to be done, the tree should be reconstructed.

2.3.2 Anchors Hierarchy

Moore proposed the anchors hierarchy as a tree-like geometric structure for spatial datasets [28]. The tree is built 'middle-out' by choosing $\sqrt{n}$ initial points known as anchors, and then merging them (to form the top of the tree) and subdividing them (to form the leaves). Each anchor maintains a list of the points for which it is the closest anchor, sorted by their distances from the anchor. Using the triangle inequality, adding additional anchors can be done efficiently. Moore showed that the anchors hierarchy is effective at eliminating many distance computations in k-means. However, Elkan showed it is less effective at eliminating distance computations than his method for even moderate values of k than his method [12].

While tree-structured acceleration methods of the k-means algorithm are interesting and often useful, they are often slower and less effective in practice than the methods we turn to now. One reason pointed out by Elkan is that tree-structured methods must build a static structure before clustering begins, without knowledge of the number of clusters. Because the number of clusters is not fixed, and their positions change during clustering, static trees are less able to reduce the number of k-means distance calculations.

2.4 Triangle Inequality Approaches

The triangle inequality is a simple but very powerful tool from geometry. If $a, b, c \in \mathbb{R}^d$, then the triangle inequality states that

$$\|a - c\| \leq \|a - b\| + \|b - c\| \tag{2.2}$$

for Euclidean vector norm $\|a\| = \sqrt{a^T a}$. Intuitively, this means that the length of line segment (a, c) is at most the sum of the lengths of line segments (a, b) and (b, c). In other words, the shortest path between two points a and c is a straight line; taking a path that goes through an intermediate point b cannot reduce the distance.

The triangle inequality is applicable in the k-means algorithm in multiple ways. Generally, we desire to use it to prove that some center must be closer to a point than all other centers. Ideally, we wish to do this with as little computation as possible. For a point x and two centers c and c', here are some of the different ways the triangle inequality can be used in k-means:

Table 2.2 This table shows different acceleration methods that use distance bounds, sorting, and the triangle inequality in various k-means algorithms

	Accelerations used					
Algorithm	1	2	3	4	5	6
Lloyd						
Compare-means [32]	X					
Sort-means [32]		X			X	
Elkan [12]	X	X	X	k		
Hamerly [15]	X		X	1		
Drake [11]	X		X	$b < k$		
Annular (this chapter)	X		X	1		X
Heap (this chapter)	X		X	0		

1. Distance from a center to its closest other center
2. Distance from a center to all other centers
3. Upper bound on point-center distances
4. Lower bound on point-center distances
5. For each center, sort all centers by distance from it
6. Sort all centers by their vector norm

The column numbers correspond to the list on the right, which lists contexts where the triangle inequality can help k-means avoid point-center distance calculations. Please see the algorithm descriptions for more details. Column 4 lists the number of lower bounds used per point

1. To prove that c' is closer to x than c, given only distances $\|c' - x\|$ and $\|c' - c\|$.
2. To prove that c' is closer to x than c, given only norms $\|x\|$ and $\|c\|$ and distance $\|x - c'\|$.
3. To maintain an upper bound on $\|x - c\|$ when c is moving.
4. To maintain a lower bound on $\|x - c\|$ when c is moving.

Table 2.2 shows the ways the triangle inequality is used in many of the algorithms described in this chapter.

2.4.1 Using the Triangle Inequality for Center-Center and Center-Point Distances

Phillips [32] demonstrated two ways to use the triangle inequality to accelerate k-means. Both of them use the triangle inequality to prove that centers that are far from a point's assigned center are also far from the point, and therefore can be excluded from distance computations with that point. He called his two algorithms compare-means and sort-means.

Compare-means uses the triangle inequality to prove that if center c' is close to point x, and some other center c is far away from another center c', then c' must be closer than c to x. Given the already-computed distances $\|x - c'\|$ and $\|c - c'\|$, and applying the triangle inequality we can show:

$$\|c - c'\| \leq \|x - c\| + \|x - c'\| \qquad \text{By the triangle inequality.}$$

$$\|c - c'\| - \|x - c'\| \leq \|x - c\|$$

Thus if we also know that $2\|x - c'\| \leq \|c - c'\|$ (which is trivial to calculate given the distances are already known), we can show

$$2\|x - c'\| - \|x - c'\| \leq \|x - c\|$$
$$\|x - c'\| \leq \|x - c\|$$

which proves that c is not closer than c' to x, without measuring the distance $\|x - c\|$. Phillips' compare-means algorithm uses this inequality in the innermost loop of k-means to prove that some point-center distances need not be computed. The algorithm computes and caches the center-center distances each time the centers move (once per iteration).

Sort-means computes a $k \times k$ matrix of center-center distances each time the centers move, and then sorts each row of matrix by distance. This gives each center a ranking of the other centers by distance. Whenever the algorithm wants to find the closest center for some point x, it searches the centers in the order of increasing distance from its currently-assigned center c. Then, if it can ever prove that the distance $\|c - c'\|$ to some other center c' is greater than twice the distance $\|x - c\|$, it can stop searching. Thus sort-means uses the same inequality as compare-means, but it searches the centers in a different order. Because of the search order it may avoid examining some far-away centers. Of course, sort-means has the extra overhead of sorting the center-center distance matrix each time the centers move.

2.4.2 Maintaining Distance Bounds with the Triangle Inequality

We can use the triangle inequality to cheaply maintain an upper bound on the distance between points, after one point has moved. Suppose $x, c, c' \in \mathbb{R}^d$, where x is a point to cluster, c is a cluster center and c' is its new position after an iteration of k-means. If we know $\|x - c\|$ (from a previous iteration of k-means) and $\|c - c'\|$ (calculated when we move the cluster centers), we can provide an upper bound on $\|x - c'\|$ without explicitly calculating its exact value:

$$\|x - c'\| \leq \|x - c\| + \|c - c'\|. \tag{2.3}$$

Intuitively, this upper bound assumes that in the worst case, c moved directly away from x a distance of $\|c - c'\|$, along the vector $x - c$.

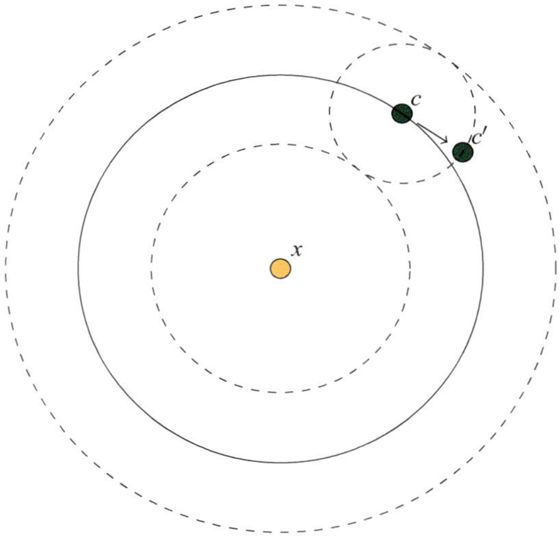

Fig. 2.1 Using the triangle inequality to bound the distance between x and c', the new location of the center c. Assume the distance $\|x - c\|$ has been calculated. It is illustrated by the *solid circle* centered on x and going through c. After c moves to c', we measure $\|c - c'\|$ (illustrated by the *dashed circle* centered on c). The upper and lower bounds on $\|x - c'\|$ are then given by $\|x - c\| - \|c - c'\| \leq \|x - c'\| \leq \|x - c\| + \|c - c'\|$, which are illustrated by the *two dashed circles* centered on x. Thus, c' must be inside the region bounded by these *two dashed circles*

Another way to apply the triangle inequality is to form a lower bound on the distance between two points. Again considering x, c, and c' as a point to cluster and the old and new positions of a center, we can form the lower bound on $\|x - c'\|$:

$$\|x - c\| \leq \|x - c'\| + \|c - c'\| \tag{2.4}$$

$$\|x - c\| - \|c - c'\| \leq \|x - c'\|. \tag{2.5}$$

Similar to the upper bound, this distance bound provides a lower bound that assumes the worst case—that c moved directly toward x a distance of $\|c - c'\|$, along the vector $x - c$.

Further, the upper (lower) bound can be updated correctly and efficiently in subsequent k-means iterations by adding (subtracting) the distance moved by a center each time it moves. This allows us to maintain both upper and lower distance bounds between a point and a moving center without explicitly calculating distances (Fig. 2.1).

2.4.3 Elkan's Algorithm: k Lower Bounds, k^2 Center-Center Distances

Elkan [12] introduced an algorithm which uses the triangle inequality multiple ways to avoid distance calculations in the k-means algorithm (see Algorithm 3 for pseudocode). For each clustered point $x(i)$, the algorithm employs one upper bound and k lower bounds. The upper bound is on the distance between $x(i)$ and its closest center $c(a(i))$; that is, $u(i) \geq \|x(i) - c(a(i))\|$. Each lower bound $\ell(i, j) \leq \|x(i) - c(j)\|$ bounds the distance between $x(i)$ and center $c(j)$. The upper (lower) bounds may be efficiently updated by adding (subtracting) the distance moved by each center after each k-means iteration.

Each time the centers move, Elkan's algorithm calculates and caches the distance between each pair of centers, as well as half the distance between each center $c(j)$ and its closest other center as $s(j)$. When it is true, the test $u(i) \leq s(a(i))$ allows Elkan's algorithm to avoid the innermost loop for $x(i)$. This is because no other center could possibly be closer to $x(i)$ than its currently-assigned center.

When Elkan's algorithm reaches the innermost loop, it may want to determine whether $c(j)$ is closer to $x(i)$ than the currently assigned center. However, if either $u(i) \leq \ell(i, j)$ or $u(i) \leq \|c(a(i)) - c(j)\|/2$, then this calculation is unnecessary, because it is not possible for $c(j)$ to be the closest center. The latter test uses the cached center-center distances.

2.4.4 Hamerly's Algorithm: 1 Lower Bound

Hamerly [15] altered Elkan's algorithm by reducing the number of bounds used (see Algorithm 4). Hamerly's algorithm uses the same upper bound $u(i)$ for each point $x(i)$—for the distance between that point and its closest center $c(a(i))$. But instead of k lower bounds, it uses only one lower bound per point, $\ell(i)$. This lower bound does not bound the distance from $x(i)$ to any particular cluster center. Instead, it represents the minimum distance that any center—except for the closest—can be to that point.

Consider the case where $u(i) \leq \ell(i)$. If this is true, it is not possible for any center to be closer to $x(i)$ than its assigned center. Thus, determining the assignment for $x(i)$ does not require knowing any exact distances, and the algorithm can skip the innermost loop that computes the distances between $x(i)$ and the k centers.

However, if $\ell(i) < u(i)$ then it might be that the closest center for $x(i)$ has changed. In this case, Hamerly's algorithm first tightens the upper bound by computing the exact distance $u(i) \leftarrow \|x(i) - c(a(i))\|$. If this reduces $u(i)$ significantly, then possibly $u(i) \leq \ell(i)$ and the algorithm can skip the innermost loop. If not, then it must compute the distances between $x(i)$ and all k cluster centers. Keeping track of the closest and second-closest allows the algorithm to

Algorithm 3 Elkan's algorithm—using k lower bounds per point and k^2 center-center distances

```
procedure ELKAN(X, C)
  a(i) ← 1, u(i) ← ∞, ∀i ∈ N {Initialize invalid bounds, all in one cluster.}
  ℓ(i, j) ← 0, ∀i ∈ N, j ∈ K
  while not converged do
    compute ‖c(j) − c(j′)‖, ∀j, j′ ∈ K
    compute s(j) ← min_{j′≠j} ‖c(j) − c(j′)‖/2, ∀j ∈ K
    for all i ∈ N do
      if u(i) ≤ s(a(i)) then continue with next i
      r ← True
      for all j ∈ K do
        z ← max(ℓ(i, j), ‖c(a(i)) − c(j)‖/2)
        if j = a(i) or u(i) ≤ z then continue with next j
        if r then
          u(i) ← ‖x(i) − c(a(i))‖
          r ← False
          if u(i) ≤ z then continue with next j
        ℓ(i, j) ← ‖x(i) − c(j)‖
        if ℓ(i, j) < u(i) then a(i) ← j
    for all j ∈ K do {Move the centers and track their movement}
      move c(j) to its new location
      let δ(j) be the distance moved by c(j)
    for all i ∈ N do {Update the upper and lower distance bounds}
      u(i) ← u(i) + δ(a(i))
      for all j ∈ K do
        ℓ(i, j) ← ℓ(i, j) − δ(j)
```

find the correct assignment, compute a tight $\ell(i)$ (for the second-closest center), and possibly tighten $u(i)$ (if the assignment happens to change).

Since $u(i)$ is the same as in Elkan's algorithm, it is updated the same way whenever the centers move. As the lower bound $\ell(i)$ is different, it is updated differently. When the centers move, the algorithm also tracks the maximum distance δ moved by any center. Then by the triangle inequality, the lower bound for each point can be updated as $\ell(i) \leftarrow \ell(i) - \delta$. This is correct since $\ell(i)$ represents the closest distance between any (non-assigned) center and $x(i)$, and no center could have moved closer toward $x(i)$ than the distance δ. In fact, a small optimization is possible. For those points assigned to the furthest-moving center, their lower bounds may be reduced by the distance moved by the second-furthest-moving center.

Why does Hamerly's algorithm not track the identity of the second-closest center? Knowing its identity would allow the algorithm to more efficiently tighten $\ell(i)$ (avoiding the innermost loop over all k centers). The reason is that the second-closest center's identity can change as the centers move, and without looking at all centers it can't be proved that the second-closest center remains the same over multiple iterations.

Hamerly's algorithm has several efficiency tradeoffs compared with Elkan's algorithm. With fewer lower bounds, Hamerly's algorithm uses less memory.

Algorithm 4 Hamerly's algorithm—using 1 lower bound per point

procedure HAMERLY(X, C)
  $a(i) \leftarrow 1, u(i) \leftarrow \infty, \ell(i) \leftarrow 0, \forall i \in N$ {Initialize invalid bounds, all in one cluster.}
  while not converged **do**
    compute $s(j) \leftarrow \min_{j' \neq j} \|c(j) - c(j')\|/2, \forall j \in K$
    for all $i \in N$ **do**
      $z \leftarrow \max(\ell(i), s(a(i)))$
      if $u(i) \leq z$ **then** continue with next i
      $u(i) \leftarrow \|x(i) - c(a(i))\|$ {Tighten the upper bound}
      if $u(i) \leq z$ **then** continue with next i
      Find $c(j)$ and $c(j')$, the two closest centers to $x(i)$, as well as the distances to each.
      if $j \neq a(i)$ **then**
        $a(i) \leftarrow j$
        $u(i) \leftarrow \|x(i) - c(a(i))\|$
      $\ell(i) \leftarrow \|x(i) - c(j')\|$
    for all $j \in K$ **do** {Move the centers and track their movement}
      move $c(j)$ to its new location
      let $\delta(j)$ be the distance moved by $c(j)$
    $\delta' \leftarrow \max_{j \in K} \delta(j)$
    for all $i \in N$ **do** {Update the upper and lower distance bounds}
      $u(i) \leftarrow u(i) + \delta(a(i))$
      $\ell(i) \leftarrow \ell(i) - \delta'$

It spends less time checking bounds (in the innermost loop) and updating bounds (when centers move). Having the single lower bound allows it to avoid entering the innermost loop more often than Elkan's algorithm. On the other hand, Elkan's algorithm computes fewer distances than Hamerly's, since Elkan's has more bounds to prune the required distance calculations. Also, Hamerly's algorithm works better in low dimension than in high dimension. Its single lower bound reduces by the maximum distance moved by any center, and in high dimension all centers tend to move a lot due to the curse of dimensionality.

2.4.5 *Drake's Algorithm:* $1 < b < k$ *Lower Bounds*

Elkan's and Hamerly's algorithms keep, respectively, k bounds and one lower bound per clustered point. Drake and Hamerly [11] bridged the gap between these two extreme values by using $1 < b < k$ lower bounds on the b closest centers to each point. The value of b can be selected in advance or adaptively learned while the algorithm runs. Drake's algorithm uses one upper bound per clustered point. Thus, Drake's algorithm uses $(b + 1)n$ total distance bounds.

For a given point, the first $b - 1$ lower bounds represent the minimal distance from the point to its $b - 1$ closest centers, excluding the currently assigned center. The last (b^{th}) lower bound is treated specially, and we will discuss it later. Since k-means is only concerned with the closest center, we can avoid distance calculations to far-away centers if we can use the bounds to prove that the closest is one of the b closest (the current assigned center plus the $b - 1$ next closest centers).

There are two minor complications that arise from keeping b bounds per point:

- For each of the b lower bound we must keep the identity of the associated center. In Hamerly's algorithm, the identity of the lower-bound center is not kept. Elkan's algorithm keeps the lower bounds in the same order as center indexes, implicitly giving the bound-center association. Keeping a center label for each bound increases the algorithm's memory footprint.
- In order to make the algorithm efficient, the lower bounds should be kept in sorted order by distance from the point. Sorting incurs overhead each time the centers move.

Nevertheless, Drake and Hamerly show that this algorithm works well in practice and can be faster than both Elkan and Hamerly's algorithms under certain conditions.

The first $b - 1$ lower bounds for a point represent the lower bounds to the associated points that are ranked 2 through b in increasing distance from the point. The last lower bound (number b, furthest from the center) represents something a bit different. Instead of being associated with one particular center, it represents the lower bound on all the furthest $k - b$ centers. This is much like Hamerly's one lower bound, but only for the outermost centers.

When searching for the closest center to a given point, we only need to search the centers whose corresponding lower bounds are less than the upper bound for that point. Moving from the center with the smallest lower bound outward, we can hopefully stop searching after just a few comparisons. For each lower-upper bound comparison that fails, we tighten the lower (and upper) bounds as necessary. If we reach the last bound and still have not proven that any of the b tracked centers are the closest, then we must search all remaining $k - b$ centers.

At the end of each k-means iteration, we must adjust the lower bounds. Each is reduced by the amount its associated center moved. The outermost bound should be reduced by the maximum amount moved by any of the $k - b$ outermost centers. However, in practice it is far more efficient, though not as tight, to reduce the last bound by the largest distance moved by *any* center, not just the furthest $k - b$ centers. When doing this, some bound might 'collapse' on (i.e. become smaller than) a lower bound for a supposedly closer center. When this happens, we reduce each bound for closer centers so they are at most the value that is collapsing.

The number of lower bounds b used by Drake's algorithm represents a tradeoff. Increasing b incurs more computational overhead to update the bound values and sort each point's closest centers by their bound, but it is also more likely that one of the bounds will prevent searching over all k centers. Drake's algorithm uses a simple way to determine a 'good' value for b adaptively. Starting with a large value for b, it may choose to reduce b each iteration based on which bounds are being

Algorithm 5 Drake's algorithm—using b lower bounds per point

procedure DRAKE(X, C, b)
  $a(i) \leftarrow 1, u(i) \leftarrow \infty, \forall i \in N$ {Initialize invalid bounds, all in one cluster.}
  $\ell(i, j) \leftarrow 0, \forall i \in N, j \in \{1, \ldots, b\}$
  while not converged **do**
    $m \leftarrow b$
    for all $i \in N$ **do**
      $j \leftarrow \arg\max_{1 \leq j' \leq b} u(i) \leq \ell(i, j')$
      if $j < b$ **then** {The bounds pruned the outer centers.}
        compute distances and reorder the j centers closest to $x(i)$
      else if $j = b$ or $\ell(i, b) < u(i)$ **then** {Bounds were ineffective.}
        compute distances from $x(i)$ to all centers and sort the b closest
      $m \leftarrow \max(m, j)$
    $b \leftarrow \max(k/8, m)$ {Reduce b if possible}
    for all $j \in K$ **do** {Move the centers and track their movement}
      move $c(j)$ to its new location
      let $\delta(j)$ be the distance moved by $c(j)$
    $\delta' \leftarrow \max_{j \in K} \delta(j)$
    for all $i \in N$ **do** {Update the upper and lower distance bounds}
      $u(i) \leftarrow u(i) + \delta(a(i))$
      $\ell(i, b) \leftarrow \ell(i, b) - \delta'$
      for $j = b - 1$ down to 1 **do**
        let $c(z)$ be the center that is the jth closest to $x(i)$
        $\ell(i, j) \leftarrow \min(\ell(i, j) - \delta(z), \ell(i, j + 1))$

used. If the algorithm is able to stop searching after only $b' < b$ lower bounds (over the entire dataset), then it reduces b down to b'. Experimentally, Drake determined that for $k > 8$, $k/8$ is a good floor for b.

2.4.6 Annular Algorithm: Sorting the Centers by Norm

Hamerly's algorithm employing one lower bound is effective at avoiding many distance computations in low dimensional spaces. But whenever the lower and upper bounds for a point cross (i.e. $\ell(i) < u(i)$), the algorithm must compute the distance between the point and all k centers. This search determines the two closest centers and tightens the upper and lower bounds. Similarly, when the last lower bound in Drake's algorithm fails to prune the search, it must search over all centers (it has already searched over b centers, and it must continue searching over the remaining $k - b$ centers represented by the last lower bound).

In this subsection we describe an efficient method that can prune such searches between a point x and all k centers. By sorting the centers by their vector norms, we can eliminate from consideration many centers whose norms are too large or too small to be closest to x. We can do this with only the knowledge of the norm of x, the norms of all k centers, and (an upper bound on) the distance between x and a (hopefully closeby) center.

Consider ordering the k cluster centers by their vector norms, $\|\cdot\|$. This order may change at most once per iteration of k-means, and is inexpensive to compute and maintain. This ordering affords a novel way to structure the search for a point's closest center and potentially avoid examining all centers.

For a point x having norm $\|x\|$, we can use the norm-ordering of the centers and the triangle inequality to prune the search over all centers. Assume that we know the exact distance $\|x - c'\|$ between x and a reasonably close center c' (such as its currently assigned center). Then consider a center c that is actually closer to x. Starting with two different statements from the triangle inequality, we have

$$\begin{aligned} \|c\| &\leq \|x - c\| + \|x\|, \\ \|x\| &\leq \|x - c\| + \|c\|. \end{aligned} \tag{2.6}$$

Then we can combine these to get

$$\begin{aligned} \|c\| - \|x\| &\leq \|x - c\|, \\ \|x\| - \|c\| &\leq \|x - c\|; \end{aligned} \tag{2.7}$$

$$\big|\, \|x\| - \|c\| \,\big| \leq \|x - c\| \qquad \text{combining the two results in (2.7).} \tag{2.8}$$

Because c is closer than c' to x, we can deduce

$$\big|\, \|x\| - \|c\| \,\big| \leq \|x - c\| \leq \|x - c'\|. \tag{2.9}$$

Thus, any center c that is closer to x than c' must satisfy Inequality (2.9). From a different perspective, consider that c is actually *farther* from x than c'. Then c must violate this inequality—i.e. $\big|\, \|x\| - \|c\| \,\big| > \|x - c'\|$. In this case, c can be eliminated from consideration because c' must be closer to x. Note that while the above discussion relied on knowing an *exact* distance between x and some close center c', the same derivation can be done with an upper bound on this distance.

We can use this knowledge to prune the search for the closest center. Given (an upper bound on) a distance $\|x - c'\|$ between x and some center c' (e.g. the currently assigned center), we can eliminate centers c where

$$\big|\, \|x\| - \|c\| \,\big| > \|x - c'\|.$$

And since we have ordered the centers by their norm, we only need to compute the distances between x and those centers c whose norm falls in the range

$$\|x - c'\| - \|x\| \leq \|c\| \leq \|x - c'\| + \|x\|$$

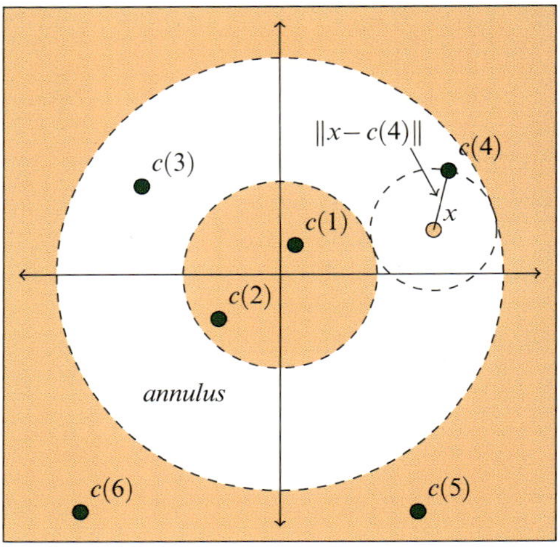

Fig. 2.2 The annular region (*white ring* centered at origin) bounds where the closest center for x might be. Centers $c(j)$ are numbered by their distance from the origin. Point x has $c(4)$ as its previously-closest center, so the width of the annulus is $2\|x - c(4)\|$ (*dashed circle* centered at x)

These bounds form an annular region centered on the origin in which potentially closer centers to x may lie. We can use binary search on the centers ordered by their norms to locate the smallest-norm center that fits this inequality, and then compute the distance between x and each center within the annulus. Figure 2.2 gives a graphical representation of this search space.

We have implemented this annular search in conjunction with Hamerly's algorithm, with one additional change. Each time Hamerly's algorithm searches over all centers, it needs to discover not just the closest center, but also the second-closest center (to tighten the lower bound). Thus, when constructing the annulus for x, we use twice the distance between x and its second-closest center as the annulus width. But since Hamerly's algorithm does not explicitly track the second-closest center, the augmented annular search algorithm attempts to do so. As mentioned above, though, the second-closest center can change. However, the annular search does not need to know which is the actual second-closest center—it only needs the index of a center which is likely to be close to x to form the search annulus, and is farther from x than the assigned center. Thus when it does find the actual second-closest center, it caches its identity for constructing the annulus later.

2.4.7 Kernelized k-Means with Distance Bounds

As with many distance-based algorithms, k-means can be 'kernelized' by applying the kernel trick [10, 34]. Starting from the definition of squared Euclidean distance which can be written using inner products,

$$\|x - y\|^2 = \langle x, x\rangle - 2\langle x, y\rangle + \langle y, y\rangle,$$

we can kernelize the distance by replacing each inner product with a call to a kernel function $K(x, y) = \langle \phi(x), \phi(y) \rangle$ which represents the inner product of x and y after they have been transformed to a new space, within the range of ϕ. Here ϕ typically represents a function that yields a higher-dimensional vector. Thus,

$$\begin{aligned}\|\phi(x) - \phi(y)\|^2 &= \langle \phi(x), \phi(x) \rangle - 2\langle \phi(x), \phi(y) \rangle + \langle \phi(y), \phi(y) \rangle \\ &= K(x, x) - 2K(x, y) + K(y, y)\end{aligned}$$

and if K is easy to compute (without explicitly using ϕ), then using a kernel is a relatively efficient way to perform k-means implicitly in the space of ϕ.

We can apply the triangle inequality algorithms to kernelized k-means. However, we must be careful to avoid inefficiencies that come from kernelizing the algorithm. In particular, since the centers live implicitly in the high-dimensional range of ϕ, and we don't represent high-dimensional feature vectors explicitly, any time we want to use a center, we instead use the kernel applied to all of the points that are in its cluster. That is, in kernel k-means we define center $c(j)$ by its member points:

$$c(j) = \frac{1}{n(j)} \sum_{i|a(i)=j} \phi(x(i))$$

so that when we want to know $\|x - c(j)\|^2$, we compute

$$\begin{aligned}\|\phi(x) - c(j)\|^2 &= K(x, x) - 2\langle \phi(x), c(j) \rangle + \langle c(j), c(j) \rangle \\ &= K(x, x) - 2\left\langle \phi(x), \frac{1}{n(j)} \sum_{i|a(i)=j} \phi(x(i)) \right\rangle \\ &\quad + \left\langle \frac{1}{n(j)} \sum_{i|a(i)=j} \phi(x(i)), \frac{1}{n(j)} \sum_{i|a(i)=j} \phi(x(i)) \right\rangle \\ &= K(x, x) - \frac{2}{n(j)} \sum_{i|a(i)=j} K(x, x(i)) \\ &\quad + \frac{1}{n(j)^2} \sum_{i|a(i)=j} \sum_{i'|a(i')=j} K(x(i), x(i'))\end{aligned}$$

Note that, naively, this single distance computation, which is the core of k-means, has a runtime of $O(n(j)^2)$ (ignoring the cost of computing the kernel). Under the reasonable assumption that clusters are roughly equal size, this is $O(n^2/k^2)$. Thus naive point-center distance computations are quite costly in kernelized k-means, especially when compared with the runtime of the non-kernelized version of k-means. However, simply caching the inner product $\langle c(j), c(j) \rangle$ for each cluster center at the beginning of each k-means iteration allows us to bring the cost down to a much better, though still very costly, $O(n(j))$ (or $O(n/k)$ under the assumption that all clusters are equal size). Either way, in kernelized k-means it is even more appealing to accelerate the algorithm by a method which avoids distance calculations altogether. We can directly apply any of the triangle inequality algorithms to kernelized k-means.

When applying an acceleration method such as Elkan's algorithm to kernelized k-means, we must additionally compute the center movement at each iteration. This is no more costly than computing the inner product for each center with itself, which is already performed each iteration as discussed previously.

2.5 Heap-Ordered k-Means: Inverting the Innermost Loops

Next we turn to a way of restructuring Hamerly's algorithm. We motivate the next algorithm in three ways:

1. it's desirable to reduce the memory use of the accelerated algorithm—in other words the number of bounds kept per point;
2. since $k < n$, the memory-efficient way of searching all point-center distances is to have a nested loop over all n on the outside and all k on the inside, but we may wish to invert this loop structure so that the outer loop is over k; and
3. thus far, the algorithms that are accelerated by the triangle inequality examine all n points every iteration, but we would like an algorithm which only investigates those points whose triangle inequality bounds have been violated.

For all these reasons, we consider an algorithm that orders all the points by their likelihood of needing cluster reassignment—in other words, those for which $\ell(i) - u(i)$ is smallest (perhaps even negative).

2.5.1 *Reducing the Number of Bounds Kept*

This new algorithm, Heap-ordered k-means, replaces each pair of bounds kept for each point $(u(i), \ell(i))$ by Hamerly's algorithm with a single value representing their difference, $\ell u(i) = \ell(i) - u(i)$. The reasoning is that the bounds avoid distance calculations whenever

$$\begin{aligned} u(i) &\leq \ell(i) \\ 0 &\leq \ell(i) - u(i) \\ 0 &\leq \ell u(i). \end{aligned}$$

Thus, we can reduce by half the number of distance bounds used by Hamerly's algorithm simply by replacing the upper and lower bounds by their difference. Whenever $\ell u(i) < 0$, the distance bound for $x(i)$ has been violated and we need to tighten its bound (possibly reassigning the point to another cluster in the process).

Each update to $\ell u(i)$ is done as follows. If in Hamerly's algorithm we would increment $u(i)$ by a and decrement $\ell(i)$ by b, then $\ell u(i)$ is decremented by $b + a$. Thus, updates to this single bound are simple.

2.5.2 Cost of Combining Distance Bounds

There is a cost of replacing two bounds with one. In the innermost loop of Hamerly's algorithm, the bounds fail whenever $u(i) > \ell(i)$ and it may have to examine all k point-center distances involving x. However, before that happens it tightens the upper bound as $u(i) = \|x(i) - c(a(i))\|$. If tightening $u(i)$ causes the bounds to become ordered again ($u(i) \leq \ell(i)$), then it has avoided $k - 1$ distance calculations. However, when the new bound $\ell u(i) < 0$, we cannot tighten it by looking at just the two centers defining the bound (the first- and second-closest centers), because we do not know the identity of the second-closest center. So when $\ell u(i) < 0$, we must examine all k point-center distances for x to determine whether $a(i)$ is still its closest center. We might reduce the search over all k by using an annular search approach (see Sect. 2.4.6).

2.5.3 Inverting the Loops Over n and k

Usually, the innermost loops in Lloyd's algorithm loop over all n points, and for each point over all k cluster centers to find its closest. We could invert these two loops, searching over each cluster and within that each point. But naively doing so requires maintaining an extra distance for each of the n points indicating its distance to the closest of the centers examined so far.

Whichever way we structure the nesting of the these two loops, the naive strategy examines all n points, and typically all nk point-center pairs. It would be an advantage to examine only those points which could possibly change cluster membership, avoiding completely those points whose assignments are provably 'safe' (via distance bounds). It turns out that we can achieve this goal by using a single (combined) lower-upper distance bound, a heap for each cluster, and making the outer loop be over the k clusters.

2.5.4 Heap-Structured Bounds

For each cluster we construct a min-heap of the points assigned to that cluster, ordered by an estimate of their distance from the cluster center. For now assume

Algorithm 6 Heap-ordered algorithm—inefficient version

procedure HEAPKMEANS-INEFFICIENT(X, C)
  construct k min-heaps: $h(j)$ for each $j \in K$
  insert $(-1, x(i))$ into $h(1)$ for each $i \in N$ {put all in the first cluster, with violated bounds}
  while not converged **do**
    for all $j \in K$ **do**
      while $h(j)$ is not empty and $(\ell u(i), i)$ at the top of $h(j)$ has $\ell u(i) < 0$ **do**
        remove $(\ell u(i), x(i))$ from $h(j)$
        find $c(j')$ and $c(j'')$, the two closest centers for $x(i)$
        compute $\ell u(i) = \|x(i) - c(j'')\| - \|x(i) - c(j')\|$ {tighten the bound}
        put $(\ell u(i), i)$ into $h(j')$
    move each center to the mean of its assigned points
    update $\ell u(i)$ for each $i \in N$, restructuring each heap as necessary

each heap entry for $x(i)$ is a pair $(lu(i), i)$. (This is not exactly the case, shortly we will adjust this definition.) But this suffices to show that the point at the top of the heap is the one whose bound is closest to failing, or has already failed (if $\ell u(i) < 0$).

This basic approach leads to Algorithm 6 for examining only those points whose bounds have failed. While this is a reasonable algorithm, it is hampered by the fact that it must visit every point to update each $\ell u(i)$, restructuring the heap as it goes. This violates a primary goal we have for this algorithm: to avoid the $O(n)$ factor of considering every point in each iteration of k-means.

We can improve this algorithm by removing the per-iteration update for $\ell u(i)$. This is possible by changing the key for the heap from $\ell u(i)$ to be a related, but static value, and keeping the updates for $\ell u(i)$ external to the heap. First we will need some new notation.

Let $\ell u(i, t)$ be the value of $\ell u(i)$ at k-means iteration t. Instead of updating $\ell u(i)$ at each iteration, we keep track of the cumulative *updates* for $\ell u(i)$, which turn out to be the same for all points assigned to the same center. Suppose $x(i)$ is assigned to center $c(j)$. Let $c(j)_t$ be its location at k-means iteration t. At iteration t, we compute the current value

$$\ell u(i, t) \leftarrow \ell u(i, t-1) - \|c(j)_t - c(j)_{t-1}\| - m(t), \qquad \text{where} \tag{2.10}$$

$$m(t) \leftarrow \max_{j' \in K} \|c(j')_t - c(j')_{t-1}\| \tag{2.11}$$

is the distance moved by the furthest-moving center for that iteration. Then for each center $c(j)$ we maintain the following structure

$$z(j, t) = m(t) + \sum_{p=1}^{t} \|c(j)_p - c(j)_{p-1}\|, \tag{2.12}$$

where $c(j)_0$ is the center's initial position. Then $z(j, t)$ is the distance center $c(j)$ has traveled since the beginning of the algorithm plus the furthest distance any center has traveled, up through iteration t. This can be computed efficiently each iteration, taking $O(kd)$ time for all centers.

Assume that at iteration t, point $x(i)$ becomes newly assigned to center $c(j)$, and has second-closest center $c(j')$. Then we can compute the (tight) value of $\ell u(i,t) = \|x(i) - c(j')_t\| - \|x(i) - c(j)_t\|$. Into heap $h(j)$ we place the pair

$$(\ell u(i,t) + z(j,t), i). \tag{2.13}$$

In other words, we order the heap by $\ell u(i,t)$ offset by the current value of $z(j,t)$. Consider now the value of $\ell u(i,t')$ at some later iteration (i.e. $t' > t$):

$$\begin{aligned}
\ell u(i,t') &= \ell u(i,t) - \|c(j)_{t+1} - c(j)_t\| - m(t+1) - \ldots - \|c(j)_{t'} - c(j)_{t'-1}\| - m(t') \\
&= \ell u(i,t) - \sum_{p=t+1}^{t'} \|c(j)_p - c(j)_{p-1}\| + m(p) \\
&= \ell u(i,t) + \sum_{p=1}^{t} \|c(j)_p - c(j)_{p-1}\| + m(p) - \sum_{p=1}^{t'} \|c(j)_p - c(j)_{p-1}\| + m(p) \\
&= \ell u(i,t) + z(j,t) - z(j,t')
\end{aligned} \tag{2.14}$$

In Algorithm 6 at iteration $t' > t$ we would check if the top of the heap has $\ell u(i,t') < 0$. Starting with this and using Eq. (2.14) and the new heap structure, we find the equivalent test

$$\begin{aligned}
\ell u(i,t') &< 0 \\
\ell u(i,t) + z(j,t) - z(j,t') &< 0 \\
\ell u(i,t) + z(j,t) &< z(j,t'),
\end{aligned} \tag{2.15}$$

which is exactly what we used as the distance key on the heap (i.e. $\ell u(i,t) + z(j,t)$) and have been updating in the intervening iterations (i.e. $z(j,t')$).

Thus, we can keep a heap structure that does not require updating as $\ell u(i)$ changes, instead accumulating updates for each center external to the heap structure, and achieve equivalent tests for $\ell u(i) < 0$. Note that we do not need to keep the history of $z(j,t)$ for all iterations; we only ever need the value for the current iteration. This leads us to the more efficient Algorithm 7.

2.5.5 Analysis of Heap-Structured k-Means

To analyze Algorithm 7, we assume a simple binary-heap implementation that takes $O(\log(n))$ time to insert and remove, and introduce two new terms. The number

Algorithm 7 Heap-ordered algorithm—using 1 bound per point, and one per cluster

procedure HEAPKMEANS(X, C)
  construct min-heap $h(j)$ for each $j \in K$
  let $z(j) \leftarrow 0$ for each $j \in K$
  insert $(-1, i)$ into $h(1)$ for each $i \in N$ {put all in the first cluster, with violated bounds}
  while not converged **do**
    for all $j \in K$ **do**
      while $h(j)$ is not empty and (y, i) at the top of $h(j)$ has $y < z(j)$ **do**
        remove (y, i) from $h(j)$
        compute the distance from $x(i)$ to each center
        let $c(j')$ and $c(j'')$ be its closest and second-closest centers
        insert $(\|x(i) - c(j'')\| - \|x(i) - c(j')\| + z(j'), i)$ into $h(j')$
    for all $j \in K$ **do**
      move $c(j)$ to the average of its assigned points
      calculate $\delta(j)$ as the distance $c(j)$ moved
    compute $\delta' = \max_{j \in K} \delta(j)$
    for all $j \in K$ **do**
      update $z(j) \leftarrow z(j) + \delta(j) + \delta'$

of iterations performed by k-means is w, and the number of bound violations per iteration is v. In other words, v is the number of points that must be removed from any heap in one iteration of k-means. Then the running time is

$$O(n + wv(\log(n) + kd)). \tag{2.16}$$

The most important thing to notice about this analysis is the lack of a wn term, which does occur in other algorithms based on the triangle inequality. While there is a term wv, and v depends on n, in general $v < n$ and highly clustered data will have $v \ll n$.

2.6 Parallelization

There are multiple ways to parallelize the k-means algorithm. While the purpose of our study is to improve the core k-means algorithm, we also want to show that such improvements are suitable for parallelization. In particular, we consider the simplest case of parallelizing the algorithm over a shared-memory, multicore machine.

In a shared-memory context with p processors, the most straightforward way to parallelize the batch k-means algorithm is to partition the n data points to be clustered into p subsets each of size n/p. The cluster centers are replicated across (or shared by) all processors.

During each iteration, each processor assigns each point in its partition to the center nearest that point. After the assignment step, each processor computes for its partition the (partial) sufficient statistics required to compute the new center locations. In particular, for each center each processor must compute the vector sum of the points assigned to that center, as well as the number of points assigned to it.

Using these partial results from all processors at the end of the iteration, the new cluster centers can be computed and shared with all processors. Thus the algorithm is embarrassingly parallel within each iteration, but requires synchronization of all processors between iterations. Note that any k-means optimization algorithm which is not batch but stochastic in nature will not be able to directly apply this type of parallelization and maintain identical output.

A few details are worth noting. For heap-based k-means presented in this chapter, the algorithm can be parallelized this way by constructing k heaps for each processor (resulting in pk total heaps). For Drake's algorithm which may reduce the number of lower bounds used as the algorithm proceeds, each thread can adaptively reduce the number of bounds that it uses without affecting other threads, allowing the potential for optimization within smaller regions of the dataset.

We have implemented multithreaded versions of the algorithms we use in experiments, and perform experiments to see how well they scale with an increasing number of available processors. Please see Sect. 2.7 for the experimental results.

2.7 Experiments and Discussion

In this section we describe the experimental results the algorithms discussed in this chapter on several real-world and synthetic datasets. Table 2.3 describes the datasets. Table 2.4 describes the algorithms tested.

Table 2.3 A list of the datasets used in the experiments

Name	Description	Number of points n	Dimension d
Uniform-2/8/32	Synthetic, uniform distribution	1,000,000	2/8/32
Clustered-2/8/32	Synthetic, 50 separated spherical Gaussian clusters	1,000,000	2/8/32
BIRCH	10×10 grid of Gaussian clusters	100,000	2
MNIST-50	Random projection from mnist784	60,000	50
Covertype	Soil cover measurements	581,012	54
KDD Cup 1998	Response rates for fundraising campaign	95,412	56
MNIST-784	Raster images of handwritten digits	60,000	784

Table 2.4 Algorithms tested. Given the same initialization, all algorithms produce the same result

Algorithm	Type of acceleration	Unique features
Lloyd	None	Baseline algorithm for batch k-means.
Compare-means [32]	Triangle inequality	avoid innermost loop when closest other center is far away.
Sort-means [32]	Triangle inequality, sorting centers	Search over centers in increasing distance from closest center.
Elkan [12]	Triangle inequality, distance bounds	1 upper bound, k lower bounds per point.
Hamerly [15]	Triangle inequality, distance bounds	1 upper bound, 1 lower bound per point.
Drake (adaptive version) [11]	Triangle inequality, distance bounds, sorting bounds	1 upper bound, b lower bounds per point; b is chosen adaptively.
Annular (this chapter)	Triangle inequality, distance bounds, sorting centers	Like Hamerly, but with norm-ordered centers.
Heap (this chapter)	Triangle inequality, distance bounds	Uses k heaps of assigned points, ordered by bounded distance from center. Upper and lower bounds are combined into one value.
Kernelized Lloyd	None	Lloyd's algorithm with kernels.
Kernelized Elkan (this chapter)	Triangle inequality, bounds	Applying Elkan's algorithm to kernelized k-means.

2.7.1 Testing Platforms

We ran our tests on two Linux 64-bit Intel platforms. One is a 128-node parallel machine with 8 processors and 16 GB of RAM per node. We used up to 8 simultaneous threads on this machine. The other is a more recent 12-core computer with 16 GB of RAM which we used for testing up to 12 simultaneous threads. We implemented all of the algorithms tested in C++ and Pthreads. For algorithms that used similar structures, we tried to use common code wherever possible to minimize differences due to implementation.

2.7.2 Speedup Relative to the Naive Algorithm

Figures 2.3 and 2.4 show the speedup of the accelerated algorithms relative to the naive algorithm. Speedup is defined as the time for the naive algorithm divided by the time for the accelerated algorithm. For each dataset, we run it with multiple k values. Speedups of up to 50x are observed, with the largest accelerations being for low-dimensional, naturally-clustered data. This is an important case for user-facing applications.

The results show a general improvement in speedup for most accelerated algorithms as the number of clusters increases. This makes sense because as the number of total possible distance calculations rises with k, so does the number of 'far away' centers that can be pruned using the acceleration techniques in this chapter. The speedup curves are not monotonic because the number of iterations varies (depending on k and the initialization), and when k-means performs very few iterations, all the algorithms take roughly the same amount of time. One reason is because the accelerations using distance bounds provide the most benefit when the centers are moving very little.

We observe generally that not all of the accelerated algorithms always outperform the naive algorithm. For example, Elkan's algorithm shows very little if any improvement when the dimension is 2. On the other end of the dimension spectrum, sort-means and compare-means perform about the same as the naive algorithm when the data is unstructured (uniform random) and of dimension 8 or 32.

In the highest dimension dataset, MNIST-784, Elkan's algorithm is the clear winner. It benefits in this high-dimension space by being the best at avoiding distance calculations, where distance calculations are very expensive. Drake's algorithm is second-best, since it uses fewer bounds than Elkan's and is unable to avoid as many distance calculations. Generally, as dimension increases the algorithm gains more benefit from caching additional (lower) distance bounds.

2.7.3 *Parallelism*

We implemented and tested multithreaded versions of each algorithm we investigate. Here we look at how well each is able to use additional computation resources, in terms of speedup and efficiency. While we try to parallelize all parts of each algorithm, the different steps of each algorithm require different amounts of thread synchronization in each iteration, and some parts are not easily parallelized (e.g. a single sort that occurs each iteration whose input depends on data from all threads, and whose output must be shared to all threads, as happens in the Annular algorithm).

Figure 2.5 shows the speedup of each algorithm with respect to the number of threads. Each algorithm's speedup is computed relative to using only one thread with that algorithm. We measure wall-clock time for using one thread and for using t threads, and divide the former by the latter to obtain the speedup. All versions of an algorithm (e.g. single-threaded versus multithreaded), given the same initialization, produce the same sequence of k-means iterations and final clustering.

Figure 2.6 shows the efficiency of each algorithm with respect to the number of threads. We define efficiency as speedup divided by the number of threads. So a perfect efficiency would be a line fixed at 1.0, and a program which cannot use multiple threads would have an efficiency curve of $1/t$ where t is the number of threads.

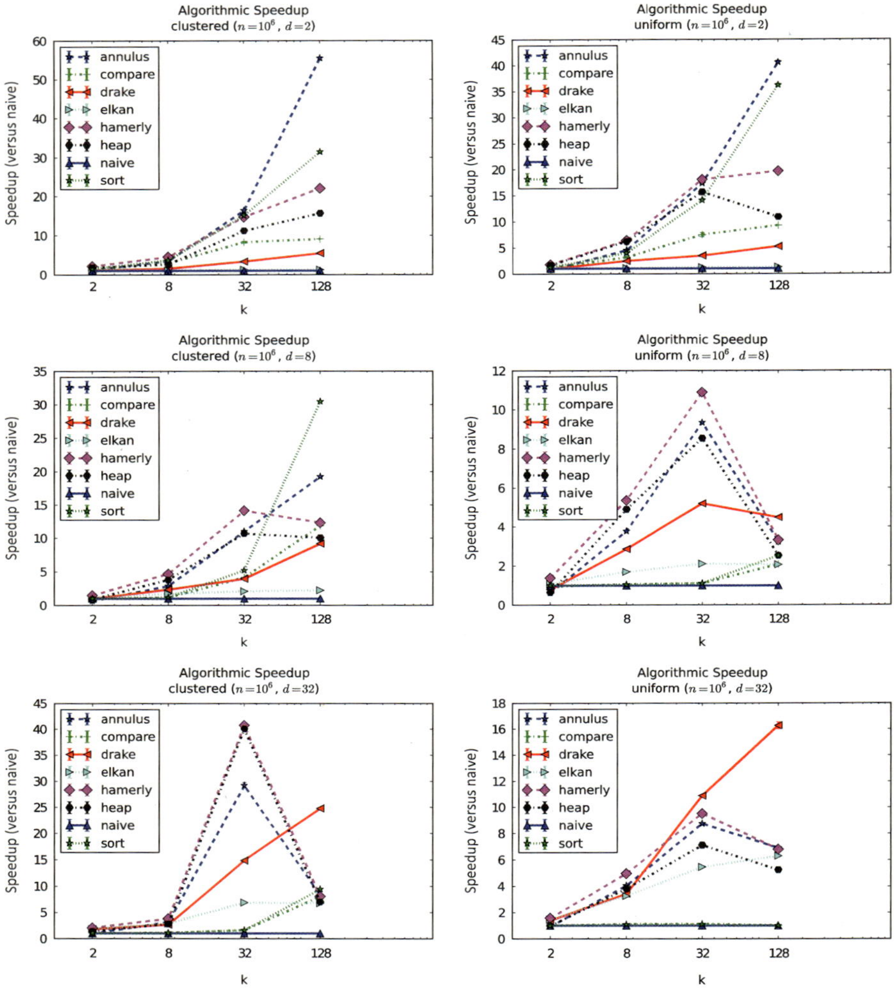

Fig. 2.3 Speedup relative to the naive algorithm for synthetic datasets (clustered and uniform). Speedup is defined as time(naive)/time(accelerated)

It is clear that not all algorithms use the additional available computation equally well. The algorithm that benefits the most from additional threads is the non-accelerated (naive) Lloyd's algorithm, which obtains nearly linear speedups. This is likely for two reasons: it has the least synchronization between threads, and its per-thread behavior is the most predictable (each thread will do approximately equal work). Accelerated algorithms require more synchronization since there is more information kept and shared between threads. For per-thread behavior, it's possible

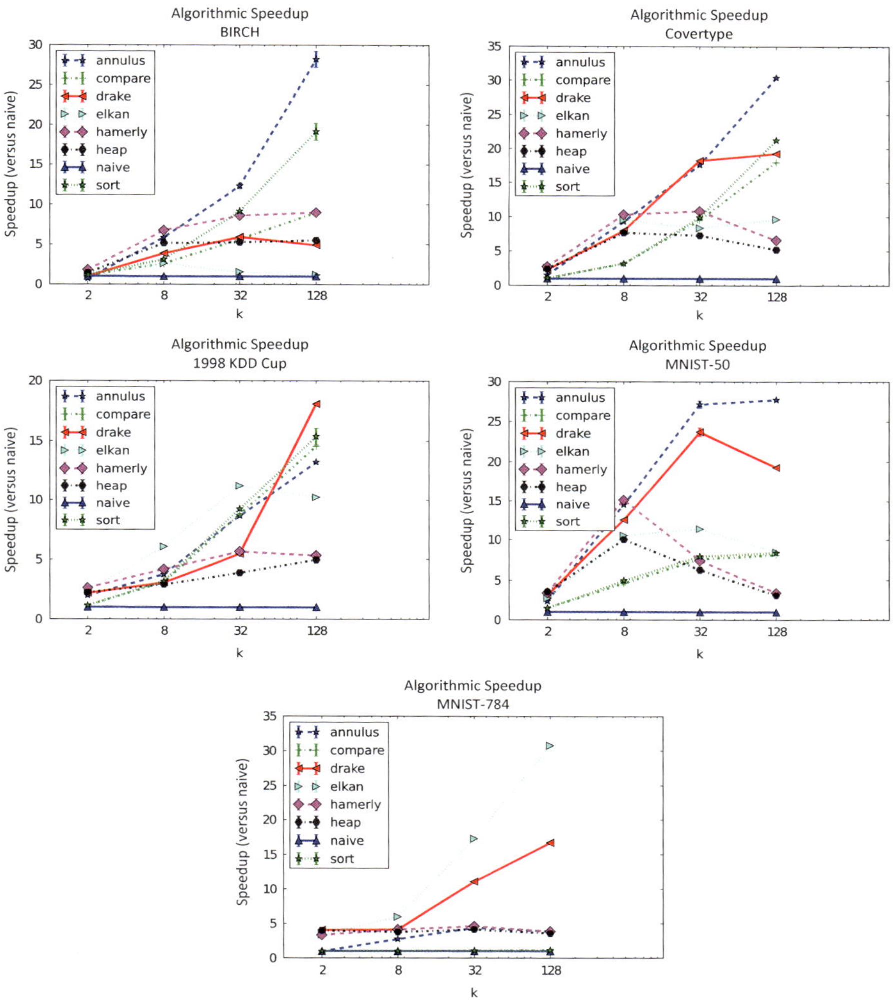

Fig. 2.4 Speedup relative to the naive algorithm for other datasets. Speedup is defined as time(naive)/time(accelerated)

that one thread will have more work to do than another due to, e.g., distance bounds being more effective for the data assigned to the thread.

2.7.4 Number of Distance Calculations

The number of distance calculations performed by *k*-means for several datasets is shown in Figs. 2.7 and 2.8 (for clustered and uniform synthetic datasets). While the datasets for the two figures are comparable in terms of the number

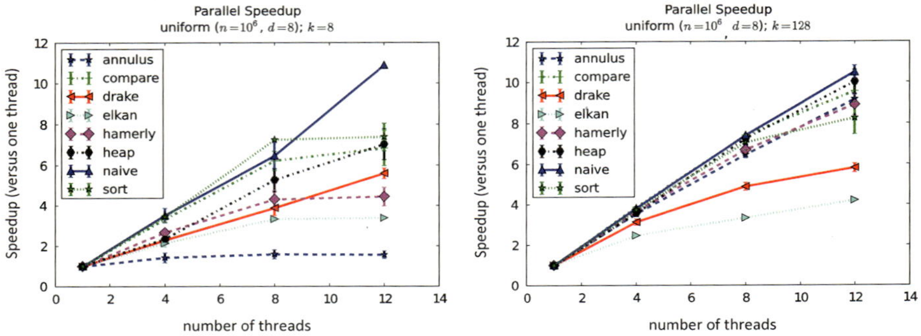

Fig. 2.5 Speedup for each algorithm as a function of the number of threads. Speedup for t threads is defined as time(single-thread)/time(t threads)

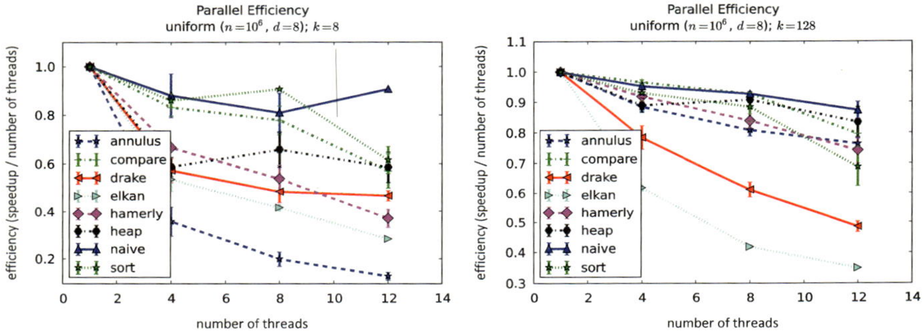

Fig. 2.6 Parallel efficiency for each algorithm as a function of the number of threads. Efficiency for t threads is defined the speedup(t threads)/t. Perfect efficiency is 1.0, and higher is better

of points, dimensions, and cluster centers used, they differ in structure (clustered versus uniform). For clustered data, all accelerated algorithms appear to compute dramatically fewer distances than the naive algorithm. However, there is a stark difference in the uniform datasets. Those accelerated algorithms that use some kind of distance bounds (Elkan, Hamerly, Annular, Drake, and Heap) all do much better than those algorithms which do not (Compare-means and Sort-means), when the dimension is 8 or higher. Thus, the distance bounds seem to be a key part of reducing distance computations in k-means.

As point-center distance calculations are especially expensive in kernelized k-means algorithms, we tested the effectiveness of Elkan's algorithm on this algorithm. Table 2.5 shows the number of distances calculated by both naive kernel k-means and Elkan's version on a small dataset. It is clear that Elkan's algorithm saves a dramatic number of distance calculations even in kernel spaces.

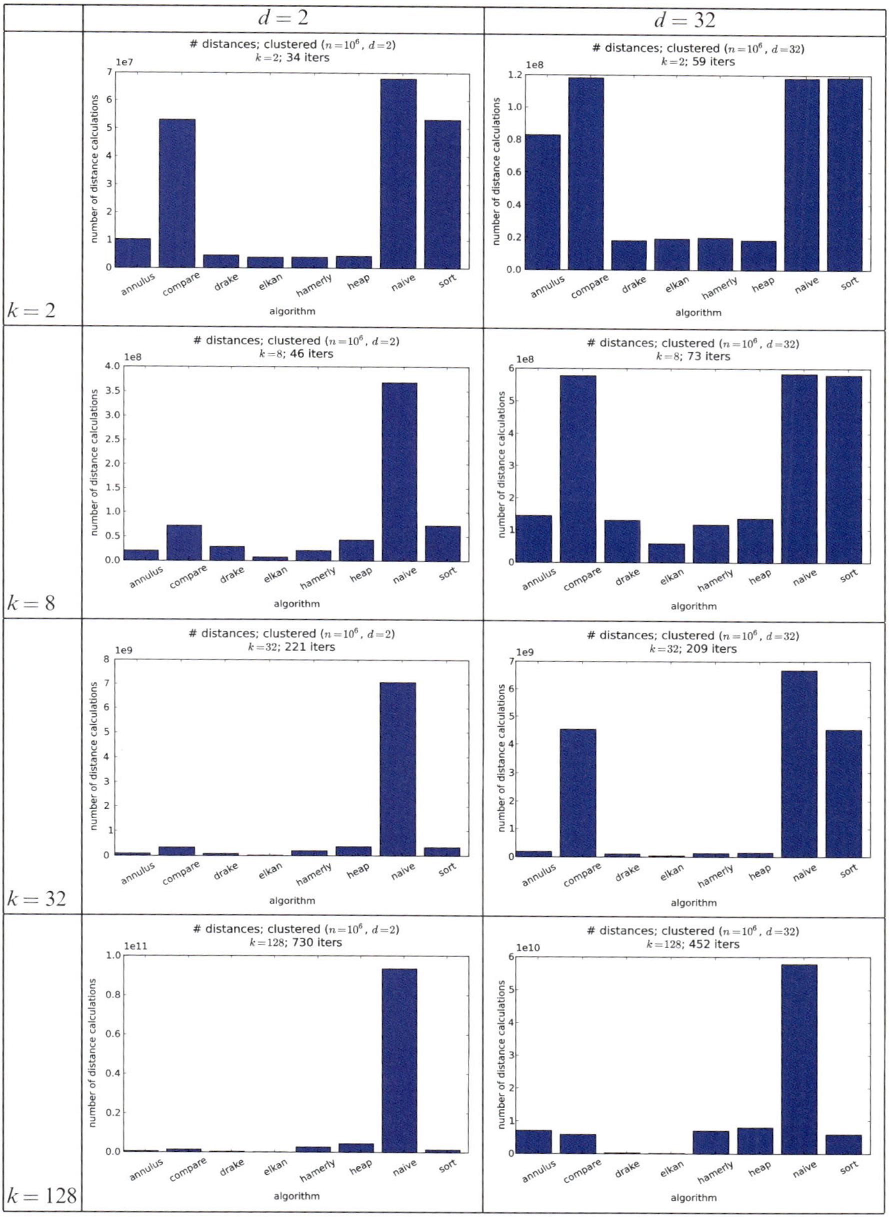

Fig. 2.7 Number of distance calculations performed for synthetic clustered data in 2 and 32 dimensions, over $k = 2, 8, 32$, and 128

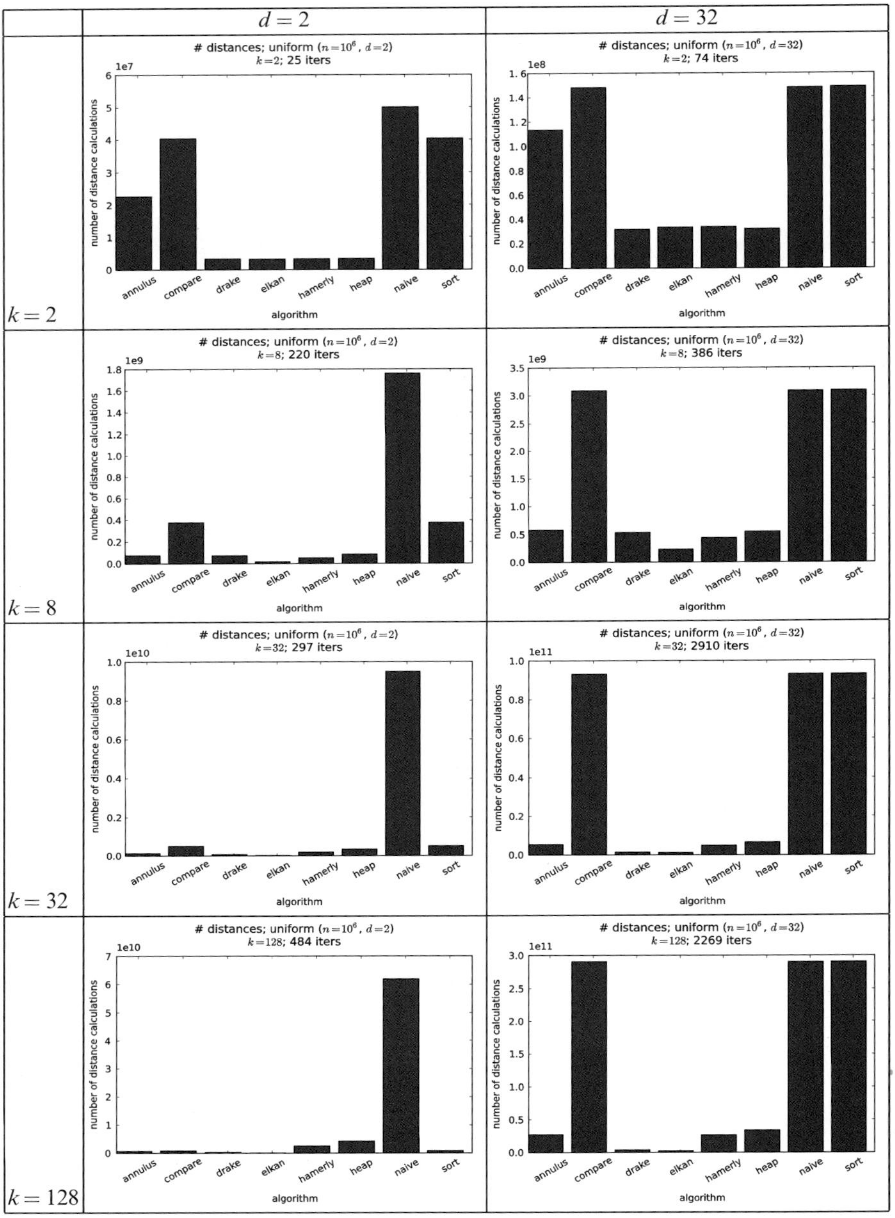

Fig. 2.8 Number of distance calculations performed for synthetic uniform data in 2, 8, and 32 dimensions, over $k = 2, 8, 32$, and 128

Table 2.5 The number of distances computed by unaccelerated kernel k-means and Elkan's kernel k-means

k	Iterations	Number of distance calculations	
		Naive kernel k-means	Elkan kernel k-means
2	24	96,467	26,766
8	39	624,750	155,130
32	43	2,752,810	708,491
128	14	3,584,506	628,437

We used a small synthetic dataset: a uniform random distribution, with 2,000 points and 8 dimensions. We use a Gaussian kernel with bandwidth τ = 10,000. The number of iterations ranged from 14 to 43

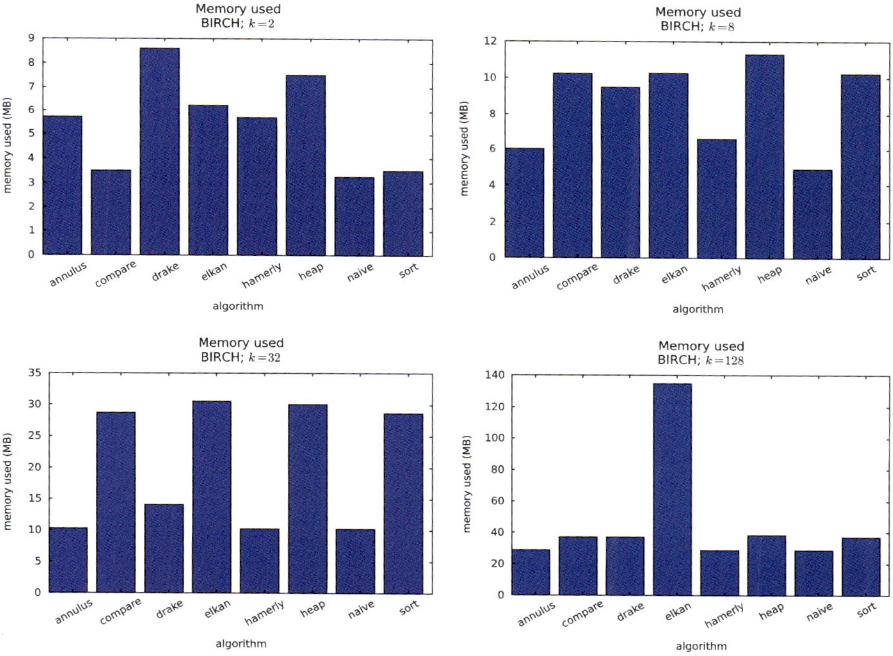

Fig. 2.9 The amount of memory used (in megabytes) for the BIRCH, over k = 2, 8, 32, and 128

2.7.5 Memory Use

Figures 2.9 and 2.10 show the amount of memory used by k-means for the small BIRCH dataset (n =100,000 and d = 2) and a large synthetic uniform dataset (n =1,000,000 and d = 32). As opposed to the amount of time used, the amount of memory used is a function of just the number of points, dimension, and number of clusters. It's clear that when k is small, the algorithms all use about the same amount

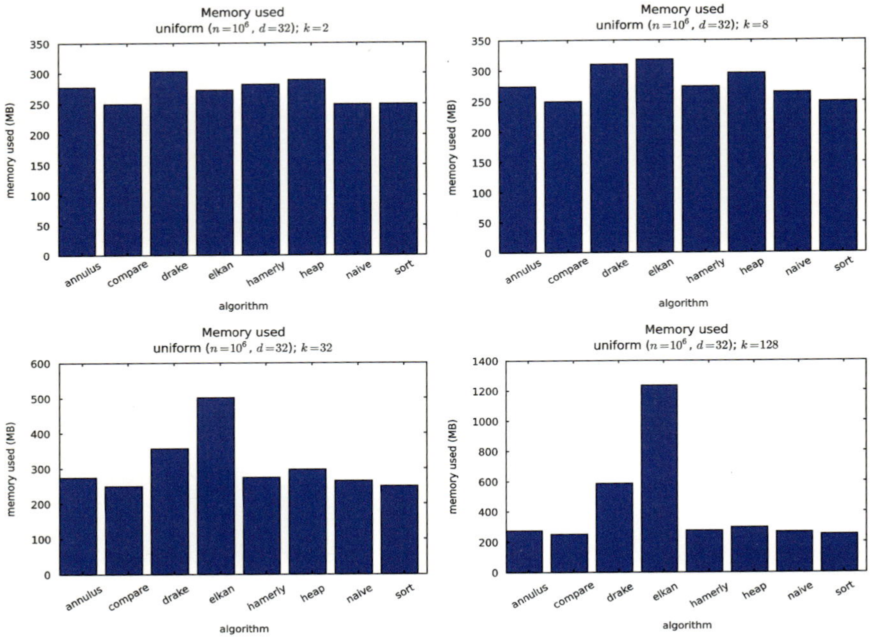

Fig. 2.10 The amount of memory used (in megabytes) for synthetic uniform data in 32 dimensions, over $k = 2, 8, 32$, and 128

of memory. When k is large, however, the large number of bounds used by Elkan's and Drake's algorithms, begin to set them apart as using a lot more memory. It is nice that for a small relative increase in memory footprint, many of these algorithms afford significant speedups.

2.8 Conclusion

This chapter presents a number of alternatives to Lloyd's very popular and widely used batch k-means algorithm. All those presented aim to provide exactly the same answer as Lloyd's (given the same initialization), but faster. Some algorithms are from the literature of the last decade (Compare-means, Sort-means; Elkan's, Hamerly's, and Drake's algorithms), and some are new (Annular, Heap).

The algorithms studied here rely on the geometric triangle inequality to avoid unnecessary and costly distance calculations. This proves to be simple to implement and can provide dramatic speedups of up to 40x in our tests. There are multiple ways to apply the triangle inequality to speed up k-means. Practically, using the triangle inequality to inexpensively maintain a set of distance bounds between points and centers is the idea with the greatest benefit.

References

1. Agarwal PK, Har-Peled S, Varadarajan KR (2005) Geometric approximation via coresets. Comb Comput Geom 52:1–30
2. Apache Mahout http://mahout.apache.org/. Version 0.8, Accessed 24 Jan 2014
3. Arthur D, Vassilvitskii S (2007) kmeans++: the advantages of careful seeding. In: ACM-SIAM symposium on discrete algorithms, pp 1027–1035
4. Arthur D, Manthey B, Röglin H (2011) Smoothed analysis of the k-means method. J ACM 58(5):19
5. Bei C-D, Gray RM (1985) An improvement of the minimum distortion encoding algorithm for vector quantization. IEEE Trans Commun 33(10):1121–1133
6. Bottou L, Bengio Y (1995) Convergence properties of the k-means algorithms. In: Advances in neural information processing systems, vol 7. MIT Press, Cambridge, 585–592
7. Celebi ME (2011) Improving the performance of k-means for color quantization. Image Vis Comput 29(4):260–271
8. Celebi ME, Kingravi HA, Vela PA (2013) A comparative study of efficient initialization methods for the k-means clustering algorithm. Expert Syst Appl 40(1):200–210
9. Coates A, Ng AY, Lee H (2011) An analysis of single-layer networks in unsupervised feature learning. In: International conference on artificial intelligence and statistics, pp 215–223
10. Dhillon I, Guan Y, Kulis B (2005) A unified view of kernel k-means, spectral clustering and graph cuts. Technical Report TR-04-25, University of Texas at Austin
11. Drake J, Hamerly G (2012) Accelerated k-means with adaptive distance bounds. In: 5th NIPS workshop on optimization for machine learning
12. Elkan C (2003) Using the triangle inequality to accelerate k-means. In: Proceedings of the twentieth international conference on machine learning (ICML), pp 147–153
13. Forgy EW (1965) Cluster analysis of multivariate data: efficiency versus interpretability of classifications. In: Biometric society meeting, Riverside
14. Fu K-S, Mui JK (1981) A survey on image segmentation. Pattern Recognit 13(1):3–16
15. Hamerly G (2010) Making k-means even faster. In: SIAM international conference on data mining
16. Hamerly G, Elkan C (2002) Alternatives to the k-means algorithm that find better clusterings. In: Proceedings of the eleventh international conference on Information and knowledge management, pp 600–607. ACM, New York
17. Hartigan JA, Wong MA (1979) Algorithm as 136: a k-means clustering algorithm. J R Stat Soc Ser C Appl Stat 28(1):100–108
18. Hochbaum DS, Shmoys DB (1985) A best possible heuristic for the k-center problem. Math Oper Res 10(2):180–184
19. Kanungo T, Mount DM, Netanyahu NS, Piatko CD, Silverman R, Wu AY (2002) An efficient k-means clustering algorithm: analysis and implementation. IEEE Trans Pattern Anal Mach Intell 24:881–892
20. Kaukoranta T, Franti P, Nevalainen O (2000) A fast exact gla based on code vector activity detection. IEEE Trans Image Process 9(8):1337–1342
21. Lai JZC, Liaw Y-C (2008) Improvement of the k-means clustering filtering algorithm. Pattern Recognit 41(12):3677–3681
22. Lai JZC, Liaw Y-C, Liu J (2008) A fast vq codebook generation algorithm using codeword displacement. Pattern Recognit 41(1):315–319
23. Linde Y, Buzo A, Gray R (1980) An algorithm for vector quantizer design. IEEE Trans Commun 28(1):84–95
24. Lloyd S (1982) Least squares quantization in PCM. IEEE Trans Inf Theory 28:129–137
25. Low Y, Gonzalez J, Kyrola A, Bickson D, Guestrin C, Hellerstein JM (2010) Graphlab: a new parallel framework for machine learning. In: Conference on uncertainty in artificial intelligence (UAI)

26. MacQueen JB (1967) Some methods for classification and analysis of multivariate observations. In: 5th Berkeley symposium on mathematical statistics and probability, vol 1. University of California Press, Berkeley, pp 281–297
27. Moore AW (1991) An introductory tutorial on kd-trees. Technical Report 209, Carnegie Mellon University
28. Moore AW (2000) The anchors hierarchy: using the triangle inequality to survive high dimensional data. In: twelfth conference on uncertainty in artificial intelligence. AAAI Press, Stanford, CA, pp 397–405
29. Ng AY, Jordan MI, Weiss Y et al (2002) On spectral clustering: analysis and an algorithm. Adv Neural Inf Process Syst 2:849–856
30. Pan J-S, Lu Z-M, Sun S-H (2003) An efficient encoding algorithm for vector quantization based on subvector technique. IEEE Trans Image Process 12(3):265–270
31. Pelleg D, Moore A (1999) Accelerating exact k-means algorithms with geometric reasoning. In: ACM SIGKDD fifth international conference on knowledge discovery and data mining, pp 277–281
32. Phillips SJ (2002) Acceleration of k-means and related clustering algorithms. In: Mount D, Stein C (eds) Algorithm engineering and experiments. Lecture notes in computer science, vol 2409. Springer, Berlin, Heidelberg, pp 61–62
33. Ra S-W, Kim JK (1993) A fast mean-distance-ordered partial codebook search algorithm for image vector quantization. IEEE Trans Circuits Syst II 40(9):576–579
34. Schölkopf B, Smola A, Müller K-R (1998) Nonlinear component analysis as a kernel eigenvalue problem. Neural Comput 10(5):1299–1319
35. Sculley D (2010) Web-scale k-means clustering. In: Proceedings of the 19th international conference on World Wide Web. ACM, New York, pp 1177–1178
36. Sherwood T, Perelman E, Hamerly G, Calder B (2002) Automatically characterizing large scale program behavior. SIGOPS Oper Syst Rev 36(5):45–57
37. Tai S-C, Lai CC, Lin Y-C (1996) Two fast nearest neighbor searching algorithms for image vector quantization. IEEE Trans Commun 44(12):1623–1628
38. Vattani A (2011) k-means requires exponentially many iterations even in the plane. Discrete Comput Geom 45(4):596–616
39. Wettschereck D, Dietterich T (1991) Improving the performance of radial basis function networks by learning center locations. In Neural Inf Process Syst 4:1133–1140
40. Wu X, Kumar V, Quinlan JR, Ghosh J, Yang Q, Motoda H, McLachlan GJ, Ng AFM, Liu B, Yu PS, Zhou Z-H, Steinbach M, Hand DJ, Steinberg D (2008) Top 10 algorithms in data mining. Knowl Inf Syst 14(1):1–37
41. Yael https://gforge.inria.fr/projects/yael/. Version v1845, Accessed 24 Jan 2014

Chapter 3
Linear, Deterministic, and Order-Invariant Initialization Methods for the K-Means Clustering Algorithm

M. Emre Celebi and Hassan A. Kingravi

Abstract Over the past five decades, k-means has become the clustering algorithm of choice in many application domains primarily due to its simplicity, time/space efficiency, and invariance to the ordering of the data points. Unfortunately, the algorithm's sensitivity to the initial selection of the cluster centers remains to be its most serious drawback. Numerous initialization methods have been proposed to address this drawback. Many of these methods, however, have time complexity superlinear in the number of data points, which makes them impractical for large data sets. On the other hand, linear methods are often random and/or sensitive to the ordering of the data points. These methods are generally unreliable in that the quality of their results is unpredictable. Therefore, it is common practice to perform multiple runs of such methods and take the output of the run that produces the best results. Such a practice, however, greatly increases the computational requirements of the otherwise highly efficient k-means algorithm. In this chapter, we investigate the empirical performance of six linear, deterministic (non-random), and order-invariant k-means initialization methods on a large and diverse collection of data sets from the UCI Machine Learning Repository. The results demonstrate that two relatively unknown hierarchical initialization methods due to Su and Dy outperform the remaining four methods with respect to two objective effectiveness criteria. In addition, a recent method due to Erişoğlu et al. performs surprisingly poorly.

Keywords Data mining • Unsupervised learning • Clustering • K-means • Cluster center initialization • Maximin

---

M.E. Celebi (✉)
Department of Computer Science, Louisiana State University, Shreveport, LA, USA
e-mail: ecelebi@lsus.edu

H.A. Kingravi
School of Electrical and Computer Engineering, Georgia Institute of Technology, Atlanta, GA, USA
e-mail: kingravi@gatech.edu

M.E. Celebi (ed.), *Partitional Clustering Algorithms*,
DOI 10.1007/978-3-319-09259-1_3

3.1 Introduction

Clustering, the unsupervised classification of patterns into groups, is one of the most important tasks in exploratory data analysis [59]. Primary goals of clustering include gaining insight into, classifying, and compressing data. Clustering has a long and rich history in a variety of scientific disciplines including anthropology, biology, medicine, psychology, statistics, mathematics, engineering, and computer science. As a result, numerous clustering algorithms have been proposed since the early 1950s [58].

Clustering algorithms can be broadly classified into two groups: hierarchical and partitional [59]. Hierarchical algorithms recursively find nested clusters either in a top-down (divisive) or bottom-up (agglomerative) fashion. In contrast, partitional algorithms find all the clusters simultaneously as a partition of the data and do not impose a hierarchical structure. Most hierarchical algorithms have time complexity quadratic or higher in the number of data points [111] and therefore are not suitable for large data sets, whereas partitional algorithms often have lower complexity.

Given a data set $\mathcal{X} = \{\mathbf{x}_1, \mathbf{x}_2, \ldots, \mathbf{x}_N\} \subset \mathbb{R}^D$, i.e., N points (vectors) each with D attributes (components), hard partitional algorithms divide $\mathcal{X}$ into K exhaustive and mutually exclusive clusters $\mathcal{C} = \{\mathcal{C}_1, \mathcal{C}_2, \ldots, \mathcal{C}_K\}$, $\bigcup_{i=1}^{K} \mathcal{C}_i = \mathcal{X}$, $\mathcal{C}_i \cap \mathcal{C}_j = \varnothing$ for $1 \leq i \neq j \leq K$. These algorithms usually generate clusters by optimizing a criterion function [48]. The most intuitive and frequently used criterion function is the Sum of Squared Error (SSE) given by

$$\mathrm{SSE} = \sum_{i=1}^{K} \sum_{\mathbf{x}_j \in \mathcal{C}_i} \left\| \mathbf{x}_j - \mathbf{c}_i \right\|_2^2, \tag{3.1}$$

where

$$\mathbf{c}_i = \frac{1}{|\mathcal{C}_i|} \sum_{\mathbf{x}_j \in \mathcal{C}_i} \mathbf{x}_j \tag{3.2}$$

and

$$\left\| \mathbf{x}_j \right\|_2 = \left(\sum_{d=1}^{D} x_{jd}^2 \right)^{1/2} \tag{3.3}$$

denote the centroid of cluster $\mathcal{C}_i$ (with cardinality $|\mathcal{C}_i|$) and the Euclidean (ℓ_2) norm of vector $\mathbf{x}_j = (x_{j1}, x_{j2}, \ldots, x_{jD})$, respectively.

The number of ways in which a set of N objects can be partitioned into K non-empty groups is given by Stirling numbers of the second kind

$$\mathcal{S}(N, K) = \frac{1}{K!} \sum_{i=0}^{K} (-1)^{K-i} \binom{K}{i} i^N, \tag{3.4}$$

which can be approximated by $K^N/K!$ It can be seen that a complete enumeration of all possible clusterings to determine the global minimum of (3.1) is clearly computationally prohibitive except for very small data sets. In fact, this non-convex optimization problem is proven to be NP-hard even for $K = 2$ [4, 30] or $D = 2$ [79, 106]. Consequently, various heuristics have been developed to provide approximate solutions to this problem [102]. Most of the early approaches [12, 39, 53, 61, 77, 78, 98, 100] were simple procedures based on the alternating minimization algorithm [28]. In contrast, recent approaches are predominantly based on various metaheuristics [29, 94] that are capable of avoiding bad local minima at the expense of significantly increased computational requirements. These include heuristics based on simulated annealing [70], evolution strategies [10], tabu search [3], genetic algorithms [85], variable neighborhood search [51], memetic algorithms [90], scatter search [89], ant colony optimization [50], differential evolution [91], and particle swarm optimization [91]. Among all these heuristics, Lloyd's algorithm [77], often referred to as the (batch) k-means algorithm, is the simplest and most commonly used one. This algorithm starts with K arbitrary centers, typically chosen uniformly at random from the data points. Each point is assigned to the nearest center and then each center is recalculated as the mean of all points assigned to it. These two steps are repeated until a predefined termination criterion is met. K-means can be expressed in algorithmic notation as follows:

1. Choose the initial set of centers $\mathbf{c}_1, \mathbf{c}_2, \ldots, \mathbf{c}_K$ arbitrarily.
2. Assign point $\mathbf{x}_j$ $(j \in \{1, 2, \ldots, N\})$ to the nearest center with respect to ℓ_2 distance, that is

$$\mathbf{x}_j \in \mathcal{C}_{\hat{\imath}} \iff \hat{\imath} = \underset{i \in \{1,2,\ldots,K\}}{\arg\min} \left\| \mathbf{x}_j - \mathbf{c}_i \right\|_2^2 .$$

3. Recalculate center $\mathbf{c}_i$ $(i \in \{1, 2, \ldots, K\})$ as the centroid of $\mathcal{C}_i$, that is

$$\mathbf{c}_i = \frac{1}{|\mathcal{C}_i|} \sum_{\mathbf{x}_j \in \mathcal{C}_i} \mathbf{x}_j .$$

4. Repeat steps 2 and 3 until convergence.

K-means is undoubtedly the most widely used partitional clustering algorithm [15, 17, 41, 48, 58, 59, 86, 110, 111]. Its popularity can be attributed to several reasons. First, it is conceptually simple and easy to implement. Virtually every data mining software includes an implementation of it. Second, it is versatile, i.e., almost every aspect of the algorithm (initialization, distance function, termination criterion, etc.) can be modified. This is evidenced by hundreds of publications over the last fifty years that extend k-means in a variety of ways. Third, it has a time complexity that is linear in N, D, and K (in general, $D \ll N$ and $K \ll N$). For this reason, it can be used to initialize more expensive clustering algorithms such as expectation maximization [82], fuzzy c-means [16, p. 35], DBSCAN [31], spectral clustering [27, 108], ant colony clustering[84], and particle swarm clustering [105]. Furthermore, numerous sequential [34,35,49,63,66,72,92] and parallel [5, 14, 26, 46, 57, 71, 74, 109] acceleration techniques are available in the literature. Fourth, it has a storage complexity that is linear in N, D, and K. In addition, there exist disk-based variants that do not require all points to be stored in memory [20, 38, 62, 88]. Fifth, it is guaranteed to converge [97] at a quadratic rate [18]. Finally, it is invariant to data ordering, i.e., random shufflings of the data points.

On the other hand, k-means has several significant disadvantages. First, it requires the number of clusters, K, to be specified in advance. The value of this parameter can be determined automatically by means of various internal/relative cluster validity measures [6, 9, 107]. Second, it can only detect compact, hyperspherical clusters that are well separated. This can be alleviated by using a more general distance function such as the Mahalanobis distance, which permits the detection of hyperellipsoidal clusters [80, 81]. Third, due its utilization of the squared Euclidean distance, it is sensitive to noise and outlier points since even a few such points can significantly influence the means of their respective clusters. This can be addressed by outlier pruning [112] or by using a more robust distance function such as the city-block (ℓ_1) distance [37, 60, 99]. Fourth, due to its gradient descent nature, it often converges to a local minimum of the criterion function [97]. For the same reason, it is highly sensitive to the selection of the initial centers [25]. Adverse effects of improper initialization include empty clusters, slower convergence, and a higher chance of getting stuck in bad local minima [23]. Fortunately, except for the first two, these drawbacks can be remedied by using an adaptive initialization method (IM).

A large number of IMs have been proposed in the literature [23, 25, 32, 54, 93]. Unfortunately, many of these have time complexity superlinear in N [1, 2, 8, 21, 53, 65, 68, 73, 75, 95], which makes them impractical for large data sets (note that k-means itself has linear time complexity). In contrast, linear IMs are often random and/or order-sensitive [7, 11, 19, 39, 61, 78, 100, 104], which renders their results unreliable. In this study, we investigate the empirical performance of six linear, deterministic (non-random), and order-invariant k-means IMs on a large and diverse collection of data sets from the UCI Machine Learning Repository.

The rest of the chapter is organized as follows. Section 3.2 presents an overview of linear, deterministic, and order-invariant k-means IMs. Section 3.3 describes the experimental setup. Section 3.4 presents and discusses the experimental results. Finally, Sect. 3.5 gives the conclusions.

3.2 Linear, Deterministic, and Order-Invariant K-Means Initialization Methods

In this study, we focus on IMs that have time complexity linear in N. This is because k-means itself has linear complexity, which is perhaps the most important reason for its popularity. Therefore, an IM for k-means should not diminish this advantage of the algorithm. Accordingly, the following six linear, deterministic, and order-invariant IMs are investigated.

The maximin (MM) method [47] chooses the first center $\mathbf{c}_1$ arbitrarily from the data points and the remaining $(K - 1)$ centers are chosen successively as follows. In iteration i $(i \in \{2, 3, \ldots, K\})$, the ith center $\mathbf{c}_i$ is chosen to be the point with the greatest minimum ℓ_2 distance to the previously selected $(i - 1)$ centers, i.e., $\mathbf{c}_1, \mathbf{c}_2, \ldots, \mathbf{c}_{i-1}$. This method can be expressed in algorithmic notation as follows:

1. Choose the first center $\mathbf{c}_1$ arbitrarily from the data points.
2. Choose the next center $\mathbf{c}_i$ $(i \in \{2, 3, \ldots, K\})$ as the point $\mathbf{x}_{\hat{j}}$ that satisfies

$$\hat{j} = \underset{j \in \{1,2,\ldots,N\}}{\arg\max} \left(\min_{k \in \{1,2,\ldots,i-1\}} \left\| \mathbf{x}_j - \mathbf{c}_k \right\|_2^2 \right).$$

3. Repeat step 2 $(K - 1)$ times.

Despite the fact that it was originally developed as a 2-approximation to the K-center clustering problem,[1] MM is commonly used as a k-means initializer.[2] In this study, the first center is chosen to be the centroid of $\mathcal{X}$ given by

$$\bar{\mathbf{x}} = \frac{1}{N} \sum_{j=1}^{N} \mathbf{x}_j. \tag{3.5}$$

Note that $\mathbf{c}_1 = \bar{\mathbf{x}}$ gives the optimal SSE when $K = 1$.

Katsavounidis et al.'s method (KK) [67] is identical to MM with the exception that the first center is chosen to be the point with the greatest ℓ_2 norm,[3] that is, the point $\mathbf{x}_{\hat{j}}$ that satisfies

$$\hat{j} = \underset{j \in \{1,2,\ldots,N\}}{\arg\max} \left\| \mathbf{x}_j \right\|_2^2. \tag{3.6}$$

The PCA-Part (PP) method [101] uses a divisive hierarchical approach based on Principal Component Analysis (PCA) [64]. Starting from an initial cluster that contains the entire data set $\mathcal{X}$, the method successively selects the cluster with the greatest SSE and divides it into two subclusters using a hyperplane that passes

[1] Given a set of N points in a metric space, the goal of K-center clustering is to find K representative points (centers) such that the maximum distance of a point to a center is minimized [52, p. 63]. A polynomial-time algorithm is said to be a *δ-approximation* algorithm for a minimization problem if for every instance of the problem it delivers a solution whose cost is at most δ times the cost of the optimal solution (δ is often referred to as the "approximation ratio" or "approximation factor") [55, p. xv].

[2] Interestingly, several authors including Thorndike [103], Casey and Nagy [22], Batchelor and Wilkins [13], Kennard and Stone [69], and Tou and Gonzalez [104, pp. 92–94] had proposed similar (or even identical) methods decades earlier. Gonzalez [47], however, was the one to prove the theoretical properties of the method.

[3] This choice was motivated by a vector quantization application.

through the cluster centroid and is orthogonal to the principal eigenvector of the cluster covariance matrix. This iterative cluster selection and splitting procedure is repeated $(K - 1)$ times. The final centers are then given by the centroids of the resulting K subclusters. This method can be expressed in algorithmic notation as follows:

1. Let $\mathcal{C}_i$ be the cluster with the greatest SSE and $\mathbf{c}_i$ be the centroid of this cluster. In the first iteration, $\mathcal{C}_1 = \mathcal{X}$ and $\mathbf{c}_1 = \bar{\mathbf{x}}$.
2. Let p be the projection of $\mathbf{c}_i$ on the principal eigenvector $\mathbf{v}_i$ of $\mathcal{C}_i$, i.e., $p = \mathbf{c}_i \cdot \mathbf{v}_i$, where '·' denotes the dot product.
3. Divide $\mathcal{C}_i$ into two subclusters $\mathcal{C}_{i_1}$ and $\mathcal{C}_{i_2}$ according to the following rule: For any $\mathbf{x}_j \in \mathcal{C}_i$, if $\mathbf{x}_j \cdot \mathbf{v}_i \leq p$, then assign $\mathbf{x}_j$ to $\mathcal{C}_{i_1}$; otherwise, assign it to $\mathcal{C}_{i_2}$.
4. Repeat steps 1–3 $(K - 1)$ times.

The Var-Part (VP) method [101] is an approximation to PP, where, in each iteration, the covariance matrix of the cluster to be split is assumed to be diagonal. In this case, the splitting hyperplane is orthogonal to the coordinate axis with the greatest variance. In other words, the only difference between VP and PP is the choice of the projection axis.

Figure 3.1 [24] illustrates VP on a toy data set with four natural clusters [96][68, p. 100]. In iteration 1, the initial cluster that contains the entire data set is split into two subclusters along the Y axis using a line (i.e., a one-dimensional hyperplane) passing through the mean point (92.026667). Between the resulting two clusters, the one above the line has a greater SSE. In iteration 2, this cluster is thus split along the X axis at the mean point (66.975000). In the final iteration, the cluster with the greatest SSE, i.e., the bottom cluster, is split along the X axis at the mean point (41.057143). In Fig. 3.1d, the centroids of the final four clusters are denoted by stars.

The maxisum (MS) method [36] is a recent modification of MM. It can be expressed in algorithmic notation as follows:

1. Determine the attribute with the greatest absolute *coefficient of variation* (ratio of the standard deviation to the mean), that is, the attribute $\mathbf{x}_{.d_1}$ that satisfies

$$d_1 = \underset{d \in \{1,2,\ldots,D\}}{\arg\max} \left| \frac{s_d}{m_d} \right|,$$

where

$$m_d = \frac{1}{N} \sum_{j=1}^{N} x_{jd}$$

and

$$s_d^2 = \frac{1}{N-1} \sum_{j=1}^{N} (x_{jd} - m_d)^2$$

denote the mean and variance of the dth attribute $\mathbf{x}_{.d}$, respectively.

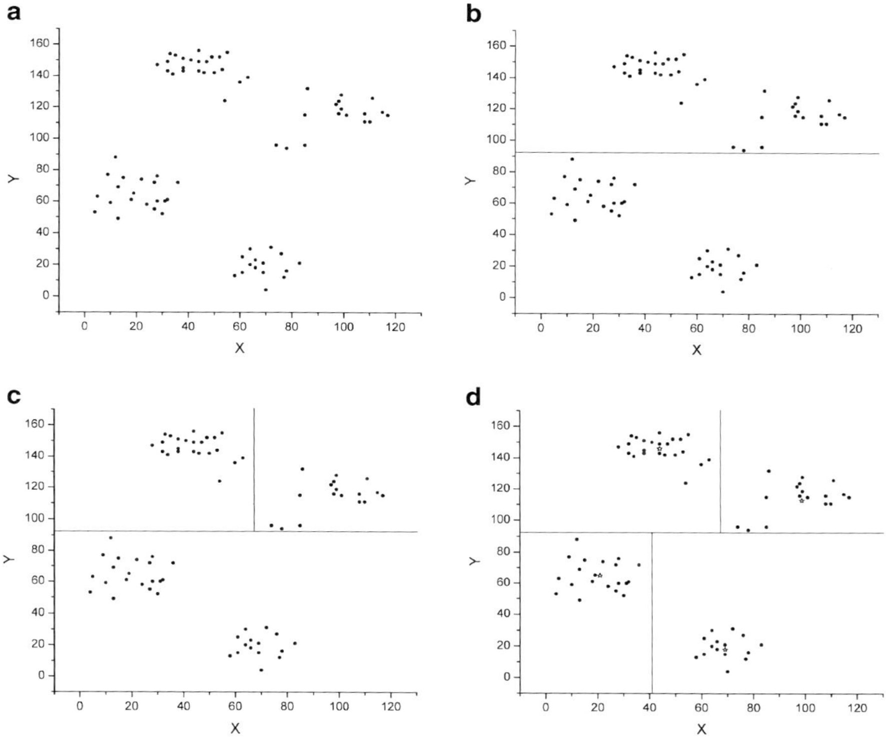

Fig. 3.1 Illustration of Var-Part on the Ruspini data set. (**a**) Input data set. (**b**) Iteration 1. (**c**) Iteration 2. (**d**) Iteration 3

2. Determine the attribute with the least *Pearson product-moment correlation* with $\mathbf{x}_{.d_1}$, that is, the attribute $\mathbf{x}_{.d_2}$ that satisfies

$$d_2 = \underset{d \in \{1,2,\ldots,D\}}{\arg\min} \frac{\sum_{j=1}^{N} (x_{jd_1} - m_{d_1})(x_{jd} - m_d)}{\sqrt{\sum_{j=1}^{N} (x_{jd_1} - m_{d_1})^2} \sqrt{\sum_{j=1}^{N} (x_{jd} - m_d)^2}}. \tag{3.7}$$

Note that since we calculated the mean and standard deviation of each attribute in step 1, the following expression can be used in place of (3.7) to save computational time:

$$d_2 = \underset{d \in \{1,2,\ldots,D\}}{\arg\min} \sum_{j=1}^{N} \left(\frac{x_{jd_1} - m_{d_1}}{s_{d_1}} \right) \left(\frac{x_{jd} - m_d}{s_d} \right). \tag{3.8}$$

3. Let $\mathcal{Y} = \{\mathbf{y}_1, \mathbf{y}_2, \ldots, \mathbf{y}_N\} \subset \mathbb{R}^2$ be the projection of $\mathcal{X} = \{\mathbf{x}_1, \mathbf{x}_2, \ldots, \mathbf{x}_N\} \subset \mathbb{R}^D$ onto the two-dimensional subspace determined in steps 1 and 2. In other words, $\mathbf{y}_j = \left(x_{jd_1}, x_{jd_2}\right)$ for $j \in \{1, 2, \ldots, N\}$.
4. Choose the first center $\mathbf{c}_1$ as the point farthest from the centroid $\bar{\mathbf{y}}$ of $\mathcal{Y}$ with respect to ℓ_2 distance, that is, the point $\mathbf{y}_{\hat{j}}$ that satisfies

$$\hat{j} = \underset{j \in \{1,2,\ldots,N\}}{\arg\max} \left\|\mathbf{y}_j - \bar{\mathbf{y}}\right\|_2^2.$$

5. Choose the next center $\mathbf{c}_i$ $(i \in \{2, 3, \ldots, K\})$ as the point with the greatest cumulative ℓ_2 distance from the previously selected $(i - 1)$ centers, that is, the point $\mathbf{y}_{\hat{j}}$ that satisfies

$$\hat{j} = \underset{j \in \{1,2,\ldots,N\}}{\arg\max} \sum_{k=1}^{i-1} \left\|\mathbf{y}_j - \mathbf{c}_k\right\|_2.$$

6. Repeat step 5 $(K - 1)$ times.

Note that steps 1 and 2 above provide rough approximations to the first two PCs and that steps 4 and 5 are performed in the two-dimensional subspace spanned by the attributes determined in steps 1 and 2.

Clearly, MS is a derivative of MM. Differences between the two methods are as follows:

- MM chooses the first center arbitrarily from the data points, whereas MS chooses it to be the point farthest from the mean of the projected data set.
- MM chooses the remaining $(K - 1)$ centers iteratively based on their minimum distance from the previously selected centers, whereas MS uses a cumulative distance criterion. Note that while the selection criterion used in MM provides an approximation guarantee of factor 2 for the K-center clustering problem (see footnote 1 on page 83), it is unclear whether or not MS offers any approximation guarantees.
- MM performs the distance calculations in the original D-dimensional space, whereas MS works in a two-dimensional subspace. A serious drawback of the projection operation employed in MS is that the method disregards all attributes but two and therefore is likely to be effective only for data sets in which the variability is mostly on two dimensions. Unfortunately, the motivation behind this particular projection scheme is not given by Erişoğlu et al.

Interestingly, MS also bears a striking resemblance to a method proposed by DeSarbo et al. [33] almost three decades earlier. The latter method differs from the former in two ways. First, it works in the original D-dimensional space. Second, it chooses the first two centers as the pair of points with the greatest ℓ_2 distance. Unfortunately, the determination of the first two centers in this method leads to a time complexity quadratic in N. Therefore, this method was not included in the experiments. More recently, Glasbey et al. [43] mentioned a very similar method within the context of color palette design.

We also experimented with a modified version of the MS method (MS+), which is identical to MS with the exception that there is no projection involved. In other words, MS+ operates in the original D-dimensional space.

For a comprehensive overview of these methods and others, the reader is referred to a recent article by Celebi et al. [25]. It should be noted that, in this study, we do not attempt to compare a mix of deterministic and random IMs. Instead, we focus on deterministic methods for two main reasons. First, these methods are generally computationally more efficient as they need to be executed only once. In contrast, random methods are inherently unreliable in that the quality of their results is unpredictable and thus it is common practice to perform multiple runs of such methods and take the output of the run[4] that produces the best objective function value. Second, several studies [24, 25, 101] demonstrated that despite the fact that they are executed only once, some deterministic methods are highly competitive with well-known and effective random methods such as Bradley and Fayyad's method [19] and k-means++ [7].

3.3 Experimental Setup

3.3.1 Data Set Descriptions

The experiments were performed on 24 commonly used data sets from the UCI Machine Learning Repository [40]. Table 3.1 gives the data set descriptions. For each data set, the number of clusters (K) was set equal to the number of classes (K'), as commonly seen in the related literature [2, 7, 21, 24, 25, 42, 54, 65, 87, 95, 101].

3.3.2 Attribute Normalization

Normalization is a common preprocessing step in clustering that is necessary to prevent attributes with large variability from dominating the distance calculations and also to avoid numerical instabilities in the computations. Two commonly used normalization schemes are linear scaling to unit range (min-max normalization) and linear scaling to unit variance (z-score normalization). Several studies revealed that the former scheme is preferable to the latter since the latter is likely to eliminate valuable between-cluster variation [44, 45, 83]. As a result, we used the min-max normalization scheme to map the attributes of each data set to the [0, 1] interval.

[4]Each 'run' of a random IM involves the execution of the IM itself followed by that of the clustering algorithm, e.g., k-means.

Table 3.1 Data set descriptions (N: # points, D: # attributes, K': # classes)

ID	Data Set	N	D	K'
01	Breast cancer Wisconsin (original)	683	9	2
02	Breast tissue	106	9	6
03	Ecoli	336	7	8
04	Steel plates faults	1,941	27	7
05	Glass identification	214	9	6
06	Heart disease (Cleveland)	297	13	5
07	Ionosphere	351	34	2
08	Iris (Bezdek)	150	4	3
09	ISOLET	7,797	617	26
10	Landsat satellite (Statlog)	6,435	36	6
11	Letter recognition	20,000	16	26
12	Multiple features (Fourier)	2,000	76	10
13	Libras movement	360	90	15
14	Optical digits	5,620	64	10
15	Page blocks classification	5,473	10	5
16	Pen digits	10,992	16	10
17	Person activity	164,860	3	11
18	Image segmentation	2,310	19	7
19	Shuttle (Statlog)	58,000	9	7
20	Spambase	4,601	57	2
21	Vertebral column	310	6	3
22	Wall-following robot navigation	5,456	24	4
23	Wine	178	13	3
24	Yeast	1,484	8	10

3.3.3 *Performance Criteria*

The performance of the IMs was quantified using two effectiveness (quality) and one efficiency (speed) criteria:

- **Initial SSE (IS)**: This is the SSE value calculated after the initialization phase, before the clustering phase. It gives us a measure of the effectiveness of an IM by itself.
- **Final SSE (FS)**: This is the SSE value calculated after the clustering phase. It gives us a measure of the effectiveness of an IM when its output is refined by k-means. Note that this is the objective function of the k-means algorithm, i.e., (3.1).
- **Number of Iterations (NI)**: This is the number of iterations that k-means requires until reaching convergence when initialized by a particular IM. It is an efficiency measure independent of programming language, implementation

style, compiler, and CPU architecture. Note that we do not report CPU time measurements since on most data sets that we tested each of the six IMs completed within a few milliseconds (gcc v4.4.5, Intel Core i7-3960X 3.30GHz).

The convergence of k-means was controlled by the disjunction of two criteria: the number of iterations reaches a maximum of 100 or the relative improvement in SSE between two consecutive iterations drops below a threshold [76], i.e., $(\text{SSE}_{i-1} - \text{SSE}_i)/\text{SSE}_i \leq \epsilon$, where SSE_i denotes the SSE value at the end of the ith ($i \in \{2, \ldots, 100\}$) iteration. The convergence threshold was set to $\epsilon = 10^{-6}$.

3.4 Experimental Results and Discussion

Tables 3.2, 3.3, and 3.4 give the performance measurements for each method (the best values are underlined). Since the number of iterations fall within [0, 100], we can directly obtain descriptive statistics such as minimum, maximum, mean, and median for this criterion over the 24 data sets. In contrast, initial/final SSE values are unnormalized and therefore incomparable across different data sets. In order to circumvent this problem, for each data set, we calculated the percent SSE of each method relative to the worst (greatest) SSE. For example, it can be seen from Table 3.2 that on the Breast Cancer Wisconsin data set the initial SSE of `MM` is 498, whereas the worst initial SSE on the same data set is 596 and thus the ratio of the former to the latter is 0.836. This simply means that on this data set `MM` obtains $100(1 - 0.836) \approx 16\,\%$ better initial SSE than the worst method, `KK`. Table 3.5 gives the summary statistics for the normalized initial/final SSE's obtained in this manner and those for the number of iterations. As usual, *min* (minimum) and *max* (maximum) represent the *best* and *worst* case performance, respectively. *Mean* represents the *average* case performance, whereas median quantifies the *typical* performance of a method without regard to outliers. For example, with respect to the initial SSE criterion, `PP` performs, on the *average*, about $100 - 21.46 \approx 79\,\%$ better than the worst method.

For convenient visualization, Fig. 3.2 shows the box plots that depict the five-number summaries (minimum, 25th percentile, median, 75th percentile, and maximum) for the normalized initial/final SSE's calculated in the aforementioned manner and the five-number summary for the number of iterations. Here, the bottom and top end of the whiskers of a box represent the *minimum* and *maximum*, respectively, whereas the bottom and top of the box itself are the 25th percentile (*Q1*) and 75th percentile (*Q3*), respectively. The line that passes through the box is the 50th percentile (*Q2*), i.e., the *median*, while the small square inside the box denotes the *mean*.

With respect to effectiveness, the following observations can be made:

- `VP` and `PP` performed very similarly with respect to both initial and final SSE.

Table 3.2 Initial SSE comparison of the initialization methods

ID	MM	KK	VP	PP	MS	MS+
01	498	596	247	240	478	596
02	19	18	8	8	50	21
03	48	76	20	19	104	68
04	2,817	3,788	1,203	1,262	5,260	4,627
05	45	117	21	20	83	132
06	409	557	249	250	773	559
07	827	1,791	632	629	3,244	3,390
08	18	23	8	8	42	42
09	221,163	298,478	145,444	124,958	368,510	318,162
10	4,816	7,780	2,050	2,116	7,685	11,079
11	5,632	7,583	3,456	3,101	12,810	14,336
12	4,485	7,821	3,354	3,266	7,129	8,369
13	1,023	1,114	628	592	1,906	1,454
14	25,291	36,691	17,476	15,714	43,169	42,213
15	635	2,343	300	230	1,328	7,868
16	12,315	16,159	5,947	5,920	17,914	16,104
17	5,940	7,196	1,269	1,468	42,475	50,878
18	1,085	1,617	472	416	3,071	1,830
19	1,818	14,824	316	309	26,778	28,223
20	772	13,155	782	783	5,101	13,155
21	37	103	23	20	83	103
22	11,004	21,141	8,517	7,805	19,986	20,122
23	87	185	51	53	153	212
24	115	261	77	63	209	658

- On 23 (out of 24) data sets, VP and PP obtained the two best initial SSE's. Therefore, in applications where an approximate clustering of the data set is desired, these hierarchical methods should be used.
- On 23 data sets, either MS or MS+ obtained the worst initial SSE. In fact, on one data set (#19, Shuttle), these methods gave respectively 86.7 and 91.3 times worse initial SSE than the best method, PP.
- On 20 data sets, VP and PP obtained the two best final SSE's. Since final SSE is the objective function of k-means, from an optimization point of view, these two methods are the best IMs.

Table 3.3 Final SSE comparison of the initialization methods

ID	MM	KK	VP	PP	MS	MS+
01	239	239	239	239	239	239
02	7	7	7	7	11	10
03	19	20	17	18	40	20
04	1,331	1,329	1,167	1,168	1,801	1,376
05	23	23	19	19	31	22
06	249	249	248	243	276	253
07	826	629	629	629	629	629
08	7	7	7	7	7	7
09	135,818	123,607	118,495	118,386	174,326	121,912
10	1,742	1,742	1,742	1,742	1,742	1,742
11	2,749	2,783	2,735	2,745	4,520	3,262
12	3,316	3,284	3,137	3,214	3,518	3,257
13	502	502	502	486	783	530
14	14,679	14,649	14,581	14,807	21,855	14,581
15	230	295	227	215	230	310
16	5,049	4,930	4,930	5,004	7,530	5,017
17	1,195	1,195	1,182	1,177	1,226	1,192
18	433	443	410	405	745	446
19	726	658	235	274	728	496
20	765	765	778	778	778	765
21	23	23	19	19	23	23
22	7,772	7,772	7,774	7,774	7,772	7,772
23	63	49	49	49	49	49
24	61	61	69	59	60	63

- On 16 data sets, MS obtained the worst final SSE. In fact, on one data set (#19, Shuttle), MS gave 3.1 times worse final SSE than the best method, VP.
- A comparison between Fig. 3.2a, b reveals that there is significantly less variation among the IMs with respect to final SSE compared to initial SSE. In other words, the performance of the IMs is more homogeneous with respect to final SSE. This was expected because, being a local optimization procedure, k-means can take two disparate initial configurations to similar (or, in some cases, even identical) local minima. Nevertheless, as Tables 3.2 and 3.3 show, VP and PP consistently performed well, whereas MS/MS+ generally performed poorly.

With respect to computational efficiency, the following observations can be made:

- An *average* (or *typical*) run of KK lead to the fastest k-means convergence.
- An *average* (or *typical*) run of PP lead to the second fastest k-means convergence.
- An *average* run of MS lead to the slowest k-means convergence.
- A *typical* run of MM lead to the slowest k-means convergence.

Table 3.4 Number of iterations comparison of the initialization methods

ID	MM	KK	VP	PP	MS	MS+
01	8	7	4	4	7	7
02	7	6	6	7	9	4
03	14	12	17	7	4	10
04	25	16	11	42	12	12
05	6	5	6	5	7	6
06	12	10	3	4	11	16
07	3	6	3	3	7	6
08	6	5	4	4	12	19
09	32	36	82	45	34	81
10	53	17	28	27	24	33
11	72	63	100	83	91	65
12	37	32	14	25	31	29
13	13	7	17	11	18	16
14	36	24	16	22	29	17
15	27	18	25	15	30	12
16	19	17	13	17	22	29
17	31	31	100	63	91	53
18	31	9	10	18	16	22
19	22	8	30	16	14	9
20	5	5	9	10	11	5
21	11	10	10	9	8	10
22	24	14	20	8	20	19
23	9	7	5	7	7	8
24	73	43	33	21	71	49

In summary, our experiments showed that VP and PP performed very similarly with respective to both effectiveness criteria and they outperformed the remaining four methods by a large margin. The former method has a time complexity of $\mathcal{O}(ND)$, whereas the latter one has a complexity of $\mathcal{O}(ND^2)$ when implemented using the power method [56]. Therefore, on high dimensional data sets, the former method might be preferable. On the other hand, on low dimensional data sets, the latter method might be preferable as it often leads to faster k-means convergence. The main disadvantage of these two methods is that they are more complicated to implement due to their hierarchical formulation. As for the remaining four methods, when compared to MM, KK was significantly worse in terms of initial SSE, slightly better in terms of final SSE, and significantly better in terms of number of iterations. Interestingly, despite its similarities with MM, the most recent method that we examined, i.e., MS, often gave the worst results. It was also demonstrated that by eliminating the two-dimensional projection step, the performance of MS can be substantially improved with respect to final SSE. This, however, comes at the expense of a performance degradation with respect to initial SSE. Consequently, in either of its forms, the MS method rediscovered recently by

Table 3.5 Summary statistics for Tables 3.2, 3.3, and 3.4

Criterion	Statistic	MM	KK	VP	PP	MS	MS+
IS	Min	5.87	14.14	1.12	1.10	16.88	42.00
	Q1	29.24	52.74	15.64	14.35	76.19	87.15
	Median	41.95	72.04	20.78	19.26	94.71	100.00
	Q3	53.57	89.42	33.07	32.69	100.00	100.00
	Max	83.56	100.00	41.44	40.27	100.00	100.00
	Mean	40.28	69.03	22.55	21.46	83.16	90.53
FS	Min	47.50	50.00	32.28	37.64	74.19	50.00
	Q1	67.11	66.25	63.87	62.85	100.00	69.03
	Median	89.31	83.09	74.69	72.75	100.00	84.34
	Q3	99.99	97.90	98.21	93.68	100.00	99.15
	Max	100.00	100.00	100.00	100.00	100.00	100.00
	Mean	83.21	81.56	76.23	75.77	96.46	82.68
NI	Min	3.00	5.00	3.00	3.00	4.00	4.00
	Q1	8.50	7.00	6.00	7.00	8.50	8.50
	Median	20.50	11.00	13.50	13.00	15.00	16.00
	Q3	31.50	21.00	26.50	23.50	29.50	29.00
	Max	73.00	63.00	100.00	83.00	91.00	81.00
	Mean	24.00	17.00	23.58	19.71	24.42	22.38

Erişoğlu et al. does not appear to outperform the classical MM method or the more recent hierarchical methods VP and PP. This is not surprising given that MS can easily choose two nearby points as centers provided that they each have a large cumulative distance to all other centers [43].

3.5 Conclusions

In this chapter we examined six linear, deterministic, and order-invariant methods used for the initialization of the k-means clustering algorithm. These included the popular maximin method and three of its variants and two relatively unknown divisive hierarchical methods. Experiments on a large and diverse collection of real-world data sets from the UCI Machine Learning Repository demonstrated that the hierarchical methods outperform the remaining four methods with respect to two objective effectiveness criteria. These hierarchical methods can be used to initialize k-means effectively, particularly in time-critical applications that involve large data sets. Alternatively, they can be used as approximate clustering algorithms without additional k-means refinement. Our experiments also revealed that the most recent variant of the maximin method performs surprisingly poorly.

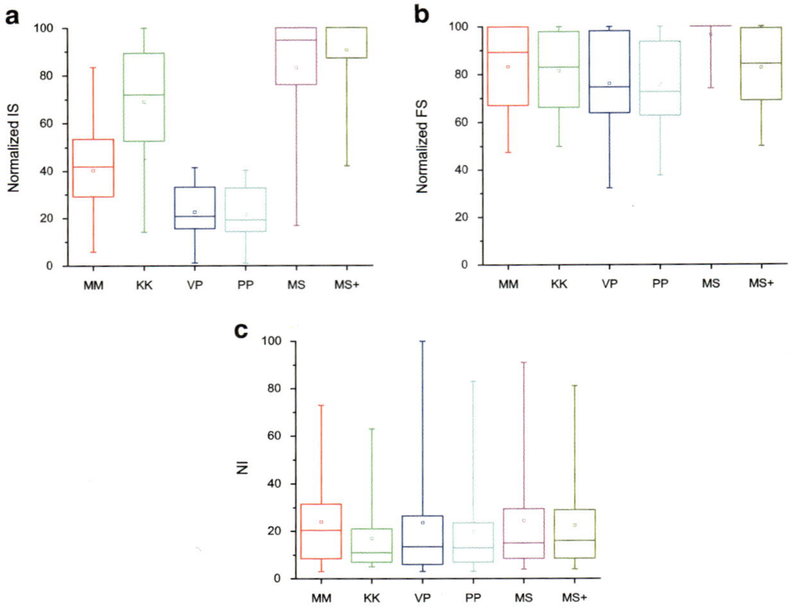

Fig. 3.2 Box plots for the performance criteria. (**a**) Normalized initial SSE. (**b**) Normalized final SSE. (**c**) Number of iterations

Acknowledgements This work was supported by a grant from the US National Science Foundation (1117457).

References

1. Al-Daoud M (2005) A new algorithm for cluster initialization. In: Proceedings of the 2nd world enformatika conference, pp 74–76
2. Al Hasan M, Chaoji V, Salem S, Zaki M (2009) Robust partitional clustering by outlier and density insensitive seeding. Pattern Recognit Lett 30(11):994–1002
3. Al-Sultan KS (1995) A Tabu search approach for the minimum sum-of-squares clustering problem. Pattern Recognit 28(9):1443–1451
4. Aloise D, Deshpande A, Hansen P, Popat P (2009) NP-hardness of Euclidean sum-of-squares clustering. Mach Learn 75(2):245–248
5. An F, Mattausch HJ (2013) K-means clustering algorithm for multimedia applications with flexible HW/SW co-design. J Syst Archit 59(3):155–164
6. Arbelaitz O, Gurrutxaga I, Muguerza J, Pérez JM, Perona I (2013) An extensive comparative study of cluster validity indices. Pattern Recognit 46(1):243–256
7. Arthur D, Vassilvitskii S (2007) K-means++: the advantages of careful seeding. In: Proceedings of the 18th annual ACM-SIAM symposium on discrete algorithms, pp 1027–1035
8. Astrahan MM (1970) Speech analysis by clustering, or the hyperphoneme method. Technical Report AIM-124, Stanford University

9. Baarsch J, Celebi ME (2012) Investigation of internal validity measures for K-means clustering. In: Proceedings of the 2012 IAENG international conference on data mining and applications, pp 471–476
10. Babu GP, Murty MN (1994) Clustering with evolution strategies. Pattern Recognit 27(2): 321–329
11. Ball GH, Hall DJ (1967) A clustering technique for summarizing multivariate data. Behav Sci 12(2):153–155
12. Banfield CF, Bassill LC (1977) A transfer algorithm for non-hierarchical classification. Appl Stat 26(2):206–210
13. Batchelor BG, Wilkins BR (1969) Method for location of clusters of patterns to initialise a learning machine. Electron Lett 5(20):481–483
14. Bekkerman R, Bilenko M, Langford J (eds) (2012) Scaling up machine learning: parallel and distributed approaches. Cambridge University Press, Cambridge
15. Berkhin P (2006) A survey of clustering data mining techniques. In: Kogan J, Nicholas C, Teboulle M (eds) Grouping multidimensional data: recent advances in clustering. Springer, Berlin, pp 25–71
16. Bezdek JC, Keller J, Krisnapuram R, Pal NR (1999) Fuzzy models and algorithms for pattern recognition and image processing. Kluwer, Dordecht
17. Bock HH (2007) Clustering methods: a history of K-means algorithms. In: Brito P, Cucumel G, Bertrand P, de Carvalho F (eds) Selected contributions in data analysis and classification. Springer, Berlin, pp 161–172
18. Bottou L, Bengio Y (1995) Convergence properties of the K-means algorithms. In: Tesauro G, Touretzky DS, Leen TK (eds) Advances in neural information processing systems, vol 7. MIT Press, Cambridge, pp 585–592
19. Bradley PS, Fayyad U (1998) Refining initial points for K-means clustering. In: Proceedings of the 15th international conference on machine learning, pp 91–99
20. Bradley PS, Fayyad U, Reina C (1998) Scaling clustering algorithms to large databases. In: Proceedings of the 4th international conference on knowledge discovery and data mining, pp 9–15
21. Cao F, Liang J, Jiang G (2009) An initialization method for the K-means algorithm using neighborhood model. Comput Math Appl 58(3):474–483
22. Casey RG, Nagy G (1968) An autonomous reading machine. IEEE Trans Comput 17(5): 492–503
23. Celebi ME (2011) Improving the performance of K-means for color quantization. Image Vis Comput 29(4):260–271
24. Celebi ME, Kingravi HA (2012) Deterministic initialization of the K-means algorithm using hierarchical clustering. Intern J Pattern Recognit Artif Intell 26(7):1250018
25. Celebi ME, Kingravi HA, Vela PA (2013) A Comparative study of efficient initialization methods for the K-means clustering algorithm. Expert Syst Appl 40(1):200–210
26. Chen TW, Chien SY (2010) Bandwidth adaptive hardware architecture of K-means clustering for video analysis. IEEE Trans VLSI Syst 18(6):957–966
27. Chen WY, Song Y, Bai H, Lin CJ, Chang EY (2011) Parallel spectral clustering in distributed systems. IEEE Trans Pattern Anal Mach Intell 33(3):568–586
28. Csiszar I, Tusnady G (1984) Information geometry and alternating minimization procedures. Stat Decis Suppl Issue (1):205–237
29. Das S, Abraham A, Konar A (eds) (2009) Metaheuristic clustering. Springer, Berlin
30. Dasgupta S (2008) The hardness of K-means clustering. Technical Report CS2008-0916, University of California, San Diego
31. Dash M, Liu H, Xu X (2001) '$1 + 1 > 2$': merging distance and density based clustering. In: Proceedings of the 7th international conference on database systems for advanced applications, pp 32–39
32. de Amorim RC, Komisarczuk P (2012) On Initializations for the Minkowski weighted K-means. In: Proceedings of the 11th international symposium on intelligent data analysis, pp 45–55

33. DeSarbo WS, Carroll JD, Clark LA, Green PE (1984) Synthesized clustering: a method for amalgamating alternative clustering bases with differential weighting of variables. Psychometrika 49(1):57–78
34. Drake J, Hamerly G (2012) Accelerated K-means with adaptive distance bounds. In: Proceedings of the 5th NIPS workshop on optimization for machine learning
35. Elkan C (2003) Using the triangle inequality to accelerate K-means. In: Proceedings of the 20th international conference on machine learning, pp 147–153
36. Erişoğlu et al. (2011) Pattern Recognit Lett 32(14):1701–1705
37. Estivill-Castro V, Yang J (2004) Fast and robust general purpose clustering algorithms. Data Min Knowl Discov 8(2):127–150
38. Farnstrom F, Lewis J, Elkan C (2000) Scalability for clustering algorithms revisited. SIGKDD Explor 2(1):51–57
39. Forgy E (1965) Cluster analysis of multivariate data: efficiency vs. interpretability of classification. Biometrics 21:768
40. Frank A, Asuncion A (2014) UCI machine learning repository. School of Information and Computer Sciences, University of California, Irvine. http://archive.ics.uci.edu/ml
41. Ghosh J, Liu A (2009) K-Means. In: Wu X, Kumar V (eds) The top ten algorithms in data mining. Chapman and Hall/CRC, London, pp 21–35
42. Gingles C, Celebi ME (2014) Histogram-based method for effective initialization of the K-means clustering algorithm. In: Proceedings of the 27th international Florida artificial intelligence research society conference, pp 333–338
43. Glasbey C, van der Heijden G, Toh VFK, Gray A (2006) Colour displays for categorical images. Color Res Appl 32(4):304–309
44. Gnanadesikan R, Kettenring JR (1995) Weighting and selection of variables for cluster analysis. J Classif 12(1):113–136
45. Gnanadesikan R, Kettenring JR, Maloor S (2007) Better alternatives to current methods of scaling and weighting data for cluster analysis. J Stat Plan Inference 137(11):3483–3496
46. Gokhale M, Frigo J, McCabe, K., Theiler J, Wolinski C, Lavenier D (2003) Experience with a hybrid processor: K-means clustering. J Supercomput 26(2):131–148
47. Gonzalez T (1985) Clustering to minimize the maximum intercluster distance. Theor Comput Sci 38(2–3):293–306
48. Hall LO (2012) Objective function-based clustering. WIREs Data Min Knowl Discov 2(4):326–339
49. Hamerly G (2010) Making K-means even faster. In: Proceedings of the 2010 SIAM international conference on data mining, pp 130–140
50. Handl J, Knowles J, Dorigo M (2005) Ant-based clustering and topographic mapping. Artif Life 12(1):35–61
51. Hansen P, Mladenovic N (2001) J-means: a new local search heuristic for minimum sum of squares clustering. Pattern Recognit 34(2):405–413
52. Har-Peled S (2011) Geometric approximation algorithms. American Mathematical Society, Providence
53. Hartigan JA, Wong MA (1979) Algorithm AS 136: a K-means clustering algorithm. J R Stat Soc C 28(1):100–108
54. He J, Lan M, Tan CL, Sung SY, Low HB (2004) Initialization of cluster refinement algorithms: a review and comparative study. In: Proceedings of the 2004 IEEE international joint conference on neural networks, pp 297–302
55. Hochbaum DS (ed) (1997) Approximation algorithms for NP-hard problems. PWS Publishing Company, Boston
56. Hotelling H (1936) Simplified calculation of principal components. Psychometrika 1(1): 27–35
57. Hwang WJ, Hsu CC, Li HY, Weng SK, Yu TY (2010) High speed C-means clustering in reconfigurable hardware. Microprocess Microsyst 34(6):237–246
58. Jain AK (2010) Data clustering: 50 years beyond K-means. Pattern Recognit Lett 31(8): 651–666

59. Jain AK, Murty MN, Flynn PJ (1999) Data clustering: a review. ACM Comput Surv 31(3):264–323
60. Jajuga K (1987) A clustering method based on the L_1-norm. Comput Stat Data Anal 5(4): 357–371
61. Jancey RC (1966) Multidimensional group analysis. Aust J Bot 14(1):127–130
62. Jin R, Goswami A, Agrawal G (2006) Fast and exact out-of-core and distributed K-means clustering. Knowl Inf Syst 10(1):17–40
63. Jin X, Kim S, Han J, Cao L, Yin Z (2011) A general framework for efficient clustering of large datasets based on activity detection. Stat Anal Data Min 4(1):11–29
64. Jolliffe IT (2002) Principal component analysis, 2nd edn. Springer, Berlin
65. Kang P, Cho S (2009) K-means clustering seeds initialization based on centrality, sparsity, and isotropy. In: Proceedings of the 10th international conference on intelligent data engineering and automated learning, pp 109–117
66. Kanungo T, Mount D, Netanyahu N, Piatko C, Silverman R, Wu A (2002) An efficient K-means clustering algorithm: analysis and implementation. IEEE Trans Pattern Anal Mach Intell 24(7):881–892
67. Katsavounidis I, Kuo CCJ., Zhang Z (1994) A new initialization technique for generalized Lloyd iteration. IEEE Signal Process Lett 1(10):144–146
68. Kaufman L, Rousseeuw P (1990) Finding groups in data: an introduction to cluster analysis. Wiley-Interscience, London
69. Kennard RW, Stone LA (1969) Computer aided design of experiments. Technometrics 11(1):137–148
70. Klein RW, Dubes RC (1989) Experiments in projection and clustering by simulated annealing. Pattern Recognit 22(2):213–220
71. Kohlhoff KJ, Pande VS, Altman RB (2013) K-means for parallel architectures using all-prefix-sum sorting and updating steps. IEEE Trans Parallel Distrib Syst 24(8):1602–1612
72. Lai JZC., Huang TJ, Liaw YC (2009) A fast K-means clustering algorithm using cluster center displacement. Pattern Recognit 42(11):2551–2556
73. Lance GN, Williams WT (1967) A general theory of classificatory sorting strategies - II. Clustering systems. Comput J 10(3):271–277
74. Li Y, Zhao K, Chu X, Liu J (2013) Speeding up K-means algorithm by GPUs. J Comput Syst Sci 79(2):216–229
75. Likas A, Vlassis N, Verbeek J (2003) The global K-means clustering algorithm. Pattern Recognit 36(2):451–461
76. Linde Y, Buzo A, Gray R (1980) An algorithm for vector quantizer design. IEEE Trans Commun 28(1):84–95
77. Lloyd S (1982) Least squares quantization in PCM. IEEE Trans Inf Theory 28(2):129–136
78. MacQueen J (1967) Some methods for classification and analysis of multivariate observations. In: Proceedings of the 5th Berkeley symposium on mathematical statistics and probability, pp 281–297
79. Mahajan M, Nimbhorkar P, Varadarajan K (2012) The planar k-means problem is NP-hard. Theor Comput Sci 442, 13–21
80. Mao J, Jain AK (1996) A self-organizing network for hyperellipsoidal clustering (HEC). IEEE Trans Neural Netw 7(1):16–29
81. Melnykov I, Melnykov V (2014) On K-means algorithm with the use of Mahalanobis distances. Stat Probab Lett 84, 88–95
82. Melnykov V (2013) Challenges in model-based clustering. Wiley Interdiscip Rev Comput Stat 5(2):135–148
83. Milligan G, Cooper MC (1988) A study of standardization of variables in cluster analysis. J Classif 5(2):181–204
84. Monmarché N, Slimane M, Venturini G (1999) On improving clustering in numerical databases with artificial ants. In: Proceedings of the 5th European conference on advances in artificial life, pp 626–635
85. Murthy CA, Chowdhury N (1996) In search of optimal clusters using genetic algorithms. Pattern Recognit Lett 17(8):825–832

86. Omran MGH, Engelbrecht AP, Salman A (2007) An overview of clustering methods. Intell Data Anal 11(6):583–605
87. Onoda T, Sakai M, Yamada S (2012) Careful seeding method based on independent components analysis for K-means clustering. J Emerg Technol Web Intell 4(1):51–59
88. Ordonez C, Omiecinski E (2004) Efficient disk-based K-means clustering for relational databases. IEEE Trans Knowl Data Eng 16(8):909–921
89. Pacheco JA (2005) A scatter search approach for the minimum sum-of-squares clustering problem. Comput Oper Res 32(5):1325–1335
90. Pacheco JA, Valencia O (2003) Design of hybrids for the minimum sum-of-squares clustering problem. Comput Stat Data Anal 43(2):235–248
91. Paterlini S, Krink T (2006) Differential evolution and particle swarm optimization in partitional clustering. Comput Stat Data Anal 50(5):1220–1247
92. Pelleg D, Moore A (1999) Accelerating exact K-means algorithms with geometric reasoning. In: Proceedings of the 5th ACM SIGKDD international conference on knowledge discovery and data mining, pp 277–281
93. Pena JM, Lozano JA, Larranaga P (1999) An empirical comparison of four initialization methods for the K-means algorithm. Pattern Recognit Lett 20(10):1027–1040
94. Rayward-Smith VJ (2005) Metaheuristics for clustering in KDD. In: Proceedings of the IEEE congress on evolutionary computation, vol. 3, pp 2380–2387
95. Redmond SJ, Heneghan C (2007) A method for initialising the K-means clustering algorithm using kd-trees. Pattern Recognit Lett 28(8):965–973
96. Ruspini EH (1970) Numerical methods for fuzzy clustering. Inf Sci 2(3):319–350
97. Selim SZ, Ismail MA (1984) K-means-type algorithms: a generalized convergence theorem and characterization of local optimality. IEEE Trans Pattern Anal Mach Intell 6(1):81–87
98. Sparks DN (1973) Euclidean cluster analysis. Appl Stat 22(1):126–130
99. Späth H (1976) L1 cluster analysis. Computing 16(4):379–387
100. Späth H (1977) Computational experiences with the exchange method: applied to four commonly used partitioning cluster analysis criteria. Eur J Oper Res 1(1):23–31
101. Su T, Dy JG (2007) In search of deterministic methods for initializing K-means and Gaussian mixture clustering. Intell Data Anal 11(4):319–338
102. Tarsitano A (2003) A computational study of several relocation methods for K-means algorithms. Pattern Recognit 36(12):2955–2966
103. Thorndike RL (1953) Who belongs in the family? Psychometrika 18(4):267–276
104. Tou JT, Gonzales RC (1974) Pattern recognition principles. Addison-Wesley, Reading
105. van der Merwe, D.W., Engelbrecht AP (2003) Data clustering using particle swarm optimization. In: Proceedings of the 2003 IEEE congress on evolutionary computation, pp 215–220
106. Vattani A (2011) K-means requires exponentially many iterations even in the plane. Discrete Comput Geom 45(4):596–616
107. Vendramin L, Campello RJG.B., Hruschka ER (2010) Relative clustering validity criteria: a comparative overview. Stat Anal Data Min 3(4):209–235
108. von Luxburg, U (2007) A tutorial on spectral clustering. Stat Comput 17(4):395–416
109. Wu R, Zhang B, Hsu M (2009) Clustering billions of data points using GPUs. In: Proceedings of the 6th ACM conference on computing Frontiers, pp 1–6
110. Xiao Y, Yu J (2012) Partitive clustering (K-means family). WIREs Data Min Knowl Discov 2(3):209–225
111. Xu R, Wunsch D (2005) Survey of clustering algorithms. IEEE Trans Neural Netw 16(3): 645–678
112. Zhang JS, Leung YW (2003) Robust clustering by pruning outliers. IEEE Trans Syst Man Cybern B 33(6):983–999

Chapter 4
Nonsmooth Optimization Based Algorithms in Cluster Analysis

Adil M. Bagirov and Ehsan Mohebi

Abstract Cluster analysis is an important task in data mining. It deals with the problem of organization of a collection of objects into clusters based on a similarity measure. Various distance functions can be used to define the similarity measure. Cluster analysis problems with the similarity measure defined by the squared Euclidean distance, which is also known as the minimum sum-of-squares clustering, has been studied extensively over the last five decades. However, problems with the L_1 and L_∞ norms have attracted less attention. In this chapter, we consider a nonsmooth nonconvex optimization formulation of the cluster analysis problems. This formulation allows one to easily apply similarity measures defined using different distance functions. Moreover, an efficient incremental algorithm can be designed based on this formulation to solve the clustering problems. We develop incremental algorithms for solving clustering problems where the similarity measure is defined using the L_1, L_2 and L_∞ norms. We also consider different algorithms for solving nonsmooth nonconvex optimization problems in cluster analysis. The proposed algorithms are tested using several real world data sets and compared with other similar algorithms.

Keywords Cluster analysis • Nonsmooth optimization • Nonconvex optimization • Partition clustering • Incremental algorithm • k-means algorithm • Similarity measure

4.1 Introduction

The cluster analysis deals with the problem of organization of a collection of objects into clusters based on similarity. It is among the most important tasks in data mining and also known as *unsupervised partition* of objects. The cluster analysis has found many applications in different areas such as medicine, bioinformatics, engineering, business. There are different types of clustering problems. We consider

A.M. Bagirov (✉) • E. Mohebi
Faculty of Science, School of Science, Information Technology and Engineering, Federation University Australia, Victoria, 3353, Australia
e-mail: a.bagirov@federation.edu.au; e.mohebi@federation.edu.au

M.E. Celebi (ed.), *Partitional Clustering Algorithms*,
DOI 10.1007/978-3-319-09259-1_4

the unconstrained hard clustering problem. It can be formulated as an optimization problem. Optimization models of the hard cluster analysis problems include combinatorial, mixed integer nonlinear programming and nonconvex nonsmooth optimization models [9, 12, 18]. The clustering is a global optimization problem, it has many solutions and only global solutions provide the best cluster structure of a data set.

The similarity measure in the cluster analysis can be defined using various distance functions. The use of different similarity measures allows one to find different cluster structures in a data set. The widely used similarity measure is based on the squared Euclidean distance. Such a clustering problem is also known as the minimum sum-of-squares clustering problem. Various optimization techniques have been applied to solve it. These techniques include branch and bound [26,29], cutting plane [34], interior point methods [45], the variable neighborhood search algorithm [35] and metaheuristics like simulated annealing, tabu search, genetic algorithms [2,20,51,53].

The similarity measure can also be defined using the L_1 and L_∞ norms. However, cluster analysis problems with such similarity measures have attracted significantly less attention than those with the use of the squared Euclidean norm. There are several papers where clustering problems with the similarity measures based on the L_1 and L_∞ norms were considered. The paper [22] (see, also [40]) seems to be the first paper where the L_1-norm (also known as the Manhattan norm) was used to define the similarity measure in the cluster analysis problems. In [1], the authors show that the distance metric based on the L_1-norm is consistently more preferable than the distance metric based on the L_2-norm for high dimensional data mining applications.

In papers [17, 54], the ISODATA clustering algorithm (a generalized version of the k-means) was developed using the Euclidean distance and this algorithm for the L_1-norm was introduced in [38]. In [38], the authors also outline the comparative results for clustering using both distances. These results demonstrate that the L_1-norm based algorithm is superior when the correlation coefficient is high and negative. The L_∞-norm is used as a convex surrogate constraint for clustering in [39]. In [30], the authors examine the use of a range of Minkowski norms for the clustering. The X-means algorithm introduced in [47] can use similarity measures based on different norms.

An approach for fuzzy clustering using the L_1-norm and a maximum entropy inference method is presented in [32]. In the paper [56], the authors give an alternative fuzzy c-means algorithm with the L_1-norm and fuzzy covariance. Results show that these algorithms are robust to noise and outliers.

In the method, called Hyperbox clustering with the Ant Colony Optimization (ACO) [48], hyperboxes, which can be defined using the L_∞ norm, are placed in the search space using the ACO and then clustered using the Nearest-Neighbor method. After running the ACO, until the stopping criteria have been achieved, overlapping hyperboxes are joined into one cluster and non-overlapping hyperboxes are assigned to their own clusters. Accuracy values from this clustering algorithm are significantly better than those from the fuzzy c-means and the ACO, and the runtime improvement when compared to the latter is substantial.

Since the clustering is a global optimization problem, the application of special procedures to generate starting cluster centers is crucial to improve the accuracy of local search algorithms. Such procedures are widely used to improve the performance of the k-means algorithm. For example, k-means++ algorithm introduced in [3], chooses centers at random from the data points, but weighs the data points according to their squared distance from the closest center already chosen. In [25], the authors investigate various initialization methods for the k-means algorithm and present their comparative analysis. Various initialization algorithms are also discussed in [19, 21, 24, 52].

Incremental approaches provide an efficient way to generate starting points for cluster centers. Note that there are two different types of incremental algorithms in cluster analysis. In the first type of incremental algorithms the input is presented as a sequence of items and can be examined in only a few passes (typically just one). At each iteration of such algorithms clusters are updated according to newly arrived data. These algorithms require limited memory and also limited processing time per item (see [33] and references therein). In the second type of incremental algorithms the data is considered as static and clusters are computed incrementally. Such algorithms compute clusters step by step starting with one cluster for the whole data set gradually adding one cluster center at each iteration [9, 15, 16, 36, 42].

In this chapter, we examine the second type of incremental clustering algorithms with the similarity measures based on the L_1, L_2 and L_∞ norms. The clustering problem is formulated as a nonsmooth nonconvex optimization problem. The algorithms considered in this chapter are based on the combination of the incremental approach, the k-means and nonsmooth optimization algorithms. Using incremental approach auxiliary clustering problems are defined and solved to generate starting points for cluster centers. The discrete gradient method of the nonsmooth optimization, smoothing techniques and heuristics based on the k-means are applied to solve optimization problems. Testing and comparison of algorithms including other similar algorithms such as the multi-start versions of various k-means (Forgy, Lloyd, McQueen and Hartigan), k-means++, X-means algorithms are presented. Visualization of results are presented to demonstrate the difference between cluster structures obtained using different similarity measures.

The rest of the chapter is organized as follows. Optimization formulations of the clustering problem are given in Sect. 4.2. The auxiliary cluster problem is introduced in Sect. 4.3. Section 4.4 presents a general scheme for an incremental clustering algorithm. An algorithm for computation of starting points for cluster centers is described in Sect. 4.5. The modified incremental clustering algorithm is given in Sect. 4.6. Three different algorithms for solving optimization problems are discussed in Sect. 4.7. Section 4.8 presents the implementation of algorithms. Numerical results with the proposed algorithms using different similarity measures are reported in Sect. 4.9. The comparison of the proposed algorithm with some other clustering algorithms are presented in Sect. 4.10. Section 4.11 concludes the chapter.

4.2 Optimization Formulations of Clustering Problems

In this section we present three different optimization formulations of the clustering problem. Assume that a finite set A of points in the n-dimensional space $\mathbb{R}^n$ is given, that is

$$A = \{a^1, \dots, a^m\}, \text{ where } a^i \in \mathbb{R}^n,\ i = 1, \dots, m.$$

The hard unconstrained clustering problem is the distribution of the points of the set A into a given number k of disjoint subsets $A^j,\ j = 1, \dots, k$ with respect to predefined criteria such that:

1) $A^j \neq \emptyset,\ j = 1, \dots, k;$
2) $A^j \bigcap A^l = \emptyset,\ j, l = 1, \dots, k,\ j \neq l;$
3) $A = \bigcup_{j=1}^{k} A^j;$
4) no constraints are imposed on the clusters $A^j,\ j = 1, \dots, k.$

The sets $A^j,\ j = 1, \dots, k$ are called clusters. Each cluster A^j can be identified by its center $x^j \in \mathbb{R}^n,\ j = 1, \dots, k$. Data points from the same cluster are similar and data points from different clusters are dissimilar to each other. The similarity between points can be measured using different distance functions. Let $d : \mathbb{R}^n \times \mathbb{R}^n \to \mathbb{R}_+$ be a distance function. Here $\mathbb{R}_+ = \{x \in \mathbb{R} : x \geq 0\}$.

In this chapter, we define the distance function d using the squared Euclidean norm, L_1-norm and L_∞-norm:

1. The distance function using the squared Euclidean norm:

$$d(x, y) = \sum_{i=1}^{n} (x_i - y_i)^2;$$

2. The distance function using the L_1-norm:

$$d(x, y) = \sum_{i=1}^{n} |x_i - y_i|;$$

3. The distance function using the L_∞-norm:

$$d(x, y) = \max_{i=1,\dots,n} |x_i - y_i|.$$

Note that the distance function d defined using the L_1 and L_∞ norms is nonsmooth. The use of different distance functions can lead to the finding of different cluster structures in the data set.

4.2.1 Combinatorial Formulation of the Clustering Problem

Denote the set of k clusters in the set A by $\bar{A} = (A^1, \dots, A^k)$ and a set of all possible k-partitions of the set A by $\tilde{A}$. Then *the combinatorial formulation* can be given as:

$$\text{minimize } \Psi_k(\bar{A}) = \frac{1}{m} \sum_{j=1}^{k} \sum_{a \in A^j} d(x^j, a) \tag{4.1}$$

subject to

$$\bar{A} = (A^1, \dots, A^k) \in \tilde{A}. \tag{4.2}$$

Here x^j is the center of the cluster A^j, $j = 1, \dots, k$ which can be found by solving the following optimization problem:

$$\text{minimize } \frac{1}{|A^j|} \sum_{a \in A^j} d(x, a) \text{ subject to } x \in \mathbb{R}^n. \tag{4.3}$$

Here $|\cdot|$ stands for the cardinality of a set. If the squared Euclidean distance is used for the similarity measure then the center x^j can be found explicitly as follows:

$$x^j = \frac{1}{|A^j|} \sum_{a \in A^j} a, \ j = 1, \dots, k. \tag{4.4}$$

Note that in this formulation decision variables are nonempty subsets of the set A and therefore optimization algorithms cannot be directly applied to solve the problem (4.1)–(4.2). The combinatorial formulation can be used to solve clustering problems only in very small data sets.

4.2.2 Mixed Integer Nonlinear Programming Formulation of the Clustering Problem

Alternatively, the problem of the finding k clusters in the set A can be reduced to the following optimization problem:

$$\text{minimize } \psi_k(x, w) = \frac{1}{m} \sum_{i=1}^{m} \sum_{j=1}^{k} w_{ij} \, d(x^j, a^i) \tag{4.5}$$

subject to

$$x = (x^1, \dots, x^k) \in \mathbb{R}^{k \times n}, \tag{4.6}$$

$$\sum_{j=1}^{k} w_{ij} = 1, \; i = 1, \ldots, m, \tag{4.7}$$

$$w_{ij} \in \{0, 1\}, \; i = 1, \ldots, m, \; j = 1, \ldots, k. \tag{4.8}$$

Here w_{ij} is the association weight of the pattern a^i with the cluster j, given by

$$w_{ij} = \begin{cases} 1, \text{ if } a^i \text{ is allocated to the cluster } j; \\ 0, \text{ otherwise.} \end{cases}$$

w is an $m \times k$ matrix.

The problem (4.5)–(4.8) is called *the mixed integer nonlinear programming formulation* of the clustering problem. It contains mn integer variables $w_{ij}, i = 1, \ldots, m, j = 1, \ldots, k$ and kn continuous variables $x^j \in \mathbb{R}^n, j = 1, \ldots, k$.

Cluster centers $x^j, \; j = 1, \ldots, k$ can be found by solving the problem (4.3). If the similarity measure is defined using the squared Euclidean norm then x^j is a centroid of the cluster $A^j, \; j = 1, \ldots, k$ and it can be found applying the formula (4.4). In this case the problem (4.5)–(4.8) becomes integer programming problem as cluster centers $x^j, j = 1, \ldots, k$ are not decision variables.

4.2.3 Nonsmooth Nonconvex Optimization Formulation of the Clustering Problem

Nonsmooth nonconvex optimization formulation of the clustering problem is as follows [12, 13, 18]:

$$\text{minimize } f_k(x) \text{ subject to } x = (x^1, \ldots, x^k) \in \mathbb{R}^{k \times n}, \tag{4.9}$$

where

$$f_k(x^1, \ldots, x^k) = \frac{1}{m} \sum_{i=1}^{m} \min_{j=1,\ldots,k} d(x^j, a^i). \tag{4.10}$$

If $k = 1$ then the function f_k is convex and it is nonconvex if $k > 1$ due to the minimum operation. If the similarity measure is defined using the squared Euclidean distance then for $k = 1$ the function f_k is smooth that is it is continuously differentiable for any $x \in \mathbb{R}^n$. However, for $k \geq 2$ this function is nonsmooth due to the minimum operation that is its gradient does not exist at all $x \in \mathbb{R}^{k \times n}$. If the similarity measure is defined using the L_1 or L_∞ norms then the function f_k is nonsmooth for all $k \geq 1$. This is due to the minimum operation and the fact that both L_1 and L_∞ norm based distance functions are nonsmooth.

4.2.4 *Comparison of Different Formulations of Clustering Problem*

In this subsection we compare objective functions in different formulations of the clustering problems. We call objective functions Ψ_k, ψ_k and f_k *cluster functions*. Note that the objective function Ψ_k explicitly depends, in particular, on clusters (or subsets of the set A). Optimization methods cannot be applied to minimize such functions and the combinatorial formulation cannot be used to solve clustering problems considered in this chapter. Therefore, we compare only objective functions ψ_k and f_k.

Comparing these two functions (also two different formulations of the clustering problem) one can note that:

1. The objective function ψ_k depends on variables $w_{ij},\ i = 1, \ldots, m,\ j = 1, \ldots, k$ (coefficients, which are binary integers) and $x^1, \ldots, x^k,\ x^j \in \mathbb{R}^n,\ j = 1, \ldots, k$ (cluster centers, which are continuous variables). However, the function f_k depends only on continuous variables $x^1, \ldots, x^k$.
2. The number of variables in Problem (4.5)–(4.8) is $(m + n) \times k$ whereas in Problem (4.9) this number is only $n \times k$. Notice that the number of variables in Problem (4.9) does not depend on the number m of instances. In many real world data sets the number of instances m is substantially greater than the number of features n.
3. Since the function f_k is represented as a sum of minima functions it is nonsmooth for $k > 1$, that is it is not differentiable everywhere. Both functions ψ_k and f_k are nonconvex for $k > 1$.
4. Problem (4.5)–(4.8) is mixed integer nonlinear programming problem and Problem (4.9) is nonsmooth global optimization problem. However, they are equivalent in the sense that their global minimizers coincide [18].

Items 1 and 2 can be considered as advantages of the nonsmooth optimization formulation (4.9) of the clustering problem. Nonsmooth optimization models of unsupervised and supervised data classification problems are also discussed in [4, 5, 23].

4.3 The Auxiliary Cluster Problem

The objective function (4.10) in Problem (4.9) is nonconvex and the problem itself is global optimization problem. However, the most of existing global optimization techniques cannot be applied to solve this problem because its size becomes large as the number of clusters increases. These techniques are extremely time consuming for solving such clustering problems. Solving clustering problems in large data sets is out of reach of most of global optimization algorithms. Any local optimization algorithm starting from a given point will end up at the closest local minimizer which can be significantly different from the global minimizer. Global minimizers provide the best cluster structure of a data set with the smallest number of clusters.

We propose to apply local search methods, including heuristic methods such as the k-means and deterministic methods of optimization, to solve clustering problems. Such methods are very fast even in large data sets however their success in finding global or near global solutions to the clustering problems highly depends on the choice of starting points. In order to generate such points we introduce the auxiliary cluster function. Starting points are found by minimizing this function.

Assume that the solution $x^1, \dots, x^{l-1}, \; l > 1$ to the $(l-1)$-clustering problem is known. Define by

$$r^i_{l-1} = \min_{j=1,\dots,l-1} d(x^j, a^i)$$

the distance between the data point $a^i, \; i = 1, \dots, m$ and its cluster center. We will also use the notation r^a_{l-1} for the data point $a \in A$. Consider the following function:

$$\bar{f}_l(y) = \frac{1}{m} \sum_{i=1}^{m} \min \left\{ r^i_{l-1}, d(y, a^i) \right\}, \; y \in \mathbb{R}^n. \tag{4.11}$$

We call this function *the l-th auxiliary cluster function*. One can see that the function $\bar{f}_l$ is represented as a sum of minima of constants r^i_{l-1} and the distance function $d(y, a^i), i = 1, \dots, m$ which is convex. Therefore the function $\bar{f}_l$ is nonsmooth and in general, nonconvex. *The l-th auxiliary clustering problem* is formulated as follows:

$$\text{minimize } \bar{f}_l(y) \text{ subject to } y \in \mathbb{R}^n. \tag{4.12}$$

This is a nonsmooth global optimization problem and it may have many local minimizers. The success of any local method for solving this problem heavily depend on the choice of starting points. Therefore it is important to develop an algorithm to generate good starting points for its solution. One such algorithm will be designed in Sect. 4.5.

4.4 An Incremental Clustering Algorithm

In this section we describe a general scheme for an incremental clustering algorithm. Incremental algorithms are becoming popular in data mining and in particular, in the cluster analysis. There are two types of incremental algorithms. In algorithms of the first type new data points are dynamically added at each iteration of the algorithm to the data set and clusters are updated accordingly. In algorithms of the second type a data set is static and an algorithm computes clusters incrementally. Such incremental clustering algorithms build clusters dynamically adding one cluster center at a time. Therefore this type of incremental algorithms can also be called *sequential clustering algorithms*. In this chapter we consider only the second type of incremental clustering algorithms. The general scheme of such algorithms for finding the k-partition of the set A is as follows.

Algorithm 1 *An incremental clustering algorithm.*
Input: *The data set A and the number k of clusters to be computed.*
Output: *The l-partition of the set A with $l = 1, \dots, k$.*

Step 1. *(Initialization). Compute the center $x^1 \in \mathbb{R}^n$ of the set A. Set $l := 1$.*
Step 2. *(Stopping criterion) Set $l := l + 1$. If $l > k$ then stop. The k-partition problem has been solved.*
Step 3. *(Computation of the next cluster center). Find a starting point $\bar{y} \in \mathbb{R}^n$ for the l-th cluster center by solving Problem (4.12).*
Step 4. *(Refinement of all cluster centers). Select $(x^1, \dots, x^{l-1}, \bar{y})$ as a new starting point to solve the l-partition problem (4.9) (or the problem (4.5)–(4.8)) (when $k = l$). Let $y^1, \dots, y^l$ be a solution to this problem.*
Step 5. *(Solution to the l-partition problem). Set $x^j := y^j$, $j = 1, \dots, l$ as a solution to the l-partition problem and go to Step 2.*

Remark 1. To date incremental algorithms were designed for solving the minimum sum-of-squares clustering problems [9, 15, 41, 42, 46]. In this chapter, we develop incremental algorithms also for solving clustering problems where the similarity measure is defined using L_1 and L_∞ norms.

Remark 2. One can see that Algorithm 1 in addition to the k-partition problem solves also all intermediate l-partition problems, where $l = 1, \dots, k - 1$. Step 3, where one finds a starting point for the l-th cluster center, and Step 4 are the most important steps of Algorithm 1. In the next section we design an algorithm for finding starting points for the l-th cluster center.

4.5 Computation of Starting Points for Cluster Centers

Various initial seeding procedures were introduced to improve the efficiency of the k-means algorithms. Such procedures were considered to design different versions of the k-means algorithm such as the X-means [47] and k-means++ [3] algorithms. In the paper [25], a comparative study of various initialization methods for the k-means algorithm is presented. In this section, we discuss an initial seeding procedure based on nonsmooth optimization formulation of the clustering problem. This procedure is designed for the incremental algorithm described in the previous section. This means that in order to find initial points for the l-th cluster center, $l > 1$, it is assumed that cluster centers for the $(l - 1)$-partition problem are known.

Given the solution $x^1, \dots, x^{l-1}$, $l > 1$ to the $(l-1)$-clustering problem one can divide the whole space $\mathbb{R}^n$ into two subsets as follows:

$$S_1 = \{y \in \mathbb{R}^n : d(y,a) \geq r^a_{l-1}, \ \forall a \in A\},$$
$$S_2 = \{y \in \mathbb{R}^n : \exists a \in A \text{ such that } d(y,a) < r^a_{l-1}\}.$$

All cluster centers $x^1, \dots, x^{l-1} \in S_1$. It is clear that $S_1 \bigcup S_2 = \mathbb{R}^n$ and $S_1 \bigcap S_2 = \emptyset$. Moreover,

$$\bar{f}_l(y) = f_{l-1}(x^1, \dots, x^{l-1}) = \frac{1}{m} \sum_{a \in A} r^a_{l-1}, \ \forall y \in S_1, \tag{4.13}$$

that is the l-th auxiliary cluster function is constant on the set S_1 and any point from this set is a global maximizer of the function $\bar{f}_l$. In general, a local search method will terminate at any of these points. Hence, points from the set S_1 cannot be considered as starting points to solve the problem (4.12). Therefore, starting points should be chosen from the set S_2.

For any $y \in S_2$ one can divide the set A into two subsets as follows:

$$B_1(y) = \{a \in A : d(y,a) \geq r^a_{l-1}\},$$
$$B_2(y) = \{a \in A : d(y,a) < r^a_{l-1}\}.$$

It is obvious that the set $B_2(y)$ contains all data points which are closer to the point y than their cluster centers. Since $y \in S_2$ the set $B_2(y) \neq \emptyset$. Furthermore, $B_1(y) \bigcap B_2(y) = \emptyset$ and $A = B_1(y) \bigcup B_2(y)$. Then

$$\bar{f}_l(y) = \frac{1}{m} \left(\sum_{a \in B_1(y)} r^a_{l-1} + \sum_{a \in B_2(y)} d(y,a) \right).$$

The difference $v_l(y)$ between the value $f_{l-1}(x^1, \dots, x^{l-1})$ (see (4.13)) and the value of the l-th auxiliary cluster function at y is:

$$v_l(y) = \frac{1}{m} \sum_{a \in B_2(y)} \left(r^a_{l-1} - d(y,a) \right)$$

which can be rewritten as

$$v_l(y) = \frac{1}{m} \sum_{a \in A} \max\{0, r^a_{l-1} - d(y,a)\}. \tag{4.14}$$

It is obvious that data points $a \in A$, which are not among cluster centers $x^1, \dots, x^{l-1}$, belong to the set S_2. Because these points attract at least themselves.

Therefore in order to compute starting points for solving the auxiliary clustering problem (4.12) we use data points $a \in A \setminus S_1$. We introduce the following number:

$$v^1_{max} = \max_{a \in A \setminus S_1} v_l(a). \tag{4.15}$$

The number v^1_{max} is the largest decrease of the auxiliary cluster function $\bar{f}_k(y)$ comparing to the value $f_{l-1}(x^1, \ldots, x^{l-1})$ for all $y \in A \setminus S_1$. Among all data points $a \in A$, a point $\bar{a} \in A \setminus S_1$, satisfying the condition $v_l(\bar{a}) = v^1_{max}$, provides the largest decrease of the cluster function f_l comparing with the value $f_{l-1}(x^1, \ldots, x^{l-1})$ if $\bar{a}$ is chosen as the l-th cluster center.

Let $\gamma_1 \in [0, 1]$ be a given number. Define the following subset of A:

$$\bar{A}_1 = \left\{a \in A \setminus S_1 : v_l(a) \geq \gamma_1 v^1_{max}\right\}. \tag{4.16}$$

The set $\bar{A}_1$ contains points $a \in A \setminus S_1$ which provide sufficient decrease of the cluster function f_l comparing with the value $f_{l-1}(x^1, \ldots, x^{l-1})$ if these points are chosen as the l-th cluster center. If $\gamma_1 = 1$ then we choose only points with largest decrease and if $\gamma_1 = 0$ then $\bar{A}_1 = A \setminus S_1$.

For each $a \in \bar{A}_1$ we compute the set $B_2(a)$ and its center $c(a)$. In this stage we replace points $a \in \bar{A}_1$ by points $c(a)$, because the center $c(a)$ is the better representative of the set $B_2(a)$ than the point a. If the similarity measure is defined using the squared Euclidean norm then $c(a)$ is the centroid of the set $B_2(a)$. In all other cases $c(a)$ is defined as the solution to the following convex minimization problem:

$$\text{minimize} \quad \frac{1}{|B_2(a)|} \sum_{b \in B_2(a)} d(x, b) \quad \text{subject to } x \in \mathbb{R}^n.$$

As a result we get the following set:

$$\overline{C}_1 = \{c \in \mathbb{R}^n : \exists a \in \bar{A}_1 \text{ such that } c = c(a)\}.$$

Then using (4.14) we compute $v^2_l(c) = v_l(c)$ for each $c \in \overline{C}_1$. Finally, we compute the maximum of all numbers $v^2_l(c),\ c \in \overline{C}_1$:

$$v^2_{max} = \max_{c \in \overline{C}_1} v^2_l(c). \tag{4.17}$$

Let $\gamma_2 \in [0, 1]$ be a given number. We define the following subset of $\overline{C}_1$:

$$\bar{A}_2 = \left\{c : c \in \overline{C}_1 \text{ and } v^2_l(c) \geq \gamma_2 v^2_{max}\right\}. \tag{4.18}$$

The set $\bar{A}_2$ contains all points $c \in \overline{C}_1$ which provide sufficient decrease of the cluster function f_l comparing with the value $f_{l-1}(x^1, \dots, x^{l-1})$ if these points are chosen as the l-th cluster center. If $\gamma_2 = 1$ then we choose points with the largest decrease and if $\gamma_2 = 0$ then $\bar{A}_2 = \overline{C}_1$.

All points from the set $\bar{A}_2$ are considered as starting points for solving the auxiliary clustering problem (4.12).

Applying a local search algorithm, Problem (4.12) is solved starting from each point of the set $\bar{A}_2$. Such local search algorithms will be discussed in Sect. 4.7. A local search algorithm generates the same number of solutions as the number of starting points. The set of these solutions is denoted by $\bar{A}_3$. Since the local method can arrive to the same solution starting from different points we remove from the set $\bar{A}_3$ solutions which are sufficiently close to each other keeping only one of them. Sufficiently close solutions can be defined using some predefined tolerance. It is clear that $\bar{A}_3 \neq \emptyset$.

Next we define

$$\bar{f}_l^{min} = \min_{y \in \bar{A}_3} \bar{f}_l(y). \tag{4.19}$$

Let $\gamma_3 \in [1, \infty)$ be a given number. Compute the following set:

$$\bar{A}_4 = \left\{ y \in \bar{A}_3 : \bar{f}_l(y) \leq \gamma_3 \bar{f}_l^{min} \right\}. \tag{4.20}$$

If $\gamma_3 = 1$ then $\bar{A}_4$ contains stationary points of the problem (4.12) with the lowest value $\bar{f}_l^{min}$. If γ_3 is sufficiently large then $\bar{A}_4 = \bar{A}_3$.

Points from the set $\bar{A}_4$ are considered as a starting point for the l-th cluster center in Step 4 of Algorithm 1. Summarizing all steps for computing the set $\bar{A}_4$ we can design the following algorithm for finding starting points to solve the clustering problem (4.9).

Algorithm 2 *Computation of starting points.*
Input: *The data set A and the solution* $x^1, \dots, x^{l-1}$, $l > 1$ *to the* $(l-1)$*-clustering problem.*
Output: *The set of starting points for the* l*-th cluster center.*

Step 0. *(Initialization). Select numbers* $\gamma_1, \gamma_2 \in [0, 1]$ *and* $\gamma_3 \in [1, \infty)$.
Step 1. *Compute* v^1_{max} *using (4.15) and the set* $\bar{A}_1$ *using (4.16).*
Step 2. *Compute* v^2_{max} *using (4.17) and the set* $\bar{A}_2$ *using (4.18).*
Step 3. *Apply a local search algorithm to compute a set* $\bar{A}_3$ *of solutions to the auxiliary clustering problem (4.12) using points from the set* $\bar{A}_2$ *as starting points.*
Step 4. *Compute* $\bar{f}_l^{min}$ *using (4.19) and the set* $\bar{A}_4$ *using (4.20).* $\bar{A}_4$ *is the set of starting points to solve the clustering problem (4.9).*

Thus, we use more than one starting points to solve the clustering problem (4.9) in Step 4 of Algorithm 1. Moreover, these points always guarantee decrease of the clustering function at each iteration of the incremental algorithm and they are away from each other in the search space. Such an approach allows us to apply local search methods to solve the global optimization problem (4.9).

4.6 The Modified Incremental Clustering Algorithm

In this section the incremental clustering Algorithm 1 is modified by applying Algorithm 2 in Step 3. Then Algorithm 1 for solving the k-partition Problem (4.9) can be rewritten as follows:

Algorithm 3 *A multistart incremental algorithm.*
Input: *The data set* A *and the number* k *of clusters to be computed.*
Output: *The* l*-partition of the set* A *with* $l = 1, \dots, k$.

Step 1. *(Initialization). Compute the center* $x^1 \in \mathbb{R}^n$ *of the set* A*. Set* $l := 1$.
Step 2. *(Stopping criterion) Set* $l := l + 1$*. If* $l > k$ *then stop. The* k*-partition problem has been solved.*
Step 3. *(Computation of the set of starting points for the next cluster center). Apply Algorithm 2 to find a set* $\bar{A}_4$ *of starting points for the* l*-th cluster center.*
Step 4. *(Refinement of all cluster centers). For each* $\bar{y} \in \bar{A}_4$ *select* $(x^1, \dots, x^{l-1}, \bar{y})$ *as a starting point to solve the* l*-partition problem (4.9) (when* $k = l$*). Let* $(y^1(\bar{y}), \dots, y^l(\bar{y}))$ *be a solution to this problem and* $f_{l\bar{y}} = f_l(y^1(\bar{y}), \dots, y^l(\bar{y}))$ *be a value of the cluster function at this solution.*
Step 5. *(Computation of the best solution). Compute*

$$f_l^{min} = \min_{\bar{y} \in \bar{A}_4} f_{l\bar{y}}$$

and the set of best solutions

$$C = \left\{(y^1(\bar{y}), \dots, y^l(\bar{y})) : f_{l\bar{y}} = f_l^{min}\right\}.$$

Step 6. *(Solution to the* l*-partition problem). Take any* $(y^1(\bar{y}), \dots, y^k(\bar{y})) \in C$*, set* $x^j := y^j(\bar{y}),\ j = 1, \dots, l$ *as a solution to the* l*-th partition problem and go to Step 2.*

4.7 Solving Optimization Problems

In this section we discuss local search algorithms for solving both the auxiliary clustering problem (4.12) and the clustering problem (4.9). We will consider three different algorithms for this purpose: (a) k-means type heuristic algorithm; (b) the discrete gradient method of nonsmooth optimization and (c) an algorithm based on smoothing techniques.

4.7.1 k-Means Type Heuristic Algorithm

This algorithm was discussed in [9] (see, also [15]). In order to solve the auxiliary clustering problem (4.12) the algorithm fixes the first $(l-1)$ cluster centers and updates only the l-th center. The algorithm proceeds as follows.

Algorithm 4 *Heuristic algorithm for minimizing the auxiliary cluster function.*
Input: *The data set A and the starting point* $y \in \bar{A}_4$.
Output: *Local minimizer of Problem (4.12).*

Step 1. *Compute the set* $B_2(y)$ *and its center* c.
Step 2. *Compute the set* $B_2(c)$ *and its center.*
Step 3. *Recompute the set* $B_2(c)$ *and its center until no more data points escape or return to this set. The last center* c *is the solution to Problem (4.12).*

It is proved in [9] that Algorithm 4 converges to the local minimizer of Problem (4.12) after finite number of iterations.

We apply the k-means algorithm for solving Problem (4.9). This algorithm converges to the local solutions of the problem (4.9).

4.7.2 The Discrete Gradient Method

The objective functions in problems (4.9) and (4.12) are nonsmooth nonconvex and nonsmooth optimization algorithms can be applied to solve them. Various subdifferential generalizations can be applied to design such algorithms. Among these generalizations the Clarke subdifferential [27] is most widely used. However, the Clarke subdifferential calculus cannot be applied to compute subgradients of the functions f_k and $\bar{f}_k$ since this calculus exists in the form of "inclusions" not "equalities". Therefore, derivative free methods are better choice than subgradient methods for solving Problems (4.9) and (4.12).

We apply the discrete gradient method to solve these problems. This method is a derivative-free method and uses discrete gradients which are approximations to subgradients. Discrete gradients are computed using only values of a function. The convergence of the discrete gradient method is proved for quite general nonsmooth and in particular, nonconvex optimization problems under very mild conditions. The detailed description of this method can be found in [11] (see, also [7,8]).

Since the objective function $\bar{f}_k$ in Problem (4.12) and the objective function f_k in Problem (4.9) are piecewise partially separable (see [14] for the definition of the piecewise partial separability), we use the version of the discrete gradient method described in [14]. The use of the special structure of these problems such as the piecewise partial separability allows one to significantly reduce the number of function evaluations in the discrete gradient method which is crucial for any derivative free method. It is worth to note that all starting points to solve

Problem (4.12) are chosen from the set S_2 and therefore the discrete gradient method will never terminate at local maximizers of this problem and will always find its local minimizers.

4.7.3 An Algorithm Based on Smoothing Techniques

In this subsection we will demonstrate how smoothing techniques can be applied to approximate the objective functions in Problems (4.9) and (4.12) by smooth functions. This will allow us to apply powerful smooth optimization algorithms such as the conjugate gradient and quasi-Newton methods to solve cluster analysis problems. We will use the hyperbolic smoothing for this purpose. Other smoothing techniques can be applied in a similar way.

The hyperbolic smoothing functions were originally introduced to approximate the following nonsmooth function (see, for example, [55]):

$$\varphi(x) = \max\{0, x\}, \ x \in \mathbb{R}. \tag{4.21}$$

The hyperbolic function smoothing the function (4.21) is as follows:

$$\phi_\tau(x) = \frac{x + \sqrt{x^2 + \tau^2}}{2}. \tag{4.22}$$

Here $\tau > 0$ is called the *precision or smoothing parameter*.

Proposition 1. *The function $\phi_\tau(x)$ has the following properties:*

1. *$\phi_\tau(\cdot)$ is an increasing convex C^∞ function;*
2. *$\varphi(x) < \phi_\tau(x) \le \varphi(x) + \frac{\tau}{2}, \ \forall \tau > 0.$*

The hyperbolic smoothing functions for maximum functions

$$\psi(x) = \max_{i=1,\dots,p} \psi_i(x)$$

were studied in [10]. Here $p \ge 1$ and functions $\psi_i, \ i = 1, \dots, p$ are continuously differentiable. Using an additional variable $t \in \mathbb{R}$ we introduce the following function:

$$\Psi(x,t) = t + \sum_{i=1}^{p} \max(0, \psi_i(x) - t).$$

It is clear that $\psi(x) = \Psi(x, \psi(x))$. Then the hyperbolic smoothing function $H(x, \tau)$ for ψ can be written as

$$H(x,\tau) = t + \sum_{i=1}^{p} \frac{\psi_i(x) - t + \sqrt{(\psi_i(x)-t)^2 + \tau^2}}{2}, \; t = \psi(x). \tag{4.23}$$

See [55] and [10] for details.

4.7.3.1 Hyperbolic Smoothing of the Cluster Function

In this subsection we define hyperbolic smoothing of the clustering function f_k defined by (4.10). It is clear that

$$\min_{j=1,\dots,k} d(x^j, a^i) = -\max_{j=1,\dots,k} -d(x^j, a^i), \; i = 1, \dots, m.$$

Then

$$f_k(x^1, \dots, x^k) = -\frac{1}{m} \sum_{i=1}^{m} \max_{j=1,\dots,k} -d(x^j, a^i).$$

Consider the function:

$$\Psi_k(x^1, \dots, x^k) = -\frac{1}{m} \sum_{i=1}^{m} \left(t_i + \sum_{j=1}^{k} \max(0, -d(x^j, a^i) - t_i) \right).$$

Here

$$t_i = \max_{j=1,\dots,k} -d(x^j, a^i) = -\min d(x^j, a^i) \leq 0, \; i = 1, \dots, m.$$

It follows from (4.23) that the hyperbolic smoothing $U_k(x^1, \dots, x^k, \tau)$ of the function f_k is as follows:

$$U_k(x^1, \dots, x^k, \tau) = -\frac{1}{m} \sum_{i=1}^{m} \left(t_i + \sum_{j=1}^{k} \frac{-d(x^j, a^i) - t_i + \sqrt{(d(x^j, a^i) + t_i)^2 + \tau^2}}{2} \right)$$

$$= \frac{1}{m} \sum_{i=1}^{m} \left(-t_i + \sum_{j=1}^{k} \frac{t_i + d(x^j, a^i) - \sqrt{(d(x^j, a^i) + t_i)^2 + \tau^2}}{2} \right). \tag{4.24}$$

4.7.3.2 Hyperbolic Smoothing of the Auxiliary Cluster Function

In this subsection we define the hyperbolic smoothing of the auxiliary cluster function $\bar{f}_k$ defined by (4.11) (here we take $l = k$). This function can be rewritten as:

$$\bar{f}_k(y) = \frac{1}{m}\sum_{i=1}^{m} r_{k-1}^i + \frac{1}{m}\sum_{i=1}^{m} \min\ (0, d(y, a^i) - r_{k-1}^i).$$

Then we have

$$\bar{f}_k(y) = \frac{1}{m}\sum_{i=1}^{m} r_{k-1}^i - \frac{1}{m}\sum_{i=1}^{m} \max\ (0, r_{k-1}^i - d(y, a^i)).$$

Applying (4.22) we can approximate the auxiliary cluster function $\bar{f}_k$ by the following function:

$$\bar{U}_k(y,\tau) = \frac{1}{m}\sum_{i=1}^{m} r_{k-1}^i - \frac{1}{m}\sum_{i=1}^{m} \frac{r_{k-1}^i - d(y,a^i) + \sqrt{(r_{k-1}^i - d(y,a^i))^2 + \tau^2}}{2}$$

$$= \frac{1}{m}\sum_{i=1}^{m} \frac{r_{k-1}^i + d(y,a^i) - \sqrt{(r_{k-1}^i - d(y,a^i))^2 + \tau^2}}{2}, \quad y \in \mathbb{R}^n. \tag{4.25}$$

Since the squared Euclidean distance is differentiable there is no need to smooth it. However, two other distance functions using L_1 and L_∞ norms are nondifferentiable. Therefore we use the hyperbolic smoothing technique to approximate them with the smooth functions.

4.7.3.3 Hyperbolic Smoothing of L_1-Norm

Here we describe the hyperbolic smoothing of the L_1-norm. It is obvious that for $x, a \in \mathbb{R}^n$

$$d(x,a) = \|x - a\|_1 = \sum_{q=1}^{n} \max\{x_q - a_q, a_q - x_q\}$$

$$= \sum_{q=1}^{n} \left[(x_q - a_q) + 2\max\{0, a_q - x_q\}\right].$$

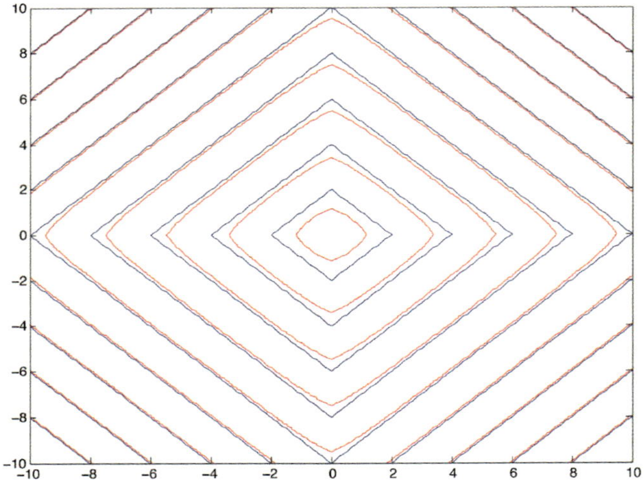

Fig. 4.1 Smooth approximation of L_1-norm

Then applying (4.22) we can approximate $\|x - a\|_1$ by the following smooth function:

$$\eta_1(x, a, \tau) = \sum_{q=1}^{n} \left[(x_q - a_q)^2 + \tau^2\right]^{1/2}. \tag{4.26}$$

Figure 4.1 illustrates the shape of the L_1-norm (blue) and its hyperbolic smooth approximation (red).

Thus, replacing the distance function d in (4.24) and (4.25) by its approximation η_1 defined by (4.26) we get the following smooth approximations for the cluster and the auxiliary cluster functions, respectively:

$$H_k^1(x^1, \ldots, x^k, \tau) = \frac{1}{m} \sum_{i=1}^{m} \left(-t_i + \sum_{j=1}^{k} \frac{t_i + \eta_1(x^j, a^i, \tau) - \sqrt{(\eta_1(x^j, a^i, \tau) + t_i)^2 + \tau^2}}{2} \right), \tag{4.27}$$

$$\bar{H}_k^1(y, \tau) = \frac{1}{m} \sum_{i=1}^{m} \frac{r_{k-1}^i + \eta_1(y, a^i, \tau) - \sqrt{(r_{k-1}^i - \eta_1(y, a^i, \tau))^2 + \tau^2}}{2}, \quad y \in \mathbb{R}^n. \tag{4.28}$$

4.7.3.4 Hyperbolic Smoothing of L_∞-Norm

Now consider the smoothing of the distance function using L_∞-norm. In this case

$$d(x,a) = \|x-a\|_\infty = \max_{q=1,\dots,n} |x_q - a_q|. \tag{4.29}$$

Then

$$\|x-a\|_\infty = \max_{q=1,\dots,n} \max\{\theta_q(x), -\theta_q(x)\}. \tag{4.30}$$

where $\theta_q(x) = x_q - a_q$.

The hyperbolic smoothing $\Omega_q(x,\tau)$ of the function $\beta_q(x) = \max\{\psi_q(x), -\psi_q(x)\}$ is

$$\Omega_q(x,\tau) = \beta_q(x) + \frac{\theta_q(x) - \beta_q(x) + \sqrt{(\theta_q(x)-\beta_q(x))^2 - \tau^2}}{2} + \frac{-\theta_q(x) - \beta_q(x) + \sqrt{(-\theta_q(x)-\beta_q(x))^2 - \tau^2}}{2}.$$

Then the hyperbolic smoothing of the distance function $d(x,a)$ can be expressed as

$$\eta_2(x,a,\tau) = d(x,a) + \sum_{q=1}^{n} \frac{\Omega_q(x,\tau) - d(x,a) + \sqrt{(\Omega_q(x,\tau) - d(x,a))^2 - \tau^2}}{2}. \tag{4.31}$$

Figure 4.2 illustrates the shape of the L_∞-norm (blue) and its hyperbolic smooth approximation (red).

Replacing the distance function d in (4.24) and (4.25) by its approximation η_2 defined by (4.31) we get the following smooth approximations for the cluster and the auxiliary cluster functions, respectively:

$$H_k^2(x^1,\dots,x^k,\tau) = \frac{1}{m}\sum_{i=1}^{m}\left(-t_i + \sum_{j=1}^{k} \frac{t_i + \eta_2(x^j,a^i,\tau) - \sqrt{(\eta_2(x^j,a^i,\tau)+t_i)^2 + \tau^2}}{2}\right), \tag{4.32}$$

$$\bar{H}_k^2(y,\tau) = \frac{1}{m}\sum_{i=1}^{m} \frac{r_{k-1}^i + \eta_2(y,a^i,\tau) - \sqrt{(r_{k-1}^i - \eta_2(y,a^i,\tau))^2 + \tau^2}}{2}, \quad y \in \mathbb{R}^n. \tag{4.33}$$

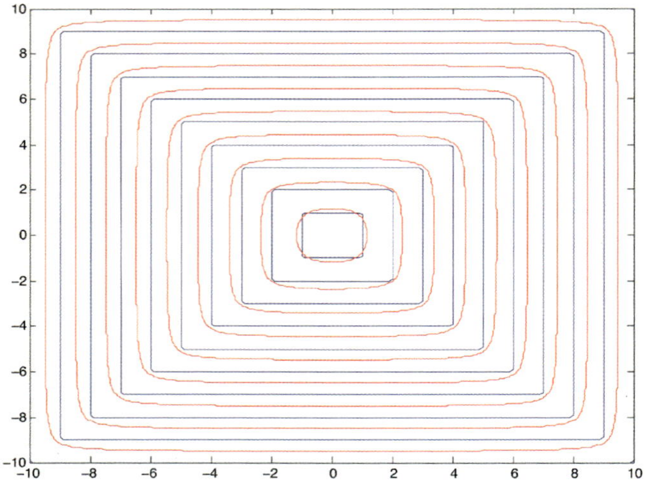

Fig. 4.2 Smooth approximation of L_∞-norm

4.7.3.5 Smooth Clustering Problems

Now we can replace nonsmooth optimization problems (4.9) and (4.12) by their smooth approximations as follows. Smooth auxiliary clustering and clustering problems with the similarity measure defined by the squared Euclidean distance are, respectively:

$$\text{minimize } \bar{U}_k(y, \tau) \text{ subject to } y \in \mathbb{R}^n. \tag{4.34}$$

$$\text{minimize } U_k(x^1, \dots, x^k, \tau) \text{ subject to } x^1, \dots, x^k \in \mathbb{R}^n. \tag{4.35}$$

Smooth auxiliary clustering and clustering problems with the similarity measure defined by the L_1-norm are, respectively:

$$\text{minimize } \bar{H}_k^1(y, \tau) \text{ subject to } y \in \mathbb{R}^n. \tag{4.36}$$

$$\text{minimize } H_k^1(x^1, \dots, x^k, \tau) \text{ subject to } x^1, \dots, x^k \in \mathbb{R}^n. \tag{4.37}$$

Finally, smooth auxiliary clustering and clustering problems with the similarity measure defined by the L_∞-norm are, respectively:

$$\text{minimize } \bar{H}_k^2(y, \tau) \text{ subject to } y \in \mathbb{R}^n. \tag{4.38}$$

$$\text{minimize } H_k^2(x^1, \dots, x^k, \tau) \text{ subject to } x^1, \dots, x^k \in \mathbb{R}^n. \tag{4.39}$$

In order to solve these problems we take any sequence $\{\tau_l\}$ such that $\tau_l > 0$ and $\tau_l \to 0$ as $l \to \infty$. Then applying any smooth optimization algorithms we can get sequences of solutions which converge to the solutions of problems (4.9) and (4.12).

4.8 Implementation of Incremental Clustering Algorithm

In this section, we discuss the implementation of Algorithm 3. The most important steps in Algorithm 3 are Steps 3 and 4. Other steps of this algorithm are easy to implement. In Step 3 we apply Algorithm 2 to find the set of starting points for the l-th cluster center. Algorithm 2 contains three parameters, namely, γ_1, γ_2 and γ_3. The choice of these parameters depends on the size of a data set.

One can see from (4.16) that if γ_1 is close to 0 then the set $\bar{A}_1$ will contain most of data points which may lead to a large number of starting points and make the algorithm very time-consuming. Therefore in order to avoid having a large number of starting points the choice of γ_1 and γ_2 should depend on the number of points in a data set. Since the difference between the values of the auxiliary clustering problem (4.12) and the clustering problem (4.9) is not expected to be large one can choose the parameter γ_3 from $[1, 2]$. These parameters are chosen as follows:

1. $\gamma_1 = 0.3, \gamma_2 = 0.5$ and $\gamma_3 = 2$ for data sets with the number of data points $m \le 200$;
2. $\gamma_1 = 0.6, \gamma_2 = 0.8$ and $\gamma_3 = 1.25$ for data sets with the number of data points $200 < m \le 2500$;
3. $\gamma_1 = 0.8, \gamma_2 = 0.99$ and $\gamma_3 = 1.05$ for data sets with the number of data points $m > 2500$.

In order to determine the neighboring points from the set $\bar{A}_3$ we use the following tolerance:

$$\varepsilon = \frac{10^{-4} f_1^{min}}{l}$$

where f_1^{min} is the optimal value of the cluster function f_k when $k = 1$ and l is the number of clusters. If the distance between two points in $\bar{A}_3$ is less than ε then we remove one of them (any) and keep another one.

Since the proposed algorithm is an incremental we do not consider data points which are very close to previous cluster centers as candidates to be starting points. In order to do so we apply a scheme already discussed in [15].

We apply the Quasi-Newton method with BFGS update to solve smooth clustering problems (4.34), (4.35), (4.36), (4.37), (4.38) and (4.39).

In general, Algorithm 3 can find only stationary points of the clustering problem (4.9). However, the use of the special procedure to generate good starting points allows us to find either global or near global solutions to this problem which will be confirmed by the computational results presented in the next section.

4.9 Computational Results: Evaluation of the Incremental Algorithm

We tested the proposed algorithm using a number of real-world data sets. Numerical experiments were carried out on a PC with Processor Intel(R) Core(TM) i5-3470S CPU 2.90 GHz and RAM 8 GB. Algorithm 3 was implemented in Lahey Fortran 95. 12 data sets were used in numerical experiments. The brief description of these data sets is given in Table 4.1. The more detailed description can be found in [6, 49].

We consider three different versions of Algorithm 3. These versions differ from each other on optimization algorithms used to solve both the auxiliary clustering and clustering problems.

1. Incremental Nonsmooth k-means algorithm: INKA—in this algorithm both the auxiliary clustering and clustering problems are solved using the k-means algorithm.
2. Incremental Nonsmooth Optimization Clustering algorithm: INCA—in this algorithm both the auxiliary clustering and clustering problems are solved using the discrete gradient method of nonsmooth optimization.
3. Incremental Smooth Optimization Clustering algorithm: ISCA—in this algorithm both the auxiliary clustering and clustering problems are solved using their smooth approximations.

We compute up to 10 clusters in small data sets (German towns, Bavaria postal 1, Bavaria postal 2 and Iris Plant), 20 clusters in medium size data sets (Heart Disease, Breast Cancer, TSPLIB1060 and Image segmentation) and 25 clusters in large data sets (TSPLIB3038, Page Blocks, D15112 and Pla85900). Results with different similarity measures are presented separately. In all tables we use the following notation:

Table 4.1 The brief description of data sets

Data sets	Number of instances	Number of attributes
German towns	59	2
Bavaria postal 1	89	3
Bavaria postal 2	89	4
Fisher's Iris Plant	150	4
Heart Disease	297	13
Breast Cancer	683	9
TSPLIB1060	1060	2
Image Segmentation	2310	19
TSPLIB3038	3038	2
Page Blocks	5473	10
D15112	15112	2
Pla85900	85900	2

- k is the number of clusters;
- f_{best} is the best known value of the cluster function (4.10) (multiplied by m) with the corresponding number of clusters;
- E is the error in %;
- N is the number of the distance function evaluations for the computation of the corresponding number of clusters;
- t is the CPU time.

The error E is computed as

$$E = \frac{(\bar{f} - f_{best}^k)}{f_{best}^k} \cdot 100$$

where $\bar{f}$ is the optimal value of the function f_k given by (4.10) obtained by an algorithm and f_{best}^k is the best known value of f_k for given k.

4.9.1 Results for the Similarity Measure Based on the Squared L_2-Norm

Results for small data sets are given in Table 4.2. These results demonstrate that all three algorithms are efficient for finding global or near global solutions to the clustering problem in small data sets. Furthermore, optimization based clustering algorithms: INCA and ISCA are more accurate than the INKA algorithm. On the other hand, the INKA requires significantly less computational effort (both distance function evaluations and CPU time) than other two algorithms. Although the INCA algorithm is slightly more accurate than the ISCA algorithm however the former requires more computational effort than the latter.

Table 4.2 Results with the similarity measure based on the L_2-norm

k	f_{best}	INKA			INCA			ISCA		
		E	N	t	E	N	t	E	N	t
		German town								
	$\times 10^3$		$\times 10^4$			$\times 10^6$			$\times 10^5$	
2	121.4257	0.00	1.84	0.00	0.00	0.43	0.00	0.00	0.26	0.00
3	77.0086	0.00	4.21	0.02	0.00	0.94	0.00	0.00	0.53	0.02
4	49.6006	0.24	5.72	0.02	0.00	1.18	0.00	0.00	0.97	0.02
5	38.7160	0.00	8.75	0.02	0.00	2.34	0.01	0.00	1.35	0.03
6	30.5354	0.00	11.02	0.02	0.00	3.38	0.01	0.00	1.79	0.03
7	24.4326	0.08	12.51	0.02	0.00	4.13	0.03	0.00	2.23	0.03
8	21.4830	0.00	15.76	0.02	0.00	7.39	0.04	0.00	3.00	0.05
9	18.5504	2.13	18.99	0.02	0.00	10.54	0.06	0.00	3.66	0.06
10	16.3080	1.81	23.37	0.02	0.00	13.44	0.09	0.00	4.36	0.08

(continued)

Table 4.2 (continued)

k	f_{best}	INKA			INCA			ISCA		
		E	N	t	E	N	t	E	N	t
		Bavaria postal 1								
	$\times 10^6$	$\times 10^4$			$\times 10^6$			$\times 10^6$		
2	602547.2132	0.00	1.20	0.00	0.00	0.28	0.00	0.00	0.09	0.00
3	294506.5545	0.00	3.34	0.00	0.00	0.76	0.01	0.00	0.21	0.00
4	104474.6551	0.00	5.57	0.00	0.00	1.27	0.01	0.00	0.51	0.00
5	59761.5150	0.00	8.36	0.00	0.00	2.27	0.01	0.00	0.88	0.02
6	35908.5267	0.00	11.53	0.00	0.00	3.96	0.03	0.00	1.21	0.03
7	21983.1969	0.61	16.45	0.00	0.00	5.24	0.04	0.00	1.62	0.03
8	13385.4046	0.00	20.67	0.02	0.00	6.95	0.06	0.00	2.12	0.03
9	8423.7401	0.00	22.62	0.02	0.00	9.40	0.07	0.00	2.53	0.03
10	6446.4738	0.00	28.21	0.02	0.00	13.26	0.10	0.00	3.36	0.06
		Bavaria postal 2								
	$\times 10^6$	$\times 10^4$			$\times 10^6$			$\times 10^6$		
2	48631.2853	0.00	1.38	0.00	0.00	0.45	0.00	7.32	0.04	0.00
3	17398.7515	0.00	3.19	0.00	0.00	1.70	0.03	0.00	0.11	0.02
4	7559.0849	0.00	6.05	0.00	0.00	2.86	0.05	0.00	0.35	0.02
5	5342.8659	0.00	9.65	0.00	0.00	5.16	0.06	0.00	0.61	0.02
6	3187.5747	0.00	13.41	0.00	0.00	6.68	0.08	0.00	0.80	0.02
7	2215.9024	0.00	17.99	0.00	0.00	11.36	0.12	0.00	1.61	0.03
8	1704.5235	0.18	23.28	0.02	0.00	19.30	0.19	0.00	2.35	0.06
9	1401.0666	1.07	27.70	0.02	0.00	32.97	0.37	0.00	3.32	0.09
10	1181.0406	0.00	31.91	0.02	0.00	54.10	0.56	0.00	3.96	0.12
		Iris Plant								
	$\times 10^0$	$\times 10^4$			$\times 10^6$			$\times 10^6$		
2	152.3480	0.00	0.45	0.02	0.00	0.45	0.01	0.00	0.22	0.00
3	78.8510	0.00	1.29	0.02	0.00	1.36	0.01	0.00	0.55	0.00
4	57.2280	0.05	2.69	0.02	0.00	3.83	0.04	0.00	2.09	0.03
5	46.4460	0.06	3.89	0.03	0.00	6.93	0.07	0.00	3.96	0.06
6	39.0400	0.07	5.36	0.03	0.00	12.77	0.15	0.00	5.30	0.08
7	34.2980	0.01	7.74	0.05	0.00	19.24	0.21	0.00	6.88	0.11
8	29.9890	0.25	8.92	0.06	0.00	22.88	0.26	0.00	8.76	0.16
9	27.7860	0.28	12.25	0.08	0.00	52.04	0.62	0.90	15.64	0.31
10	25.8340	0.07	16.30	0.09	0.00	84.94	1.12	0.91	21.86	0.47

Table 4.3 Results with the similarity measure based on the L_2-norm (cont.)

k	f_{best}	INKA			INCA			ISCA		
		E	N	t	E	N	t	E	N	t
		Heart disease								
	$\times 10^4$		$\times 10^6$			$\times 10^8$			$\times 10^7$	
2	59.8899	0.00	0.41	0.05	0.00	0.04	0.08	0.00	0.08	0.03
5	32.7543	0.00	3.01	0.27	0.00	0.14	0.34	0.13	0.70	0.23
10	20.0521	0.00	15.26	0.94	0.00	1.86	3.62	0.01	5.87	8.46
15	14.7085	0.00	26.51	1.47	0.00	4.13	7.47	0.03	9.47	24.23
20	11.6834	1.02	37.16	1.94	0.39	7.50	14.63	0.00	12.22	45.80
		Breast cancer								
	$\times 10^4$		$\times 10^7$			$\times 10^8$			$\times 10^7$	
2	1.9323	0.00	0.10	0.09	0.00	0.03	0.04	0.00	0.23	0.05
5	1.3705	0.00	0.86	0.58	0.00	0.86	1.27	0.00	1.73	0.61
10	1.0194	0.00	2.00	1.06	0.00	2.55	3.51	0.00	8.01	2.14
15	0.8710	0.00	3.71	1.72	0.00	7.14	9.17	0.00	19.78	9.08
20	0.7677	0.00	5.82	2.48	0.00	15.07	18.29	0.00	37.10	27.36
		TSPLIB1060								
	$\times 10^6$		$\times 10^7$			$\times 10^8$			$\times 10^8$	
2	9831.9499	0.00	0.14	0.06	0.00	0.05	0.04	0.00	0.04	0.05
5	3791.0203	0.01	1.00	0.34	0.00	0.25	0.18	0.00	0.16	0.17
10	1754.8752	0.22	4.04	1.14	0.00	1.17	0.76	0.00	0.63	0.70
15	1121.9175	0.00	8.22	2.08	0.00	3.04	1.84	0.02	1.08	1.25
20	792.5267	0.14	13.79	3.19	0.00	7.48	4.46	0.03	2.27	2.70
		Image segmentation								
	$\times 10^5$		$\times 10^7$			$\times 10^8$			$\times 10^8$	
2	356.0580	0.00	0.64	0.52	0.00	0.16	0.44	0.00	0.08	0.27
5	171.4291	0.00	3.24	2.30	0.00	1.65	4.37	0.00	0.55	1.84
10	97.9546	0.64	9.22	6.00	0.00	12.41	28.42	1.15	1.93	10.26
15	65.5542	0.47	15.76	9.48	0.00	26.61	58.53	1.88	3.83	32.34
20	51.3621	0.89	25.55	14.41	0.12	45.71	98.84	0.00	6.55	89.31

Table 4.3 contains results for medium size data sets. These results show that all three algorithms are efficient for finding global or near global solutions to the clustering problem in medium size data sets. The INCA algorithm is most accurate and overall the INKA is more accurate than the ISCA algorithm. The INKA algorithm requires less distance function evaluations and CPU time than other two algorithms. The INCA algorithm uses more distance function evaluations and CPU time than the ISCA algorithm in Heart Disease and Breast Cancer data sets, however the former algorithm requires less computational effort than the latter algorithm in other two data sets.

Table 4.4 Results with the similarity measure based on the L_2-norm (cont.)

k	f_{best}	INKA			INCA			ISCA		
		E	N	t	E	N	t	E	N	t
		TSPLIB3038								
	$\times 10^6$		$\times 10^8$			$\times 10^8$			$\times 10^8$	
2	3168.8047	0.00	0.11	0.56	0.00	0.20	0.20	0.00	0.26	0.25
5	1198.1959	0.00	0.57	2.50	0.00	1.00	0.87	0.00	0.95	0.98
10	560.2569	0.57	1.53	4.92	0.00	2.81	2.23	0.00	2.47	2.64
15	356.0483	0.00	3.09	8.55	0.00	5.42	4.07	0.00	4.71	5.16
20	267.1626	0.20	4.69	11.89	0.00	11.70	7.87	0.11	7.99	8.86
25	214.4985	0.23	8.86	20.95	0.00	26.11	16.33	0.01	12.49	14.01
		Page blocks								
	$\times 10^6$		$\times 10^9$			$\times 10^9$			$\times 10^9$	
2	57936.8499	0.00	0.01	1.88	0.00	0.05	0.93	0.00	0.04	0.89
5	13218.3776	0.00	0.07	8.09	0.00	0.22	4.05	0.19	0.17	3.76
10	4532.9960	0.00	0.12	12.08	0.00	0.76	12.27	1.04	0.46	10.44
15	2493.6468	0.00	0.19	16.45	0.06	1.72	25.20	0.31	0.89	21.01
20	1720.3970	0.00	0.25	20.75	0.00	2.98	41.58	0.00	1.45	37.52
25	1210.2370	0.00	0.32	25.28	2.23	4.28	58.29	0.00	2.17	63.31
		D15112								
	$\times 10^8$		$\times 10^9$			$\times 10^9$			$\times 10^9$	
2	3684.0305	0.00	0.23	3.23	0.00	0.47	5.52	0.00	0.42	5.15
5	1327.0655	0.00	0.92	10.12	0.00	1.89	20.47	0.00	1.52	18.41
10	644.9025	1.41	2.09	17.38	0.00	4.73	47.64	0.00	3.44	41.50
15	431.3720	0.26	3.28	24.34	0.00	10.99	99.05	0.00	5.74	68.23
20	321.7737	0.01	4.52	31.11	0.00	19.72	165.97	0.00	8.65	101.82
25	253.0012	0.51	5.76	37.53	0.00	41.29	317.07	0.00	11.96	139.53
		Pla85900								
	$\times 10^{10}$		$\times 10^{11}$			$\times 10^{11}$			$\times 10^{11}$	
2	374908.4065	1.44	0.07	143.70	0.00	0.13	132.21	0.00	0.12	123.96
5	133971.8844	2.78	0.30	609.91	0.00	0.39	471.00	0.00	0.37	452.96
10	68294.1490	0.00	0.67	1358.81	0.00	0.83	1028.34	0.00	0.79	1011.17
15	46029.3761	0.17	1.04	1938.13	0.00	1.29	1608.17	0.00	1.24	1596.67
20	34985.9729	0.03	1.42	2382.82	0.00	1.78	2198.69	0.00	1.71	2210.91
25	28275.5162	0.77	1.80	2788.37	0.00	2.30	2808.12	0.00	2.21	2863.18

Table 4.4 presents results for large data sets. One can see that the accuracy of all three algorithms are similar for large data sets used in this chapter. The INKA algorithm requires least number of the distance function evaluations and CPU time among three algorithms. The INCA requires more computational effort than the ISCA algorithm in all data sets except Pla85900 data set where results are mixed.

4.9.2 Results for the Similarity Measure Based on the L_1-Norm

In this subsection we present clustering results with the similarity measure defined by the L_1-norm. Here we use only the INCA and ISCA algorithms since the INKA algorithm is not applicable with the L_1-norm.

Results for small data sets are given in Table 4.5. These results clearly demonstrate that the INCA algorithm is more accurate than the ISCA algorithm in all data sets, however in Iris Plant data set results are similar. On the other side the INCA algorithm requires significantly more distance function evaluations and CPU time than the ISCA algorithm.

Results for medium size data sets presented in Table 4.6 demonstrate that the INCA algorithm is more accurate than the ISCA algorithm in all data sets. The ISCA algorithm requires less distance function evaluations than the INCA algorithm however the former uses more CPU time than the latter algorithm.

Table 4.7 contains results with large data sets. These results show that the INCA algorithm is more accurate than the ISCA algorithm in TSPLIB3038 and Pla85900 data sets. In Page Blocks and D15112 data sets both algorithms are equally successful. Again one can see that the ISCA algorithm requires less distance function evaluations than the INCA algorithm. However results for CPU time are mixed. Overall, results presented in this subsection demonstrate that the nonsmooth optimization based INCA algorithm is more accurate, however it requires more computational effort than the smoothing technique based ISCA algorithm.

4.9.3 Results for the Similarity Measure Based on the L_∞-Norm

In this subsection we present clustering results with the INCA algorithm using the similarity measure defined by the L_∞-norm. Results with small data sets are presented in Table 4.8, with medium size data sets in Table 4.9 and with large data sets in Table 4.10. These results show that computational effort required by the INCA algorithm in these data sets is reasonable.

4.9.4 Dependence of Number of Distance Function Evaluations and CPU Time on Number of Clusters

In Figs. 4.3, 4.4 and 4.5 we present dependence of both the number of distance calculations and CPU time on the number of clusters for the INCA algorithm using the similarity measures based on the L_1, L_2 and L_∞ norms, respectively. We use two data sets with the largest number of data points, namely, D15112 and Pla85900.

Table 4.5 Results with the similarity measure based on the L_1-norm

k	f_{best}	INCA			ISCA		
		E	N	t	E	N	t
		German town					
	$\times 10^3$	$\times 10^6$			$\times 10^6$		
2	3.0740	0.00	0.82	0.00	0.00	0.31	0.01
3	2.4450	0.00	1.46	0.00	0.00	0.24	0.01
4	1.8870	0.00	2.93	0.01	0.00	0.19	0.01
5	1.6460	0.00	5.40	0.03	0.30	0.17	0.03
6	1.5060	0.00	8.34	0.04	0.60	0.15	0.03
7	1.3780	0.00	12.22	0.07	0.00	0.14	0.08
8	1.2500	0.00	15.15	0.09	1.44	0.13	0.08
9	1.1390	0.00	17.83	0.10	0.09	0.11	0.09
10	1.0440	0.00	21.60	0.14	0.38	0.10	0.11
		Bavaria postal 1					
	$\times 10^6$	$\times 10^7$			$\times 10^7$		
2	4.0249	0.00	0.42	0.03	3.21	0.32	0.09
3	2.8284	0.00	0.59	0.04	2.51	0.39	0.11
4	2.1982	0.00	1.41	0.10	0.22	0.52	0.14
5	1.7208	0.00	2.67	0.20	3.54	0.67	0.19
6	1.3003	0.00	3.29	0.24	8.72	1.13	0.31
7	1.0704	0.00	3.55	0.26	3.25	1.48	0.42
8	0.8510	0.00	3.58	0.28	2.21	1.53	0.44
9	0.7220	0.00	5.11	0.39	1.02	2.38	0.75
10	0.6037	0.00	6.02	0.45	0.33	3.40	1.09
		Bavaria postal 2					
	$\times 10^6$	$\times 10^6$			$\times 10^6$		
2	1.8600	0.00	2.36	0.03	2.38	0.52	0.03
3	1.2607	0.00	3.06	0.03	1.82	0.63	0.03
4	0.9633	0.00	5.96	0.06	1.29	0.86	0.03
5	0.7872	0.00	16.44	0.14	6.91	1.35	0.05
6	0.6736	0.00	22.76	0.20	3.64	1.79	0.08
7	0.5659	0.00	27.38	0.24	3.34	1.92	0.08
8	0.5097	0.00	36.97	0.32	1.90	2.68	0.11
9	0.4688	0.00	46.44	0.40	3.67	3.20	0.12
10	0.4340	0.00	55.42	0.48	4.63	4.75	0.23
		Iris Plant					
	$\times 10^2$	$\times 10^7$			$\times 10^7$		
2	2.1670	0.00	0.08	0.01	0.00	0.41	0.08
3	1.5920	0.00	0.93	0.09	0.00	1.64	0.28
4	1.3650	0.00	1.91	0.20	0.07	1.93	0.33
5	1.2460	0.00	3.65	0.40	0.40	2.69	0.44
6	1.1530	0.00	5.10	0.54	0.10	3.77	0.61
7	1.0620	0.00	6.38	0.70	0.01	4.06	0.65
8	1.0010	0.00	8.91	1.01	0.05	4.41	0.72
9	0.9540	0.00	13.62	1.65	0.00	5.91	0.98
10	0.9070	0.00	20.36	2.66	0.00	6.78	1.14

Table 4.6 Results with the similarity measure based on the L_1-norm (cont.)

k	f_{best}	INCA			ISCA		
		E	N	t	E	N	t
		Heart disease					
	$\times 10^4$		$\times 10^8$			$\times 10^8$	
2	2.0435	0.00	1.17	4.68	0.02	0.05	0.25
5	1.5760	0.00	5.97	27.78	0.06	1.22	6.21
10	1.3006	0.00	11.19	45.50	2.11	2.23	14.09
15	1.1158	0.00	17.17	61.65	0.00	9.97	114.00
20	1.0190	0.00	19.44	66.87	0.10	12.92	176.48
		Breast cancer					
	$\times 10^4$		$\times 10^8$			$\times 10^8$	
2	0.6401	0.00	0.57	1.04	0.02	0.13	0.50
5	0.5032	0.00	4.45	8.84	0.56	1.27	3.90
10	0.4244	0.00	10.57	21.16	1.96	1.60	5.04
15	0.3858	0.00	25.44	46.28	0.08	5.64	21.25
20	0.3583	0.00	35.96	62.29	0.11	6.28	23.52
		TSPLIB1060					
	$\times 10^6$		$\times 10^8$			$\times 10^8$	
2	3.8645	0.00	0.13	0.09	0.74	0.23	0.31
5	2.3096	0.00	1.83	1.02	0.62	0.90	1.08
10	1.5628	0.00	10.46	5.75	0.40	2.75	3.48
15	1.1983	0.00	28.89	16.33	0.73	5.21	6.75
20	1.0157	0.00	70.94	40.76	0.42	16.97	19.48
		Image segmentation					
	$\times 10^5$		$\times 10^9$			$\times 10^9$	
2	5.1916	0.00	0.04	1.35	0.01	0.14	12.28
5	3.3996	0.00	0.55	16.62	0.00	0.52	39.64
10	2.5663	0.00	8.39	221.88	0.00	3.72	289.71
15	2.1795	0.00	17.68	440.93	1.40	8.84	801.91
20	1.9411	0.00	40.85	968.79	0.05	21.48	2566.79

These graphs demonstrate that both the number of distance calculations and CPU time increases almost linearly for Pla85900 data set and for all similarity measures. This dependence is also quite close to linear one for D15112 data set. Since the INCA algorithm includes an optimization solver, it is not possible to estimate its complexity. However, results show that the number of distance calculations depends at most polynomially on the number of clusters.

Table 4.7 Results with the similarity measure based on the L_1-norm (cont.)

k	f_{best}	INCA			ISCA		
		E	N	t	E	N	t
		TSPLIB3038					
	$\times 10^6$		$\times 10^9$			$\times 10^9$	
2	3.7308	0.00	0.04	0.28	0.02	0.02	0.23
3	3.0056	0.00	0.10	0.68	0.00	0.03	0.44
5	2.2551	0.00	0.24	1.59	0.06	0.08	0.97
10	1.5502	0.00	1.70	9.95	0.14	0.99	10.33
15	1.2298	0.00	4.71	27.11	0.27	2.14	22.03
20	1.0597	0.00	9.97	56.89	0.64	4.19	42.37
25	0.9441	0.00	14.38	81.97	0.18	7.60	75.88
		Page blocks					
	$\times 10^6$		$\times 10^9$			$\times 10^9$	
2	8.4141	0.00	0.08	1.73	0.00	0.10	3.37
5	4.8821	0.00	1.18	21.43	0.00	0.45	14.88
10	3.1704	0.00	9.17	149.35	0.00	1.83	63.09
15	2.5682	0.00	12.60	197.94	0.00	6.77	241.91
20	2.2127	0.00	27.10	391.28	0.00	10.16	366.84
25	1.9737	0.00	49.83	689.86	0.00	16.43	623.74
		D15112					
	$\times 10^8$		$\times 10^9$			$\times 10^9$	
2	0.886	0.00	0.40	4.66	0.00	0.27	3.87
5	0.4998	0.00	1.40	16.08	0.00	1.15	16.08
10	0.3617	0.00	3.50	37.17	0.00	2.78	38.13
15	0.293	0.00	6.84	66.64	0.03	4.74	64.38
20	0.2501	0.00	11.86	106.82	0.00	8.62	112.29
25	0.2243	0.00	16.49	143.66	0.00	11.37	147.12
		Pla85900					
	$\times 10^{10}$		$\times 10^{10}$			$\times 10^{10}$	
2	2.0656	0.00	1.33	123.67	0.21	0.82	105.75
5	1.2569	0.00	4.38	455.53	1.84	3.12	412.28
10	0.898	0.00	9.59	1006.40	0.95	7.01	930.83
15	0.7333	0.00	14.98	1576.03	1.38	11.00	1451.20
20	0.6374	0.00	20.89	2171.28	0.91	15.10	1991.88
25	0.5693	0.00	26.67	2748.98	1.69	19.50	2557.04

Table 4.8 Results with the similarity measure based on the L_∞-norm

L_∞-norm	$N(\times 10^6)$					
k	INCA					
	f_{best}	N	t	f_{best}	N	t
	German town			Bavaria postal 1		
	$\times 10^3$			$\times 10^6$		
2	2.5573	0.64	0.00	3.9981	1.35	0.01
3	1.8719	1.22	0.00	2.7916	2.02	0.03
4	1.5069	1.78	0.01	2.1713	3.24	0.04
5	1.2476	2.15	0.01	1.7496	9.46	0.09
6	1.1316	5.03	0.03	1.3178	17.30	0.17
7	1.0415	8.81	0.06	1.0997	23.23	0.23
8	0.9595	11.54	0.09	0.9081	31.94	0.31
9	0.8875	14.62	0.10	0.7703	41.43	0.40
10	0.8196	18.32	0.14	0.6486	49.63	0.46
	Bavaria postal 2			Iris Plant		
	$\times 10^6$			$\times 10^2$		
2	0.9646	3.73	0.06	1.3651	3.87	0.06
3	0.6765	4.30	0.06	0.9656	4.53	0.06
4	0.5373	6.52	0.09	0.7922	16.52	0.23
5	0.4592	20.63	0.26	0.6750	18.70	0.24
6	0.3877	35.69	0.45	0.6197	47.87	0.65
7	0.3454	58.20	0.73	0.5650	67.40	0.92
8	0.3132	80.16	1.02	0.5187	106.01	1.46
9	0.2870	89.40	1.15	0.4953	177.23	2.51
10	0.2619	101.53	1.32	0.4729	251.74	3.60

4.9.5 Results for Purity in Data Sets with Class Labels

In order to determine which similarity measure provides better approximation of a data set one can use the notion of the *cluster purity* in situations where instances are already labeled. In this case we can compare the clusters with the "true" class labels. The purity P_i of the cluster $i,\ i = 1, \dots, k$ is defined as follows [28]:

$$P_i = \frac{1}{n_i} \max_{j=1,\dots,l} n_i^j.$$

In this expression n_i is the number of points in the cluster $i,\ i = 1, \dots, k$, n_i^j is the number of points in the cluster i that belong to the true class j and l is the number of true classes. The *total purity* P_t for the whole data set A is computed as:

Table 4.9 Results with the similarity measure based on the L_∞-norm (cont.)

L_∞-norm	$N(\times 10^8)$					
k	INCA					
	f_{best}	N	t	f_{best}	N	t
	Heart disease			Breast cancer		
	$\times 10^4$			$\times 10^4$		
2	1.0265	0.99	3.15	0.1927	0.49	1.46
3	0.8763	2.13	6.84	0.1737	0.65	1.95
5	0.7267	3.95	12.63	0.1521	1.74	5.36
7	0.6455	6.41	20.20	0.1402	3.15	9.76
10	0.5631	9.93	30.98	0.1294	8.91	27.08
12	0.5198	12.10	37.56	0.1247	15.20	45.83
15	0.4753	14.81	45.83	0.1186	21.31	64.13
18	0.4400	19.12	58.98	0.1138	29.16	87.46
20	0.4207	22.25	68.42	0.1108	34.43	103.08
	TSPLIB1060			Image segmentation		
	$\times 10^6$			$\times 10^5$		
2	2.9215	0.14	0.09	1.5901	5.14	21.68
3	2.3911	0.42	0.29	1.3954	14.03	57.34
5	1.7843	0.93	0.65	1.1563	41.75	164.11
7	1.4847	1.69	1.23	1.0269	70.93	272.14
10	1.1779	3.44	2.52	0.8852	261.00	959.59
12	1.0559	5.63	4.07	0.7998	366.10	1334.30
15	0.9299	11.21	8.19	0.7380	586.25	2110.36
18	0.8409	19.97	14.72	0.6861	831.64	2975.59
20	0.7957	43.37	32.47	0.6586	933.84	3334.88

$$P_t = \frac{1}{m} \sum_{i=1}^{k} \max_{j=1,\dots,l} n_i^j .$$

Among all data sets used in our experiments only four contain class label: Heart Disease, Breast Cancer, Image Segmentation and Page Blocks. Results obtained using the INCA algorithm are presented in Table 4.11. In this table k stands for the number of clusters and P_t for the total purity. The number of clusters where the purity achieves its maximum is presented in bold. Results show that the use of the similarity measure based on the L_1-norm provides better approximation than the use of similarity measures based on other two norms. Results for L_2 and L_∞ norms are similar in the sense of both the number of clusters and accuracy. The exception is the Image Segmentation data sets where the similarity measure based on the squared L_2-norm produced significantly better accuracy.

Table 4.10 Results with the similarity measure based on the L_∞-norm (cont.)

L_∞-norm	$N(\times 10^9)$					
k	INCA					
	f_{best}	N	t	f_{best}	N	t
	TSPLIB3038			Page blocks		
	$\times 10^6$			$\times 10^6$		
2	2.8326	0.03	0.26	5.1353	0.06	1.56
3	2.1424	0.06	0.56	4.1871	0.13	3.33
5	1.6562	0.13	1.17	2.9973	0.60	14.05
10	1.1544	0.56	4.50	1.6485	1.92	42.44
15	0.9114	1.60	12.41	1.2580	4.82	101.83
20	0.7750	3.51	26.94	1.0943	10.29	212.56
25	0.6924	7.74	58.70	0.9667	13.98	286.55
	D15112			Pla85900		
	$\times 10^8$			$\times 10^{10}$		
2	0.7018	0.35	5.42	1.5803	11.53	124.16
3	0.5178	0.64	10.12	1.2250	20.74	235.26
5	0.3952	1.27	19.76	0.9184	37.25	442.32
10	0.2663	3.01	45.05	0.6523	78.26	971.83
15	0.2148	5.17	74.30	0.5455	122.83	1507.79
20	0.1859	7.93	109.57	0.4653	172.67	2087.35
25	0.1669	14.18	179.71	0.4155	224.15	2714.76

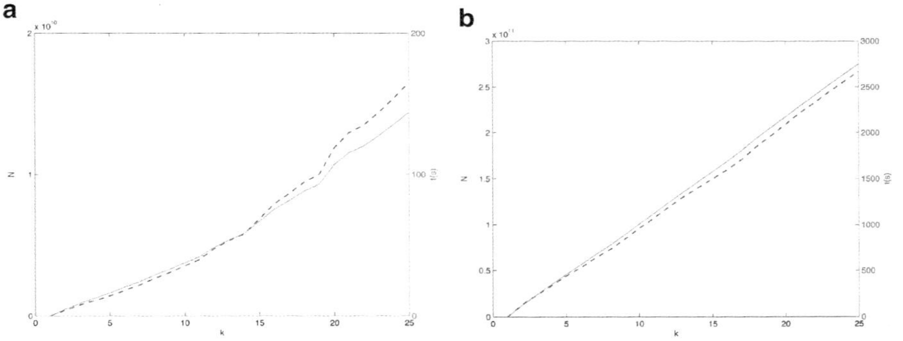

Fig. 4.3 Results for the similarity measure based on the L_1-norm. (**a**) D15112. (**b**) Pla85900

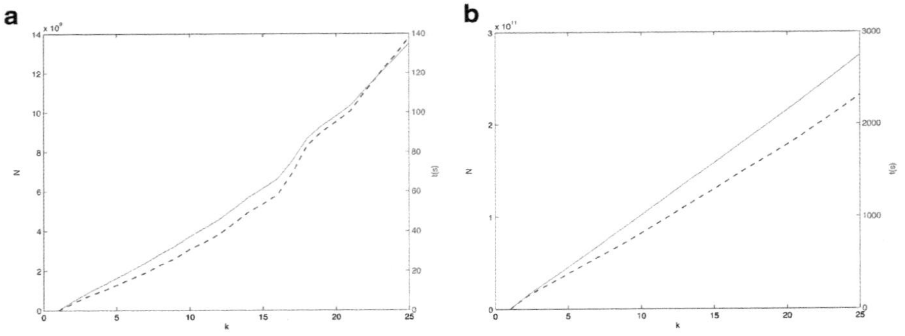

Fig. 4.4 Results for the similarity measure based on the squared L_2-norm. (**a**) D15112. (**b**) Pla85900

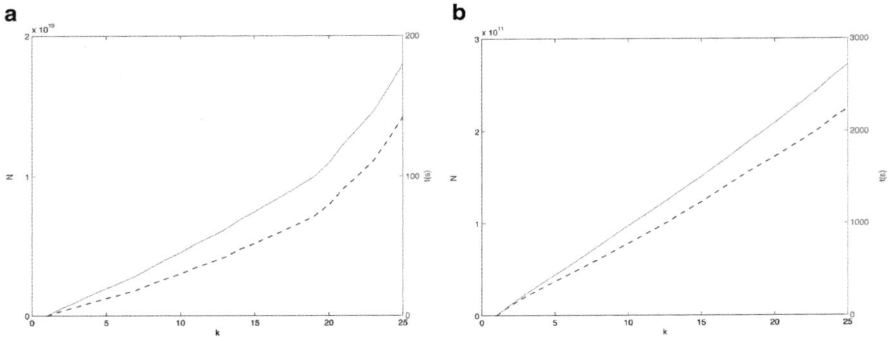

Fig. 4.5 Results for the similarity measure based on the L_∞-norm. (**a**) D15112. (**b**) Pla85900

4.9.6 Visualization of Results

We use Voronoi diagrams to visualize results obtained by the INCA algorithm for different similarity measures. In order to draw these diagrams we used the software available from [50]. Figures 4.6, 4.7, 4.8 present Voronoi diagrams for three data sets: German town, TSPLIB1060 and TSPLIB3038 data sets. In all data sets these diagrams are illustrated for five clusters. One can see that cluster structures for different similarity measures are different in all data sets, however the distributions of cluster centers for different norms are similar in TSPLIB1060 data set. These results confirm that the use of different similarity measures allows one to explore different cluster structures in data sets.

Table 4.11 Cluster purity for different similarity measures

L_1		L_2		L_∞	
k	P_t	k	P_t	k	P_t
Heart Disease					
2	59.2593	2	57.5758	2	57.5758
5	65.6566	5	62.6263	5	61.6162
10	68.0135	10	63.2997	10	65.6566
15	69.3603	**14**	**68.0135**	**13**	**68.6869**
18	**71.3805**	15	67.0034	15	68.3502
20	70.7071	20	68.0135	20	67.6768
Breast Cancer					
2	94.4363	2	96.0469	**2**	**97.0717**
5	96.3397	**3**	**97.0717**	3	96.7789
6	**97.6574**	5	96.3397	5	96.9253
10	96.4861	10	96.9253	10	96.6325
15	97.0717	15	96.9253	15	96.3397
20	96.9253	20	96.9253	20	96.3397
Image Segmentation					
2	28.5714	2	28.5714	2	20.8225
5	60.9091	5	43.8095	5	40.0866
10	74.2857	10	61.342	10	46.1472
15	77.4026	15	71.6883	15	50.4329
19	77.7489	19	74.7619	19	51.6883
20	**78.0087**	**20**	**75.671**	**20**	**54.2424**
Page Blocks					
2	89.768	2	89.8593	2	89.768
5	89.9872	5	89.9141	5	89.8776
10	90.0603	10	90.0968	10	90.0238
15	90.1517	15	90.1517	15	90.0786
20	90.1517	20	90.1882	20	90.1151
23	**90.8642**	**23**	**90.2065**	**24**	**90.1334**
25	90.8642	25	90.2065	25	90.1334

4.10 Computational Results: Comparison with Other Clustering Algorithms

In this section we compare the proposed three algorithms with other clustering algorithms. Comparison of algorithms with different similarity measures are presented separately.

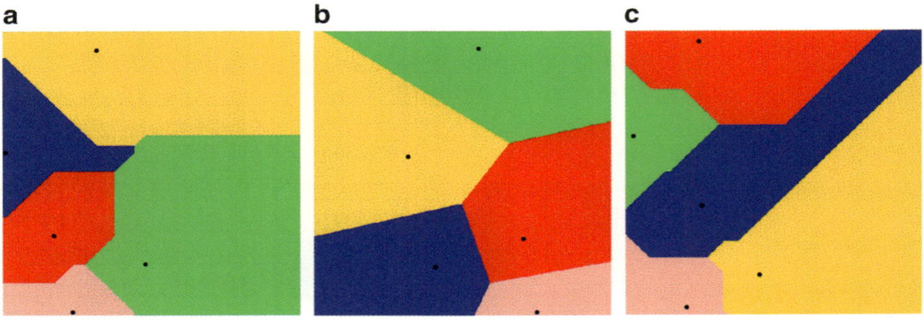

Fig. 4.6 Visualization of clusters for German towns data set. (**a**) L_1-norm. (**b**) L_2-norm. (**c**) L_∞-norm

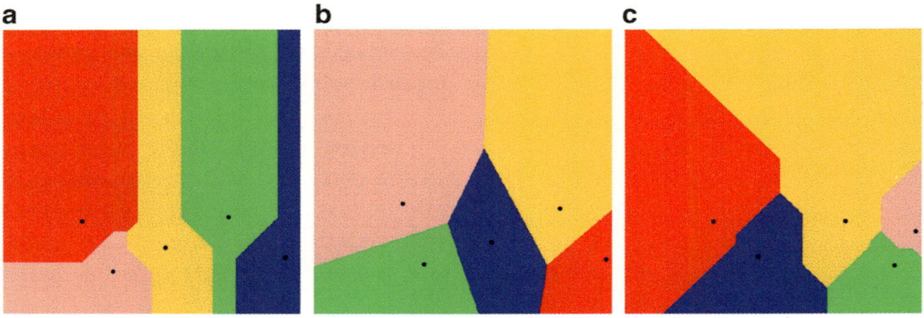

Fig. 4.7 Visualization of clusters for TSPLIB1060 data set. (**a**) L_1-norm. (**b**) L_2-norm. (**c**) L_∞-norm

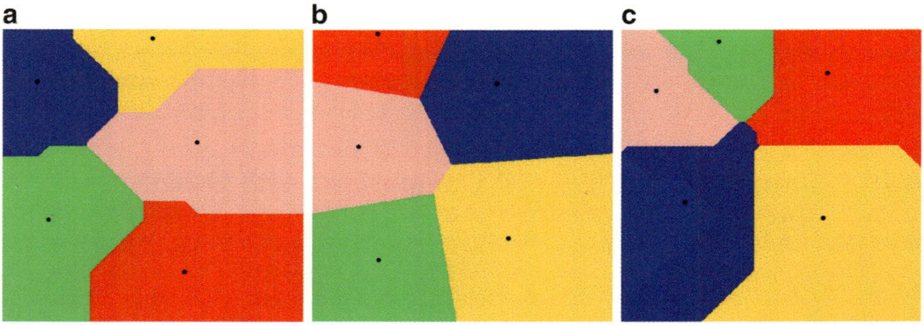

Fig. 4.8 Visualization of clusters for TSPLIB3038 data set. (**a**) L_1-norm. (**b**) L_2-norm. (**c**) L_∞-norm

4.10.1 Comparison of Algorithms Using the Similarity Measure Based on the Squared L_2-Norm

In this subsection we use the following algorithms for comparison with the INCA algorithm:

1. *Lloyd algorithm.* This clustering algorithm is the version of the k-means algorithm. It was introduced in [43].
2. *Forgy algorithm.* This algorithm is a simple alternative of least-squares algorithm [31].
3. *MacQueen algorithm.* This clustering algorithm introduced in [44] is similar to the Forgy's algorithm. The difference is in the last stage where the MacQueen algorithm moves the center points to the mean of their Voronoi set.
4. *Hartigan algorithm.* This algorithm was introduced in [37].
5. *K-means++ algorithm.* This algorithm is the version of the k-means algorithm and was introduced in [3]. It uses a special procedure for initialization of cluster centers.
6. *X-means algorithm.* This algorithm is an improvement of the original k-means algorithm [47]. It uses a special procedure for initialization of cluster centers.
7. *Global k-means algorithm (GKM).* This is an incremental algorithm [42].
8. *Modified Global k-means algorithm (MGKM).* This is the modified version of the global k-means algorithm [9].

All algorithms listed above, except the GKM and MGKM algorithms, use randomly generated starting points for cluster centers. In order to have a fair comparison with the INCA algorithm we use large number of starting points in these algorithms so that the CPU time used by them is almost the same that used by the INCA algorithm. More specifically, we used 500 starting points in all algorithms and we chose the best solution and compared it to that of found by the INCA algorithm. Results are presented in Tables 4.12, 4.13, 4.14. In these tables we include the error of a solution found by an algorithm.

Results for small data sets are presented in Table 4.12. These results show that the Hartigan, k-means++ and INCA algorithms are most accurate among all algorithms. The error by the GKM and MGKM algorithms are not large in comparison with other algorithms.

Table 4.13 contains results with medium size data sets. Again we can see that the Hartigan, k-means++ and INCA algorithms are among most accurate in these data sets. The GKM and MGKM algorithms also show good performance in most data sets.

Results for large data sets presented in Table 4.14 demonstrates that overall the INCA algorithm is most successful in these data sets. Forgy, Lloyd, McQueen, Hartigan, X-means and k-means++ algorithms failed to solve the clustering problem in Page Blocks data set.

Table 4.12 Comparison of algorithms using the similarity measure based on the L_2-norm

k	Forgy	Lloyd	McQueen	Hartigan	X-means	K-means++	GKM	MGKM	INCA
	German town								
2	0.00	0.00	0.00	0.00	8.76	0.00	0.00	0.00	0.00
3	0.00	0.00	0.00	0.00	11.63	0.00	1.45	1.45	0.00
4	0.00	0.00	0.00	0.00	15.35	0.00	0.72	0.72	0.00
5	0.00	0.00	0.00	0.00	11.44	0.00	0.00	0.00	0.00
6	0.00	0.00	0.00	0.00	27.26	0.00	0.00	0.27	0.00
7	0.00	0.00	0.00	0.00	31.00	0.00	0.09	0.00	0.00
8	0.00	0.00	0.00	0.00	46.65	0.00	1.33	1.23	0.00
9	0.00	0.00	0.00	0.00	51.81	0.68	2.13	4.41	0.00
10	0.00	0.00	0.28	0.00	80.32	0.00	1.79	1.51	0.00
	Bavaria postal 1								
2	0.00	0.00	0.00	0.00	242.83	0.00	7.75	0.00	0.00
3	0.00	0.00	0.00	0.00	579.00	0.00	0.00	0.00	0.00
4	0.00	0.00	0.00	0.00	312.15	0.00	0.00	0.00	0.00
5	0.00	0.00	0.00	0.00	763.62	0.00	0.00	0.00	0.00
6	27.65	0.00	0.00	27.65	1223.59	0.00	0.00	0.00	0.00
7	0.61	0.61	1.32	0.00	2010.27	0.00	1.50	1.50	0.00
8	0.00	0.00	0.00	0.00	3296.56	0.00	0.00	0.00	0.00
9	0.00	15.43	0.00	0.00	5312.35	0.00	0.00	0.00	0.00
10	0.00	0.00	0.00	0.00	6811.38	0.00	0.00	0.00	0.00
	Bavaria postal 2								
2	0.00	0.00	0.00	0.00	0.00	0.00	7.32	7.32	7.32
3	0.00	0.00	0.00	0.00	110.55	0.00	0.00	0.00	0.00
4	0.00	0.00	0.00	0.00	317.87	0.00	0.00	0.00	0.00
5	1.14	1.14	0.00	0.00	482.46	0.00	1.86	1.86	0.00
6	39.04	39.02	26.72	0.00	868.32	0.00	1.21	1.21	0.00
7	0.04	0.04	0.04	0.00	1260.32	0.00	0.55	0.55	0.04
8	4.68	12.57	6.06	0.00	1668.73	0.00	0.73	0.73	0.00
9	11.90	1.34	1.34	0.00	2046.36	0.14	0.14	0.14	0.00
10	3.91	11.64	5.54	0.00	2422.10	0.22	1.00	1.00	0.00
	Iris Plant								
2	0.00	0.00	0.00	0.00	1.71	0.00	0.00	0.00	0.00
3	0.00	0.00	0.00	0.00	8.92	0.00	0.01	0.01	0.00
4	0.00	0.00	0.00	0.00	37.30	0.00	0.05	0.05	0.05
5	0.00	0.00	0.00	0.00	11.76	0.00	0.54	0.54	0.00
6	0.00	0.00	0.00	0.00	19.81	0.00	1.44	1.44	0.00
7	0.00	0.00	0.00	0.00	30.49	0.00	3.17	3.17	0.00
8	0.00	0.00	0.00	0.00	47.80	0.00	1.71	1.71	0.00
9	0.00	0.00	0.01	0.00	56.56	0.00	2.85	2.85	0.00
10	0.20	0.41	0.00	0.00	60.86	0.06	3.55	3.55	0.00

Table 4.13 Comparison of algorithms using the similarity measure based on the L_2-norm (cont.)

k	Forgy	Lloyd	McQueen	Hartigan	X-means	k-means++	GKM	MGKM	INCA
	Heart disease								
2	0.00	0.00	0.00	0.00	75.01	0.00	0.00	0.00	0.00
5	0.00	0.00	0.00	0.00	216.02	0.00	0.58	0.58	0.00
10	0.05	0.06	0.14	0.00	389.14	0.09	2.61	1.33	0.03
15	0.28	0.97	0.61	0.00	531.38	1.17	0.96	1.13	0.16
20	2.90	2.65	0.92	0.00	667.67	1.77	2.18	1.35	0.41
25	3.33	5.44	5.56	0.31	781.51	4.33	3.77	0.75	0.00
30	5.45	4.32	5.66	0.03	845.62	4.68	3.76	1.38	0.00
	Breast cancer								
2	0.00	0.00	0.00	0.00	0.00	0.00	0.00	0.00	0.00
5	0.00	0.00	0.00	0.00	2.97	0.00	2.28	0.62	0.02
10	0.10	0.03	0.04	0.00	9.96	0.14	0.25	0.20	0.04
15	0.87	0.84	0.65	0.00	11.87	1.07	1.55	1.49	0.73
20	2.88	2.68	2.59	0.00	23.75	1.92	4.08	1.84	0.82
25	2.10	2.28	3.93	0.50	20.23	3.38	5.04	1.19	0.00
30	4.46	3.68	4.07	0.00	20.98	3.23	4.67	1.06	0.46
	TSPLIB1060								
2	0.00	0.00	0.00	0.00	149.58	0.00	0.00	0.00	0.00
5	0.00	0.00	0.00	0.00	24.48	0.00	0.01	0.01	0.01
10	0.00	0.00	0.00	0.00	46.36	0.00	0.23	0.04	0.03
20	0.05	0.04	0.03	0.00	45.94	0.39	0.39	0.39	0.08
25	0.11	0.26	0.00	0.08	68.25	0.09	1.81	1.81	0.00
30	0.13	0.00	0.48	0.23	73.92	1.47	2.82	2.83	0.75
40	1.57	0.93	1.89	0.00	76.85	2.39	4.00	2.82	0.27
50	3.77	3.75	3.69	0.61	88.96	4.49	2.07	1.50	0.00
	Image Segmentation								
2	0.00	0.00	0.00	0.00	6.65	0.00	0.00	0.00	0.00
5	0.00	0.00	0.00	0.00	54.85	0.00	0.00	0.00	0.00
10	1.53	1.16	0.00	0.63	106.25	0.00	1.76	1.76	1.75
20	12.08	0.18	8.87	2.38	126.52	0.77	0.00	1.40	27.72
25	11.48	15.35	17.29	4.09	173.86	4.18	0.05	0.00	23.39
30	14.65	12.37	17.01	9.35	202.64	5.75	0.07	0.06	0.00
40	14.96	12.26	17.65	15.37	260.99	6.99	1.08	1.07	0.00
50	19.96	18.87	22.55	18.82	323.81	8.13	2.22	2.21	0.00

Table 4.14 Comparison of algorithms using the similarity measure based on the L_2-norm (cont.)

k	Forgy	Lloyd	McQueen	Hartigan	X-means	k-means++	GKM	MGKM	INCA
	TSPLIB3038								
2	0.00	0.00	0.00	0.00	30.17	0.00	0.00	0.00	0.00
5	0.00	0.00	0.00	0.00	1.77	0.00	0.00	0.00	0.00
10	0.00	0.00	0.00	0.00	12.43	0.00	2.78	0.58	0.57
20	0.02	0.03	0.02	0.00	10.11	0.14	2.00	0.48	0.14
25	0.05	0.02	0.03	0.00	8.20	0.22	0.75	0.23	0.56
30	0.05	0.03	0.08	0.00	13.22	0.45	1.44	1.04	0.82
40	0.00	0.76	0.66	0.00	20.78	1.26	1.49	1.72	1.01
50	0.34	0.41	0.51	0.33	19.00	1.15	0.68	0.20	0.00
	Page blocks								
5	38.99	38.99	38.99	38.99	933.67	0.00	0.00	0.00	0.00
10	208.96	46.37	216.19	38.71	2787.71	5.20	1.53	0.73	0.00
20	247.53	248.08	678.67	200.22	2209.30	36.81	0.64	0.00	1.02
30	497.05	482.87	496.96	417.76	4644.91	31.56	2.16	1.40	0.00
50	1118.04	1110.42	1064.06	943.63	7602.65	69.65	0.11	0.07	0.00
70	1883.18	1905.86	1911.44	1708.40	13742.13	97.83	0.45	0.15	0.00
80	1572.73	1794.92	1428.82	1408.72	16659.75	86.06	1.97	0.51	0.00
100	2069.75	2415.00	2074.10	2045.52	20261.88	107.94	1.03	1.14	0.00
	D15112								
5	0.00	0.00	0.00	0.00	4.91	0.00	0.00	0.00	0.00
10	0.00	0.00	0.00	0.00	1.51	0.00	1.41	1.41	0.62
20	0.00	0.00	0.00	0.00	4.44	0.00	0.25	0.25	0.24
30	0.00	0.01	0.00	0.00	8.55	0.00	0.07	0.10	0.06
50	0.19	0.22	0.31	0.23	5.80	0.00	0.00	0.00	0.08
70	0.51	0.34	0.49	0.28	6.23	0.87	0.96	0.95	0.00
80	0.70	0.97	0.55	0.40	10.85	0.81	0.28	0.27	0.00
100	1.52	1.28	1.34	1.08	10.48	0.82	0.13	0.13	0.00
	Pla85900								
5	0.00	0.00	0.00	0.00	0.09	0.00	0.00	0.00	0.00
10	0.00	0.00	0.00	0.00	0.53	0.00	0.00	0.00	0.00
20	0.00	0.00	0.00	0.00	1.06	0.00	0.31	0.31	0.54
30	0.00	0.00	0.00	0.00	3.13	0.00	0.13	0.13	0.01
50	0.14	0.14	0.00	0.07	0.99	0.04	0.35	0.35	0.70
70	0.18	0.32	0.10	0.00	2.83	0.25	0.24	0.72	1.23
80	0.29	0.00	0.04	0.11	2.78	0.02	0.93	0.87	1.15
100	0.29	0.33	0.13	0.00	3.48	0.16	0.30	0.70	0.68

Table 4.15 Comparison of algorithms using the similarity measure based on the L_1-norm

k	k-means	X-means	INCA	k-means	X-means	INCA
	German Towns			Bavaria postal 1		
2	4.23	2.68	0.00	18.58	44.21	0.00
3	16.11	4.09	0.00	53.31	34.44	0.00
4	1.64	29.36	0.00	98.27	48.73	0.00
5	6.38	14.27	0.00	146.01	102.81	0.00
6	10.23	11.24	0.00	223.08	149.68	0.00
7	14.15	18.49	0.00	284.29	148.85	0.00
8	10.16	19.12	0.00	384.27	188.39	0.00
9	10.80	26.51	0.00	441.99	338.77	0.00
10	8.14	33.78	0.00	491.15	404.42	0.00
	Bavaria postal 1			Iris Plant		
2	1.44	13.31	0.00	0.28	2.05	0.00
3	35.65	15.29	0.00	0.06	2.07	0.00
4	57.83	27.51	0.00	0.51	12.42	0.00
5	90.57	14.30	0.00	2.97	5.34	0.00
6	119.23	10.23	0.00	6.76	7.15	0.00
7	151.72	20.40	0.00	8.00	11.85	0.00
8	178.83	32.68	0.00	9.19	17.12	0.00
9	201.37	39.58	0.00	9.12	21.59	0.00
10	128.65	42.21	0.00	12.35	23.72	0.00

4.10.2 Comparison of Algorithms Using the Similarity Measure Based on the L_1-Norm

In this subsection we present the comparison of the INCA algorithm with two clustering algorithms: the k-means and the X-means algorithms. All algorithms use the similarity measure defined by the L_1-norm. Results are presented in Tables 4.15, 4.16, 4.17. We include the error of each algorithm for given number of clusters at these tables.

Table 4.15 contains results on small data sets. These results clearly demonstrate the superiority of the INCA algorithm over other two algorithms in these data sets. Moreover, its superiority over the k-means and X-means algorithms in Bavaria Postal 1 data set is significant.

Results for medium size data sets are given in Table 4.16. Again we can see that the INCA algorithm is much more superior than other two algorithms in these data sets. In most cases results obtained by the k-means and X-means algorithms deteriorate as the number of clusters increases.

Table 4.16 Comparison of algorithms using the similarity measure based on the L_1-norm (cont.)

k	k-means	X-means	INCA	k-means	X-means	INCA
	Heart disease			Breast cancer		
2	18.31	18.63	0.00	0.05	14.46	0.00
3	26.60	30.37	0.00	5.93	15.35	0.00
5	44.06	49.02	0.00	5.03	18.16	0.00
7	52.37	59.50	0.00	6.28	20.73	0.00
10	69.75	73.73	0.00	8.15	18.87	0.00
12	77.73	84.08	0.00	11.61	19.82	0.00
15	92.18	97.24	0.00	9.18	19.20	0.00
18	101.43	108.08	0.00	8.66	21.68	0.00
20	105.50	112.45	0.00	9.24	23.27	0.00
	TSPLIB1060			Image segmentation		
2	37.69	0.17	0.00	1.68	8.78	0.00
3	17.61	18.37	0.00	16.66	5.62	0.00
5	7.30	13.17	0.00	0.13	7.59	0.00
7	6.86	9.35	0.00	2.10	17.72	0.00
10	11.56	11.79	0.00	9.42	21.76	0.00
12	12.58	9.13	0.00	12.80	25.18	0.00
15	13.13	14.00	0.00	18.87	30.73	0.00
18	14.14	9.44	0.00	20.21	33.01	0.00
20	9.17	12.54	0.00	25.86	28.25	0.00

Table 4.17 presents results for large data sets. One can see that the INCA algorithm achieves significantly better solutions than other two algorithms in most cases, except the case $k = 3$ for Pla85900 data set. The k-means and X-means algorithms are quite efficient in TSPLIB3038, D15112 and Pla85900 data sets. All these data sets have two attributes. This means that the k-means and X-means algorithms are efficient when the number of attributes is very small (2 or 3).

4.10.3 Comparison of Algorithms Using the Similarity Measure Based on the L_∞-Norm

In this subsection we present the comparison of the INCA algorithm with two clustering algorithms: the k-means and the X-means algorithms. All algorithms use the similarity measure defined by the L_∞-norm. Results are presented in Tables 4.18, 4.19, 4.20. In these tables the error of each algorithm for given number of clusters is included.

Table 4.17 Comparison of algorithms using the similarity measure based on the L_1-norm (cont.)

k	k-means	X-means	INCA	k-means	X-means	INCA
	TSPLIB3038			Page blocks		
2	4.91	0.27	0.00	27.42	29.10	0.00
3	2.05	0.62	0.00	33.09	58.19	0.00
5	1.86	4.22	0.00	83.22	111.94	0.00
10	3.24	6.99	0.00	142.63	151.02	0.00
15	2.28	1.95	0.00	164.98	158.37	0.00
20	2.93	3.40	0.00	202.92	141.11	0.00
25	2.07	4.45	0.00	238.26	148.37	0.00
	D15112			Pla85900		
2	1.16	0.64	0.00	0.51	0.46	0.00
3	8.45	3.52	0.00	0.00	0.13	0.25
5	0.44	1.05	0.00	0.21	0.37	0.00
10	1.31	0.73	0.00	0.15	0.88	0.00
15	1.03	1.91	0.00	0.27	1.28	0.00
20	2.65	3.27	0.00	0.94	0.85	0.00
25	1.61	5.10	0.00	2.01	2.12	0.00

Table 4.18 Comparison of algorithms using the similarity measure based on the L_∞-norm

k	k-means	X-means	INCA	k-means	X-means	INCA
	German Towns			Bavaria postal 1		
2	0.00	24.00	5.41	21.62	44.43	0.00
3	0.00	4.69	1.57	69.89	111.24	0.00
4	0.47	0.00	3.41	106.20	63.50	0.00
5	7.21	2.72	0.00	137.93	96.54	0.00
6	5.03	10.58	0.00	225.35	158.52	0.00
7	2.11	14.55	0.00	285.78	207.73	0.00
8	15.37	15.67	0.00	356.89	271.70	0.00
9	10.31	17.74	0.00	413.54	327.02	0.00
10	6.09	25.57	0.00	338.41	409.24	0.00
	Bavaria postal 1			Iris Plant		
2	404.07	498.62	0.00	1.08	0.00	39.10
3	601.05	771.69	0.00	0.00	2.45	25.24
4	733.30	560.73	0.00	0.00	2.64	12.37
5	806.53	648.82	0.00	4.94	0.00	4.17
6	1005.88	778.72	0.00	2.47	11.94	0.00
7	1128.26	879.76	0.00	2.74	5.85	0.00
8	1224.71	977.71	0.00	9.41	13.52	0.00
9	1278.33	1046.10	0.00	11.75	17.35	0.00
10	985.72	1161.13	0.00	14.61	15.78	0.00

Table 4.19 Comparison of algorithms using the similarity measure based on the L_∞-norm (cont.)

k	k-means	X-means	INCA	k-means	X-means	INCA
	Heart disease			Breast cancer		
2	29.51	30.08	0.00	39.34	0.32	0.00
3	50.94	51.55	0.00	12.09	5.70	0.00
5	81.01	80.94	0.00	25.71	4.84	0.00
7	102.73	101.13	0.00	22.18	8.00	0.00
10	129.34	128.72	0.00	26.43	12.78	0.00
12	148.06	145.52	0.00	23.26	11.66	0.00
15	165.77	166.08	0.00	27.23	13.10	0.00
18	184.73	186.18	0.00	28.08	13.18	0.00
20	195.60	197.98	0.00	29.56	12.67	0.00
	TSPLIB1060			Image segmentation		
2	30.88	45.57	0.00	12.77	11.62	0.00
3	17.93	23.59	0.00	9.52	13.88	0.00
5	16.21	18.04	0.00	24.66	23.08	0.00
7	12.97	11.35	0.00	25.50	20.02	0.00
10	18.52	17.35	0.00	21.36	29.97	0.00
12	15.58	17.15	0.00	34.01	38.56	0.00
15	16.58	13.69	0.00	38.66	34.67	0.00
18	13.34	13.87	0.00	43.64	45.01	0.00
20	19.86	18.54	0.00	49.17	41.86	0.00

Results for small data sets are given in Table 4.18. These results demonstrate the significant superiority of the INCA algorithm over other two algorithms in all data sets, except the cases when the number of clusters is small.

Table 4.19 contains results for medium size data sets. One can see that the INCA algorithm is much more superior than other two algorithms in these data sets. In most cases results obtained by the k-means and X-means algorithms deteriorate as the number of clusters increases.

Results for large data sets are reported in Table 4.20. The INCA algorithm achieves significantly better solutions than other two algorithms in Page Blocks data set. However in other three data sets the k-means and X-means algorithms are more successful than the INCA algorithm. These results confirm that both the k-means and X-means algorithm are efficient when the number of attributes is very small.

Table 4.20 Comparison of algorithms using the similarity measure based on the L_∞-norm (cont.)

k	k-means	X-means	INCA	k-means	X-means	INCA
	TSPLIB3038			Page blocks		
2	1.06	4.19	0.00	8.97	7.10	0.00
3	2.74	1.62	0.00	6.21	30.99	0.00
5	0.41	0.00	1.52	47.22	79.15	0.00
10	0.00	2.16	2.82	137.85	205.06	0.00
15	0.00	0.45	2.61	193.57	293.60	0.00
20	3.43	0.00	2.20	235.09	343.56	0.00
25	0.19	0.60	0.00	272.51	329.67	0.00
	D15112			Pla85900		
2	1.62	0.00	13.17	0.46	0.00	7.80
3	0.40	0.00	2.96	0.07	0.00	6.74
5	7.17	0.00	5.27	0.29	0.00	5.08
10	0.32	0.55	0.00	0.07	0.00	3.52
15	1.63	0.00	0.38	0.00	0.01	5.86
20	0.00	0.76	0.76	0.03	0.00	2.81
25	0.00	1.67	1.43	1.41	0.00	3.17

4.11 Conclusions

In this chapter, we presented different optimization models for the hard clustering problem. We demonstrated that the use of the nonsmooth nonconvex optimization formulation has some advantages over other two formulations. This formulation allows one to easily apply various similarity measures to solve clustering problems. In particular, one can use similarity measures based on the L_1, the squared L_2 and L_∞ norms. Based on this approach algorithms for solving clustering problems are designed. The proposed clustering algorithms are based on the incremental approach and involves a special procedure for finding starting points for cluster centers. Starting points are found by minimizing the so-called auxiliary cluster function. Three different algorithms are introduced to minimize both the auxiliary cluster and cluster functions. These algorithms include the k-means type heuristic algorithm, the discrete gradient method of the nonsmooth optimization and the algorithm based on the smoothing of the cluster functions.

The proposed algorithms are tested on 12 real world data sets using different similarity measures. Results demonstrate that these algorithms are efficient in finding global or near global solutions to clustering problems with different similarity measures. The comparison of the algorithm with the use of the discrete gradient method with a number of other clustering algorithms is also presented. This comparison demonstrate that the proposed algorithm is more accurate than other algorithms in many data sets. In some other data sets the performance of the proposed algorithm, the Hartigan and the k-means++ algorithms are quite similar.

References

1. Aggarwal C, Hinneburg A, Keim D (2001) On the surprising behavior of distance metrics in high dimensional space. In: ICDT '01 Proceedings of the 8th international conference on database theory, pp 420–434
2. Al-Sultan K (1995) A tabu search approach to the clustering problem. Pattern Recogn 28(9):1443–1451
3. Arthur D, Vassilvitskii S (2007) k-means++: the advantages of careful seeding. In: Bansal N, Pruhs K, Stein C (eds) SODA '07 Proceedings of the eighteenth annual ACM-SIAM symposium on discrete algorithms. SIAM, Miami, pp 1027–1035
4. Astorino A, Fuduli A (2007) Nonsmooth optimization techniques for semi-supervised classification. IEEE Trans Pattern Anal Mach Intell 29:2135–2142
5. Astorino A, Fuduli A, Gorgone E (2008) Non-smoothness in classification problems. Optim Methods Software 23:675–688
6. Bache K, Lichman M (2013) UCI machine learning repository. URL http://archive.ics.uci.edu/ml
7. Bagirov AM (1999) Minimization methods for one class of nonsmooth functions and calculation of semi-equilibrium prices. In: Eberhard A, et al (eds) Progress in optimization: contribution from Australasia. Kluwer Academic, Norwell, pp 147–175
8. Bagirov AM (2003) Continuous subdifferential approximations and their applications. J Math Sci 115(5):2567–2609
9. Bagirov AM (2008) Modified global k-means algorithm for minimum sum-of-squares clustering problems. Pattern Recogn 41(10):3192–3199
10. Bagirov AM, Al Nuaimat A, Sultanova N (2013) Hyperbolic smoothing function method for minimax problems. Optimization 62(6):759–782
11. Bagirov AM, Karasozen B, Sezer M (2008) Discrete gradient method: Derivative-free method for nonsmooth optimization. J Optim Theory Appl 137:317–334
12. Bagirov AM, Rubinov AM, Soukhoroukova N, Yearwood J (2003) Unsupervised and supervised data classification via nonsmooth and global optimization. Top 11:1–93
13. Bagirov AM, Rubinov AM, Yearwood J (2002) A Global Optimization Approach to Classification. Optim Eng 3(2):129–155
14. Bagirov AM, Ugon J (2006) Piecewise partially separable functions and a derivative-free algorithm for large scale nonsmooth optimization. J Global Optim 35:163–195
15. Bagirov AM, Ugon J, Webb D (2011) Fast modified global k-means algorithm for incremental cluster construction. Pattern Recogn 44(4):866–876
16. Bagirov AM, Yearwood J (2006) A new nonsmooth optimization algorithm for minimum sum-of-squares clustering problems. Eur J Oper Res 170(2):578–596
17. Ball GH, Hall DJ (1967) A clustering technique for summarizing multivariate data. Behav Sci 12(2):153–155
18. Bock HH (1998) Clustering and neural networks. In: Rizzi A, Vichi M, Bock HH (eds) Advances in data science and classification. Springer, Berlin and Heidelberg, pp 265–277
19. Bradley P, Fayyad U (1998) Refining initial points for k-means clustering. In: Proc. of the 15th int. conf. on machine learning, pp 91–99
20. Brown D, Huntley C (1992) A practical application of simulated annealing to clustering. Pattern Recogn 25(4):401–412
21. Cao F, Liang j, Jiang G (2009) An initialization method for the k-means algorithm using neighborhood model. Comput Math Appl 58(3):474–483
22. Carmichael J, Sneath P (1969) Taxometric maps. Syst Zool 18:402–415
23. Carrizosa E, Romero Morales D (2013) Supervised classification and mathematical optimization. Comput Oper Res 40:150–165
24. Celebi ME, Kingravi H (2012) Deterministic initialization of the k-means algorithm using hierarchical clustering. Int J Pattern Recogn Artif Intell 26(7):1250,018

25. Celebi ME, Kingravi HA, Vela PA (2013) A comparative study of efficient initialization methods for the k-means clustering algorithm. Expert Syst Appl 40:200–210
26. Cheng CH (1995) A branch and bound clustering algorithm. IEEE Trans Syst Man Cybern 25(5):895–898
27. Clarke F (1983) Optimization and nonsmooth analysis. Canadian mathematical society series of monographs and advanced texts. Wiley, New York
28. Dhillon IS, James F, Guan Y (2001) Efficient clustering of very large document collections. In: Grossman R, Kamath C, Kegelmeyer P, Kumar V, Namburu R (eds) Data mining for scientific and engineering applications. Kluwer Academic, Norwell, pp 357–382
29. Diehr G (1985) Evaluation of a branch and bound algorithm for clustering. SIAM J Sci Stat Comput 6(2):268–284
30. Doherty K, Adams R, Davey N (2004) Non-Euclidean norms and data normalisation. In: Proceedings of ESANN, pp 181–186
31. Forgy EW (1965) Cluster analysis of multivariate data: efficiency versus interpretability of classifications. Biometrics 21:768–769
32. Ghorbani M (2005) Maximum entropy-based fuzzy clustering by using l1-norm space. Turk J Math 29:431–438
33. Guha S, Meyerson A, Mishra N, Motwani R, O'Callaghan L (2003) Clustering data streams: Theory and practice. IEEE Trans Knowl Data Eng 15(3):515–528
34. Hansen P, Jaumard B (1997) Cluster analysis and mathematical programming. Math Programm 79(1-3):191–215
35. Hansen P, Mladenovic N, Perez-Britos D (2001) Variable neighborhood decomposition search. J Heuristics 7(4):335–350
36. Hansen P, Ngai E, Cheung B, Mladenovic N (2005) Analysis of global k-means an incremental heuristic for minimum sum-of-squares clustering. J Classification 22(2):287–310
37. Hartigan JA, Wong MA (1979) Algorithm as 136: A k-means clustering algorithm. J Roy Stat Soc C (Appl Stat) 28(1):100–108
38. Jajuga K (1987) A clustering method based on the L_1-norm. Comput Stat Data Anal 5(4): 357–371
39. Jalali A, Srebro N (2012) Clustering using Max-norm Constrained Optimization. CoRR abs/1202.5
40. Kaufman L, Rousseeuw P (1990) Finding groups in data: an introduction to cluster analysis. Wiley series in probability and statistics. Wiley
41. Lai JZC, Huang TJ (2010) Fast global k-means clustering using cluster membership and inequality. Pattern Recogn 43(5):1954–1963
42. Likas A, Vlassis N, Verbeek JJ (2003) The global k-means clustering algorithm. Pattern Recogn 36(2):451–461
43. Lloyd S (1982) Least squares quantization in pcm. IEEE Trans Inform Theory 28(2):129–137
44. MacQueen JB (1967) Some methods for classification and analysis of multivariate observations. In: Cam LML, Neyman J (eds) Proc. of the fifth Berkeley symposium on mathematical statistics and probability, University of California Press, vol 1, pp 281–297
45. du Merle O, Hansen P, Jaumard B, Mladenovic N (1999) An interior point algorithm for minimum sum-of-squares clustering. SIAM J Sci Comput 21(4):1485–1505
46. Ordin B, Bagirov AM (2014) A heuristic algorithm for solving the minimum sum-of-squares clustering problems. J Global Optim URL 10.1007/s10898-014-0171-5
47. Pelleg D, Moore A (2000) X-means: Extending k-means with efficient estimation of the number of clusters. In: Langley P (ed) ICML'00 Proceedings of the seventeenth international conference on machine learning. Morgan Kaufmann, San Francisco, pp 727–734
48. Ramos G, Hatakeyama Y, Dong F, Hirota K (2009) Hyperbox clustering with Ant Colony Optimization (HACO) method and its application to medical risk profile recognition. Appl Soft Comput 9(2):632–640
49. Reinelt G (1991) TSPLIB- a traveling salesman problem library. ORSA J Comput 3(4): 376–384

50. Sedgewick R, Wayne K (2007) Introduction to programming in Java. Addison-Wesley, URL http://introcs.cs.princeton.edu/java/36inheritance/Voronoi.java.html
51. Selim SZ, Al-Sultan K (1991) A simulated annealing algorithm for the clustering problem. Pattern Recogn 24(10):1003–1008
52. Su T, Dy JG (2007) In search of deterministic methods for initializing k-means and gaussian mixture clustering. Intell Data Anal 11(4):319–338
53. Sun L, Xie Y, Song X, Wang J, Yu R (1994) Cluster analysis by simulated annealing. Comput Chem 18(2):103–108
54. Venkateswarlu N, Raju P (1992) Fast isodata clustering algorithms. Pattern Recogn 25(3): 335–342
55. Xavier AE, Oliveira AAFD (2005) Optimal covering of plane domains by circles via hyperbolic smoothing. J Global Optim 31(3):493–504
56. Yang M, Hung W (2006) Alternative fuzzy clustering algorithms with l1-norm and covariance matrix. Adv Concepts Intell Vis 4179:654–665

Chapter 5
Fuzzy Clustering Algorithms and Validity Indices for Distributed Data

L. Vendramin, M.C. Naldi, and R.J.G.B. Campello

Abstract This chapter presents a unified framework to generalize a number of fuzzy clustering algorithms to handle distributed data in an *exact* way, i.e., with no approximation of results with respect to their original centralized versions. The same framework allows the exact distribution of relative validity indices used to evaluate the quality of fuzzy clustering solutions. Complexity analyses for each distributed algorithm and index are reported in terms of space, time, and communication aspects. A general procedure to estimate the number of clusters in a non-centralized fashion using the proposed framework is also described. Such a procedure is directly applicable not only to distributed data, but to parallel data processing scenarios as well. Experimental results illustrate the speedup obtained when running algorithms under the proposed framework in multiple cores of a processor, when compared to their traditional, centralized counterparts running in a single core. Additionally, the quality of the results and amount of data transmitted are assessed and compared among different fuzzy clustering algorithms.

Keywords Clustering • Fuzzy partitions • Validity indices • Distributed data • Parallel computing

5.1 Introduction

Clustering techniques can be broadly divided into three main types [1]: overlapping, partitional, and hierarchical. The two latter are related to each other in that a hierarchical clustering is a nested sequence of partitional clusterings, each of which represents a hard partition of a data set **X** into a different number of mutually disjoint

L. Vendramin • R.J.G.B. Campello
Institute of Mathematics and Computer Sciences, University of São Paulo (USP), Av. Trabalhador São-Carlense, 400 Centro, Caixa Postal: 668, CEP: 13560-970, São Carlos, SP, Brazil
e-mail: vendra@icmc.usp.br; campello@icmc.usp.br

M.C. Naldi (✉)
Federal University of Viçosa (UFV)s, Rodovia BR 354–km 310, Caixa Postal: 22, CEP: 38.810-000, Rio Paranaíba, MG, Brazil
e-mail: murilocn@ufv.br

M.E. Celebi (ed.), *Partitional Clustering Algorithms*,
DOI 10.1007/978-3-319-09259-1_5

subsets (clusters). A *hard* (or *crisp*) partition of N objects $\mathbf{X} = \{\mathbf{x}_1, \ldots, \mathbf{x}_N\}$ into a certain number k of non-null disjoint clusters $\mathbf{C}_i$ is such that $\mathbf{C}_1 \cup \ldots \cup \mathbf{C}_k = \mathbf{X}$, $\mathbf{C}_i \neq \oslash$, $\mathbf{C}_i \cap \mathbf{C}_m = \oslash \; \forall i \neq m$. It can be represented by means of a $k \times N$ partition matrix $\mathbf{U} = [u_{ij}]_{k \times N}$ whose element u_{ij} is either 1 if the jth object belongs to the ith cluster or 0 otherwise. Formally, the space of hard partitions is

$$\mathbf{M}_h = \left\{ \mathbf{U} = [u_{ij}] \in \mathbb{R}^{k \times N} \;\middle|\; u_{ij} \in \{0, 1\} \;\; \forall i, j; \;\; \sum_{i=1}^{k} u_{ij} = 1 \;\; \forall j \right\}. \tag{5.1}$$

Overlapping techniques search for *soft* or *fuzzy* partitions [2]. A soft partition of a data set can be described in a way similar to (5.1). The only difference refers to the condition of mutual disjointness, which is relaxed, i.e. the space of soft partitions of N objects into k overlapping clusters is

$$\mathbf{M}_s = \left\{ \mathbf{U} = [u_{ij}] \in \mathbb{R}^{k \times N} \;\middle|\; u_{ij} \in \{0, 1\} \;\; \forall i, j; \;\; \sum_{i=1}^{k} u_{ij} \geq 1 \;\; \forall j \right\}. \tag{5.2}$$

A *possibilistic* fuzzy partition goes even further by relaxing the constraint that each object either belongs or does not belong to a given cluster in a binary fashion, i.e. the elements of a possibilistic fuzzy partition matrix can take any value within the continuous membership interval [0,1]. Formally, the space of possibilistic fuzzy partitions is [3]:

$$\mathbf{M}_f = \left\{ \mathbf{U} = [u_{ij}] \in \mathbb{R}^{k \times N} \;\middle|\; u_{ij} \in [0, 1] \;\; \forall i, j; \;\; \forall j, \exists i, u_{ij} > 0 \right\}. \tag{5.3}$$

The definition in (5.3) is partially used by *probabilistic* fuzzy partitions, but with a constraint that allows the elements of the partition matrix to be interpreted as probabilities [4, 5]:

$$\mathbf{M}_{fp} = \left\{ \mathbf{U} = [u_{ij}] \in \mathbb{R}^{k \times N} \;\middle|\; u_{ij} \in [0, 1] \;\; \forall i, j; \;\; \sum_{i=1}^{k} u_{ij} = 1 \;\; \forall j \right\}. \tag{5.4}$$

Note that hard partitions can be seen as a special case of fuzzy partitions, in which each object is assigned a membership value 1 to the cluster the object belongs to and 0 to the other clusters. Once a fuzzy partition is available, it is easy to derive a hard partition from it by changing, for each object, its maximum membership value to 1 and the other values to 0. However, when the structure in the data contains overlapping categories, fuzzy partitions provide more precise information.

The literature on fuzzy clustering is extensive and several studies have been carried out with different characteristics and for different purposes during the past years [2, 6–9]. One of the most used fuzzy clustering algorithms is the Fuzzy c-Means (FCM) [5, 10]. FCM finds hyper-spherical clusters and partitions lying in

$\mathbf{M}_{fp}$ as defined in (5.4). Many other fuzzy clustering algorithms have been proposed to find clusters with different shapes and orientations, e.g., Gustafson-Kessel [11], Gath and Geva [2, 12], Fuzzy c-Varieties [13], and Fuzzy c-Elliptotypes [2, 6, 14]. Many others have been developed to deal with partitions lying in $\mathbf{M}_f$ as defined in (5.3), e.g., Possibilistic c-Means [3], Possibilistic Gustafson-Kessel [3], Fuzzy-Possibilistic c-Means [15], and Possibilistic-Fuzzy c-Means [16]. However, these algorithms have been originally conceived to process *centralized* databases, i.e., they assume that all data objects to be clustered can be accessed in a single storage and processing site.

Despite the large number of existing centralized algorithms, distributed environments like the internet, intranets and parallel processing systems have changed many aspects of computing. In this context, physically distributed databases have been increasingly used for knowledge discovery [17–20], giving rise to an area of study called *Distributed Knowledge Discovery in Databases* [21–23]. There is also an increasing tendency to distribute large databases across multiple processing units and then use appropriate data mining techniques to process the data [24], which is so-called *Parallel and Distributed Data Mining* [22, 25–29].

Although there is a conceptual overlap between the areas of Distributed and Parallel Data Mining [30, 31], these terms are usually used to refer to algorithms having different characteristics. "Parallel" often refers to algorithms applied to strongly coupled systems, whereas "distributed" often refers to algorithms applied to loosely coupled systems. In other words, parallel algorithms are often used in multiprocessor architectures in which the data are fully available in a single shared memory, whereas distributed algorithms usually subsume that the data set is not available in a shared memory. Typically, parallel data mining aims to increase the computational performance and distributed data mining is intended to solve problems in which there is no suitable or feasible way to centralize the data set [32]. In this chapter, we develop a framework that is applicable to both parallel and distributed scenarios.

When there are multiple databases distributed across different sites, one might argue that the data could be centralized before performing any computation. However, centralizing distributed data may not be appropriate due to several reasons [26, 29, 33, 34]. Essentially, the large amount of data may incur in high costs to transfer and store, besides probably increasing the time of the mining process. In some cases, data centralization is not possible due to computational limitations, such as working memory capacity or prohibitive processing time. In other cases, there are confidentiality and security issues.

In the realm of Distributed and Parallel Data Mining, there is an area of study called *Distributed Data Clustering* [27, 29, 35], which refers to clustering methods able to find structures in distributed databases. Many different clustering algorithms have been proposed to find structures in parallel and distributed environments [36–43]. Some of them were developed as a generalization of the centralized version of a specific algorithm [36, 37, 40, 44], whereas others run a (centralized) algorithm in each data site independently and then use some kind of ensemble to improve the results in a specific site—using information from the others [45, 46].

Another approach involves collaborative clustering, where the algorithm runs in every data site and iteratively shares information across sites trying to find a global structure [43, 47]. Despite the existence of many algorithms for parallel and distributed data, only a few of them have been developed for fuzzy clustering as a generalization of the original (centralized) ones, e.g., see [48–50]. In other words, just a few fuzzy clustering algorithms have been generalized to deal with parallel and distributed data in an *exact* way.

In our study, we are particularly interested in variants of centralized fuzzy clustering algorithms conceived to handle distributed data in an exact way. Formally, let $\mathbf{X}[ii] = \{\mathbf{x}[ii]_1, \ldots, \mathbf{x}[ii]_{N_{ii}}\}$ be a data set stored at the iith data site and composed of N_{ii} objects, $\mathbf{x}[ii]_j$ $(j = 1, \ldots, N_{ii})$, each of which is described in an n-dimensional attribute space. In addition, consider a scenario where there exists a distributed collection of these data sets, $\mathbf{X}[1], \ldots, \mathbf{X}[P]$. The goal of a distributed clustering algorithm, as an exact generalization of a centralized one, consists of finding a solution that simultaneously represents an existing structure common to all data sites. In other words, it searches for a partition matrix $\mathbf{U}$ with k clusters that represents the memberships of each of the $N = N_1 + N_2 + \ldots + N_P$ objects to the k clusters. The final result ($\mathbf{U}$) is supposed to consist exactly of the same result that would be obtained by the original (centralized) algorithm if such an algorithm could be run in a single data set containing all objects of all data sites ($\mathbf{X} = \mathbf{X}[1] \cup \mathbf{X}[2] \cup \ldots \cup \mathbf{X}[P]$).

Having appropriate algorithms in hand, a fundamental problem concerns the validation of the clustering results. A common approach to quantitatively evaluate a data partition is based on the use of relative validity indices [51–53], which make it possible to compare different partitions in a relative manner [1]. Each candidate partition obtained by a clustering algorithm can be quantitatively evaluated by a relative index and compared to other partitions of the same data set [1, 52, 54], thereby also making it possible to estimate the number of clusters from data. This includes partitions produced by fuzzy clustering algorithms, which can be evaluated by means of fuzzy relative validity indices [2, 54]. However, as such indices have been originally designed to operate when the data are centralized, they cannot be directly used to evaluate partitions obtained by a distributed data clustering procedure. To overcome this limitation, we additionally show in this chapter that the proposed framework for distributed computation of fuzzy clustering algorithms can also be extended to generalize several existing validity indices so that they can be exactly computed in distributed and paralleled scenarios.

The remainder of this chapter is organized as follows. Sections 5.2 and 5.3 review fuzzy clustering algorithms and validity indices originally developed to handle centralized data. In Sects 5.4 and 5.5 such algorithms and indices are generalized by designing a unified framework to compute them in an exact way in distributed scenarios; an original procedure to estimate the number of clusters in distributed data, which makes use of the proposed framework, is also described in Sect. 5.5. The theoretical contributions are compiled in Sect. 5.6, and experimental results are reported in Sect. 5.7. The conclusions and final remarks are given in Sect. 5.8. A number of additional, well-established fuzzy clustering algorithms that can also be generalized within the proposed framework are described in the Appendix.

5.2 Fuzzy Clustering Algorithms

When a partitioning-type fuzzy clustering algorithm is applied to a set of N data objects $\mathbf{X} = \{\mathbf{x}_1, \ldots, \mathbf{x}_N\}$, of which each is composed of n attributes (features), $\mathbf{x}_j = [x_{j1}, \ldots, x_{jn}]^T \in \mathbb{R}^n$, the final result is a fuzzy partition matrix of these data into a certain number k of clusters, such that $\mathbf{U} = [u_{ij}]_{k \times N}$, where $\mathbf{U}$ is a $k \times N$ fuzzy partition matrix whose element u_{ij} represents the membership (belongness or pertinence) of the jth object to the ith fuzzy cluster. Most of the clustering algorithms experimentally compared in this chapter minimize the function

$$J = \sum_{j=1}^{N} \sum_{i=1}^{k} (u_{ij})^m D_{ij}, \tag{5.5}$$

constrained with $u_{ij} \in [0, 1]$ and

$$\sum_{i=1}^{k} u_{ij} = 1, \; 1 \leq j \leq N, \tag{5.6}$$

i.e., $\mathbf{U} = [u_{ij}] \in \mathbf{M}_{fp}$ as defined in (5.4), where $m \in (1, \infty)$ is the fuzzification exponent (usually $m = 2$), D_{ij} is the distance between the jth object and the ith cluster, and J is a measure of intra-cluster dissimilarity. The differences among the algorithms arise in the definition of D_{ij} and the variables considered to minimize (5.5).

The most popular method to solve (5.5) is a simple Picard iteration through the first-order conditions for stationary points of (5.5) [6]. Thus, considering that D_{ij} is constant and u_{ij} are variables, a necessary condition is[1]:

$$u_{ij} = \left(\sum_{c=1}^{k} \left(\frac{D_{ij}}{D_{cj}} \right)^{1/(m-1)} \right)^{-1}, \; 1 \leq i \leq k, \; 1 \leq j \leq N. \tag{5.7}$$

[1]We do not present the mathematical techniques to find the necessary conditions throughout the chapter. They are found by using Lagrangian multipliers and setting the gradient equal to zero. For details see [2, 5] and references in the subsequent sections.

Therefore, considering that u_{ij} is constant, the other necessary conditions depend on the definition of D_{ij} and, for this reason, they are presented in the following along with the different algorithms associated with them.

5.2.1 FCM: Fuzzy c-Means

The widely used FCM algorithm [5, 10] provides a partition matrix $\mathbf{U} = [u_{ij}]_{k \times N} \in \mathbf{M}_{fp}$ by minimizing (5.5) with $(1 \leq i \leq k,\ 1 \leq j \leq N)$

$$D_{ij} = ||\mathbf{x}_j - \mathbf{v}_i||^2_{\mathbf{A}} = (\mathbf{x}_j - \mathbf{v}_i)^T \mathbf{A}(\mathbf{x}_j - \mathbf{v}_i), \tag{5.8}$$

being any squared inner-product distance norm (e.g. squared Euclidean) between the jth object, $\mathbf{x}_j$, and the ith cluster prototype, $\mathbf{v}_i$. The norm-inducing matrix $\mathbf{A}$ (a positive-definite $n \times n$ matrix) in (5.8) defines the shape of the clusters and must be defined a priori by the user. However, in most application scenarios, the user does not know the shape of the clusters present in the data, therefore $\mathbf{A}$ is usually set as the identity matrix $\mathbf{I}_{n \times n}$ and D_{ij} is defined as the squared Euclidean distance.

Holding u_{ij} constant, a necessary condition to minimize (5.5) is $(1 \leq i \leq k)$

$$\mathbf{v}_i = \frac{\sum_{j=1}^{N} (u_{ij})^m \mathbf{x}_j}{\sum_{j=1}^{N} (u_{ij})^m}. \tag{5.9}$$

The complete procedure to minimize (5.5) with D_{ij} given by (5.8) is a simple Picard iteration through (5.7) and (5.9), as given by Algorithm 1.

Some important observations based on FCM (Algorithm 1) that also hold for other algorithms discussed in later sections are:

Algorithm 1 FCM

Require: Data set $\mathbf{X} = \{\mathbf{x}_1, \ldots, \mathbf{x}_N\}$, number of clusters $k \in \{2, \ldots, N-1\}$, fuzzification exponent $m \in (1, \infty)$, initial (random) partition matrix $\mathbf{U} = [u_{ij}] \in \mathbf{M}_{fp}$, maximum number of iterations t_{max} and/or termination tolerance $\epsilon > 0$.

1: $t = 0$.
2: **repeat**
3: Compute the cluster prototypes by using (5.9).
4: Compute the distance between objects and cluster prototypes by using (5.8).
5: Update the partition matrix by using (5.7) (Save the previous partition matrix as $\hat{\mathbf{U}}$ to analyze the convergence).
6: $t = t + 1$
7: **until** $||\hat{\mathbf{U}} - \mathbf{U}|| < \epsilon$ or $t = t_{max}$.
8: **return** $\mathbf{U} = [u_{ij}]_{k \times N}$ and $\mathbf{V} = [\mathbf{v}_i]_{k \times n}$.

---

1. The initial partition matrix ($\mathbf{U}$) can be replaced by the initial cluster prototypes ($\mathbf{V} = [\mathbf{v}_i]_{k \times n} \in \mathbb{R}^{k \times n}$). In this case, step 3 is no longer executed in the first iteration.
2. The convergence criterion $||\hat{\mathbf{U}} - \mathbf{U}||$ (comparison between previous and current partition matrix) can use any norm (e.g. max) and can be replaced by other criteria (e.g., $||\hat{\mathbf{V}} - \mathbf{V}||$—comparison between previous and current cluster prototypes).
3. Step 5 requires that $D_{ij} > 0$ for all $i \in \{1, \ldots, k\}$ and $j \in \{1, \ldots, N\}$. For every j, if $D_{ij} = 0$ for $i \in I \subseteq \{1, \ldots, k\}$, then define u_{ij} in such a way that: a) $u_{ij} = 0$ for $i \notin I$; and b) $\sum_{i \in I} u_{ij} = 1$.

5.2.2 GK: Gustafson Kessel

The ordinary FCM uses the squared Euclidean distance and, therefore, induces hyper-spherical clusters[2]. To identify hyper-ellipsoidal clusters with different spreads and orientations, Gustafson and Kessel [11] generalized (5.5) and (5.8) by using independent covariance matrices for each cluster and considering these matrices as variables to be optimized. In this case, (5.5) is minimized (with respect to u_{ij}, $\mathbf{v}_i$ and $\mathbf{A}_i$) with the distances given by ($1 \leq i \leq k,\ 1 \leq j \leq N$)

$$D_{ij} = ||\mathbf{x}_j - \mathbf{v}_i||^2_{\mathbf{A}_i} = (\mathbf{x}_j - \mathbf{v}_i)^T \mathbf{A}_i (\mathbf{x}_j - \mathbf{v}_i). \tag{5.10}$$

Now, holding u_{ij} and $\mathbf{A}_i$ constant while considering $\mathbf{v}_i$ as variables, the necessary conditions to minimize (5.5) also lead to (5.9). Considering $\mathbf{A}_i$ as variables, since (5.5) becomes linear with respect to $\mathbf{A}_i$, it could be made as small as desired by making $\mathbf{A}_i$ less positive definite. To avoid this behavior, $\mathbf{A}_i$ is constrained in such a way that $\det(\mathbf{A}_i) = \rho_i$. The necessary optimality condition is then given by ($1 \leq i \leq k$)

$$\mathbf{A}_i = (\rho_i \cdot \det(\mathbf{F}_i))^{1/n} (\mathbf{F}_i)^{-1}, \tag{5.11}$$

$$\mathbf{F}_i = \frac{\sum_{j=1}^{N} (u_{ij})^m (\mathbf{x}_j - \mathbf{v}_i)(\mathbf{x}_j - \mathbf{v}_i)^T}{\sum_{j=1}^{N} (u_{ij})^m}. \tag{5.12}$$

The procedure to minimize (5.5) with the distances defined in (5.10) is given by Algorithm 2.

[2] Actually, it can be seen from (5.8) that FCM works with any inner-product distance, which must be fixed a priori by the user.

Algorithm 2 GK

Require: Besides all requirements of Algorithm 1, the cluster volumes, ρ_i.

$t = 0$.
repeat
    Compute the cluster prototypes (means) by using (5.9).
    Compute the covariance matrix for each cluster by using (5.12).
    Compute the distances by using (5.10) and (5.11).
    Update the partition matrix by using (5.7) (Save the previous partition matrix as $\hat{\mathbf{U}}$ to analyze the convergence).
    $t = t + 1$.
until $||\hat{\mathbf{U}} - \mathbf{U}|| < \epsilon$ or $t = t_{max}$.
return $\mathbf{U} = [u_{ij}]_{k \times N}$, $\mathbf{V} = [\mathbf{v}_i]_{k \times n}$ and $\mathbf{F}_i$.

Note that the cluster volumes ρ_i must be fixed a priori by the user. A common approach consists in defining these parameters as $\rho_i = 1 \; \forall i$, which induces the algorithm to find clusters with approximately the same volume.

5.2.3 *Other Fuzzy Clustering Algorithms*

Many other fuzzy clustering algorithms have been proposed in the literature. Like FCM and GK, a number of these algorithms can also be generalized to deal with distributed data by means of the unified framework proposed in this chapter, such as: Gath and Geva (GG) [2, 12], Fuzzy c-Varieties (FCV) [13], Fuzzy c-Elliptotypes (FCE) [14], Possibilistic c-Means (PCM) [3], Possibilistic Gustafson-Kessel (PGK) [3], Fuzzy-Possibilistic c-Means (FPCM) [15], and Possibilistic-Fuzzy c-Means (PFCM) [16]. Their computational time and space complexities are compared in Sect. 5.2.4. For a detailed description of these algorithms, we refer the reader to the Appendix.

5.2.4 *Complexity Analysis: Summary*

The time and space complexity analyses of the studied fuzzy clustering algorithms are summarized in Table 5.1 (see [55] for details). These analyses presume the use of the efficient implementation of the FCM algorithm as described in [56], which can also be extended to its variants. They also presume the use of a general norm-inducing matrix **A** (rather than just the identity matrix) in FCM and related variants. Finally, it is assumed that all data inputs and outputs are stored in main memory (no I/O access to secondary storage devices).

Table 5.1 Complexities for the fuzzy clustering algorithms: t, k, n, and N are the numbers of iterations, clusters, attributes, and objects, respectively.

Algorithm	Time	Space
FCM	$O(tkNn^2)$	$O(Nn + kN + n^2)$
GK	$O(tkn^3 + tkn^2N)$	$O(Nn + kN + kn^2)$
GG	$O(tkn^3 + tkn^2N)$	$O(Nn + kN + kn^2)$
FCV	$O(tkn^3 + tkn^2Nr)$	$O(Nn + kN + kn^2)$
FCE	$O(tkn^3 + tkn^2Nr)$	$O(Nn + kN + kn^2)$
PCM	$O(tkNn^2)$	$O(Nn + kN + n^2)$
PGK	$O(tkn^3 + tkn^2N)$	$O(Nn + kN + kn^2)$
FPCM	$O(tkNn^2)$	$O(Nn + kN + n^2)$
PFCM	$O(tkNn^2)$	$O(Nn + kN + n^2)$

5.3 Clustering Validation

Fuzzy clustering algorithms usually require that the number of clusters, k, be previously defined by the user [2]. This is quite restrictive in practice since k is generally unknown, especially in real-world data involving overlapping clusters and many attributes. A widely used approach to overcome this drawback—described in Sect. 5.3.12—consists in obtaining a set of data partitions with different k and then selecting that particular partition that provides the best result according to a specific quality criterion [2,6]. This section reviews several fuzzy clustering validity indices that can be used for this task. These indices can also be used to compare partitions with the same k, which is also of interest especially because most fuzzy clustering algorithms (including those reviewed in Sect. 5.2 and in the Appendix) may produce different solutions with the same k, depending on their initializations [57,58].

5.3.1 XB: Xie-Beni

The XB index is defined as [59,60]:

$$V_{XB} = \frac{\sum_{i=1}^{k}\sum_{j=1}^{N}(u_{ij})^m ||\mathbf{x}_j - \mathbf{v}_i||^2}{N \min_{l \neq s} ||\mathbf{v}_l - \mathbf{v}_s||^2}. \tag{5.13}$$

The numerator of (5.13) is the total within-cluster distance, which is equivalent to the objective function J in (5.5) with D_{ij} being the squared Euclidean distance. The ratio J/N is the normalized compactness of the fuzzy partition. The minimum squared distance between prototypes in the denominator of (5.13) is a cluster separability measure. Therefore, good partitions are distinguished by low values of V_{XB}, i.e., XB is a minimization index.

5.3.2 FSS: Fuzzy Simplified Silhouette

The FSS index [61] is a fuzzy version of the Simplified Silhouette (for hard partitions) [62] defined as

$$V_{FSS} = \frac{\sum_{j=1}^{N}(u_{pj} - u_{qj})^{\alpha} s_j}{\sum_{j=1}^{N}(u_{pj} - u_{qj})^{\alpha}}, \tag{5.14}$$

where u_{pj} and u_{qj} are the first and second largest elements of the jth column of the fuzzy partition matrix, respectively, $\alpha \geq 0$ is a user-defined weighting coefficient, and s_j is the hard silhouette for object $\mathbf{x}_j$, defined as

$$s_j = \frac{b_{pj} - a_{pj}}{max\{a_{pj}, b_{pj}\}}, \tag{5.15}$$

where a_{pj} is the distance between object $\mathbf{x}_j$ and the nearest prototype $\mathbf{v}_p$ (corresponding to the cluster to which $\mathbf{x}_j$ has the largest membership value) and b_{pj} is the distance between object $\mathbf{x}_j$ and the second nearest cluster prototype. Good partitions are expected to yield larger values of s_j (larger values of b_{pj} and smaller values of a_{pj}), hence a larger value of V_{FSS} in (5.14). Thus, FSS is a maximization index.

5.3.3 K: Kwon

The Kwon (K) index [63] is a modification of XB designed to eliminate its tendency to monotonically decrease when k approaches N and is given by

$$V_K = \frac{\sum_{i=1}^{k}\sum_{j=1}^{N}(u_{ij})^2||\mathbf{x}_j - \mathbf{v}_i||^2 + \frac{1}{k}\sum_{i=1}^{k}||\mathbf{v}_i - \overline{\mathbf{x}}||^2}{\min_{l \neq s}||\mathbf{v}_l - \mathbf{v}_s||^2}. \tag{5.16}$$

where

$$\overline{\mathbf{x}} = \frac{1}{N}\sum_{j=1}^{N}\mathbf{x}_j \tag{5.17}$$

is the grand mean of the data set. The first term of the numerator in (5.16) measures the intra-cluster similarity. The second term of the numerator in (5.16) is a penalty factor intended to counterbalance the decreasing tendency when k approaches N. The denominator in (5.16) measures the inter-cluster distances. Thus, K is a minimization index.

5.3.4 TSS: Tang-Sun-Sun

Similarly to the Kwon index, the TSS minimization index [64] modifies the XB index to eliminate its tendency to monotonically decrease when k approaches to N and it is given by

$$V_{TSS} = \frac{\sum_{i=1}^{k}\sum_{j=1}^{N}(u_{ij})^2||\mathbf{x}_j - \mathbf{v}_i||^2 + \frac{1}{k(k-1)}\sum_{i=1}^{k}\sum_{\substack{l=1\\ l\neq i}}^{k}||\mathbf{v}_i - \mathbf{v}_l||^2}{\min_{l\neq s}||\mathbf{v}_l - \mathbf{v}_s||^2 + 1/k}. \tag{5.18}$$

Again, the first term of the numerator in (5.18) measures the intra-cluster similarity. The second term of the numerator in (5.18) is a penalty factor intended to counterbalance the decreasing tendency when k approaches N. The denominator in (5.18) measures the inter-cluster distances.

5.3.5 FS: Fukuyama-Sugeno

The Fukuyama and Sugeno (FS) [60, 65] minimization index is defined as

$$V_{FS} = \sum_{i=1}^{k}\sum_{j=1}^{N}(u_{ij})^m||\mathbf{x}_j - \mathbf{v}_i||^2 - \sum_{i=1}^{k}\sum_{j=1}^{N}(u_{ij})^m||\mathbf{v}_i - \overline{\mathbf{x}}||^2, \tag{5.19}$$

where $\overline{\mathbf{x}}$ is given by (5.17). The first term in (5.19) is the intra-cluster dissimilarity. The second term in (5.19) ($\sum_{i=1}^{k}\sum_{j=1}^{N}(u_{ij})^m||\mathbf{v}_i - \overline{\mathbf{x}}||^2$) tends to increase as k increases. On the other hand, this second term tends to decrease as the degree of overlap (fuzziness of $\mathbf{U}$) between the clusters increases.

5.3.6 FHV: Fuzzy Hypervolume

The FHV minimization index [12] is based on the concepts of hypervolume and density, and is defined as

$$V_{FHV} = \sum_{i=1}^{k}[\det(\mathbf{F}_i)]^{1/2}, \tag{5.20}$$

where $\mathbf{F}_i$ is given by (5.12)[3]. Since the eigenvalues of $\mathbf{F}_i$ are directly related to the variances of the ith fuzzy cluster, the determinant in (5.20), given by the product of all the eigenvalues of $\mathbf{F}_i$, provides a measure of the dispersion (hypervolume) of the cluster.

5.3.7 APD: Average Partition Density

Based on the same idea of FHV, the APD [12] maximization index is defined as

$$V_{APD} = \frac{1}{k} \sum_{i=1}^{k} \frac{S_i}{[\det(\mathbf{F}_i)]^{1/2}}, \tag{5.21}$$

where

$$S_i = \sum_{\mathbf{x}_j \in \mathbf{R}_i} u_{ij}, \tag{5.22}$$

and $\mathbf{R}_i$ is the set of data objects that are within a pre-specified region around the cluster prototype $\mathbf{v}_i$, i.e., the objects such that $(\mathbf{x}_j - \mathbf{v}_i)^T \mathbf{F}_i^{-1} (\mathbf{x}_j - \mathbf{v}_i) < 1$.

5.3.8 PD: Partition Density

Based on APD, the PD maximization index [12] is defined as

$$V_{PD} = \frac{\sum_{i=1}^{k} S_i}{V_{FHV}}, \tag{5.23}$$

where S_i is given by (5.22) and V_{FHV} is given by (5.20).

5.3.9 SCG

The SCG maximization index [68] is defined as

$$V_{SCG} = \frac{Sep}{GComp}, \tag{5.24}$$

[3] Some authors use $\mathbf{F}_i$ in (5.12) with $m = 1$ [61,66] while others use it with a different value of m [54,67].

where $GComp$ is the global compactness of the partition defined as

$$GComp = \sum_{i=1}^{k} \text{trace}(\mathbf{F}_i), \tag{5.25}$$

with $\mathbf{F}_i$ given by (5.12), and Sep is the separation of fuzzy clusters, defined as

$$Sep = \text{trace}(\mathbf{S}_B). \tag{5.26}$$

$\mathbf{S}_B$ is the between-cluster fuzzy scatter matrix given by

$$\mathbf{S}_B = \sum_{i=1}^{k} \sum_{j=1}^{N} (u_{ij})^m (\mathbf{v}_i - \overline{\mathbf{x}})(\mathbf{v}_i - \overline{\mathbf{x}})^T, \tag{5.27}$$

where $\overline{\mathbf{x}}$ is given by (5.17).

5.3.10 PBMF

The PBMF maximization index [69] is given by

$$V_{PBMF} = \left(\frac{1}{k} \cdot \frac{E_1}{J_m} \cdot D_c \right)^2, \tag{5.28}$$

where $E_1 = \sum_{j=1}^{N} ||\mathbf{x}_j - \overline{\mathbf{x}}||$ is a constant that depends only on the database and is used to avoid small values of V_{PBMF}, $D_c = max_{i,j=1}^{k} ||\mathbf{v}_i - \mathbf{v}_j||$ measures the maximum distance between two cluster prototypes, and $J_m = \sum_{i=1}^{k} \sum_{j=1}^{N} (u_{ij})^m ||\mathbf{x}_j - \mathbf{v}_i||$ measures the intra-cluster similarity.

5.3.11 Complexity Analysis: Summary

The time and space complexity analyses of the studied fuzzy clustering validity indices are summarized in Table 5.2 (see [55] for details).

5.3.12 OMR: Ordered Multiple Runs

In practice, different approaches for determining the most appropriate k can be used [2, 6]. A common approach is based on a general procedure in which the data are

Table 5.2 Time and space complexity for the fuzzy clustering validity indices: k, n, and N are the numbers of clusters, attributes, and objects, respectively

	Time	Space		Time	Space
XB	$O(kNn)$	$O(Nn + kN)$	FHV	$O(kn^3 + kNn^2)$	$O(kn^2)$
FSS	$O(kNn)$	$O(Nn + kn)$	APD	$O(kNn^3)$	$O(kn^2)$
K	$O(kNn)$	$O(Nn + kN)$	PD	$O(kNn^3)$	$O(kn^2)$
TSS	$O(kNn)$	$O(Nn + kN)$	SCG	$O(kn^3 + kNn^2)$	$O(Nn + kN + n^2)$
FS	$O(kNn)$	$O(Nn + kN)$	PBMF	$O(kNn)$	$O(Nn + kN)$

Algorithm 3 OMR

Require: The clustering algorithm to be used, $\mathscr{A}$ (and all parameters of such algorithm), the validity index to be used, $\mathscr{I}$, the minimum number of clusters, k_{min} (usually $k_{min} = 2$), the maximum number of clusters, k_{max}, and the number of runs of $\mathscr{A}$ from random initializations of partition matrices for each k, M.

1: Initialize the best validity value: $\mathscr{V}_{best} = -Inf$ for maximization indices or $\mathscr{V}_{best} = +Inf$ for minimization ones.
2: **for** $k = k_{min}$ to k_{max} **do**
3:     **for** i=1 to M **do**
4:         Execute the clustering algorithm $\mathscr{A}$ with k clusters and obtain the clustering result $\mathscr{R}$.
5:         Evaluate the solution obtained in Step 4 using the validity index $\mathscr{I}$ and obtain the validity value $\mathscr{V}$.
6:         **if** $\mathscr{V}$ is better than $\mathscr{V}_{best}$ ("greater" for maximization indices or "lower" for minimization ones) **then**
7:             Update the best value: $\mathscr{V}_{best} \leftarrow \mathscr{V}$.
8:             Save the current solution as the best one: $\mathscr{R}_{best} \leftarrow \mathscr{R}$.
9:             Save the current number of clusters: $k_{best} \leftarrow k$.
10:         **end if**
11:     **end for**
12: **end for**
13: **return** $\mathscr{V}_{best}$, $\mathscr{R}_{best}$, and k_{best}.

first partitioned into different values of k and, then, a clustering validity index is used to assess the quality of the obtained partitions. In its simplest form, named here *Ordered Multiple Runs* (OMR), the clustering algorithm is run repeatedly for an increasing k [2, 6]. For each value of k, a number of partitions achieved by the clustering algorithm is assessed by means of some validity index (Sect. 5.3), for which the best value is kept for further reference. After doing this for each value of k in a given range, the best obtained partition (according to the validity index) is chosen. This procedure is detailed in Algorithm 3.

It is important to note that the validity index ($\mathscr{I}$) in the OMR procedure must be compatible with the clustering algorithm ($\mathscr{A}$). For example, the (dis)similarity measure used by an index may eventually need to be changed in order to work with different algorithms [61]. Another alternative is simply the use of the own objective function J of the algorithm (with its respective metric) as a validity index to find the best solution [70, 71], e.g., by plotting a curve of such objective function values against the number of clusters and then looking for a knee in the chart [52].

5.4 Distributed Fuzzy Clustering Algorithms

Let $\mathbf{X}[ii] = \{\mathbf{x}[ii]_1, \ldots, \mathbf{x}[ii]_{N_{ii}}\}$ be a data set at the iith data site, composed of N_{ii} objects ($\mathbf{x}[ii]_j$, $j = 1, \ldots, N_{ii}$), each of which is described in an n-dimensional attribute space. In addition, consider that there exists a distributed collection of these data sets, $\mathbf{X}[1], \ldots, \mathbf{X}[P]$. In the present context, the goal of a distributed fuzzy clustering algorithm consists in finding a fuzzy partition matrix $\mathbf{U}$ with k clusters that represents the memberships of each of the $N = N_1 + N_2 + \ldots + N_P$ objects to the k clusters. The final result ($\mathbf{U}$) is supposed to consist exactly of the same result that would be obtained by the original (centralized) algorithm if such an algorithm could be run in a single data set containing all objects of all data sites ($\mathbf{X} = \mathbf{X}[1] \cup \mathbf{X}[2] \cup \ldots \cup \mathbf{X}[P]$).

To circumvent the drawbacks imposed by the distributed scenarios, the data sites should be handled separately. Take, for instance, Fig. 5.1, in which the data are distributed across two data sites (DS1 and DS2). In Fig. 5.1a, the data are centralized and the clustering procedure is performed at a central node (computer). In Fig. 5.1b, the (distributed) clustering procedure is performed over the data distributed across DS1 and DS2, and the same structure found in Fig. 5.1a is found once again without the need of data centralization[4].

The next section presents the distributed/parallel versions of clustering algorithms and validity indices to work with distributed/parallel scenarios.

5.4.1 *DFCM: Distributed Fuzzy c-Means*

The algorithm called here Distributed Fuzzy c-Means (DFCM) is based on the ideas on the parallelization of computations in the FCM algorithm [48–50] and is a formal generalization of FCM to handle distributed data. Note that Algorithm 1 (FCM in Sect. 5.2.1) basically consists of three steps: (1) computation of the prototypes, (2) computation of the distances, and (3) computation of the partition matrix. The prototypes are computed according to (5.9), which needs all data objects. Rewriting (5.9) to handle objects distributed across P data sites we obtain ($1 \leq i \leq k$)

$$\mathbf{v}_i = \frac{\sum_{ii=1}^{P} \sum_{j=1}^{N_{ii}} (u[ii]_{ij})^m \mathbf{x}[ii]_j}{\sum_{ii=1}^{P} \sum_{j=1}^{N_{ii}} (u[ii]_{ij})^m}, \tag{5.29}$$

where N_{ii} is the number of objects in the iith data site, $\mathbf{x}[ii]_j$ is the jth object of the iith data site, and $u[ii]_{ij}$ is the membership of the jth object of the iith data site to

[4]The distributed scenario illustrated in Fig. 5.1b can be intentionally implemented (e.g. in many processors) in order to obtain a better performance, e.g., in a parallel context.

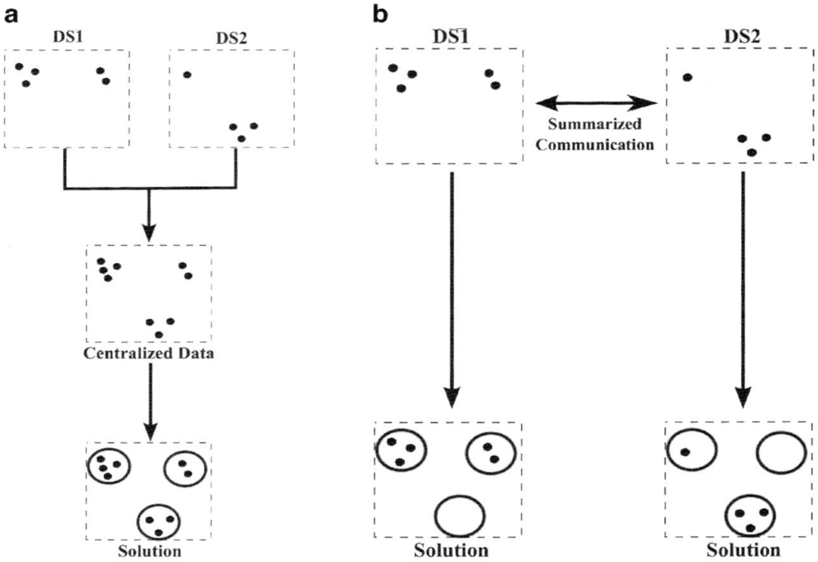

Fig. 5.1 Data clustering (**a**) with and (**b**) without data centralization. (**a**) Solution obtained with data centralization. (**b**) Same solution as in (**a**), but obtained without the need for data centralization

the ith fuzzy cluster. $\mathbf{v}_i$ is the ith fuzzy cluster prototype, referred to here as *global* prototype because it is obtained from all objects of all data sites and it is exactly the same prototype that would be obtained by (5.9) if the objects were centralized.

The computation of the *global* prototypes in (5.29) requires all objects of all data sites. Notice, however, that each data site can compute partial (local) prototypes considering only its objects, as follows ($1 \leq ii \leq P,\ 1 \leq i \leq k$)

$$\mathbf{v}[ii]_i = \frac{\sum_{j=1}^{N_{ii}} (u[ii]_{ij})^m \mathbf{x}[ii]_j}{\sum_{j=1}^{N_{ii}} (u[ii]_{ij})^m}. \tag{5.30}$$

Here, $\mathbf{v}[ii]_i$ is called the ith *local* prototype of the iith data site because it is obtained strictly from the objects of the iith data site. We call the denominator of (5.30) the "sum of memberships" and denote it as ($1 \leq i \leq k$)

$$\mu[ii]_i = \sum_{j=1}^{N_{ii}} (u[ii]_{ij})^m. \tag{5.31}$$

Now, if all sites send their *local* prototypes and sums of memberships to a central data site[5], it is possible to compute the *global* prototypes as

$$\mathbf{v}_i = \frac{\sum_{ii=1}^{P} \mathbf{v}[ii]_i \cdot \mu[ii]_i}{\sum_{ii=1}^{P} \mu[ii]_i}. \tag{5.32}$$

Once this computation has been performed, the result ($\mathbf{v}_i$, $i = 1, \cdots, k$) is communicated from the central data site to all other sites so that every data site has access to the same *global* prototypes.

The next step of FCM consists in computing the distances between objects and prototypes. Note that, having access to the *global* prototypes, every data site can compute the distances between its objects and each cluster prototype as ($1 \leq ii \leq P$, $1 \leq i \leq k$, $1 \leq j \leq N$.)

$$D[ii]_{ij} = ||\mathbf{x}[ii]_j - \mathbf{v}_i||^2_{\mathbf{A}} = (\mathbf{x}[ii]_j - \mathbf{v}_i)^T \mathbf{A} (\mathbf{x}[ii]_j - \mathbf{v}_i). \tag{5.33}$$

This computation does not require any extra communication among the data sites. Once it has been performed, every data site has the distances between the *global* prototypes and its objects.

The final step of FCM consists in updating the partition matrix. Note that the computation of the memberships of a given object $\mathbf{x}_j$ in (5.7) depends solely on the distances from this object to the cluster prototypes. In other words, the distances between the prototypes and other objects (even in the same data site) are not required to compute the membership values of a particular object $\mathbf{x}[ii]_j$. Thus, the partition matrix can be updated as

$$u[ii]_{ij} = \left(\sum_{c=1}^{k} \left(\frac{D[ii]_{ij}}{D[ii]_{cj}} \right)^{1/(m-1)} \right)^{-1}, \tag{5.34}$$

without the need of any extra communications among the data sites.

In summary, only the computation of prototypes requires communications among the data sites. The computation of the distances and the update of the partition matrix do not require any communication at all. The distributed/parallel version of FCM is presented in Algorithm 4 (every data site, including the central one, must execute the same procedure).

---

[5]A central data site is considered here to be one of the data sites, which is chosen to be also responsible for additional processing and redistribution of global information.

Algorithm 4 DFCM—Execution at the iith data site

Require: Local data set $\mathbf{X}[ii] = \{\mathbf{x}[ii]_1, \ldots, \mathbf{x}[ii]_{N_{ii}}\}$, number of clusters $k \in \{2, \ldots, N-1\}$, fuzzification exponent $m \in (1, \infty)$, initial (random) local partition matrix $\mathbf{U}[ii] = [u[ii]_{ij}] \in \mathbf{M}_{fp}$, number of iterations t_{max} and/or termination tolerance $\epsilon > 0$.

1: $t = 0$.
2: **repeat**
3: Compute the local cluster prototypes (means) and sums of memberships using (5.30) and (5.31) (Save the previous prototypes as $\hat{\mathbf{V}}$ to analyze the convergence).
4: **if** this (ii) is the central data site **then**
5: (Wait and) Receive the local prototypes ($\mathbf{v}[jj]_i$) and sums of memberships ($\mu[jj]_i$) from the other data sites.
6: Compute the global prototypes using (5.32).
7: Send the global prototypes ($\mathbf{v}_i$) to all data sites.
8: **else**
9: Send the local prototypes ($\mathbf{v}[ii]_i$) and sums of memberships ($\mu[ii]_i$) to the central data site.
10: (Wait and) Receive the global prototypes ($\mathbf{v}_i$) from the central data site.
11: **end if**
12: Compute the distances between objects and prototypes using (5.33).
13: Update the partition matrix using (5.34).
14: $t = t + 1$.
15: **until** $||\hat{\mathbf{V}} - \mathbf{V}|| < \epsilon$ or $t = t_{max}$.
16: **return** $\mathbf{U}[ii] = [u[ii]_{ij}]_{k \times N_{ii}}$ and $\mathbf{V} = [\mathbf{v}_i]_{k \times n}$.

Some important notes about this algorithm are:

1. DFCM communicates prototypes and the sums of memberships. No information about the original objects is communicated among the sites.
2. When using the same initial conditions and parameters, DFCM produces the same partition FCM would produce if all objects were centralized.
3. The convergence $||\hat{\mathbf{V}} - \mathbf{V}|| < \epsilon$ analyzes the (global) cluster prototypes. It is possible to check the convergence by using the partition matrix as $||\hat{\mathbf{U}}[ii] - \mathbf{U}[ii]|| < \epsilon$. However, by doing so, an one-bit additional communication step for each data site is required to check if $||\hat{\mathbf{U}}[ii] - \mathbf{U}[ii]|| < \epsilon \; \forall ii$.
4. Steps 3 to 11 of Algorithm 4 are related to Step 3 of Algorithm 1, Step 12 of Algorithm 4 is related to Step 4 of Algorithm 1, and Step 13 of Algorithm 4 is related to Step 5 of Algorithm 1.

5.4.2 *Framework for Distributed Data*

All algorithms discussed in Sect. 5.2 can be generalized to deal with distributed data in a way similar to DFCM, i.e., the steps of the original (centralized) algorithm are executed and, when a particular step requires communications among the data sites, the *local* values are computed and sent to the central data site, which computes the *global* values and sends them back to all other data sites.

We propose a unified framework that generalizes the algorithms cited in Sect. 5.2 to deal with distributed data. The framework is based on the execution of every step of the original (centralized) algorithm and, if a communication step is required, then it proceeds as follows:

1. Every data site computes the respective *local* values $V[ii]$.
2. Every data site sends the local values to a central data site.
3. The central data site computes the *global* values as

$$V = \frac{\left[\sum_{ii=1}^{P} \alpha[ii] \cdot (V[ii] - \Delta) \cdot \phi[ii]\right] + \Delta}{\sum_{ii=1}^{P} \alpha[ii] \cdot \psi[ii]}, \tag{5.35}$$

4. The central data site sends the *global* values back to the other sites.

Step 3 requires further clarifications. In (5.35), V is the global value to be computed by the central data site (which is different from the prototypes $\mathbf{V}$ returned from the DFCM) and sent to the other sites, P is the number of data sites, $V[ii]$ is the local value computed by the iith data site, Δ is a value which can be computed by the central data site without any communication, $\phi[ii]$ and $\psi[ii]$ are additional values that must be sent by the iith data site to the central site, and $\alpha[ii]$ is an optional user-defined weight that allows assigning a certain relative importance to the iith data site. For simplicity, in the following developments we have adopted $\alpha[ii] = 1 \; \forall ii$ (all sites have the same relative importance). Thus, (5.35) can be rewritten as

$$V = \frac{\left[\sum_{ii=1}^{P} (V[ii] - \Delta) \cdot \phi[ii]\right] + \Delta}{\sum_{ii=1}^{P} \psi[ii]}. \tag{5.36}$$

Bearing the above considerations in mind, we can summarize the proposed framework as follows. When a communication step is required, the respective step of the original (centralized) algorithm must be replaced by the steps in Algorithm 5. For example, DFCM (Sect. 5.4.1) consists of three steps. Two of them do not require any communication, namely: the computation of the distances between objects and prototypes and the update of the partition matrix. The computation of the prototypes requires communications among the sites and can be performed using Algorithm 5 by setting:

- $V = \mathbf{v}_i$ (global prototypes).
- $V[ii] = \mathbf{v}[ii]_i$ (local prototypes).
- $\phi[ii] = \psi[ii] = \sum_{j=1}^{N_{ii}} (u[ii]_{ij})^m$ (sum of memberships).
- $\Delta = 0$.

In this case, Algorithm 5 is equivalent to Steps 3–11 of Algorithm 4.

Algorithm 5 Distributed routine — Execution at the iith data site

Require: The local values $V[ii]$, $\phi[ii]$, $\psi[ii]$, and Δ.
1: **if** this is the central data site **then**
2: (Wait and) Receive the local values $V[jj]$, $\phi[jj]$, and $\psi[jj]$ from the other data sites.
3: Compute the global values by using (5.36).
4: Send the global values V to the other data sites.
5: **else**
6: Send the local values $V[ii]$, $\phi[ii]$, and $\psi[ii]$ to the central data site.
7: (Wait and) Receive the global values V from the central data site.
8: **end if**
9: **return** the global values V.

5.4.3 Other Distributed Fuzzy Clustering Algorithms

The same idea discussed above with respect to DFCM can also be applied to other fuzzy clustering algorithms. For example, the distributed version of the GK algorithm (Sect. 5.2.2) needs to compute both the global cluster prototypes, by using (5.32), and the global covariance matrices in (5.12), which can be rewritten in the distributed scenario as ($1 \leq i \leq k$)

$$\mathbf{F}_i = \frac{\sum_{ii=1}^{P} \sum_{j=1}^{N_{ii}} (u[ii]_{ij})^m (\mathbf{x}[ii]_j - \mathbf{v}_i)(\mathbf{x}[ii]_j - \mathbf{v}_i)^T}{\sum_{ii=1}^{P} \sum_{j=1}^{N_{ii}} (u[ii]_{ij})^m}. \tag{5.37}$$

The same strategy used to compute the global prototypes can also be used to compute the global covariance matrices in (5.37). Indeed, notice that setting $V = \mathbf{F}_i$, $V[ii] = \mathbf{F}[ii]$, $\Delta = 0$, and $\psi[ii] = \phi[ii] = \sum_{j=1}^{N_{ii}} (u[ii]_{ij})^m$, the global covariance matrices can be computed using (5.36) by rewriting them as

$$\mathbf{F}_i = \frac{\sum_{ii=1}^{P} \mathbf{F}[ii]_i \cdot \phi[ii]}{\sum_{ii=1}^{P} \psi[ii]}, \tag{5.38}$$

where ($1 \leq ii \leq P$, $1 \leq i \leq k$)

$$\mathbf{F}[ii]_i = \frac{\sum_{j=1}^{N_{ii}} (u[ii]_{ij})^m (\mathbf{x}[ii]_j - \mathbf{v}_i)(\mathbf{x}[ii]_j - \mathbf{v}_i)^T}{\sum_{j=1}^{N_{ii}} (u[ii]_{ij})^m}, \tag{5.39}$$

are the local covariance matrices of the iith data site (covariance matrices computed with respect to the objects in the iith data site only).

Once the global prototypes and covariance matrices have already been computed, the computation of the distances can be rewritten for the distributed scenario as:

$$D[ii]_{ij} = ||\mathbf{x}[ii]_j - \mathbf{v}_i||^2_{\mathbf{A}_i} = (\mathbf{x}[ii]_j - \mathbf{v}_i)^T \mathbf{A}_i (\mathbf{x}[ii]_j - \mathbf{v}_i) \tag{5.40}$$

Algorithm 6 DGK — Execution at the iith data site

Require: All requirements of Algorithm 4 and cluster volume constants ρ_i.
1: $t = 0$.
2: **repeat**
3: Compute the local cluster prototypes ($\mathbf{v}[ii]_i$) by using (5.30) and sums of memberships $\left(\sum_{j=1}^{N_{ii}} (u[ii]_{ij})^m\right)$.
4: Execute Algorithm 5 with $V[ii] = \mathbf{v}[ii]_i$, $\phi[ii] = \psi[ii] = \sum_{j=1}^{N_{ii}} (u[ii]_{ij})^m$, and $\Delta = 0$ to obtain the global prototypes $V = \mathbf{v}_i$ (Save the previous prototypes as $\hat{\mathbf{V}}$ to analyze the convergence.).
5: Compute the local covariance matrices by using (5.39).
6: Execute Algorithm 5 with $V[ii] = \mathbf{F}[ii]_i$, $\phi[ii]_i = \psi[ii]_i = \sum_{j=1}^{N_{ii}} (u[ii]_{ij})^m$, and $\Delta = 0$ to obtain the global covariance matrices $V = \mathbf{F}_i$.
7: Compute the norm inducing matrices ($\mathbf{A}_i$) by using (5.11).
8: Compute the distances between objects and prototypes by using (5.40).
9: Update the partition matrix by using (5.34).
10: $t = t + 1$.
11: **until** $||\hat{\mathbf{V}} - \mathbf{V}|| < \epsilon$ or $t = t_{max}$.
12: **return** $\mathbf{U}[ii] = [u[ii]_{ij}]_{k \times N_{ii}}$, $\mathbf{V} = [\mathbf{v}_i]_{k \times n}$, and $\mathbf{F}[ii]_i$.

with $\mathbf{A}_i$ given by (5.11) and the membership degrees given by (5.34). In brief, the distributed version of GK (DGK) can be summarized by Algorithm 6.

The other algorithms listed in Sect. 5.2.3 and described in the Appendix can also be generalized to handle distributed data by following the same idea. Section 5.6 summarizes the configurations of the terms in Equation (5.36) for the computation of all these algorithms.

5.4.4 *Complexity Analysis: Communication*

Since the time and space complexity analyses of the distributed algorithms follow the complexity analyses reported in Sect. 5.2.4[6], here we present only the communication complexity analysis of each algorithm. Following the distributed procedure in (5.36), it is necessary to communicate three variables: $V[ii]$, $\phi[ii]$, and $\psi[ii]$. The variables $\phi[ii]$ and $\psi[ii]$ are real numbers and require communication of a single value each. Thus, the overall communication complexity depends on the definition of $V[ii]$ and, accordingly, of its size, as addressed in the sequel.

To compute the global prototypes, it is necessary to communicate $O(n+1+1)$ values per cluster. Thus, the total communication complexity to compute the global prototypes is $O(k(n+1+1)) \rightarrow O(kn)$. To compute the global covariance matrices it is necessary to communicate $O(n^2 + 1 + 1)$ values per cluster. Thus, the total

[6]Note that the algorithm essentially runs in multiple data sites, each of which contains only a fraction of the number of objects.

Table 5.3 Communication complexities for the distributed algorithms: P, t, k, and n are the numbers of data sites, iterations, clusters, and attributes

	Elements	Complexity
DFCM	$\mathbf{v}_i$	$O(Ptkn)$
DGK	$\mathbf{v}_i$ and $\mathbf{F}_i$	$O(Pt(kn + kn^2)) \rightarrow O(Ptkn^2)$
DGG	$\mathbf{v}_i$, P_i, and $\mathbf{F}_i$	$O(Pt(kn + k + kn^2)) \rightarrow O(Ptkn^2)$
DFCV	$\mathbf{v}_i$ and $\mathbf{F}_i$	$O(Pt(kn + kn^2)) \rightarrow O(Ptkn^2)$
DFCE	$\mathbf{v}_i$ and $\mathbf{F}_i$	$O(Pt(kn + kn^2)) \rightarrow O(Ptkn^2)$
DPCM	$\mathbf{v}_i$	$O(Ptkn)$
DPGK	$\mathbf{v}_i$ and $\mathbf{F}_i$	$O(Pt(kn + kn^2)) \rightarrow O(Ptkn^2)$
DFPCM	$\mathbf{v}_i$ and D_{il}	$O(Pt(kn + k)) \rightarrow O(Ptkn)$
DPFCM	$\mathbf{v}_i$	$O(Ptkn)$

communication complexity to compute the global covariance matrices is $O(k(n^2 + 1 + 1)) \rightarrow O(kn^2)$.

To compute typicalities in (5.67), it is necessary to communicate only $O(k)$ values if the typicalities are computed in the more efficient way given by (5.70). Analogously, it can be shown that only $O(k)$ values need to be communicated in order to compute all global prior probabilities in (5.59) [55].

From the complexity analyses just reported, one can obtain the communication complexity for the distributed versions of all the fuzzy clustering algorithms described in Sect. 5.2 and in the Appendix. These complexities are summarized in Table 5.3, which also illustrates the elements that each algorithm requires to be communicated[7]. Note that the complexities do not depend on the numbers of objects in the data sites.

5.5 Distributed Clustering Validation

The indices addressed in Sect. 5.3 require centralized data. However, they depend essentially on (1) the membership of the objects to the clusters, (2) the distances between prototypes, and (3) the distances between objects and prototypes. As discussed in Sect. 5.4, (1) and (3) can be computed locally in each data site and then communicated to a central data site, whereas (2) can be obtained because each data site has access to the global prototypes. Again, when necessary, (5.36) can be used to compute the global values. Thus, the distributed version of each validity

[7]We do not present the $\phi[ii]$ and $\psi[ii]$ values that each element in Table 5.3 requires to be communicated because they are real numbers, which do not change the communication complexity. For details on the computation of such values, see Sect. 5.6.

index of Sect. 5.3 can be computed, allowing them to work with distributed data as well.

5.5.1 DXB: Distributed Xie-Beni

The XB index in (5.13) can be rewritten as

$$V_{DXB} = \frac{\sum_{ii=1}^{P} \sum_{i=1}^{k} \sum_{j=1}^{N_{ii}} (u[ii]_{ij})^m ||\mathbf{x}[ii]_j - \mathbf{v}_i||^2}{\sum_{ii=1}^{P} N_{ii} min_{l \neq s} ||\mathbf{v}_l - \mathbf{v}_s||^2}. \tag{5.41}$$

It can be computed for distributed data with (5.36) by setting $V = V_{DXB}$, $\Delta = 0$, $\phi[ii] = \psi[ii] = N_{ii}$, and $V[ii] = V_{XB}[ii]$, where $V_{XB}[ii]$ is the index computed locally with (5.13) for the iith data site:

$$V_{XB}[ii] = \frac{\sum_{i=1}^{k} \sum_{j=1}^{N_{ii}} (u[ii]_{ij})^m ||\mathbf{x}[ii]_j - \mathbf{v}_i||^2}{N_{ii} min_{l \neq s} ||\mathbf{v}_l - \mathbf{v}_s||^2}.$$

Therefore, the distributed version of the XB index can be calculated as [72]

$$V_{DXB} = \frac{\sum_{ii=1}^{P} V_{XB}[ii] \cdot N_{ii}}{\sum_{ii=1}^{P} N_{ii}}. \tag{5.42}$$

5.5.2 DFSS: Distributed Fuzzy Simplified Silhouette

The FSS index in (5.14) can be rewritten as

$$V_{DFSS} = \frac{\sum_{ii=1}^{P} \sum_{j=1}^{N_{ii}} (u[ii]_{pj} - u[ii]_{qj})^{\alpha} s[ii]_j}{\sum_{ii=1}^{P} \sum_{j=1}^{N_{ii}} (u[ii]_{pj} - u[ii]_{qj})^{\alpha}}, \tag{5.43}$$

where $s[ii]_j$ is the (prototype-based) silhouette of the jth object of the iith data site, which can be computed without any communication. Now, let $V = V_{DFSS}$, $\Delta = 0$, $\phi[ii] = \psi[ii] = \sum_{j=1}^{N_{ii}} (u[ii]_{pj} - u[ii]_{qj})^{\alpha}$, and $V[ii] = V_{FSS}[ii]$, where $V_{FSS}[ii]$ is the index computed locally with (5.14) as

$$V_{FSS}[ii] = \frac{\sum_{j=1}^{N_{ii}} (u[ii]_{pj} - u[ii]_{qj})^{\alpha} s[ii]_j}{\sum_{j=1}^{N_{ii}} (u[ii]_{pj} - u[ii]_{qj})^{\alpha}}, \tag{5.44}$$

Then, the distributed version of the FSS index can be calculated as

$$V_{DFSS} = \frac{\sum_{ii=1}^{P} V_{FSS}[ii] \cdot \phi[ii]}{\sum_{ii=1}^{P} \psi[ii]} = \frac{\sum_{ii=1}^{P} V_{FSS}[ii] \cdot \sum_{j=1}^{N_{ii}} (u[ii]_{pj} - u[ii]_{qj})^{\alpha}}{\sum_{ii=1}^{P} \sum_{j=1}^{N_{ii}} (u[ii]_{pj} - u[ii]_{qj})^{\alpha}}. \tag{5.45}$$

5.5.3 DK: Distributed Kwon

The K index in (5.16) can be rewritten as

$$V_{DK} = \frac{\sum_{ii=1}^{P} \sum_{i=1}^{k} \sum_{j=1}^{N_{ii}} (u[ii]_{ij})^m ||\mathbf{x}[ii]_j - \mathbf{v}_i||^2}{min_{l \neq s} ||\mathbf{v}_l - \mathbf{v}_s||^2} + \frac{\frac{1}{k} \sum_{i=1}^{k} ||\mathbf{v}_i - \overline{\mathbf{x}}||^2}{min_{l \neq s} ||\mathbf{v}_l - \mathbf{v}_s||^2}, \tag{5.46}$$

where $\overline{\mathbf{x}} = \frac{\sum_{ii=1}^{P} \sum_{j=1}^{N_{ii}} \mathbf{x}[ii]_j}{\sum_{ii=1}^{P} N_{ii}}$ is the grand mean, which can be computed for distributed data with (5.36) by setting $V = \overline{\mathbf{x}}$, $V[ii] = \overline{\mathbf{x}}[ii]$ (the grand mean computed locally only with the objects of the iith data site), and $\phi[ii] = \psi[ii] = N_{ii}$. Now, let $V = V_{DK}$ and $V[ii] = V_K[ii]$ being the index computed locally with (5.16) as

$$V_K[ii] = \frac{\sum_{i=1}^{k} \sum_{j=1}^{N_{ii}} (u[ii]_{ij})^m ||\mathbf{x}[ii]_j - \mathbf{v}_i||^2}{min_{l \neq s} ||\mathbf{v}_l - \mathbf{v}_s||^2} + \frac{\frac{1}{k} \sum_{i=1}^{k} ||\mathbf{v}_i - \overline{\mathbf{x}}||^2}{min_{l \neq s} ||\mathbf{v}_l - \mathbf{v}_s||^2}, \tag{5.47}$$

$$\Delta = \frac{\frac{1}{k} \sum_{i=1}^{k} ||\mathbf{v}_i - \overline{\mathbf{x}}||^2}{min_{l \neq s} ||\mathbf{v}_l - \mathbf{v}_s||^2},$$

$\phi[ii] = 1$, and $\psi[ii] = 1/P$. Then, the distributed version of the K index can be computed with (5.36), because (5.46) can be rewritten as

$$V_{DK} = \sum_{ii=1}^{P} \left[V_K[ii] - \frac{\frac{1}{k} \sum_{i=1}^{k} ||\mathbf{v}_i - \overline{\mathbf{x}}||^2}{min_{l \neq s} ||\mathbf{v}_l - \mathbf{v}_s||^2} \right] + \frac{\frac{1}{k} \sum_{i=1}^{k} ||\mathbf{v}_i - \overline{\mathbf{x}}||^2}{min_{l \neq s} ||\mathbf{v}_l - \mathbf{v}_s||^2}. \tag{5.48}$$

5.5.4 DTSS: Distributed Tang-Sun-Sun

The TSS index in (5.18) can be rewritten as

$$V_{DTSS} = \frac{\sum_{ii=1}^{P} \sum_{i=1}^{k} \sum_{j=1}^{N_{ii}} (u[ii]_{ij})^m ||\mathbf{x}[ii]_j - \mathbf{v}_i||^2}{min_{l \neq s} ||\mathbf{v}_l - \mathbf{v}_s||^2 + 1/k}$$

$$+\frac{\frac{1}{k(k-1)}\sum_{i=1}^{k}\sum_{\substack{l=1\\l\neq i}}^{k}||\mathbf{v}_i-\mathbf{v}_l||^2}{\min_{l\neq s}||\mathbf{v}_l-\mathbf{v}_s||^2+1/k}.$$

It can be computed for distributed data with (5.36) by setting $V = V_{DTSS}$,

$$\Delta=\frac{\frac{1}{k(k-1)}\sum_{i=1}^{k}\sum_{\substack{l=1\\l\neq i}}^{k}||\mathbf{v}_i-\mathbf{v}_l||^2}{\min_{l\neq s}||\mathbf{v}_l-\mathbf{v}_s||^2+1/k},$$

$\phi[ii] = 1$, $\psi[ii] = 1/P$, and $V[ii] = V_{TSS}[ii]$, which is the index computed locally with (5.18):

$$V_{TSS}[ii]=\frac{\sum_{i=1}^{k}\sum_{j=1}^{N_{ii}}(u[ii]_{ij})^m||\mathbf{x}[ii]_j-\mathbf{v}_i||^2}{min_{l\neq s}||\mathbf{v}_l-\mathbf{v}_s||^2+1/k}+\frac{\frac{1}{k(k-1)}\sum_{i=1}^{k}\sum_{\substack{l=1\\l\neq i}}^{k}||\mathbf{v}_i-\mathbf{v}_l||^2}{\min_{l\neq s}||\mathbf{v}_l-\mathbf{v}_s||^2+1/k}. \tag{5.49}$$

Therefore, the distributed version of the TSS index can be calculated as

$$V_{DTSS}=\sum_{ii=1}^{P}\left[V_T[ii]-\frac{\frac{1}{k(k-1)}\sum_{i=1}^{k}\sum_{\substack{l=1\\l\neq i}}^{k}||\mathbf{v}_i-\mathbf{v}_l||^2}{\min_{l\neq s}||\mathbf{v}_l-\mathbf{v}_s||^2+1/k}\right]$$
$$+\frac{\frac{1}{k(k-1)}\sum_{i=1}^{k}\sum_{\substack{l=1\\l\neq i}}^{k}||\mathbf{v}_i-\mathbf{v}_l||^2}{\min_{l\neq s}||\mathbf{v}_l-\mathbf{v}_s||^2+1/k}.$$

5.5.5 DFS: Distributed Fukuyama-Sugeno

The FS index in (5.19) can be rewritten as

$$V_{DFS}=\sum_{ii=1}^{P}\sum_{i=1}^{k}\sum_{j=1}^{N_{ii}}(u[ii]_{ij})^m||\mathbf{x}[ii]_j-\mathbf{v}_i||^2-\sum_{ii=1}^{P}\sum_{i=1}^{k}\sum_{j=1}^{N_{ii}}(u[ii]_{ij})^m||\mathbf{v}_i-\overline{\mathbf{x}}||^2, \tag{5.50}$$

where $\overline{\mathbf{x}} = \frac{\sum_{ii=1}^{P}\sum_{j=1}^{N_{ii}}\mathbf{x}[ii]_j}{\sum_{ii=1}^{P}N_{ii}}$ can be computed for distributed data with (5.36) by setting $V = \overline{\mathbf{x}}$, $V[ii] = \overline{\mathbf{x}}[ii]$, and $\phi[ii] = \psi[ii] = N_{ii}$. Now, let $V = V_{DFS}$ and $V[ii] = V_{FS}[ii]$ be the index computed locally with (5.19), i.e.,

$$V_{FS}[ii] = \sum_{i=1}^{k}\sum_{j=1}^{N_{ii}}(u[ii]_{ij})^m \|\mathbf{x}[ii]_j - \mathbf{v}_i\|^2 - \sum_{i=1}^{k}\sum_{j=1}^{N_{ii}}(u[ii]_{ij})^m \|\mathbf{v}_i - \overline{\mathbf{x}}\|^2, \quad (5.51)$$

$\Delta = 0$, $\phi[ii] = 1$, and $\psi[ii] = 1/P$. Then, the distributed version of FS can be computed with (5.36) because (5.50) can be rewritten as

$$V_{DFS} = \sum_{ii=1}^{P} V_{FS}[ii]. \quad (5.52)$$

5.5.6 DFHV: Distributed Fuzzy Hypervolume

The FHV index in (5.20) only needs the (global) covariance matrices. These matrices and, as a consequence, the distributed version of the index, V_{DFHV}, can be readily computed by following the steps in (5.37)–(5.39).

5.5.7 DAPD: Distributed Average Partition Density

The distributed version of APD index, V_{APD}, can be computed from (5.21) inasmuch as global values for S_i and $\mathbf{F}_i$ can be obtained from the distributed environment. As for FHV (Sect. 5.5.6), $\mathbf{F}_i$ can be computed for distributed data by following the steps in Eqs. (5.37)–(5.39). Analogously, S_i can be computed with (5.36) by setting $V = S_i$,

$$V[ii] = S[ii]_i = \sum_{\mathbf{x}_j \in \mathbf{R}_i} u[ii]_{ij},$$

$\phi[ii] = 1$, and $\psi[ii] = 1/P$, in such a way that

$$S_i = \sum_{ii=1}^{P} \sum_{\mathbf{x}[ii]_j \in \mathbf{R}[ii]_i} u[ii]_{ij} = \sum_{ii=1}^{P} S[ii]_i, \quad (5.53)$$

where $\mathbf{R}[ii]_i$ is the set of data objects at the iith data site that are within a pre-specified region around the cluster prototype $\mathbf{v}_i$, i.e., the objects such that $(\mathbf{x}[ii]_j - \mathbf{v}_i)^T \mathbf{F}_i^{-1}(\mathbf{x}[ii]_j - \mathbf{v}_i) < 1$.

5.5.8 DPD: Distributed Partition Density

The distributed version of the PD index in (5.23) can be obtained as

$$V_{DPD} = \frac{\sum_{i=1}^{k} S_i}{V_{DFHV}}, \tag{5.54}$$

where V_{DFHV} is described in Sect. 5.5.6, and S_i is given by (5.53).

5.5.9 DSCG: Distributed SCG

The SCG index in (5.24) requires the covariance matrices so that the term $GComp$ in (5.25) can be computed. As already discussed, the (global) covariance matrices can be computed for distributed data with (5.37)–(5.39). The SCG index also requires the scatter matrices $\mathbf{S}_B$ in (5.27) so that the term Sep in (5.26) can be computed. The computation of $\mathbf{S}_B$ in a distributed environment can be obtained from (5.36) by setting $V = \mathbf{S}_B$,

$$V[ii] = \mathbf{S}_B[ii] = \sum_{i=1}^{k} \sum_{j=1}^{N_{ii}} (u[ii]_{ij})^m (\mathbf{v}_i - \overline{\mathbf{x}})(\mathbf{v}_i - \overline{\mathbf{x}})^T, \tag{5.55}$$

$\phi[ii] = 1$, and $\psi[ii] = 1/P$, in such a way that

$$\mathbf{S}_B = \sum_{ii=1}^{P} \sum_{i=1}^{k} \sum_{j=1}^{N_{ii}} (u[ii]_{ij})^m (\mathbf{v}_i - \overline{\mathbf{x}})(\mathbf{v}_i - \overline{\mathbf{x}})^T. \tag{5.56}$$

These equations require the grand mean ($\overline{\mathbf{x}}$), which can also be obtained from (5.36) as described in Sect. 5.5.3. Once these terms have been computed, V_{DSCG} can be obtained from (5.24).

5.5.10 DPBMF: Distributed PBMF

The PBMF index can be rewritten as

$$V_{PBMF} = \left(\frac{1}{k} \cdot E_1 \cdot D_c\right)^2 \cdot \left(\frac{1}{J_m}\right)^2. \tag{5.57}$$

Since E_1 is constant, it does not need to be computed for relative comparison purposes [73][8]. The term $\left(\frac{1}{k} \cdot E_1 \cdot D_c\right)^2$ can be computed in the central site without any communication. The term J_m can be computed with (5.36) by setting $V = J_m$, $V[ii] = J_m[ii]$ (the value of the objective function computed in the iith site, i.e., $J_m[ii] = \sum_{i=1}^{k} \sum_{j=1}^{N} (u[ii]_{ij})^m ||\mathbf{x}[ii]_j - \mathbf{v}_i||$), $\Delta = 0$, $\phi[ii] = 1$, and $\psi[ii] = 1/P$. Then, DPBMF can be readily computed [50].

5.5.11 Complexity Analysis: Communication

Since the time and space complexity analyses of the distributed validity indices follow the complexity analyses reported in Sect. 5.3.11, we here present the complexity analysis of each index in terms of communication only. Following the distributed procedure in (5.36), it is necessary to communicate three variables: $V[ii]$, $\phi[ii]$, and $\psi[ii]$. The size of the first one depends on what is being communicated. The other variables ($\phi[ii]$ and $\psi[ii]$) are real numbers and require communication of a single value each. Thus, the communication complexity depends on the definition of $V[ii]$. For the DFHV index, the communication of the fuzzy covariance matrices is required, taking $O(kn^2)$. For the DAPD and DPD indices, it is necessary to compute the covariance matrices and S_i for each cluster, which takes $O(kn^2 + k)$. For DSCG, it is necessary to compute the covariance matrices, the grand mean, and $\mathbf{S}_B$, which takes $O(kn^2 + n + n^2)$. The DPBMF index requires the computation of the grand mean, which takes $O(n)$, and the global cluster prototypes, which takes $O(kn)$. The communication complexities of the distributed versions of the indices reviewed in Sect. 5.3 are summarized in Table 5.4. Note that the complexities do not depend on the numbers of objects in the data sites.

Table 5.4 Communication complexity for the distributed fuzzy indices: P, k, and n are the numbers of data sites, clusters, and attributes, respectively

	Complexity		Complexity
DXB	$O(P)$	DFHV	$O(Pkn^2)$
DFSS	$O(P)$	DAPD	$O(P(kn^2 + k)) \rightarrow O(Pkn^2)$
DK	$O(P)$	DPD	$O(P(kn^2 + k)) \rightarrow O(Pkn^2)$
DTSS	$O(P)$	DSCG	$O(P(kn^2 + n^2)) \rightarrow O(Pkn^2)$
DFS	$O(P)$	DPBMF	$O(P(n + kn)) \rightarrow O(Pkn)$

[8]If desired, this term can be computed from distributed data by calculating the grand mean $\bar{\mathbf{x}}$ by using (5.36) with $V = \bar{\mathbf{x}}$, $V[ii] = \bar{\mathbf{x}}[ii]$ (the grand mean computed in iith site), $\Delta = 0$, and $\phi[ii] = \psi[ii] = N_{ii}$, then computing E_1 by using (5.36) with $V = E_1$, $V[ii] = E_1[ii]$, $\Delta = 0$, $\phi[ii] = 1$, and $\psi[ii] = 1/P$.

Algorithm 7 DOMR

Require: The distributed clustering algorithm to be used, $\mathscr{A}^{\mathscr{D}}$ (and all parameters of such an algorithm), the distributed validity index to be used, $\mathscr{I}^{\mathscr{D}}$, the minimum number of clusters, k_{min} (usually $k_{min} = 2$), the maximum number of clusters, k_{max}, and the number of runs for each k, M.

1: Initialize $\mathscr{V}_{best} = -Inf$ and $\mathscr{V}_{best} = +Inf$ for maximization and minimization indices, respectively.
2: **for** $k = k_{min}$ to k_{max} **do**
3:     **for** $i = 1$ to M **do**
4:         Execute the distributed clustering algorithm $\mathscr{A}^{\mathscr{D}}$ with k clusters and obtain the clustering result $\mathscr{R}$.
5:         Evaluate the solution obtained in Step 4 with the distributed validity index $\mathscr{I}^{\mathscr{D}}$ and get the validity value $\mathscr{V}$.
6:         **if** $\mathscr{V}$ is better than $\mathscr{V}_{best}$ ("greater" for maximization indices or "lower" for minimization ones) **then**
7:             Update the best value: $\mathscr{V}_{best} \leftarrow \mathscr{V}$.
8:             Save the current solution as the best one: $\mathscr{R}_{best} \leftarrow \mathscr{R}$.
9:             Save the current number of clusters: $k_{best} \leftarrow k$.
10:         **end if**
11:     **end for**
12: **end for**
13: **return** $\mathscr{V}_{best}$, $\mathscr{R}_{best}$, and k_{best}.

5.5.12 DOMR: Distributed Ordered Multiple Runs

Once the generalizations of clustering algorithms and indices to handle distributed data are available, the procedure to estimate the number of clusters (k) described in Sect. 5.3.12 can also be generalized to work with distributed scenarios. To do so, the user must run a distributed clustering algorithm (Sect. 5.4) and evaluate the quality of each partition with a distributed validity index (Sect. 5.5). The procedure to estimate k in distributed scenarios is summarized in Algorithm 7.

5.6 Summary of Results

The elements used in the algorithms and validity indices to compute (5.36)—with particular settings for Δ, $\phi[ii]$, and $\psi[ii]$—are shown in Tables 5.5 and 5.6. Note that, in all cases, $V[ii]$ and V in (5.36) are the local quantity (computed at the iith data site) and the global quantity, respectively. For the sake of simplicity, they are omitted in the tables.

Table 5.5 Elements used by the distributed algorithms and validity indices to compute (5.36) with $\Delta = 0$

Parameter values		
	$\phi[ii]$	$\psi[ii]$
$\mathbf{v}_i$ (5.9)	$\sum_{j=1}^{N_{ii}}(u[ii]_{ij})^m$	$\sum_{j=1}^{N_{ii}}(u[ii]_{ij})^m$
$\mathbf{v}_i$ (5.68)	$\sum_{j=1}^{N_{ii}}((u[ii]_{ij})^m + (p[ii]_{ij})^\gamma)$	$\sum_{j=1}^{N_{ii}}((u[ii]_{ij})^m + (p[ii]_{ij})^\gamma)$
$\mathbf{v}_i$ (5.73)	$\sum_{j=1}^{N_{ii}}(a \cdot (u[ii]_{ij})^m + b \cdot (p[ii]_{ij})^\gamma)$	$\sum_{j=1}^{N_{ii}}(a \cdot (u[ii]_{ij})^m + b \cdot (p[ii]_{ij})^\gamma)$
$\mathbf{F}_i$ (5.12)	$\sum_{j=1}^{N_{ii}}(u[ii]_{ij})^m$	$\sum_{j=1}^{N_{ii}}(u[ii]_{ij})^m$
P_i (5.59)	$\sum_{j=1}^{N_{ii}}\sum_{c=1}^{k}(u_{cj})^m$	$\sum_{j=1}^{N_{ii}}\sum_{c=1}^{k}(u_{cj})^m$
φ_i (5.69)	1	1/P
η_i (5.64)	$\sum_{j=1}^{N_{11}}(u[ii]_{ij})^m$	$\sum_{j=1}^{N_{11}}(u[ii]_{ij})^m$
J (5.5),(5.62), (5.65),(5.71)	1	1/P
$\bar{\mathbf{x}}$ (5.17)	N_{ii}	N_{ii}
S_i (5.22)	1	1/P
$\mathbf{S}_B$ (5.27)	1	1/P

Table 5.6 Elements used by the validity indices to compute (5.36)

Parameter values			
	Δ	$\phi[ii]$	$\psi[ii]$
PC	0	N_{ii}	N_{ii}
PE	0	N_{ii}	N_{ii}
FS	0	1	$1/P$
XB	0	N_{ii}	N_{ii}
K	$\dfrac{\frac{1}{k}\sum_{i=1}^{k}\|\|\mathbf{v}_i - \bar{\mathbf{x}}\|\|^2}{min_{l\neq s}\|\|\mathbf{v}_l - \mathbf{v}_s\|\|^2}$	1	$1/P$
TSS	$\dfrac{\frac{1}{k(k-1)}\sum_{i=1}^{k}\sum_{l=1, l\neq i}^{k}\|\|\mathbf{v}_i - \mathbf{v}_l\|\|^2}{\min_{l\neq s}\|\|\mathbf{v}_l - \mathbf{v}_s\|\|^2 + 1/k}$	1	$1/P$
FSS	0	$\sum_{j=1}^{N_{ii}}(u_{pj} - u_{qj})^\alpha$	$\sum_{j=1}^{N_{ii}}(u_{pj} - u_{qj})^\alpha$

5.7 Experimental Evaluation

In this section we present an empirical evaluation of the proposed framework, which allowed us to generalize fuzzy clustering algorithms, as well as relative validity criteria, so that they can handle distributed data in an exact way, i.e., with no approximations in their results. In order to determine the most appropriate number of clusters, the DOMR procedure (Sect. 5.5.12) was adopted with the distributed versions of FCM, GK, GG, PCM, FPCM, and PFCM, described in Sect. 5.2 and in the Appendix. The DFSS validation index (Sect. 5.5.2) was chosen as $\mathscr{I}^{\mathscr{D}}$ to evaluate the quality of the partitions produced by the DOMR procedure. DFSS is the distributed fuzzy version of the Simplified Silhouette index [62], which scored the best among 40 indices in a recent comparative study [52, 74]. The experiments were performed in a parallel scenario, where the responsibility for processing predetermined sub-collections of the data is distributed across the different cores of a single processor, which can then be seen as virtually distributed data sites, even though in this parallel computing setup the data is stored in a shared memory that is accessible by all the cores.

The experiments were run in a computer equipped with an AMD 3.6GHz processor with three processing modules, six cores and 16 GB-RAM using Linux 64 bits, Matlab 2009 and Matlab Parallel Computing Toolbox. Three data sets were generated for the experiments, using the same data set generator described in [52, 75]. Each data set contains $k = 8$ clusters embedded in an $n = 8$ dimensional attribute space. The first data set contains $N = 2^{13}$ objects, the second contains $N = 2^{17}$ objects, and the third contains $N = 2^{20}$ objects. The true distribution of the objects within clusters follows a (mildly truncated) multivariate normal distribution, such that the resulting structure can be considered to consist of natural clusters that exhibit the properties of external isolation and internal cohesion.

DOMR was run 10 times for each clustering algorithm, with $M = 10$, $k_{min} = 2$, and $k_{max} = 8$, using from one to six cores of the computer. In each case, the data sets were randomly distributed across the cores (evenly). The parameters of the algorithms were set as $m = \gamma = 2$, $\rho_i = 1$, η_i as defined in (5.64), and $a = b = 1$. Two different settings were considered for the number of iterations, namely, $t = 10$ and $t = 100$.

The obtained speedups (averaged over 10 DOMR runs) are depicted in Fig. 5.2, 5.3, and 5.4. It is important to note that the architecture of the processor used is composed by three processing modules with six cores, i.e., each pair of cores share resources from the processing module they belong to. Thus, the performance per node of the processor naturally degrades if four or more cores are used in parallel, as the cores compete for shared resources. This type of architecture and hyper-threading are adopted by the vast majority of the current desktop processors and, for this reason, was chosen for this experiment.

The results in Fig. 5.2 (data set with 2^{13} objects) show that when the clustering algorithms use more cores in a small data set, particularly for few iterations, the performance is not good due to their communication overhead in comparison with the execution in a single core and due to the initial setup required to start the environment. In fact, the centralized algorithm runs very fast in small data and the time necessary to setup the parallel/distributed environment may be longer then the algorithm itself (depending on the number of iterations the algorithm runs). The setup time tends not to be significant as the database becomes larger and/or the clustering algorithm runs longer. In this context, Fig. 5.2a shows that the clustering algorithms, except for DGK and DGG (which compute covariance matrices)[9], spent

[9]Computing the covariance matrices and their inverses are computationally demanding, therefore such algorithms have a larger ratio between processing time and data communication.

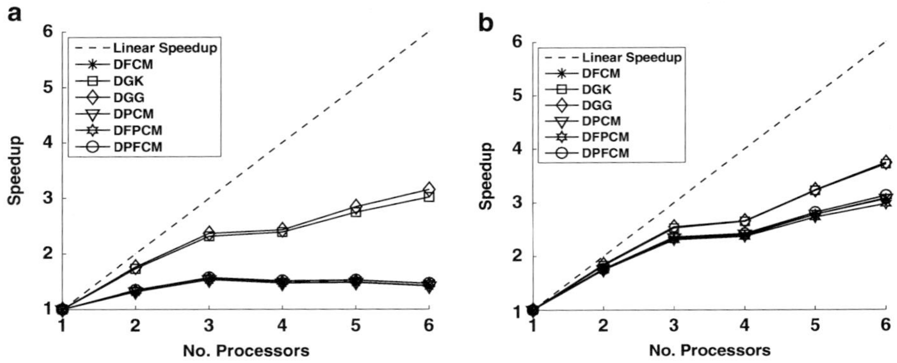

Fig. 5.2 Speedups obtained in a data set with 2^{13} objects over 10 DOMR runs. (**a**) Speedup after 10 iterations. (**b**) Speedup after 100 iterations

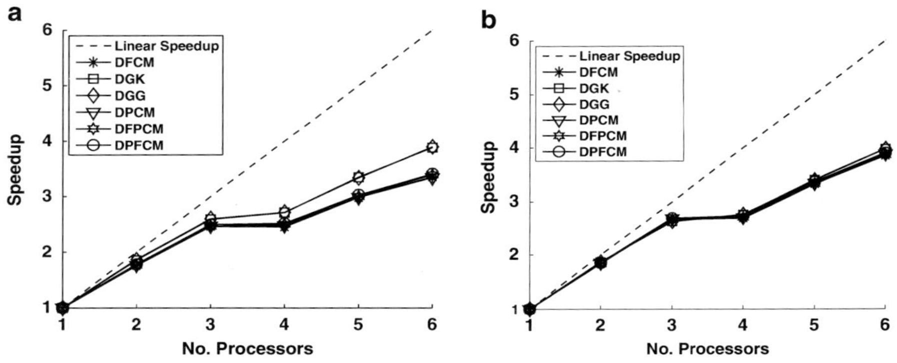

Fig. 5.3 Speedups obtained in a data set with 2^{17} objects over 10 DOMR runs. (**a**) Speedup after 10 iterations. (**b**) Speedup after 100 iterations

more time to process the data in multiple cores compared to a single one (speedup lower than 1). However, as the size of the data set increases, the speedup also increases (see Figs. 5.3 and 5.4) and the use of multiple cores becomes attractive. Additionally, the higher is the number of iterations of the algorithms, the better the resulting speedup tends to be (compare Figs. 5.2b, 5.3b, 5.4b against 5.2a, 5.3a, 5.4a, respectively).

Table 5.7 describes the amount of data transfer (in MBs) among the data sites after the execution of 10 runs of DOMR with $t = 100$ iterations of each clustering algorithm. These quantities are proportional to the number of iterations. Thus, one tenth of the quantities displayed in Table 5.7 have been observed for $t = 10$ iterations. Note that the size of the data set does not affect the amount of data communication because the algorithms transfer summarized information (such as

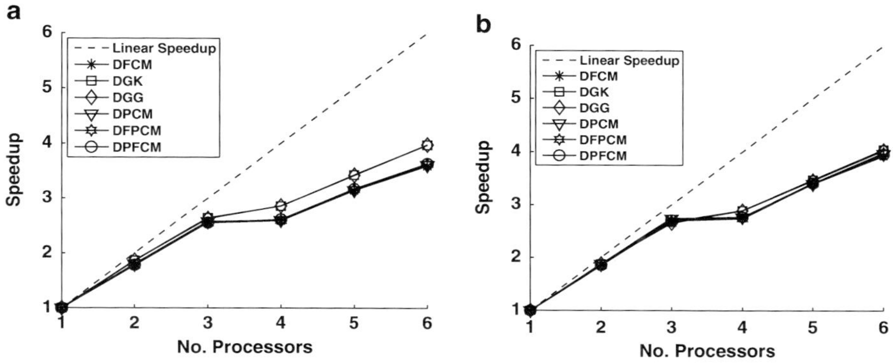

Fig. 5.4 Speedups obtained in a data set with 2^{20} objects over 10 DOMR runs. (**a**) Speedup after 10 iterations. (**b**) Speedup after 100 iterations

Table 5.7 Amount of data communication (in MBs) among the data sites using two, three, four, five and six processor cores for $t = 100$ iterations

Number of cores	Two	Three	Four	Five	Six
DFCM	156.796	241.171	325.546	409.921	494.296
DGK	969.296	1466.171	1963.046	2459.921	2956.796
DGG	994.296	1509.921	2025.546	2541.171	3056.796
DPCM	156.796	241.171	325.546	409.921	494.296
DFPCM	181.796	284.921	388.046	491.171	594.296
DPFCM	156.796	241.171	325.546	409.921	494.296

prototypes, sums of memberships, etc.). Thus, the same amount has been observed for data sets with $N = 2^{13}$, $N = 2^{17}$ e $N = 2^{20}$ objects.

The values presented in Table 5.7 indicate that the growth of the amount of data communication results from the increase of the number of cores and reflects the communication complexities presented in Sects. 5.4.4 and 5.5.11. The algorithms DGK and DGG demand the highest data transfer rates due to the distributed calculation of their covariance matrices.

Our main goal in this work is not to assess the quality of clustering algorithms. In spite of this and just for illustration purposes, the DOMR best evaluated partition obtained for each clustering algorithm, $\mathscr{R}_{best}$, was compared to the known clusters or "golden truth" with the well-known Corrected Rand (CR) external index [76]. Note that the known clusters of each data set form a hard partition, thus a hard partition was derived from $\mathscr{R}_{best}$ by changing its maximum membership value to 1 and the other values to 0 before using the CD index. The index's upper bound is 1 and its lower bound depends on the data set [77]. Zero value means chance agreement and negative values mean less than chance agreement. The CR values resulting from 10 runs of DOMR with $t = 10$ are presented in Table 5.8.

Table 5.8 CR values of the DOMR $\mathcal{R}_{best}$ partition when compared to the know clusters of data sets with sizes $N = 2^{13}$, $N = 2^{17}$ and $N = 2^{20}$ for 10 iterations of the clustering algorithms

Data set size	$N = 2^{13}$	$N = 2^{17}$	$N = 2^{20}$
DFCM	0.2317	0.2312	0.2106
DGK	0.7622	0.2499	0.2499
DGG	0.8708	1	0.8710
DPCM	0.1800	0.2179	0.0227
DFPCM	1	0.2222	0.2496
DPFCM	0.2497	0.2132	0.2048

The CR values displayed in Table 5.8 indicate that DGG was the only algorithm capable of finding high quality partitions for all data sets, when using the DOMR procedure with $M = t = 10$. However, by changing the number of iterations of the clustering algorithms to $t = 100$, the CR value is maximum (unitary) for all algorithms and data sets, which indicates that the hard partitions derived from $\mathcal{R}_{best}$ are identical to the known clusters in every case. This result indicates that the clustering algorithms are capable of finding the correct number of clusters and high quality partitions despite running in a distributed fashion by using the proposed framework.

5.8 Final Remarks

We introduced a framework that generalizes a number of fuzzy clustering algorithms so that they can handle distributed data in an exact way, i.e. with no approximations in the results. In addition, we showed that the proposed framework can be adapted to automatically estimate the number of clusters from data in both distributed and parallel processing scenarios, by using one of the ten relative fuzzy clustering validity indices that have also been extended to the non-centralized domain. Specifically, our framework allows computing global values representing a distributed collection of data as a function of local values that are computed at each data site, processor or core, and then reduced into a single solution. The complexity analyses of the resulting algorithms and validation indices were reported in terms of space, time, and communication requirements. Experimental results showed the speedup obtained by the algorithms when running in a parallel architecture, when compared to the traditional, centralized counterparts, for different sized data sets. These results also indicate that the compared algorithms are capable of finding high quality partitions when combined with a validity index by a procedure to estimate the number of clusters in distributed data. Both the clustering algorithms and the validity indices were distributed using the proposed framework. Since real-world applications in which the data are naturally or intentionally distributed across different sites or computers are becoming increasingly common (distributed/parallel scenarios), we believe that the proposed framework can be very useful in practice.

In summary, the main contributions of our work are: (a) the development of distributed versions of several well-known fuzzy clustering algorithms that so far could only be computed in a centralized way; (b) a unified framework that comprises all such new distributed algorithms and a few previously existing ones as particular parameterization cases; (c) the extension of such a framework to the computation of fuzzy clustering validity indices, which so far could only be computed in a centralized way; (d) a procedure for the estimation of the number of clusters in parallel and distributed scenarios that makes use of the distributed algorithms and validity indices that have been proposed.

It is important to stress that we are not claiming that the algorithms and indices surveyed in this chapter are all-inclusive. There is a plethora of fuzzy clustering algorithms and indices out there, many of them developed for specific application scenarios. Figuring out if the proposed framework can be applied to other algorithms/indices not explicitly discussed in this chapter and, if not, how to make a distributed version of the corresponding algorithms/indices possible, are interesting subjects for further investigations.

Another relevant topic for future research involves different possible computer architectures for distributed/parallel processing. In particular, the architecture adopted in this chapter assumes that one of the local data sites also works as a central processing node that is responsible for communicating with the other sites, thereby aggregating their local statistics and distributing back the global ones. This architecture is simple and at the same time it allows that the communication of the information from and to all data sites be made in parallel. Other architectures, however, may also be possible (e.g., involving networks with arbitrary node-to-node communication [78]). A theoretical or experimental comparison between different possible architectures is an interesting subject for future studies.

Acknowledgements The authors acknowledge the Brazilian Research Agencies CNPq, FAPESP, and FAPEMIG for the financial support.

Appendix: Additional Fuzzy Clustering Algorithms

GG: Gath and Geva

The GG algorithm [2, 12] induces hyper-ellipsoidal clusters in such a way that the distances are not linear with respect to the covariance matrices [79] and the cluster volumes do not need to be fixed a priori. In this case, (5.5) is minimized with the distances given by ($1 \leq i \leq k,\ 1 \leq j \leq N$)

$$D_{ij} = \frac{[\det(\mathbf{F}_i)]^{1/2}}{P_i} \exp\left[\frac{1}{2}(\mathbf{x}_j - \mathbf{v}_i)^T \mathbf{F}_i^{-1}(\mathbf{x}_j - \mathbf{v}_i)\right], \tag{5.58}$$

Algorithm 8 GG

Require: All requirements of Algorithm 1.
1: $t = 0$.
2: **repeat**
3: Compute the cluster prototypes by using (5.9).
4: Compute the covariance matrix and the prior probability by using (5.12) and (5.59).
5: Compute the distances by using (5.58).
6: Update the partition matrix by using (5.7) (Save the previous partition matrix as $\hat{\mathbf{U}}$ to analyze the convergence).
7: $t = t + 1$.
8: **until** $||\hat{\mathbf{U}} - \mathbf{U}|| < \epsilon$ or $t = t_{max}$.
9: **return** $\mathbf{U} = [u_{ij}]_{k \times N}$, $\mathbf{V} = [\mathbf{v}_i]_{k \times n}$ and $\mathbf{F}_i$.

where $\mathbf{F}_i$ denotes the fuzzy covariance matrix for cluster $\mathbf{C}_i$, given by (5.12) [2] and P_i is the prior probability of selecting cluster $\mathbf{C}_i$, given by Hoppner et al. [2] $(1 \leq i \leq k)$

$$P_i = \frac{\sum_{j=1}^{N} (u_{ij})^m}{\sum_{j=1}^{N} \sum_{c=1}^{k} (u_{cj})^m}. \tag{5.59}$$

Again, minimizing (5.5) with respect to $\mathbf{v}_i$ leads to (5.9). The complete procedure to minimize (5.5) with the distances defined in (5.58) is given by Algorithm 8.

FCV: Fuzzy c-Varieties

Unlike previous algorithms, the FCV algorithm [13] measures the distances among data from r-dimensional linear varieties (lines when $r = 1$, planes when $r = 2$ or hyper-planes when $2 < r < n$). Thus, it was developed for the recognition of clusters formed by lines, planes or hyper-planes, which may be useful in image processing for the identification of border lines in image recognition [2]. The FCV algorithm minimizes (5.5) with the distances given by $(1 \leq i \leq k,\ 1 \leq j \leq N)$

$$\begin{aligned} D_{ij} &= ||\mathbf{x}_j - \mathbf{v}_i||_{\mathbf{A}}^2 - \sum_{l=1}^{r} < \mathbf{x}_j - \mathbf{v}_i, \mathbf{s}_{il} > \\ &= (\mathbf{x}_j - \mathbf{v}_i)^T \mathbf{A} (\mathbf{x}_j - \mathbf{v}_i) - \sum_{l=1}^{r} ((\mathbf{x}_j - \mathbf{v}_i)^T \mathbf{A}^{1/2} \mathbf{s}_{il})^2 \end{aligned} \tag{5.60}$$

where $\mathbf{v}_i$ is a point through which the variety passes, $< \cdot, \cdot >$ denotes a scalar product, and $(\mathbf{s}_{i1}, \ldots, \mathbf{s}_{ir})$ is an r-tuple of linearly independent vectors spanning variety v_{ir} and corresponds to the r principal eigenvectors (normalized eigenvectors corresponding to the r largest eigenvalues) of the ith fuzzy within cluster scatter matrix $\mathbf{S}_i = \mathbf{A}^{1/2}\mathbf{F}_i\mathbf{A}^{1/2}$ (in which $\mathbf{F}_i$ is given by (5.12)). Thus, the varieties are defined according to the stretching of its corresponding cluster. The distances from each object to each cluster are then influenced by how close the corresponding variety passes to it. Note that if $r = 0$ then (5.60) reduces to (5.8) and FCV reduces to FCM. In addition, minimizing (5.5) by using the distance in (5.60) with respect to $\mathbf{v}_i$ leads to (5.9). FCV is summarized by Algorithm 9.

Algorithm 9 FCV

Require: All requirements of Algorithm 1 and the dimensionality r of the varieties $(0 < r < n)$.
1: $t = 0$.
2: **repeat**
3: Compute the cluster prototypes (means) by using (5.9).
4: Compute the covariance matrix for each cluster by using (5.12).
5: Extract (from each $\mathbf{S}_i = \mathbf{A}^{1/2}\mathbf{F}_i\mathbf{A}^{1/2}$) the r principal eigenvectors $\mathbf{s}_{il}$, $l = 1, \ldots, r$ (normalized eigenvectors corresponding to the r largest eigenvalues).
6: Compute the distances by using (5.60).
7: Update the partition matrix by using (5.7) (Save the previous partition matrix as $\hat{\mathbf{U}}$ to analyze the convergence).
8: $t = t + 1$.
9: **until** $||\hat{\mathbf{U}} - \mathbf{U}|| < \epsilon$ or $t = t_{max}$.
10: **return** $\mathbf{U} = [u_{ij}]_{k \times N}$, $\mathbf{V} = [\mathbf{v}_i]_{k \times n}$ and $\mathbf{F}_i$.

FCE: Fuzzy c-Elliptotypes

The FCE algorithm finds clusters in lines, planes, or hyper-planes, which have infinite length and may cluster widely separated collinear clusters together [2, 6, 14]. To avoid this problem, FCE [14] uses a distance given by a convex combination of (5.8) and (5.60) as $(1 \leq i \leq k,\ 1 \leq j \leq N)$

$$\begin{aligned} D_{ij} &= (1-\alpha)||\mathbf{x}_j - \mathbf{v}_i||_{\mathbf{A}}^2 + \alpha(||\mathbf{x}_j - \mathbf{v}_i||_{\mathbf{A}}^2 - \sum_{l=1}^{r} < \mathbf{x}_j - \mathbf{v}_i, \mathbf{s}_{il} >_{\mathbf{A}}^2) \qquad (5.61) \\ &= ||\mathbf{x}_j - \mathbf{v}_i||_{\mathbf{A}}^2 - \alpha \sum_{l=1}^{r} < \mathbf{x}_j - \mathbf{v}_i, \mathbf{s}_{il} > \\ &= (\mathbf{x}_j - \mathbf{v}_i)^T \mathbf{A}(\mathbf{x}_j - \mathbf{v}_i) - \alpha \sum_{l=1}^{r} ((\mathbf{x}_j - \mathbf{v}_i)^T \mathbf{A}^{1/2}\mathbf{s}_{il})^2, \end{aligned}$$

where $0 \leq \alpha \leq 1$ defines the stretching of the hyper-ellipsoids in the directions of $\mathbf{s}_{il}$. When $\alpha = 0$, D_{ij} reduces to (5.8). On the other hand, if $\alpha = 1$ then D_{ij} reduces to (5.60). Again, minimizing (5.5) using the distances in (5.61) with respect to $\mathbf{v}_i$ leads to (5.9). FCE is summarized in Algorithm 10.

Algorithm 10 FCE

Require: Besides all the requirements of Algorithm 1, the dimensionality r of the varieties ($0 < r < n$), and the weighting parameter $0 \leq \alpha \leq 1$.

$t = 0$.
repeat
  Compute the cluster prototypes by using (5.9).
  Compute the covariance matrix for each cluster by using (5.12).
  Extract (from each $\mathbf{S}_i = \mathbf{A}^{1/2}\mathbf{F}_i\mathbf{A}^{1/2}$) the r principal eigenvectors $\mathbf{s}_{il}$, $l = 1, \ldots, r$ (normalized eigenvectors corresponding to the r largest eigenvalues).
  Compute the distances by using (5.61).
  Update the partition matrix by using (5.7) (Save the previous partition matrix as $\hat{\mathbf{U}}$ to analyze the convergence).
  $t = t + 1$.
until $||\hat{\mathbf{U}} - \mathbf{U}|| < \epsilon$ or $t = t_{max}$.
return $\mathbf{U} = [u_{ij}]_{k \times N}$, $\mathbf{V} = [\mathbf{v}_i]_{k \times n}$ and $\mathbf{F}_i$.

PCM: Possibilistic c-Means

All previous algorithms work with partitions in $\mathbf{M}_{fp}$—see (5.4). However, the constraint $\sum_{i=1}^{k} u_{ij} = 1$ may lead to unrealistic results. Take as an example two clusters $\mathbf{C}_1$ and $\mathbf{C}_2$ with some degree of overlapping, as displayed in Fig. 5.5. The membership degree of object $\mathbf{x}_1$ is equal to 0.5 for both clusters (i.e., $u_{11} = u_{21} = 0.5$) because the distances from $\mathbf{x}_1$ to the cluster prototypes $\mathbf{v}_1$ and $\mathbf{v}_2$ (D_{11} and D_{21}) are the same ($D_{11} = D_{21}$). This is expected since $\mathbf{x}_1$ is in the overlapping area. Note, however, that the membership degree of the outlier object $\mathbf{x}_2$ is also equal to 0.5 for both clusters ($u_{11} = u_{21} = 0.5$). The reason is that although it is an outlier far from the clusters, it is equidistant from both cluster prototypes (i.e., $D_{12} = D_{22}$). Therefore, the membership calculation in (5.7) leads to an equal membership degree for both clusters.

In an attempt to avoid this problem, the PCM algorithm [3] deals with partitions lying in $\mathbf{M}_f$—see (5.3). Because now $0 \leq u_{ij} \leq 1$ (called typicalities) may be as small as possible (close to 0), the objective function in (5.5) must be modified in order to prevent the trivial solution $u_{ij} = 0$ for all i, j. Thus, PCM aims at minimizing the function

$$J = \sum_{i=1}^{k}\sum_{j=1}^{N}(u_{ij})^m D_{ij} + \sum_{i=1}^{k}\eta_i \sum_{j=1}^{N}(1 - u_{ij})^m, \tag{5.62}$$

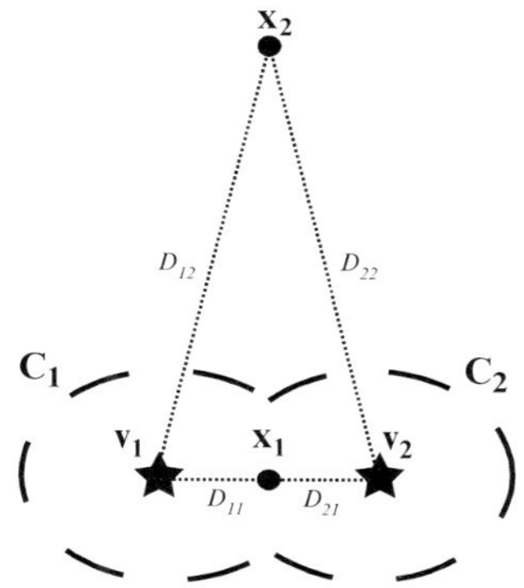

Fig. 5.5 Example of membership of two objects to two clusters

where D_{ij} is given by (5.8), and $\eta_i > 0$ is a weighting parameter defined by the user. The first term of (5.62) is identical to (5.5). The second term forces the typicalities u_{ij} to be as large as possible, avoiding the trivial solution $\mathbf{U} = \mathbf{0}$ (recall that $\mathbf{U} \in \mathbf{M}_f$).

The most popular method to solve (5.62) is a simple Picard iteration through the first-order conditions for stationary points of (5.62). Thus, holding D_{ij} constant, one finds that a necessary condition is[10] $(1 \leq i \leq k, 1 \leq j \leq N)$

$$u_{ij} = \left[1 + \left(\frac{D_{ij}}{\eta_i}\right)^{1/(m-1)}\right]^{-1}. \tag{5.63}$$

Holding $\mathbf{v}_i$ constant, one finds that another necessary condition to minimize (5.62) is given by (5.9). Note that η_i is an important user-defined parameter. When η_i approximates to 0, the solution tends to be trivial ($\mathbf{U} = \mathbf{0}$). If η_i is large, u_{ij} will be as large as possible. In other words, η_i determines the relative degree to which the second term in (5.62) is important in comparison with the first one. If the two terms are to be weighted roughly equally, then η_i should be of the order of D_{ij}. In practice, Krishnapuram and Keller [3] found that the following definition works well for a given initial partition[11]

$$\eta_i = \frac{\sum_{j=1}^{N}(u_{ij})^m D_{ij}}{\sum_{j=1}^{N}(u_{ij})^m}. \tag{5.64}$$

The complete procedure is given by Algorithm 11.

[10] The columns in $\mathbf{U}$ are independent and, therefore, it is possible to decompose the objective function into k individual components.

[11] For example, a partition given by a previous run of FCM.

Algorithm 11 PCM

Require: Data set $\mathbf{X} = \{\mathbf{x}_1, \ldots, \mathbf{x}_N\}$, number of clusters $k \in \{2, \ldots, N-1\}$, fuzzification exponent $m \in (1, \infty)$, initial (random) partition matrix $\mathbf{U} = [u_{ij}] \in \mathbf{M}_f$, weighting parameters $\eta_i \geq 0$, number of iterations t_{max} and/or termination tolerance $\epsilon > 0$.

1: $t = 0$.
2: **repeat**
3: Compute the cluster prototypes by using (5.9).
4: Compute the distance between objects and prototypes by using (5.8).
5: Update the partition matrix by using (5.63) (Save the previous partition matrix as $\hat{\mathbf{U}}$ to analyze the convergence).
6: $t = t + 1$
7: **until** $||\hat{\mathbf{U}} - \mathbf{U}|| < \epsilon$ or $t = t_{max}$.
8: **return** $\mathbf{U} = [u_{ij}]_{k \times N}$ and $\mathbf{V} = [\mathbf{v}_i]_{k \times n}$.

Algorithm 12 PGK

Require: Besides all the requirements of Algorithm 11, the cluster volumes, ρ_i.

1: $t = 0$.
2: **repeat**
3: Compute the cluster prototypes (means) using (5.9).
4: Compute the covariance matrix for each cluster using (5.12).
5: Compute the distances using (5.10) and (5.11).
6: Update the partition matrix using (5.63) (Save the previous partition matrix as $\hat{\mathbf{U}}$ to analyze the convergence).
7: $t = t + 1$
8: **until** $||\hat{\mathbf{U}} - \mathbf{U}|| < \epsilon$ or $t = t_{max}$.
9: **return** $\mathbf{U} = [u_{ij}]_{k \times N}$, $\mathbf{V} = [\mathbf{v}_i]_{k \times n}$, and $\mathbf{F}_i$.

PGK: Possibilistic Gustafson-Kessel

The PGK algorithm [3] is a possibilistic version of the GK algorithm (Sect. 5.2.2). It minimizes (5.62) with the distances given by (5.10). In this case, holding D_{ij} constant, the necessary conditions to minimize (5.62) also lead to (5.63). Holding u_{ij} and $\mathbf{A}_i$ constant, the necessary conditions lead to (5.9). Finally, holding u_{ij} and v_{ij} constant and constraining $\mathbf{A}_i$ in such a way that $\det(\mathbf{A}_i) = \rho_i$, the necessary conditions lead to (5.11) and (5.12). PGK is summarized in Algorithm 12.

FPCM: Fuzzy-Possibilistic c-Means

Since the columns of partitions $\mathbf{U} \in \mathbf{M}_f$ are independent, PCM-like algorithms tend to obtain concentrical clusters when the global minimum has been achieved [80,81]. To avoid this problem, the FPCM algorithm [15] finds memberships and typicalities simultaneously between the objects and clusters (i.e., one matrix $\mathbf{U} \in \mathbf{M}_{fp}$ and another matrix $\mathbf{P} \in \mathbf{M}_f$). The objective function to be minimized is then given by

$$J = \sum_{i=1}^{k} \sum_{j=1}^{N} ((u_{ij})^m + (p_{ij})^\gamma) D_{ij}, \tag{5.65}$$

where D_{ij} is given by (5.8), $m, \gamma > 0$ are fuzzification exponents, $\mathbf{U} = [u_{ij}] \in \mathbf{M}_{fp}$ is the membership matrix, and $\mathbf{P} = [p_{ij}] \in \mathbf{M}_f$ is the typicality matrix. In contrast with memberships (u_{ij}), a given typicality p_{ij} should not depend on the typicalities assigned to the other clusters ($p_{cj}, c \neq i$). Because $\mathbf{P} = [p_{ij}] \in \mathbf{M}_f$, p_{ij} could be as small as desired in order to minimize (5.65). To avoid it, in [15] the authors constrained the typicality matrix by[12] [15] ($1 \leq i \leq k$)

$$\sum_{j=1}^{N} p_{ij} = 1. \tag{5.66}$$

The usual method to solve (5.65) is a simple Picard iteration through the first-order conditions for stationary points of (5.65). Holding D_{ij} and p_{ij} constant, the necessary conditions to minimize (5.65) lead to (5.7). Holding D_{ij} and u_{ij} constant, the necessary conditions lead to ($1 \leq i \leq k,\ 1 \leq j \leq N$)

$$p_{ij} = \left(\sum_{l=1}^{N} \left(\frac{D_{ij}}{D_{il}} \right)^{1/(\gamma-1)} \right)^{-1}. \tag{5.67}$$

Finally, holding u_{ij} and p_{ij} constant, the necessary conditions lead to

$$\mathbf{v}_i = \frac{\sum_{j=1}^{N} ((u_{ij})^m + (p_{ij})^\gamma) \mathbf{x}_j}{\sum_{j=1}^{N} ((u_{ij})^m + (p_{ij})^\gamma)}. \tag{5.68}$$

The complete procedure is given in Algorithm 13.

It is worth noticing that the typicalities in (5.67) can be computed more efficiently by following a similar idea as in [56]. More precisely, note that

$$\varphi_i = \sum_{l=1}^{N} \left(\frac{1}{D_{il}} \right)^{1/(\gamma-1)}, \quad 1 \leq i \leq k, \tag{5.69}$$

can be computed for each cluster before computing p_{ij} in (5.67), which now reduces to

$$p_{ij} = \left((D_{ij})^{1/(\gamma-1)} \cdot \varphi_i \right)^{-1} \tag{5.70}$$

[12] Note that, since $\mathbf{U} \in \mathbf{M}_{fp}$, it is constrained by (5.6).

Algorithm 13 FPCM

Require: Besides all the requirements of Algorithm 1, the fuzzification exponent $\gamma \in (1, \infty)$ and the initial (random) partition matrix $\mathbf{P} = [p_{ij}] \in \mathbf{M}_f$.
1: $t = 0$.
2: **repeat**
3: Compute the cluster prototypes by using (5.68).
4: Compute the distances between objects and prototypes by using (5.8).
5: Update the possibilistic partition matrix by using (5.67).
6: Update the probabilistic partition matrix by using (5.7) (Save the previous partition matrix as $\hat{\mathbf{U}}$ to analyze the convergence).
7: $t = t + 1$
8: **until** $||\hat{\mathbf{U}} - \mathbf{U}|| < \epsilon$ or $t = t_{max}$.
9: **return** $\mathbf{U} = [u_{ij}]_{k \times N}$, $\mathbf{P} = [p_{ij}]_{k \times N}$, and $\mathbf{V} = [\mathbf{v}_i]_{k \times n}$.

and takes only $O(kN)$ operations to be computed for all objects and clusters.

PFCM: Possibilistic-Fuzzy c-Means

The constraint (5.66) used by FPCM related to the possibilistic matrix, **P**, may induce unreal (lower) typicality values when N is large. To circumvent this problem, the PFCM algorithm [16] rules out constraint (5.66) and aims at minimizing the following objective function

$$J = \sum_{i=1}^{k} \sum_{j=1}^{N} (a \cdot (u_{ij})^m + b \cdot (p_{ij})^{\gamma}) D_{ij} + \sum_{i=1}^{k} \eta_i \sum_{j=1}^{N} (1 - p_{ij})^{\gamma}, \tag{5.71}$$

where $\mathbf{U} = [u_{ij}] \in \mathbf{M}_{fp}$, $\mathbf{P} = [p_{ij}] \in \mathbf{M}_f$, $a > 0$ and $b > 0$ define the relative importance between memberships and typicalities, $m > 1$ and $\gamma > 1$ are fuzzification exponents, and $\eta_i > 0\ \forall i$ are user-defined weighting parameters (see Sect. 5.8). Note that if $a = 0$, then minimizing (5.71) is equivalent to minimizing (5.62) (PCM). On the other hand, if $b = 0$ and $\eta_i = 0\ \forall i$, then (5.71) reduces to (5.5) (FCM). In fact, if $b = 0$ then it follows from (5.71) that p_{ij} is not affected by D_{ij} and by objects $\mathbf{x}_j$ and centers $\mathbf{v}_i$. Therefore, minimizing (5.71) is equivalent to minimizing (5.5) (FCM).

The method to solve (5.71) is a simple Picard iteration through the first-order conditions for stationary points. Holding D_{ij} and p_{ij} constant the necessary conditions to minimize (5.71) lead to (5.7). Holding D_{ij} and u_{ij} constant, the necessary conditions to minimize (5.71) lead to ($1 \le i \le k,\ 1 \le j \le N$)

$$p_{ij} = \left(1 + \left(\frac{b}{\eta_i} D_{ij}\right)^{1/(\gamma-1)}\right)^{-1}. \tag{5.72}$$

Algorithm 14 PFCM

Require: Besides all the requirements of Algorithm 1, the weighting parameters $\eta_i \geq 0$, fuzzification exponent $\gamma \in (1, \infty)$, initial (random) partition matrix $\mathbf{P} = [p_{ij}] \in \mathbf{M}_f$, and importances between memberships and typicalities $a > 0$ and $b > 0$.

1: $t = 0$.
2: **repeat**
3: Compute the cluster prototypes by using (5.73).
4: Compute the distance between objects and prototypes by using (5.8).
5: Update the possibilistic partition matrix by using (5.72).
6: Update the probabilistic partition matrix by using (5.7) (Save the previous partition matrix as $\hat{\mathbf{U}}$ to analyze the convergence).
7: $t = t + 1$
8: **until** $||\hat{\mathbf{U}} - \mathbf{U}|| < \epsilon$ or $t = t_{max}$.
9: **return** $\mathbf{U} = [u_{ij}]_{k \times N}$, $\mathbf{P} = [p_{ij}]_{k \times N}$, and $\mathbf{V} = [\mathbf{v}_i]_{k \times n}$.

Then, holding u_{ij} and p_{ij} constant, the necessary conditions lead to

$$\mathbf{v}_i = \frac{\sum_{j=1}^{N}(a \cdot (u_{ij})^m + b \cdot (p_{ij})^\gamma)\mathbf{x}_j}{\sum_{j=1}^{N}(a \cdot (u_{ij})^m + b \cdot (p_{ij})^\gamma)}. \tag{5.73}$$

The complete procedure is given in Algorithm 14.

References

1. Jain AK, Dubes RC (1988) Algorithms for clustering data. Prentice Hall, Englewood Cliffs
2. Hoppner F, Klawonn F, Kruse R, Runker T (1999) Fuzzy cluster analysis: methods for classification, data analysis and image recognition. Wiley, London
3. Krishnapuram R, Keller JM (1993) A possibilistic approach to clustering. IEEE Trans Fuzzy Syst 1(2):98–110
4. Ruspini EH (1970) Numerical methods for fuzzy clustering. Inf Sci 2:319–350
5. Bezdek JC (1981) Pattern recognition with fuzzy objective function algorithms. Plenum Press, New York
6. Babuska R (1998) Fuzzy modeling for control. Kluwer, Norwell
7. Yang MS (1993) A survey of fuzzy clustering. Math Comput Model 18(11):1–16
8. Baraldi A, Blonda P (1999) A survey of fuzzy clustering algorithms for pattern recognition. I. IEEE Trans Syst Man Cybern B Cybern 29(6):778–785
9. Baraldi A, Blonda P (1999) A survey of fuzzy clustering algorithms for pattern recognition. II. IEEE Trans Syst Man Cybern B Cybern 29(6):786–801
10. Dunn JC (1973) A fuzzy relative of the isodata process and its use in detecting compact well-separated clusters. J Cybern 3:32–57
11. Gustafson DE, Kessel WC (1979) Fuzzy clustering with a fuzzy covariance matrix. In: 1978 IEEE conference on decision and control including the 17th symposium on adaptive processes, 10–12 January 1979, pp 761–766. doi:10.1109/CDC.1978.268028
12. Gath I, Geva AB (1989) Unsupervised optimal fuzzy clustering. IEEE Trans Pattern Anal Mach Intell 11(7):773–780

13. Bezdek JC, Coray C, Gunderson R, Watson J (1981) Detection and characterization of cluster substructure I. Linear structure: fuzzy c-Lines. SIAM J Appl Math 40(2):339
14. Bezdek JC, Coray C, Gunderson R, Watson J (1981) Detection and characterization of cluster substructure II. Fuzzy c-Varieties and convex combinations thereof. SIAM J Appl Math 40(2):358
15. Pal NR, Pal K, Bezdek JC (1997) A mixed c-means clustering model. In: Proceedings of 6th international fuzzy systems conference, Barcelona, pp 11–21
16. Pal NR, Pal K, Keller JM, Bezdek JC (2005) A possibilistic fuzzy c-means clustering algorithm. IEEE Trans Fuzzy Syst 13(4):517–530
17. Park B, Kargupta H, Johnson E, Sanseverino E, Hershberger D, Silvestre L (2001) Distributed, collaborative data analysis from heterogeneous sites using a scalable evolutionary technique. Appl Intell 16:19–42
18. Tsoumakas G, Angelis L, Vlahavas I (2004) Clustering classifiers for knowledge discovery from physically distributed databases. Data Knowl Eng 49:223–242
19. Datta S, Bhaduri K, Giannella C, Wolff R, Kargupta H (2006) Distributed data mining in peer-to-peer networks. IEEE Internet Comput 10:18–26
20. Zhang S, Zhang C, Wu X (2004) Knowledge discovery in multiple databases. Springer, Berlin
21. Park B-H, Kargupta H (2002) Distributed data mining: algorithms, systems, and applications. In: Data mining handbook, pp 341–358
22. Zaki MJ, Pan Y (2002) Introduction: recent developments in parallel and distributed data mining. Distrib Parallel Databases 11:123–127. doi:10.1023/A:1013918601668
23. Congiusta A, Talia D, Trunfio P (2007) Distributed data mining services leveraging WSRF. Future Gener Comput Syst 23(1):34–41
24. Kargupta H, Chan P (eds) (2000) Advances in distributed and parallel knowledge discovery. MIT Press, Cambridge
25. Johnson E, Kargupta H (2000) Collective, hierarchical clustering from distributed, heterogeneous data. In: Zaki M, Ho C-T (eds) Large-scale parallel data mining. Lecture notes in computer science, vol 1759. Springer, Berlin, Heidelberg, pp 803–803
26. Kargupta H, Huang W, Sivakumar K, Johnson E (2001) Distributed clustering using collective principal component analysis. Knowl Inf Syst 3:422–448
27. Klusch M, Lodi S, Moro G (2003) Distributed clustering based on sampling local density estimates. In: Proceedings of the 18th International Joint Conference on artificial Intelligence (IJCAI'03). Morgan Kaufmann Publishers Inc., San Francisco, 485–490
28. da Silva J, Giannella C, Bhargava R, Kargupta H, Klusch M (2005) Distributed data mining and agents. Eng Appl Artif Intell 18:791–807
29. S. Merugu, Ghosh J (2005) A privacy-sensitive approach to distributed clustering. Pattern Recognit Lett 26(4):399–410
30. Freitas A, Lavington SH (1997) Mining very large databases with parallel processing. Kluwer, Boston
31. Zaki MJ (2000) Parallel and distributed data mining: an introduction. In: Large-scale parallel data mining, vol 1759. Springer, Berlin, Heidelberg, pp. 1–23
32. Zaki MJ (1999) Parallel and distributed association mining: a survey. IEEE Concurrency 7:14–25
33. Pedrycz W (2005) Knowledge-based clustering from data to information granules. Wiley, Hoboken
34. da Silva JC, Klusch M (2007) Privacy preserving pattern discovery in distributed time series. In: IEEE 23rd international conference on data engineering workshop, pp 207–214
35. da Silva JC, Klusch M (2006) Inference in distributed data clustering. Eng Appl Artif Intell 19(4):363– 369
36. Olson C (1995) Parallel algorithms for hierarchical clustering. Parallel Comput 21:1313–1325
37. Dhillon IS, Modha DS (2000) A data-clustering algorithm on distributed memory multiprocessors. In: Revised papers from large-scale parallel data mining, workshop on large-scale parallel KDD systems, SIGKDD, London. Springer, Berlin, pp 245–260

38. Tian J, Zhu L, Zhang S, Liu L (2005) Improvement and parallelism of k-Means clustering algorithm. Tsinghua Sci Technol 10:277–281
39. Z Du, Lin F (2005) A novel parallelization approach for hierarchical clustering. Parallel Comput 31:523–527
40. Garg A, Mangla A, Gupta N, Bhatnagar V (2006) PBIRCH: a scalable parallel clustering algorithm for incremental data. In: 10th International database engineering and applications symposium, pp 315–316
41. Hammouda KM, Kamel MS (2009) Hierarchically distributed Peer-to-Peer document clustering and cluster summarization. IEEE Trans Knowl Data Eng 21:681–698
42. Datta S, Giannella CR, Kargupta H (2009) Approximate distributed K-means clustering over a peer-to-peer network. IEEE Trans Knowl Data Eng 21:1372–1388
43. Coletta LFS, Vendramin L, Hruschka ER, Campello RJGB, Pedrycz W (2012) Collaborative fuzzy clustering algorithms: some refinements and design guidelines. IEEE Trans Fuzzy Syst 20(3):444–462
44. Forman G, Zhang B (2000) Distributed data clustering can be efficient and exact. ACM SIGKDD Explor Newslett 2:34–38
45. Strehl A, Ghosh J (2003) Cluster ensembles — a knowledge reuse framework for combining multiple partitions. J Mach Learn Res 3:583–617
46. Pedrycz W, Hirota K (2008) A consensus-driven fuzzy clustering. Pattern Recognit Lett 29:1333–1343
47. Pedrycz W, Rai P (2008) Collaborative clustering with the use of fuzzy c-means and its quantification. Fuzzy Sets Syst. 159(18):2399–2427
48. Kwok T, Smith K, Lozano S, Taniar D (2002) Parallel fuzzy c-means clustering for large data sets. In: Monien B, Feldmann R (eds) Euro-Par 2002 parallel processing. Lecture notes in computer science, vol. 2400. Springer, Berlin, pp 365–374. 10.1007/3-540-45706-2_48
49. Rahimi S, Zargham M, Thakre A, Chhillar D (2004) A parallel fuzzy c-mean algorithm for image segmentation. In: Processing NAFIPS '04. IEEE annual meeting of the fuzzy information, vol 1, pp 234–237
50. Modenesi M, Costa M, Evsukoff A, Ebecken N (2007) Parallel fuzzy c-means cluster analysis. In: Daydé M, Palma JMLM, Coutinho ÁLGA, Pacitti E, Lopes JC (eds) High performance computing for computational science - VECPAR 2006. Lecture notes in computer science, Springer Berlin, vol 4395, pp 52–65. doi:10.1007/978-3-540-71351-7_5
51. Milligan GW, Cooper MC (1985) An examination of procedures for determining the number of clusters in a data set. Psychometrika 50(2):159–179
52. Vendramin L, Campello RJGB, Hruschka ER (2010) Relative clustering validity criteria: a comparative overview. Stat Anal Data Min 3(4):209–235
53. Arbelaitz O, Gurrutxaga I, Muguerza J, Pérez JM, Perona I (2013) An extensive comparative study of cluster validity indices. Pattern Recognit 46(1):243–256
54. Halkidi M, Batistakis Y, Vazirgiannis M (2001) On clustering validation techniques. J Intell Inf Syst 17:107–145
55. Vendramin L (2012) Study and development of fuzzy clustering algorithms in centralized and distributed scenarios (in Portuguese). Master's Dissertation, University of São Paulo, Brazil. http://www.teses.usp.br/teses/disponiveis/55/55134/tde-10092012-163429/en.php
56. Kolen JF, Hutcheson T (2002) Reducing the time complexity of the fuzzy c-means algorithm. IEEE Trans Fuzzy Syst 10(2):263–267
57. Celebi ME, Kingravi HA (2012) Deterministic initialization of the k-means algorithm using hierarchical clustering. Intern J Pattern Recognit Artif Intell 26(07):1250018
58. Celebi ME, Kingravi HA, Vela PA (2013) A comparative study of efficient initialization methods for the k-means clustering algorithm. Expert Syst Appl 40(1):200–210
59. Xie XL, Beni G (1991) A validity measure for fuzzy clustering. IEEE Trans Pattern Anal Mach Intell 13:841–847
60. Pal NR, Bezdek JC (1995) On cluster validity for the fuzzy c-means model. IEEE Trans Fuzzy Syst 3(3):370–379

61. Campello RJGB, Hruschka ER (2006) A fuzzy extension of the silhouette width criterion for cluster analysis. Fuzzy Sets Syst 157(21):2858–2875
62. Hruschka ER, Campello RJGB, de Castro, LN (2006) Evolving clusters in gene-expression data. Inf Sci 176:1898–1927
63. Kwon SH (1998) Cluster validity index for fuzzy clustering. Electron Lett 34(22):2176–2177
64. Tang Y, Sun F, Sun Z (2005) Improved validation index for fuzzy clustering. In: Proceedings of the 2005 American control conference, vol. 2, pp 1120–1125
65. Fukuyama Y, Sugeno M (1989) A new method of choosing the number of clusters for the fuzzy c-means method. In: Proceedings of 5th fuzzy systems symposium, pp 247–250
66. Bouguessa M, Wang S, Sun H (2006) An objective approach to cluster validation. Pattern Recognit Lett 27:1419–1430
67. Wang W, Zhang Y (2007) On fuzzy cluster validity indices. Fuzzy Sets Syst 158(19): 2095–2117
68. Bouguessa M, Wang SR (2004) A new efficient validity index for fuzzy clustering. In: Proceedings of 2004 international conference on machine learning and cybernetics, vol 3, pp 26–29
69. Pakhira MK, Bandyopadhyay S, Maulik U (2004) Validity index for crisp and fuzzy clusters. Pattern Recognit 37:487–501
70. Yoshinari Y, Pedrycz W, Hirota K (1993) Construction of fuzzy models through clustering techniques. Fuzzy Sets Syst 54(2):157–165
71. Nikhil R, Pal K, Bezdek JC, Runkler TA (1997) Some issues in system identification using clustering. In: International conference on neural networks, vol 4, pp 2524–2529
72. Vendramin L, Campello RJGB, Coletta LFS, Hruschka ER (2011) Distributed fuzzy clustering with automatic detection of the number of clusters. In: International symposium on distributed computing and artificial intelligence. Advances in intelligent and soft computing, vol 91. Springer, Berlin, Heidelberg, pp 133–140
73. Pakhira MK, Bandyopadhyay S, Maulik U (2005) A study of some fuzzy cluster validity indices, genetic clustering and application to pixel classification. Fuzzy Sets Syst 155(2): 191–214
74. Vendramin L, Campello RJGB, Hruschka ER (2009) On the comparison of relative clustering validity criteria. In: SIAM international conference on data mining, pp 733–744
75. Milligan GW, Cooper MC (1981) A monte carlo study of thirty internal criterion measures for cluster analysis. Psychometrika 46(2):187–199
76. Hubert L, Arabie P (1985) Comparing partitions. J Classif 2:193–218
77. Milligan GW, Schilling DA (1985) Asymptotic and finite sample characteristics of four external criterion measures. Multivar Behav Res 20(1):97–109
78. Kowalczyk W, Vlassis N (2005) Newscast EM. Adv Neural Inf Process Syst 17:713–720
79. Bezdek JC, Dunn JC (1975) Optimal fuzzy partitions: a heuristic for estimating the parameters in a mixture of normal distributions. IEEE Trans Comput C-24(8):835–838
80. Barni M, Cappellini V, Mecocci A (1996) Comments on "A possibilistic approach to clustering". IEEE Trans Fuzzy Syst 4(3):393–396
81. Timm H, Borgelt C, Kruse R, Doring C (2004) An extension to possibilistic fuzzy cluster analysis. Fuzzy Sets Syst 147(1):3–16

Chapter 6
Density Based Clustering: Alternatives to DBSCAN

Christian Braune, Stephan Besecke, and Rudolf Kruse

Abstract Clustering data has been an important task in data analysis for years as it is now. The de facto standard algorithm for density-based clustering today is DBSCAN. The main drawback of this algorithm is the need to tune its two parameters ϵ and *minPts*. In this paper we explore the possibilities and limits of two novel different clustering algorithms. Both require just one DBSCAN-like parameter. Still they perform well on benchmark data sets. Our first approach just uses a parameter similar to DBSCAN's *minPts* parameter that is used to incrementally find protoclusters which are eventually merged while discarding those that are too sparse. Our second approach only uses a local density without any minimum number of points to be specified. It estimates clusters by seeing them from spectators watching the data points at different angles. Both algorithms lead to results comparable to DBSCAN. Our first approach yields similar results to DBSCAN while being able to assign multiple cluster labels to a points while the second approach works significantly faster than DBSCAN.

Keywords Density based clustering • Multi-class clustering • Spike train analysis • Prototype based clustering • Kernel based clustering

6.1 Introduction

Clustering is notably one of the most important tools in exploratory data analysis [19]. It helps detect structures and relations within a data set of unordered, unlabeled data and provides data analysts with means to draw conclusion or discover knowledge in large amounts of data. Clustering techniques can be used in huge variety of fields spanning from simple point-based data to text [1, 15, 20] or image analysis [12, 13, 18, 26].

Formally clustering can be seen as a partitioning of a data set into several non-empty, (not necessarily) disjoint subsets.

C. Braune (✉) • S. Besecke • R. Kruse
University Magdeburg, Magdeburg, Germany
e-mail: christian.braune@ovgu.de; stephan.besecke@st.ovgu.de; rudolf.kruse@ovgu.de

M.E. Celebi (ed.), *Partitional Clustering Algorithms*,
DOI 10.1007/978-3-319-09259-1_6

Definition 1. A data set is a set of points $x_i \in \mathbb{R}^d$, where d is the dimensionality of the data set $\mathscr{D} = \{x_1, x_2, \ldots, x_n\}$ with n points.

Definition 2. We denote a simple clustering of $\mathscr{D}$ as $\mathscr{C} = \{C_1, C_2, \ldots, C_k\}$ with

$$\bigcup_{i=1}^{k} C_i = \mathscr{D} \text{ and } C_i \cap C_j = \emptyset \leftrightarrow C_i \neq C_j .$$

That is, every point in $\mathscr{D}$ belongs to one cluster and all clusters are disjoint.

This is commonly understood as a clustering of a data set. We will see later that there are circumstances under which we may want to allow overlapping clusters and have to weaken the cluster definition in the sense that they do not have to be disjoint. The result of a clustering algorithm may either be the clusters themselves in the form of sets of points or as a list of labels $\{\lambda_1, \ldots, \lambda_n\}$ such that $\lambda_i = j$ if and only if $x_i \in C_j$. Usually the cluster label (or index) $\lambda(x)$ will be a positive integer or a set of labels:

$$\lambda_{\mathscr{C}}(x) = \{i \,|\, x \in C_i, C_i \in \mathscr{C}\}$$

If the clustering algorithm applied is able to recognize *noise* (points that do form any cluster) the cluster label for such points is usually set to -1 or $\emptyset$.

Classical clustering algorithms such as k-means [21] require the user to specify the number of clusters they want to find. The clustering algorithm will then report exactly this number of clusters, whether or not the clustering makes any sense or not. Some clustering algorithm that are implemented as optimization problems can be run several times with different parameters to choose the best result according to the objective function. In the case of k-means this poses two major problems:

1. The choice of the initial cluster centers' locations may lead to totally different clusters [2, 6, 7]. Thus, the algorithm needs to run several times for each value of k chosen, to ensure that an (nearly) optimal solution for this parameter has been found.
2. With increasing k the clustering becomes better and better – at least in terms of the objective function used by k-means.

$$J(U, C)_{\mathscr{D}} = \sum_{i=1}^{n} \sum_{j=1}^{|C|} u_{ij} d^2\left(x_i, c_j\right).$$

In this function U describes a membership matrix with values u_{ij} being either 1 (if x_i belongs to cluster c_j) or 0 (otherwise). $d^2(x, y)$ describes the distance (resulting from any metric applicable) between two points x and y. It is easy to see that the function will reach a global optimum if $k = n$. Some penalty terms would be required (such as minimum description length or Bayesian information criterion) to ensure the clustering does not get too complex. Alternatively fuzzy

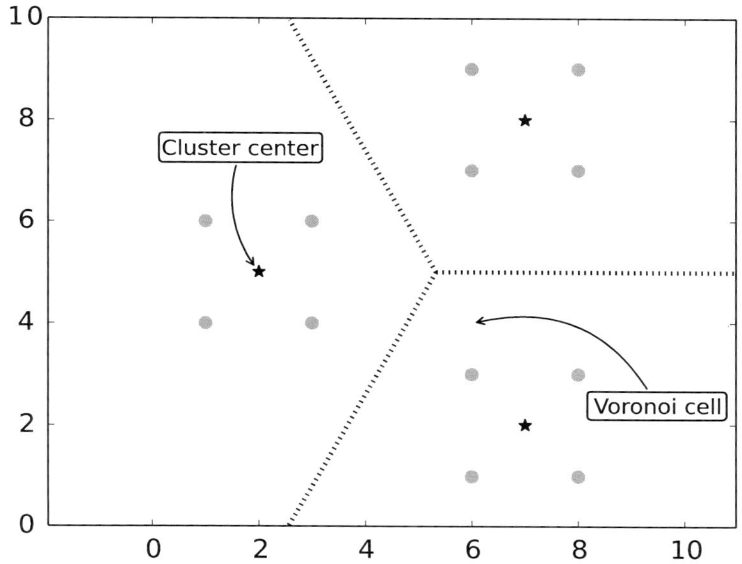

Fig. 6.1 k-Means clustering of a simple data set consisting of three clusters. Voronoi cells of a point c are the areas in which no other cluster center is closer to said point

clustering (fuzzy c-means) [8] can be applied. Points that have an equally high degree of membership in several clusters might indicate that these clusters can be merged.

Another drawback is k-means' limitation to convex-shaped clusters. Due to the nature of the Voronoi cells clusters that are non-convex can only be captured in a single cluster if the other data points' locations allow this to happen (see Fig. 6.1). More often than not the cluster will be split apart and the resulting clustering becomes meaningless [24].

The last problem of k-means we want to mention here is its vulnerability against noise. As noise we consider data points that do not fit well into any cluster. k-means assigns every point to a cluster and moves the cluster prototype to the center of all points assigned, making it susceptible for outliers. Such points may even distort the clustering into complete nonsense if the amount of noise is sufficiently high.

From these disadvantages classical cluster algorithms possess, we define some properties we expect a clustering algorithm should have:

- Ideally a clustering algorithm would be able to choose an appropriate number of clusters itself without the additional need of user interaction or penalty terms.
- It should also be able to detect clusters of arbitrary shape.
- The algorithms result should at most change slightly in the presence of noise or outliers.
- The number of parameters that need to be specified by the user should be as small as possible.

As there is *no free lunch* [25] for clustering, we will hardly see an algorithm that completely fulfills all of these requirements. Luckily there exist algorithms that fulfill most of these. These are density-based clustering algorithms. One of the most prominent representatives of this class of clustering algorithms will be described in the following section.

Other alternatives for clustering a data set include hierarchical clustering [5], where each data point forms its own cluster and in each step of the algorithm the two clusters that are closest to each other (according to a chosen linkage criterion) are merged until only one cluster remains. The distances at which clusters are merged can be visualized into a dendrogram and then be analyzed for finding a suitable threshold. Instead of merging clusters there is also a divisive procedure that implements a top-down approach [16].

6.2 Related Work

Some of the problems mentioned in the previous section can be traced back to a limited definition of what a cluster actually is. Clusters are implicitly defined by the k-means algorithms as sets of data points that where each point within this set is close to each other point within the set and distant each point not in the same set. This is a very broad notion of density which can be refined in several ways as we will see in the following.

6.2.1 DBSCAN

DBSCAN is possibly the most prominent density-based clustering algorithm as of today. Points are categorized into so-called core points which have many neighbouring points in their direct vicinity, border points which lie in the neighborhood of at least one core point, and noise points which are neither of the first.

The following part describes the DBSCAN algorithm in more detail, following the original definitions and structure of [10]. First, we need to formally define when a point should become a core point of a cluster. For this we need to define the term ϵ-neighborhood.

Definition 3. The ϵ-neighborhood of a point x_i is the set of points, which have a distance of at most ϵ to that point:

$$\mathcal{N}_\epsilon(x_i) = \{x \in \mathcal{D} \mid d(x_i, x) \leq \epsilon\}.$$

Then we can define core points with respect to a local density measure, i.e. the number of points within a point's ϵ-neighborhood.

Definition 4. To the set of core points belong all points whose ϵ-neighborhood contains at least *minPts* other points:

$$\mathscr{P} = \{x \in \mathscr{D} \mid \|\mathscr{N}_\epsilon(x)\| \geq minPts\}.$$

Now, a point that lies within an ϵ-neighborhood of a core point might itself be no core point because it lies on the border of a cluster and thus does not fulfill the density criterion posed by the *minPts* parameter.

Definition 5. A point p is called *directly density-reachable* from a core point x if $p \in \mathscr{N}_\epsilon(x)$.

Definition 6. A point p is called *density-reachable* from a core point x if there is a series of points $p_1, p_2, \ldots, p_n$ with $p = p_n$ and $x = p_1$ and for every (p_i, p_{i+1}) holds: p_{i+1} is directly density-reachable from p_i.

Definition 7. Two points p and q are called *density-connected* if there is a point o such that p and q are both density-reachable from o.

With these notions of density we can finally define the clusters DBSCAN will find:

Definition 8. A set $C \subseteq \mathscr{D}$ is called *cluster* if for every $p, q \in C$ holds: p and q are density-connected and there is no superset of C that is also a *cluster*.

Every point that is not density-connected to any other point (and thus does not belong to any cluster) is considered to be noise. Figure 6.2 visualizes the different concepts of connectivity explained before. The set of points $C = \{p, x_1, x_2, x_3, q\}$ forms a cluster in this example.

With this notion of what a cluster actually is, we can find almost arbitrarily-shaped clusters in also high-dimensional data. Even clusters whose convex hulls are

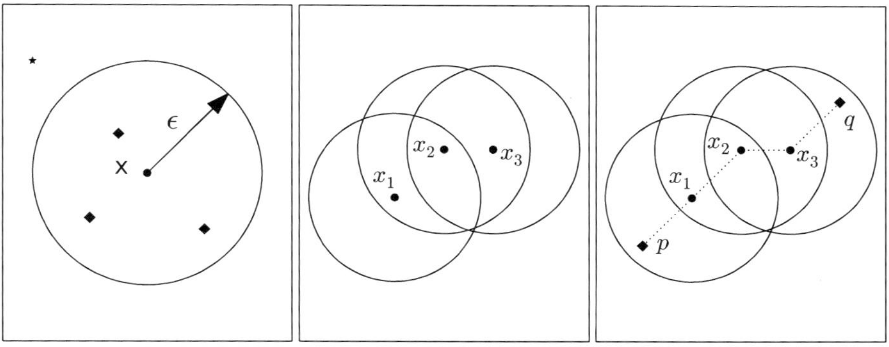

Fig. 6.2 The *left plot* shows a point x and its ϵ-neighborhood. The *star-shaped point* lies not in $\mathscr{N}_\epsilon(x)$, the *diamond-shaped points* do. The *center plot* shows three points x_1, x_2, x_3. x_1 and x_3 are directly density-reachable from x_2 and thus x_3 is density-reachable from x_1. The *right plot* shows two additional points p and q which are density-connected by x_1, x_2 and x_3

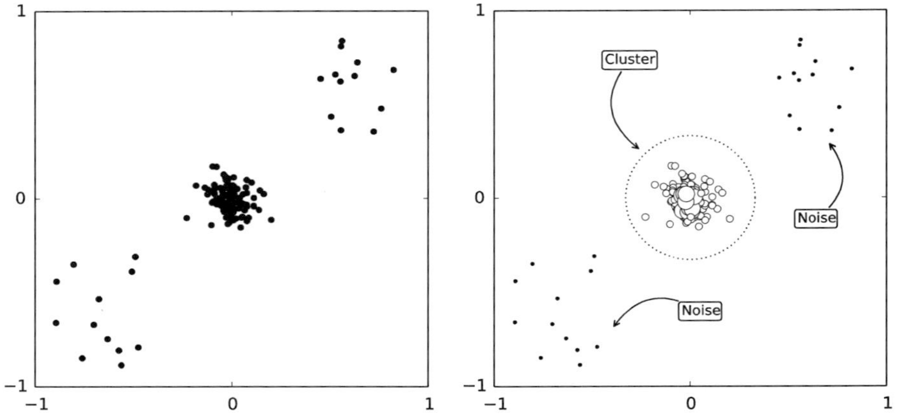

Fig. 6.3 A single cluster with two patches of noise in the *bottom left* and *top right corner* in the *left plot*. In the *right plot* core points are plotted as *bigger circles*, density-connected non-core points smaller

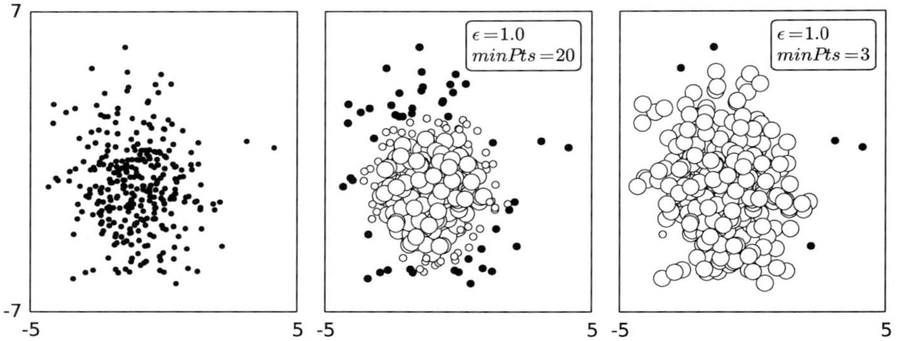

Fig. 6.4 *Left plot* shows the data set of 300 points, originally all drawn from the same multivariate Gaussian distribution. The other plots show the clustering with the given parameters. *minPts* has been chosen too large. *White points* belong to the same cluster, *black points* are considered noise

intersecting (which are not linear separable) can be distinguished with this algorithm. With the help of suitable data structures the region queries (asking for the points in a certain neighborhood of a point) can be computed fairly fast [17] leading to an overall fast computation of the clustering result (Fig. 6.3).

Yet, there are still some drawbacks that should not go unmentioned in this algorithm. First, the choice of *minPts* greatly influences the outcome of the algorithm. A value too low will turn the result close to what could have been achieved by hierarchical clustering. In fact, the resulting clustering is equal to that of hierarchical clustering with a cutoff threshold of ϵ if we chose $minPts = 2$. However, if the value for *minPts* is chosen too large then only few to none points might fulfill the core point condition and no clusters will be formed at all. This effect is illustrated in Fig. 6.4.

The second parameter – ϵ – influences the outcome as well [9]. Since it is used to calculate the neighborhood of a point it is crucial in determining whether a point should become a core point or not. Smaller values of ϵ require also smaller values of *minPts*, if the same density is to be used, but this might lead to overall smaller clusters as the border points of a cluster (the ones which are density-connected but not core points) may not be captured by any ϵ-neighborhood. These points would become noise then.

Lastly, DBSCAN is very bad at differentiating between clusters with different densities [10] or clusters with locally varying density. Clusters with higher densities require a smaller ϵ or a smaller *minPts*, while low density clusters need these values increased. If now in a data sparse and tight clusters occur at the same time, DBSCAN runs into a situation where either the sparse clusters would not be recognized at all or might be merged with the tight cluster, making it impossible to distinguish these. Figure 6.5 shows this for a sample data set of two clusters generated with different densities.

As for all clustering algorithms that are based on distance measures DBSCAN does not cope very well with increasing dimensionality of the data set. The curse of dimensionality is here only one problem. The *minPts*-ϵ combination of parameters scales not very intuitively with increasing number of data set attributes. While a parameter setting of $(minPts, \epsilon) = (20, 2)$ means that we hope to see approximately 1.59 data points per unit of hypervolume in a two-dimensional data set, the same setting leads to 0.25 points per hypervolume unit in a data set as simple as only containing four attributes.

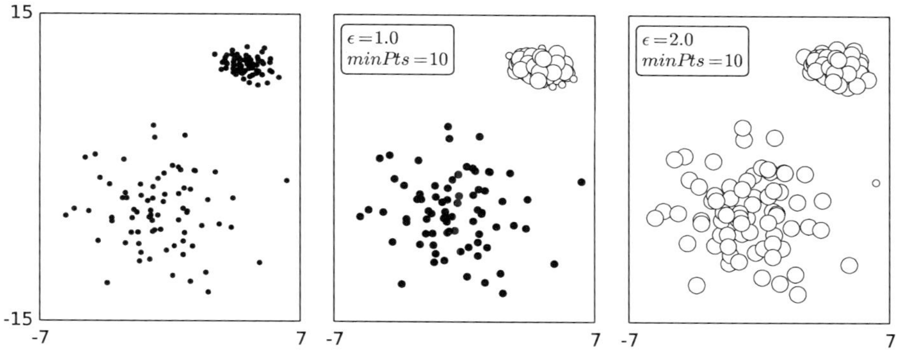

Fig. 6.5 Effects of different choices for ϵ if clusters of different densities are present. *Left plot* shows the data set of 150 points, distributed across two clusters of different densities. The other plots show the clustering with the given parameters. *White points* belong to the same cluster, *black points* are considered noise

6.3 Clustering with a Different Notion of Density

In this section we will describe to density-based clustering methods that circumvent the necessity to both specify a parameter for ϵ and *minPts*. Both algorithms use a slightly different notion of density, which thus enables them to cluster data sets with only one of the parameters required by DBSCAN. Of course this comes at a cost as we will explain later.

6.3.1 Black Hole Clustering

The first algorithm we want to describe uses only an ϵ as parameter to estimate clusters in an iterative way. The basic idea behind this algorithm is that if data are organized in different clusters and if we placed *black holes* outside of the bounding box of the data each single data point would be dragged towards the black hole. Depending on the original distance to the black hole, it would take a different amount of time for a data point to finally fall into the black hole and disappear. Clusters of data points should disappear at roughly the same time and could be identified in such a way.

Of course, if two clusters had the same distance to the black hole they would also disappear at the same time and could not be distinguished. Thus, we need to place several black holes around the data and observe each of it individually, recording the time (or equivalently the distance) from each point to each black hole. Clusters that could already be separated by one of the previous black holes can be processed separately. The algorithm identifies points on rings around each black hole as belonging to the same cluster. By processing the rings separately we are in a sense triangulating the clusters (in the two-dimensional case).

Figure 6.6 shows the pseudocode for this clustering algorithm. One of its advantages is that it does not require any special implementation of data structures other than an array or a list and that the overall runtime is considerably fast compared to other clustering algorithms. To calculate the complexity of this algorithm we take a look at the pseudocode:

Line 3 To construct the black holes we need to know the bounding box of the data set which takes $\mathbf{O}(n)$ and construct the location of each black hole in $\mathbf{O}(k)$ $k+1$ times. For proper triangulation we need $k+1$ observers in k-dimensional space to ensure that we correctly locate a point.
Line 4 Calculating the distance array needs $\mathbf{O}((k+1)k \cdot n)$ as distance calculation use $\mathbf{O}(k)$ and we need to perform this $(k+1)$ times for each data point.
Line 10 Sorting the array of points according to one of the distance can be done in $\mathbf{O}(n \cdot \log n)$.
Line 9 to 21 Each point is copied exactly once: $\mathbf{O}(n)$.
Line 8 to 24 This loop is executed $(k+1)$ times.

```
 1: procedure BHC(𝒟, ε)                              ▷ Cluster 𝒟 w.r.t. ε into disjoint subgroups
 2:     b ← k+1 for 𝒟 ⊆ ℝ^k                          ▷ Each data point has k attributes
 3:     bh ← Array of b points on a k-simplex surrounding 𝒟
 4:     ds ← {d⃗ = (d_1, d_2, ..., d_b) | d_i = dist(x, bh[i]) ∀x ∈ 𝒟}   ▷ Array of vectors,
 5:     j ← 0                                  ▷ each containing the distance to the respective black hole
 6:     cs ← [ds]                              ▷ Array containing the array of all distances
 7:     c2 ← []                                                          ▷ Empty list
 8:     while j < b do                                 ▷ Process every black hole seperately
 9:         for each c ∈ cs do
10:             Sort c by its j^th component in ascending order.
11:             cur ← [c[0]]                    ▷ Insert first element of c into current cluster
12:             for i = 1 to len(c) do
13:                 if c[i+1][j] − c[i][j] < ε then   ▷ Either the two points are close to each other
14:                     Append c[i+1] to cur     ▷ then the current point belongs to the current
    cluster
15:                 else                                                 ▷ or they are not,
16:                     Append cur to c2          ▷ then the current cluster is finished and
17:                     cur ← [c[i+1]]             ▷ the point already starts a new one.
18:                 end if
19:             end for
20:             Append cur to c2                  ▷ The last cluster has to be appended, too
21:         end for
22:         cs ← c2                                        ▷ Repeat with the splitted list
23:         c2 ← []                                        ▷ and empty the temporary
24:     end while
25:     return cs                  ▷ cs is now an array of arrays each containing the cluster points
26: end procedure
```

Fig. 6.6 Pseudocode for the black hole clustering algorithm

The complexity of the black hole clustering algorithm is then

$$\mathbf{O}\left(\overbrace{n + k\cdot(k+1)}^{\text{Line 3}} + \underbrace{(k+1)k\cdot n}_{\text{Line 4}} + \overbrace{(k+1)\cdot(\underbrace{n\cdot\log n}_{\text{Line 10}} + n)}^{\text{Line 8 to 24}}\right)$$

$$= \mathbf{O}\left(n + k^2 + k + k^2 n + kn + kn\log n + kn + n\log n + n\right)$$

$$= \mathbf{O}\left(2n + k + 2kn + n\cdot\log n + k^2 + k^2 n + k\cdot n\cdot\log n\right)$$

And since we can assume that n is much larger than k:

$$= \mathbf{O}\left(k\cdot n\cdot\log n\right).$$

Although this algorithm beats DBSCAN's runtime for large data set and can estimate the number of clusters present quite well it is limited to finding mainly convex clusters that are clearly linearly separable. It also has no built-in, explicit noise detection like DBSCAN has, instead noise will be kept in very small lists of cluster, each only containing few (most often only one) data points. Section 6.4 will give a more thorough evaluation of the algorithms capabilities.

6.3.2 Protoclustering

Our second algorithm abstains from using a measure for local density and rather uses only a *minPts* parameter. It works by iteratively finding subspaces of high density, forming the points contained within these into a set of so-called protoclusters which are then further analyzed for cluster structure. The idea behind this is that clusters are regions in space that are denser than their surrounding and can be identified as such.

In Fig. 6.7 we can see the process of forming the protoclusters for a very small data set of 9 points. We choose $minPts = 3$ for obvious reasons. The first step in forming the protoclusters is identifying dense regions in data space. For this we consider the whole data set to be a single cluster and calculate its center point. The point that lies furthest away from this center is the point must susceptible to being noise and is thus removed from the data set. This, however, changes the mean point, which has to be recalculated. We then proceed to remove the farthest point again and repeat the center calculation. With every step the bounding (hyper-)box becomes smaller and smaller while the number of points within it only decreases by one in each iteration. If only less than or equal to *minPts* points are left, the remaining points are returned as first protocluster.

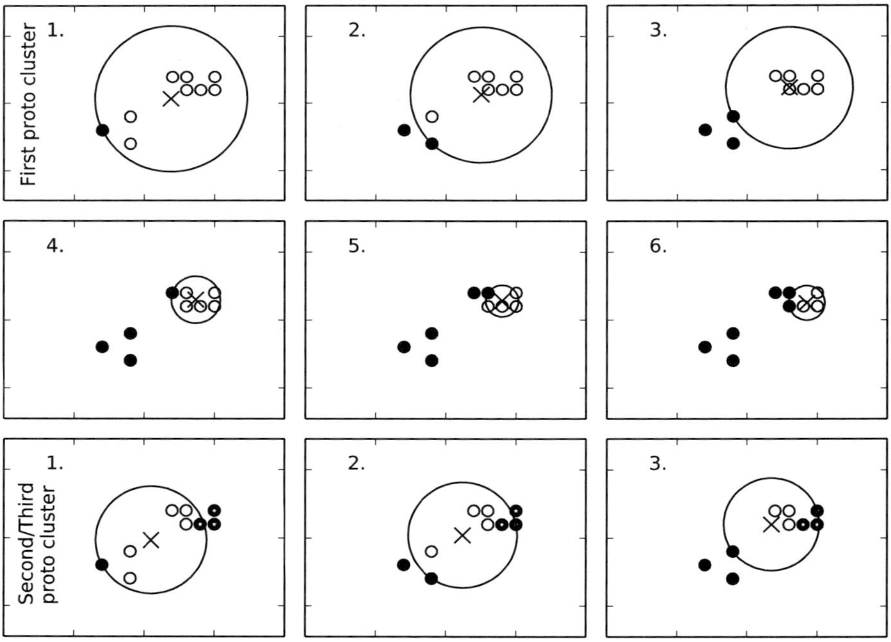

Fig. 6.7 The *first two rows* show how points are removed one by one and how the center point changes to find the first protocluster. The *bottom row* shows the generation of the second protocluster and since only *minPts* points are left they automatically turn into the third protocluster

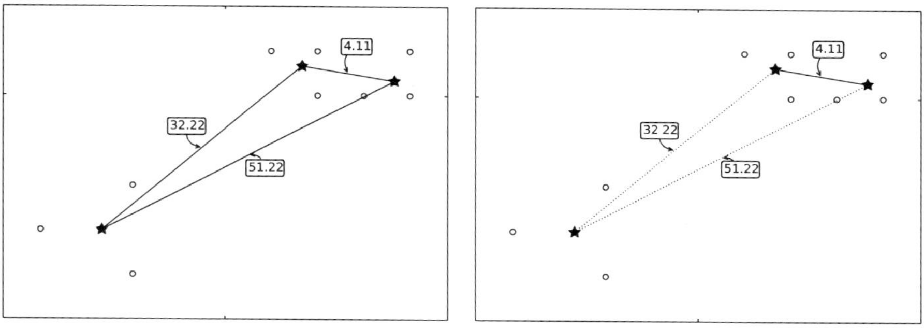

Fig. 6.8 The distance graph resulting from the protoclusters. The edges are labeled with the Euclidean distances between the protoclusters' centers The *right plot* shows which edges can be removed since their weights are significantly higher than the remaining one

The protocluster generation proceeds with the remaining data points until eventually all points are grouped into $c = \lceil \frac{n}{minPts} \rceil$ protoclusters ($c-1$ of size *minPts* and 1 of size at most *minPts*). Each cluster within the data set is now represented by several protoclusters and their structure can be used as a skeleton for the real clusters. To find, which protoclusters can be merged, we can look at the pairwise distances between the protoclusters' center points. Figure 6.8 shows the distance graph for the previous example. Obviously the distances to the protocluster in the bottom left corner of the data set are significantly higher than the distance between the top right protoclusters. If we remove all the edges from this graph whose distances (weights) are too large, we see (in Fig. 6.8 on the right) that only the protoclusters that originate from points that belong to the same cluster stay connected. So, clusters can be seen as connected components in the reduced distance graph and the final clustering would look like the one in Fig. 6.9.

The pseudocode of the algorithm so far can be found in Fig. 6.10.

With the implicit iterative adaptation to the local densities we can also find clusters with varying densities as Fig. 6.11 shows. With a proper threshold chosen the connected components even become cliques in the reduced distance graph which might give us another application for this algorithm: overlapping clusters. If a point may not only belong to one cluster but to several (similar to multi-label classification, where one is interested in assigning not only one but possibly several labels to a data point) such overlapping cluster structures could be identified as cliques in the distance graph structure. In such a graph, vertices could also belong to several cliques. Such a (admittedly constructed) example of a data set where we might assign two cluster labels to the central four points can be found in Fig. 6.12 and of course it would be also valid to consider the whole data set to be only a single cluster.

Since the protoclusters' size is fixed, we might have skipped some of the border points of a noisier cluster and such a point would later be part of a less dense protocluster. The center of such low-density protoclusters is probably a bad representation for the points contained in the cluster.

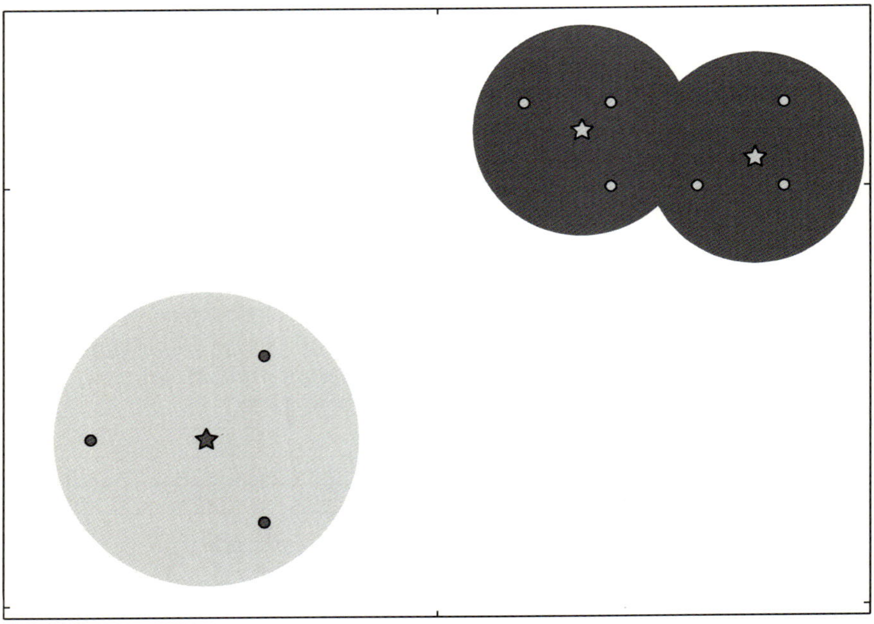

Fig. 6.9 The resulting clustering after connected components are identified

procedure PROTOCLUSTERING($\mathscr{D}, minPts$) ▷ Cluster $\mathscr{D}$ into $\lceil \frac{n}{minPts} \rceil$ disjoint protoclusters
  $ps \leftarrow []$ ▷ Initialize with an empty list
  while $\sum_{p \in ps} |p| < |\mathscr{D}|$ **do** ▷ As long as not every point is part of a protocluster
    $cand \leftarrow \{x \in \mathscr{D} | x$is not part of any protocluster yet$\}$ ▷ Get all points that are not yet in a protocluster
    while $|cand| \geq minPts$ **do** ▷ If there are more than *minPts* points left as candidates
      $c \leftarrow mean(cand)$ ▷ Calculate their mean
      $o = \arg\max_{x \in cand} d(x, c)$ ▷ Find the one farthest from the center
      $cand \setminus \{o\}$ ▷ And remove it from the set of candidates
    end while
    Append *cand* to *ps* ▷ The remaining form a protocluster
  end while
  return *ps* ▷ *ps* is now a list of arrays each containing the points that belong to the same protocluster
end procedure

Fig. 6.10 Pseudocode for the protoclustering algorithm

In Fig. 6.13 you can actually see two protocluster centers which are placed in a rather odd way. This is, because the points contained in these protoclusters lie on different sides of the circle. Since the process generating the protoclusters organized the prototypes in such a way that less than *minPts* points remained close to each other, a protocluster was formed that spans a larger hypervolume. Thus, these points

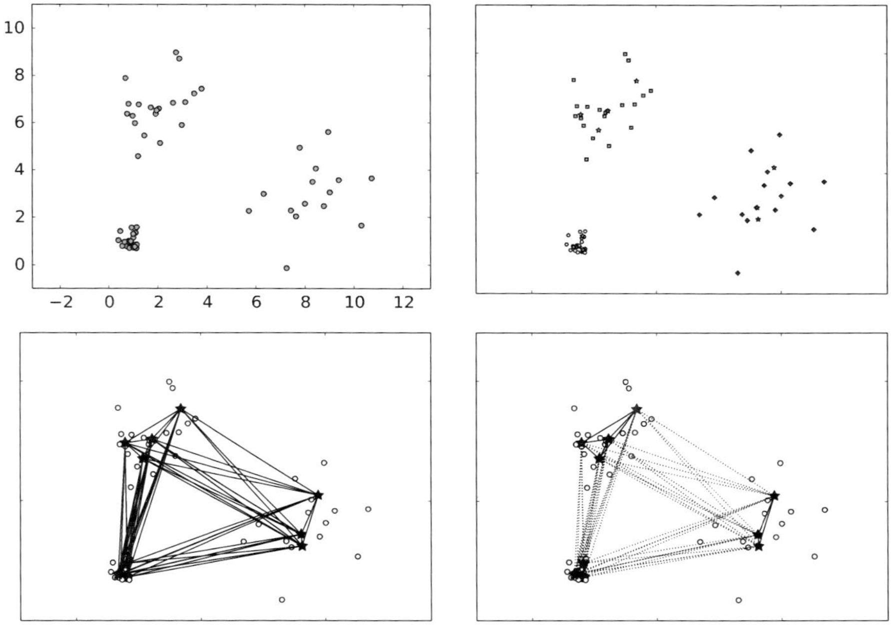

Fig. 6.11 A data set with three clusters of different densities, the final clustering and the distance graph (full and reduced)

never formed a dense region again until the very end of the protocluster generation. In Fig. 6.14 this is even more extreme as here some points are actually assigned to the wrong cluster.

To cope with these problems a *clean-up* step should be performed once after the protoclusters have been generated. For every point we check whether there is a protocluster that fits the point better than the one it is currently assigned to. This can be done simply by choosing the protocluster center that lies nearest to each point. If $\mathscr{P}$ is the set of protoclusters (represented by their center points $p_i \in \mathscr{P}$), the new protocluster for a point $x \in \mathscr{D}$ is:

$$\lambda(x) = \arg\min_{p \in \mathscr{P}} d(x, p)$$

6.4 Evaluation

To evaluate the algorithm we described in the previous section, we use two different scenarios. Since our black hole clustering approach produces mainly the same results as the DBSCAN algorithm (given that we have convex, non-overlapping clusters) and cannot properly identify clusters whose convex hulls are intersecting, we generated several data sets of different size and with a different number

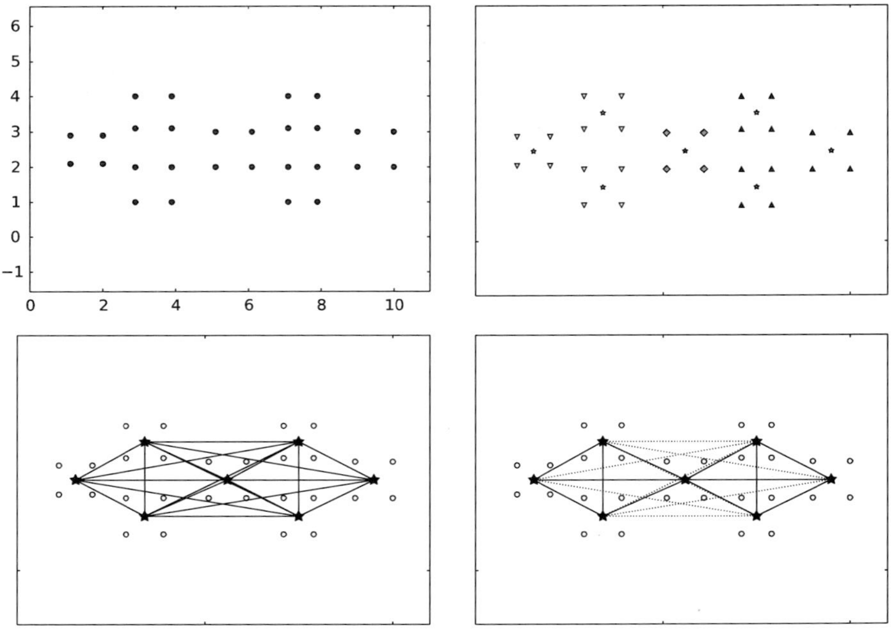

Fig. 6.12 A data set with two overlapping clusters. The final clustering assigns two different cluster labels to the central four points

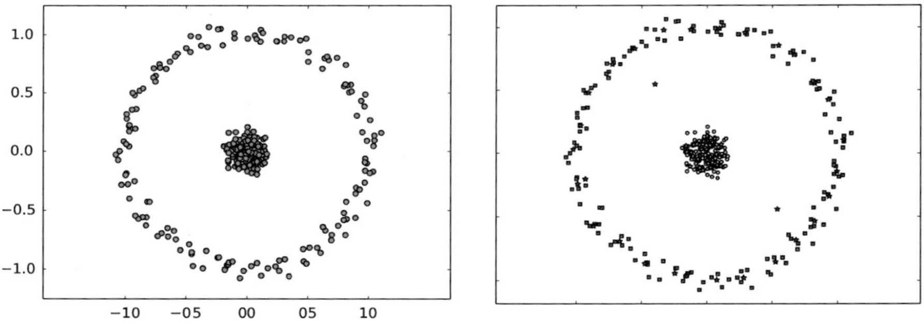

Fig. 6.13 A data set with *two circles*, one cluster is completely surrounded by the other

of clusters. To evaluate the clusterings we use the Adjusted Rand Index [14]. The Rand Index is a measure of agreement between two labelings of points and measure (independent of permutation of label indices) how well two different clusterings overlap. Given that we know the original labels of the data points from the generating process we can use this measure to assess the quality of the clustering. Values near 1.0 indicate strong agreement between the two sets of labels while values close to −1.0 indicate the opposite. Values near 0.0 indicate random label distribution. The Adjusted Rand Index used for our evaluation also corrects for guessing the correct label just by chance and can be seen as a normalized version of the original Rand Index.

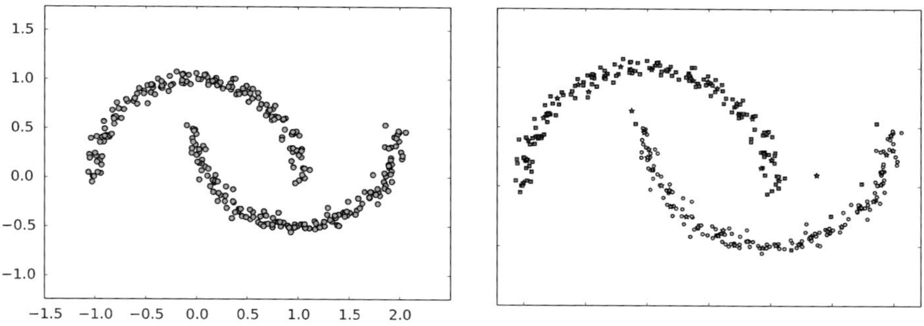

Fig. 6.14 A data set with *two half moons*, intruding each others convex hulls

For the protoclustering algorithm we resort to an application of the algorithm where we already tested its validity: spike train analysis (see [3,4,11,23] for a more in-depth explanation on the task). Neurons emit signals when stimulated by enough neurotransmitters. According to one of many competing hypotheses in neurobiology information is processed by several neurons and one piece of information that is processed by the same set of neurons excites these in a way that they emit signals at the same time. Signals recorded from a neuron over time are called spike train and can be analyzed in several ways. One of them is to turn the train into a vector space representation which then in turn can be clustered. Since the stochastical behaviour of a spike can roughly be described by a poisson process, we can simulate spike trains with the desired properties (coincident emitation of signals) and see, if our clustering algorithm can identify these groups of synchronized spike trains within a larger base of noise. Since it is unclear whether neurons actually only participate in one group or can be part of several groups at the same time, we also generated spike trains which overlap on some neurons (very similar to the data set from Fig. 6.12).

6.4.1 Black Hole Clustering Evaluation

To show the quality of the black hole clustering algorithm, we generated several data sets of one to four clusters. Each cluster was generated with the same number of points. The cluster centers' coordinates were each drawn from uniform distribution on the interval $[-10, 10]$. Around the centers points were drawn from a bivariate Gaussian distribution. Cluster indices were stored for each point to evaluate the resulting clustering against a ground truth. The size of the data set was $s \in \{100, 250, 500, 1{,}000, 2{,}500, 5{,}000, 10{,}000, 25{,}000, 50{,}000, 100{,}000\}$ data points in two dimensions. The tests themselves were repeated one hundred times for each size. DBSCAN was initialized with the same parameter for epsilon as our algorithm.

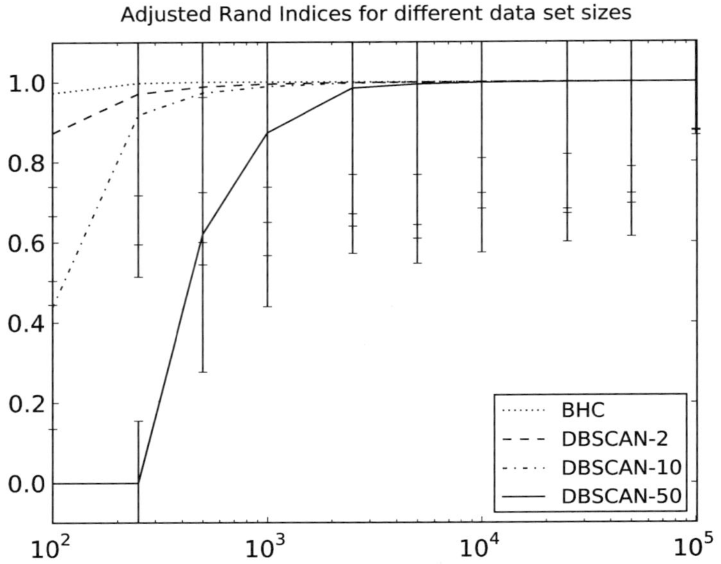

Fig. 6.15 Evaluation results for black hole clustering vs. different instances of DBSCAN. Please note the logarithmic x axis. Error bars indicate one standard deviation

For *minPts* we chose values from $\{2, 10, 50\}$ and ran the algorithms on the same data set. The computer used to perform the tests was a `HP Z400` workstation (8*GB* RAM, Six cores, each 3.3 GHz, hyperthreading enabled but not used) running with *MacOS Lion*. No parallelization was used during the tests, each algorithm could only use one of the cores available. The DBSCAN implementation from `scikit-learn` [22] was used. The results for the different test setups can be seen in Fig. 6.15 and the results for the runtime evaluation in Fig. 6.16. As we can see from the Fig. 6.15 the more points are contained in the data set, the better the clustering result. Naturally, if only a little more than 100 points are contained in the data set and they are distributed across several clusters, we expect DBSCAN to fail when it requires at least 50 points in any ϵ-neighborhood. Thus, the quality of the clustering for DBSCAN increases with the number of data points as this requirement can be met more easily – especially for blob-shaped clusters. The same but to a lesser degree holds true for the other DBSCAN instances that can meet the *minPts* requirement more easily right at the beginning of our tests.

On the other hand, our method surpasses DBSCAN in any of our chosen settings. Nearly perfect results are achieved even for small datasets, where the only errors originate from extreme outliers that were drawn from the same distribution but do not really fit into the ideal cluster shape. Since they are still labeled as belonging to a certain cluster they generate an error if labeled as noise. Still, the variation of the results and the results themselves are always better than for any instance of the DBSCAN algorithm.

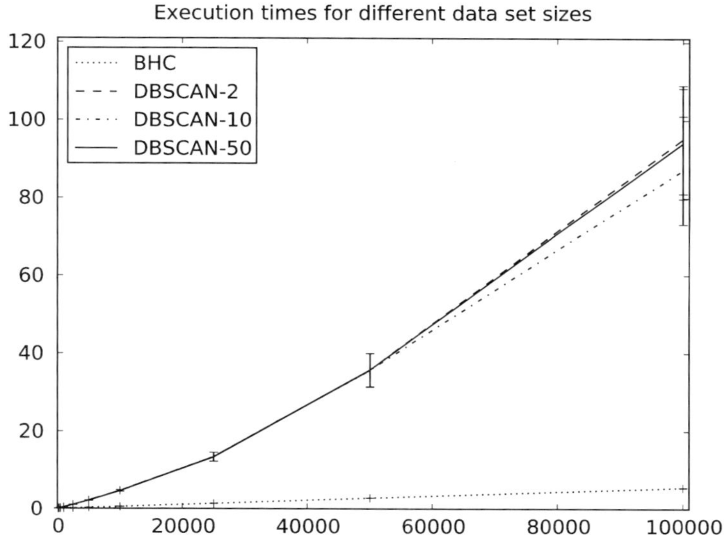

Fig. 6.16 Runtime evaluation results for black hole clustering vs. different instances of DBSCAN. The differences between the different setups of the DBSCAN are so little that they cannot be distinguished up to $x = 50{,}000$ in this plot. Error bars indicate one standard deviation (nearly zero for black hole clustering)

The runtime of all DBSCAN instances is nearly the same and increases approximately linearly with the number of points in the dataset as Fig. 6.16 shows. On the other hand nearly no increase in runtime is measurable for our clustering method (approximately 90 s for DBSCAN vs. less than 5 s for BHC).

6.4.2 *Protoclustering*

As for the protoclustering algorithm we generated 10,000 data sets with 80 % noise and only 20 % data points (spike trains). The main challenge here is to properly distinguish between the noise and non-noise data points. Of all data available from a single spike train again at least another 75 % are just noise and no usable information can be gained from those spikes within a train. But still these signals may coincide by chance with signals from other spike trains. On yet another level information about coincidences across synchronized spike trains may be lost (due to biological constraints) and up to 40 % of the coincidence information is lost again.

In Fig. 6.17 the accuracy over all the trials can be seen. Since the data was generated artificially we can again evaluate against a ground truth.

If we assume that spike trains can be part of more than one cluster we can model the information loss described earlier by different overlapping groups of spike trains. Hence we do not need to model the information loss as in the previous

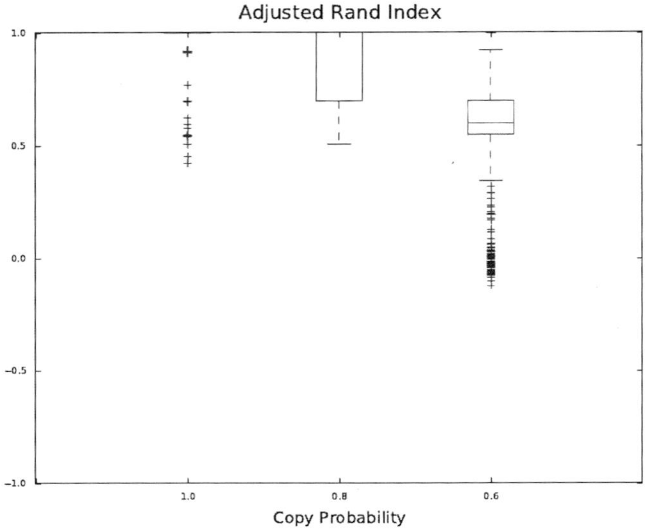

Fig. 6.17 This plot shows cluster quality for different spike train setups

setup but rather hope to find the points belonging to the overlapping subgroups. Evaluation becomes a bit more complicated as cluster evaluation measures usually only apply to single label clusterings. For this evaluation setup we generated two groups of spike trains that overlap on a subgroup of 5 points. The *ground truth* label for the overlapping group is different from the other two labels. Points that were clustered into two clusters were automatically labeled with this extra labels. The evaluation could then be performed against this ground truth. The results are shown in Figs. 6.17 and 6.18. As we might expect from the data generation process, the results show that clusters can be identified for different copy probabilities as long as the probability does not drop too much. For $c = 0.6$ most of the results are still valid but the quality detoriates a lot, yet the vast majority of all results still indicates a lot of the labels were actually correct. If we are looking for overlapping clusters (cf. Fig. 6.18) we see better results than we would be expecting under such a difficult setting. Comparing the results to the previous ones we have to know, that any setting of different copy probabilities can be modelled by overlapping clusters of different sizes and a suitable overlap structure. Because of that we do not need to test anything beside a copy probability of $c = 1.0$ and we can see that the multiple assignment of labels works in our example.

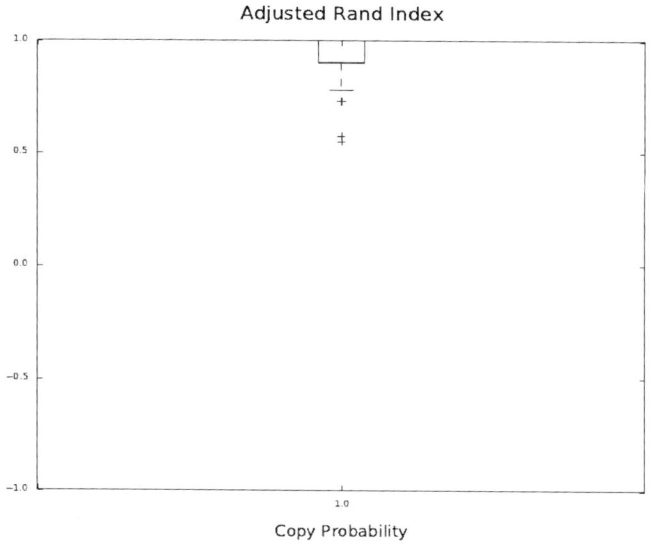

Fig. 6.18 This plot shows cluster quality for overlapping clusters of spike trains

6.5 Conclusion

In the previous sections we described how density-based clustering like DBSCAN overcomes some of the major problems of classical partitional clustering algorithms. We also described two method which – under certain restrictions – produce similar or better results than DBSCAN does while using less parameters that need to be estimated by the data analyst. Our two algorithms – the *black hole clustering* and the *protoclustering* algorithm – both use only one version of the parameters used by DBSCAN. This comes at the cost of some flexibility or speed but allow either a quick explorative analysis on hundreds of thousands of points even on desktops machines or allow multilabel clustering on smaller data sets.

Still, some problems are yet unsolved. Especially for the multilabel clustering scenario the choice of the threshold to reduce its distance graph is crucial and not easily determined. Spherical clusters can be found with ease since many protoclusters will lie in the same cluster and there will be many small distances within the clusters. The threshold can then be determined by clustering the one-dimensional data set of weights in the graph and remove the edges with the larger distances. This leads to good results yet is a rather unsatisfying method. Here some more sophisticated methods should be explored.

References

1. Beil F, Ester M, Xu X (2002) Frequent term-based text clustering. In: Proceedings of the eighth ACM SIGKDD international conference on knowledge discovery and data mining. ACM, New York
2. Bradley PS, Fayyad UM (1998) Refining initial points for K-means clustering. In: Proceedings of the fifteenth international conference on machine learning (ICML), vol 98, pp 91–99
3. Braune C, Borgelt C, Grün S (2012) Assembly detection in continuous neural spike train data. In: Advances in intelligent data analysis XI. Springer, Berlin, Heidelberg, pp 78–89
4. Braune C, Borgelt C, Kruse R (2013) Behavioral clustering for point processes. In: Advances in intelligent data analysis XII. Springer, Berlin, Heidelberg, pp 127–137
5. Bridges CC Jr (1966) Hierarchical cluster analysis. Psychol Rep 18(3):851–854
6. Celebi ME, Kingravi H (2012) Deterministic initialization of the K-means algorithm using hierarchical clustering. Int J Pattern Recognit Artif Intell 26(7):1250018
7. Celebi ME, Kingravi H, Vela PA (2013) A comparative study of efficient initialization methods for the K-means clustering algorithm. Expert Syst Appl 40(1):200–2010
8. Döring C, Lesot MJ, Kruse R (2006) Data analysis with fuzzy clustering methods. Comput Stat Data Anal 51(1):192–214
9. Esmaelnejad J, Habibi J, Yeganeh SH (2010) A novel method to find appropriate ϵ for DBSCAN. In: Intelligent information and database systems. Springer, Berlin, Heidelberg, pp 93–102
10. Ester M, Kriegel HP, Sander J, Xu X (1966) A density-based algorithm for discovering clusters in large spatial databases with noise. In: Proceedings of 2nd international conference on knowledge discovery and data mining (KDD), vol 96, pp 226–231
11. Gerstein, GL, Perkel DH (1969) Simultaneously recorded trains of action potentials: analysis and functional interpretation. Science 164(3881):828–830
12. Hall LO, Bensaid AM, Clarke LP, Velthuizen RP, Silbiger MS, Bezdek JC (1992) A comparison of neural network and fuzzy clustering techniques in segmenting magnetic resonance images of the brain. IEEE Trans Neural Netw 3(5):672–682
13. Hentschel C, Stober S, Nürnberger A, Detyniecki M (2008) Automatic image annotation using a visual dictionary based on reliable image segmentation. In: Adaptive multimedia retrieval: retrieval, user, and semantics. Springer, Berlin, Heidelberg, pp 45–56
14. Hubert L, Arabie P (1985) Comparing partitions. J Classif 2(1):193–218
15. Jing L, Ng MK, Xu J, Huang JZ (2005) Subspace clustering of text documents with feature weighting k-means algorithm. In: Advances in knowledge discovery and data mining. Springer, Berlin, Heidelberg, pp 802–812
16. Kaufman L, Rousseeuw PJ (1990) Finding groups in data: an introduction to cluster analysis. Wiley, New York
17. Katayama N, Satoh S (1997) The SR-tree: an index structure for high-dimensional nearest neighbor queries. ACM SIGMOD Record 26(2):440–447
18. Krinidis S, Chatzis V (2010) A robust fuzzy local information C-means clustering algorithm. IEEE Trans Image Process 19(5):1328–1337
19. Kruse R, Borgelt C, Klawonn F, Moewes C, Steinbrecher M, Held P (2013) Computational intelligence: a methodological introduction. Springer, Berlin
20. Li Y, Luo C, Chung SM (2008) Text clustering with feature selection by using statistical data. IEEE Trans Knowl Data Eng 20(5):641–652
21. MacQueen JB (1967) Some methods for classification and analysis of multivariate observations. In: Proceedings of 5th Berkeley symposium on mathematical statistics and probability, vol 1. University of California Press, Berkeley, pp 281–297
22. Pedregosa F, Varoquaux G, Gramfort A, Michel V, Thirion B, Grisel O, Blondel M, Prettenhofer P, Weiss R, Dubourg V, Vanderplas J, Passos A, Cournapeau D, Brucher M, Perrot M, Duchesnay E (2011) Scikit-learn: machine learning in python. J Mach Learn Res 12:2825–2830

23. Perkel DH, Gerstein GL, Moore GP (1967) Neuronal spike trains and stochastic point processes: II. Simultaneous spike trains. Biophys J 7(4):419–440
24. Sugar CA, James GM (2003) Finding the number of clusters in a dataset. J Am Stat Assoc 98(463):750–763
25. Wolpert DH, Macready WG (1997) No free lunch theorems for optimization. IEEE Trans Evol Comput 1.1:67–82
26. Zhang X, Jiao L, Liu F, Bo L, Gong M (2008) Spectral clustering ensemble applied to SAR image segmentation. IEEE Trans Geosci Remote Sens 46(7):2126–2136

Chapter 7
Nonnegative Matrix Factorization for Interactive Topic Modeling and Document Clustering

Da Kuang, Jaegul Choo, and Haesun Park

Abstract Nonnegative matrix factorization (NMF) approximates a nonnegative matrix by the product of two low-rank nonnegative matrices. Since it gives semantically meaningful result that is easily interpretable in clustering applications, NMF has been widely used as a clustering method especially for document data, and as a topic modeling method.

We describe several fundamental facts of NMF and introduce its optimization framework called block coordinate descent. In the context of clustering, our framework provides a flexible way to extend NMF such as the sparse NMF and the weakly-supervised NMF. The former provides succinct representations for better interpretations while the latter flexibly incorporate extra information and user feedback in NMF, which effectively works as the basis for the visual analytic topic modeling system that we present.

Using real-world text data sets, we present quantitative experimental results showing the superiority of our framework from the following aspects: fast convergence, high clustering accuracy, sparse representation, consistent output, and user interactivity. In addition, we present a visual analytic system called UTOPIAN (User-driven Topic modeling based on Interactive NMF) and show several usage scenarios.

Overall, our book chapter cover the broad spectrum of NMF in the context of clustering and topic modeling, from fundamental algorithmic behaviors to practical visual analytics systems.

Keywords Nonnegative matrix factorization • Document clustering • Topic modeling • Block coordinate descent • Interactive visual analytics

D. Kuang • J. Choo • H. Park (✉)
Georgia Institute of Technology, Atlanta, GA, USA
e-mail: da.kuang@cc.gatech.edu; jaegul.choo@cc.gatech.edu; hpark@cc.gatech.edu

M.E. Celebi (ed.), *Partitional Clustering Algorithms*,
DOI 10.1007/978-3-319-09259-1_7

7.1 Introduction to Nonnegative Matrix Factorization

Nonnegative matrix factorization (NMF) is a dimension reduction method and factor analysis method. Many dimension reduction techniques are closely related to the low-rank approximations of matrices, and NMF is special in that the low-rank factor matrices are constrained to have only nonnegative elements. The nonnegativity reflects the inherent representation of data in many application areas, and the resulting low-rank factors lead to physically natural interpretations [33]. NMF was first introduced by Paatero and Tapper [43] as positive matrix factorization and subsequently popularized by Lee and Seung [33]. Over the last two decades, NMF has received enormous attention and has been successfully applied to a broad range of important problems in the areas including text mining [44, 52], computer vision [21, 36], bioinformatics [6, 11, 22], spectral data analysis [45], and blind source separation [10], and many others.

Suppose a nonnegative matrix $A \in \mathbb{R}^{m \times n}$ is given. When the desired lower dimension is k, the goal of NMF is to find the two matrices $W \in \mathbb{R}^{m \times k}$ and $H \in \mathbb{R}^{k \times n}$ having only nonnegative entries such that

$$A \approx WH. \tag{7.1}$$

According to (7.1), each data point, which is represented as the column of A, can be approximated by an additive combination of the nonnegative basis vectors, which are represented as the columns of W. As the goal of dimension reduction is to discover compact representation in the form of (7.1), k is assumed to satisfy that $k < \min\{m, n\}$. The matrices W and H are found by solving an optimization problem defined with the Frobenius norm (a distance measure between two given matrices), the Kullback-Leibler (KL) divergence (a distance measure between two probability distributions) [34,38], or other divergences [12,38]. In this book chapter, we focus on NMF based on the Frobenius norm, which is the most commonly used formulation:

$$\min_{W \geq 0, H \geq 0} f(W, H) = \|A - WH\|_F^2. \tag{7.2}$$

The constraints in (7.2) indicate that all the entries of W and H are nonnegative.

NMF with the formulation (7.2) has been very successful in partitional clustering, and many variations have been proposed for different settings such as constrained clustering and graph clustering [7, 22, 30, 37]. NMF especially performs well as a document clustering and topic modeling method. Due to an ever increasing amount of document data and the complexity involved with analyzing them in practice, revealing meaningful insights and thus guiding users in their decision-making processes has long been an active area of research. Document clustering is an important task in text mining with the goal of organizing a large text collection into several semantic clusters and helping users browse documents efficiently. Topic modeling is related to soft clustering where the documents are represented as a weighted combination of topics in terms of their proximity to each topic. In addition to its soft clustering aspect, topic modeling also deals with the semantic meaning of each cluster/topic and models it as a weighted combination of keywords. Because of the nonnegativity constraints in NMF, the result of NMF can be viewed as document clustering and topic modeling results directly, which will be elaborated by theoretical and empirical evidences in this book chapter.

The goal of this book chapter is to provide an overview of NMF used as a clustering and topic modeling method for document data. We present a wide spectrum of material including the theoretical justification of NMF as a clustering method (Sect. 7.2), an algorithmic framework and extensions (Sect. 7.3), empirical performances and practical issues (Sects. 7.4–7.5), as well as a visual analytic system called UTOPIAN (Sect. 7.6). Our emphasis is placed on NMF in the context of document clustering and topic modeling; however, the presented methodology applies to data types beyond text, for example, DNA microarray and RNA sequencing data in the biological domain.

We recommend the readers be familiar with linear algebra and numerical optimization theory.

Notations: Notations used in this book chapter are as follows. A lower-case letter, such as x, denotes a scalar; an upper-case letter, such as X, denotes a matrix; a bold-face lower-case letter, such as $\mathbf{x}$, denotes a column vector. We typically use i, j as indices: For example, $i \in \{1, \cdots, n\}$. The elements of a sequence of vectors or matrices are denoted by superscripts within parentheses, such as $X^{(1)}, \cdots, X^{(n)}$, and the entire sequence is denoted by $\{X^{(i)}\}$. The entries of a matrix are denoted by subscripts, such as x_{ij} for a matrix X. $X \geq 0$ indicates that the elements of X are nonnegative, i.e., X is a *nonnegative matrix*. $\mathbb{R}$ and $\mathbb{R}_+$ denote the set of real numbers and nonnegative real numbers, respectively. $\|\cdot\|_2$ and $\|\cdot\|_F$ denotes the L_2 norm and the Frobenius norm, respectively. The operator $.*$ denotes entrywise multiplication of matrices.

7.2 Nonnegative Matrix Factorization for Clustering

Dimension reduction and clustering are closely related. Consider the low-rank approximation in (7.1), where $A \in \mathbb{R}_+^{m \times n}$, $W \in \mathbb{R}_+^{m \times k}$, $H \in \mathbb{R}_+^{k \times n}$, and $k << \min(m, n)$ is the pre-specified lower rank. The columns of A represent n data points in an m-dimensional space. Each column of H is the k-dimensional representation of a data point. If we can use H to derive an assignment of the n data points into k groups, clustering can be viewed as a special type of dimension reduction. One example is the classical K-means clustering:

$$\min \sum_{i=1}^{n} \|\mathbf{a}_i - \mathbf{w}_{g_i}\|_2^2, \tag{7.3}$$

where $\mathbf{a}_1, \cdots, \mathbf{a}_n$ are the columns of A, $\mathbf{w}_1, \cdots, \mathbf{w}_k$ are the k centroids, and $g_i = j$ when the i-th data point is assigned to the j-th cluster ($1 \le j \le k$). Consider K-means formulated as a dimension reduction problem [26]:

$$\min_{H \in \{0,1\}^{k \times n}, H^T \mathbf{1}_k = \mathbf{1}_n} \|A - WH\|_F^2, \tag{7.4}$$

where $\mathbf{1}_k \in \mathbb{R}^{k \times 1}$ and $\mathbf{1}_n \in \mathbb{R}^{n \times 1}$ are the column vectors whose elements are all 1's. In the K-means formulation for (7.4), the columns of W are the cluster centroids, and the single nonzero element in each column of H indicates the clustering assignment. Another example of dimension reduction is NMF:

$$\min_{W \ge 0, H \ge 0} \|A - WH\|_F^2.$$

In this formulation, the columns of W provide the basis of a latent k-dimensional space, and the columns of the second factor H provide the representation of $\mathbf{a}_1, \cdots, \mathbf{a}_n$ in the latent space. With only the nonnegativity constraints on H, this formulation can still be interpreted as clustering results: The columns of W are interpreted as k cluster representatives, and the i-th column of H contains the soft clustering membership of the i-th data point for the k clusters. NMF is best known for the *interpretability* of the latent space it finds [33]. In the case of document clustering and topic modeling, the basis vectors in W represent k topics, and the coefficients in the i-th column of H indicate the topic proportions for $\mathbf{a}_i$, the i-th document. To obtain a hard clustering result, we can simply choose the topic with the largest weight, i.e., the largest element in each column of H.

Historically, NMF has been extensively compared with K-means and singular value decomposition (SVD). We give several clarifications and caveats regarding using NMF as a clustering method. It has been shown that K-means and NMF have the equivalent form of an objective function, $\|A - WH\|_F^2$ [13]. However, each clustering method has its own conditions under which it performs well. K-means assumes that data points in each cluster follow a spherical Gaussian distribution [16]. In contrast, the NMF formulation (7.2) provides a better low-rank approximation of the data matrix A than the K-means formulation (7.4). If $k \le \text{rank}(A)$, the columns of W are linearly independent due to $\text{rank}(A) \le \text{nonnegative-rank}(A)$[1] [3]. Therefore, NMF performs well when different clusters correspond to linearly independent vectors [30].

One caveat is that NMF does not always perform well as a clustering method. Consider the example in Fig. 7.1, where the two cluster centers are along the same direction and thus the two centroid vectors are linearly dependent. While NMF still approximates all the data points well in this example, no two linearly independent

[1] The nonnegative rank of a matrix $X \in \mathbb{R}_+^{m \times n}$ is the smallest number $\hat{k}$ such that $X = WH$ where $W \in \mathbb{R}_+^{m \times \hat{k}}$ and $H \in \mathbb{R}_+^{\hat{k} \times n}$.

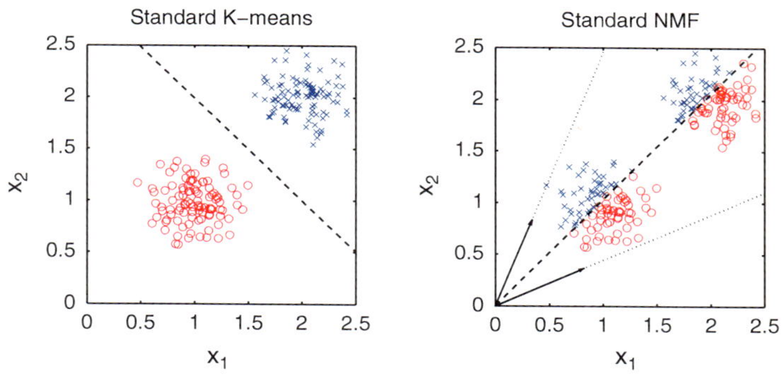

Fig. 7.1 An example with two ground-truth clusters, and the different clustering results given by K-means and NMF. The "o" and "x" markers in each figure indicate the cluster membership of the data points given by the clustering algorithm. The *left figure* shows that K-means correctly identifies the two clusters where the two centroids are linearly dependent. The *right figure* shows that NMF, on the contrary, uses two linearly independent vectors as cluster representatives marked by the two *thick arrows*, and leads to incorrect clustering results

vectors in a two-dimensional space can represent the two clusters shown in Fig. 7.1. Since K-means and NMF have different conditions under which each of them does clustering well, they may generate very different clustering results in practice.

In contrast to NMF, rank-k SVD provides the best rank-k approximation but allows negative entries:

$$\min_{U^T U = I, V^T V = I} \|A - WY\|_F = \|A - U \Sigma V^T\|_F, \tag{7.5}$$

where $U \in \mathbb{R}^{m \times k}$, $\Sigma \in \mathbb{R}^{k \times k}$, and $V \in \mathbb{R}^{n \times k}$. Thus we cannot interpret the coefficients in the lower-dimensional space spanned by the columns of U as clustering assignments. In other words, the rows of V cannot be used as cluster indicators directly. Instead, an additional clustering method such as K-means has to be applied to a lower-dimensional representation of the data such as the rows of V to generate clusters.

The success of NMF as a clustering method depends on the underlying data set, and its greatest success has been in the area of document clustering [15, 26, 37, 44, 46, 52]. In a document data set, data points are often represented as unit-length vectors [40] and embedded in a linear subspace. For a term-document matrix A, a basis vector $\mathbf{w}_j$ is interpreted as the keyword-wise distribution of a single topic. When these distributions of k topics are linearly independent, which is usually the case, NMF can properly extract the ground-truth clusters determined by the true cluster labels.

Recently, NMF has been applied to topic modeling, a task related to document clustering, and achieved satisfactory results [1, 2]. Both document clustering and topic modeling can be considered as dimension reduction processes. Compared to standard topic modeling methods such as probabilistic latent semantic indexing (p-LSI) [20] and latent Dirichlet allocation (LDA) [5], NMF essentially gives the same output types: A keyword-wise topic representation (the columns of W), and a topic-wise document representation (the columns of H). The only difference, however, is that the columns of W and H do not have a unit L_1-norm unlike the p-LSI and LDA outputs. Nonetheless, such a difference is negligible in that (7.1) can be manipulated via diagonal scaling matrices as

$$A \approx WH = (WD_W)(D_W^{-1}H) = \hat{W}\hat{H}, \tag{7.6}$$

where the diagonal component of the diagonal matrix $D_W \in \mathbb{R}_+^{k \times k}$ corresponds to the column sums of W. Now the new matrix $\hat{W}$ is column-normalized, giving an output analogous to the first outputs from p-LSI and LDA, but the second output $\hat{H}$ is still not column-normalized. The column normalization on H does not affect the interpretation of each document in terms of its relative relationships to topics. In this sense, NMF can be used as an alternative to topic modeling methods.

Note that NMF with KL-divergence is another commonly used formulation for topic modeling. It has a probabilistic interpretation and can be shown to be equivalent to p-LSI under certain constraints. However, algorithms for solving NMF with KL-divergence and p-LSI are typically much slower than those for solving NMF based on the Frobenius norm (7.2) [51]. Therefore, we focus on (7.2) in this chapter since there are many justified and efficient optimization algorithms developed for (7.2) in the literature.

7.3 Optimization Framework for Nonnegative Matrix Factorization

Although NMF is known as an NP-hard problem [49], one can still hope to find a local minimum as an approximate solution for NMF. In this section, we introduce an algorithmic framework to optimize the objective function (7.2), namely the *block coordinate descent* (BCD) framework. Multiplicative updating (MU) is another popular framework for solving NMF [33]. However, it has slow convergence and may lead to inferior solutions [18, 39]. In the later section, we will compare the solutions given by these two frameworks empirically and show the better clustering quality given by BCD.

7.3.1 Block Coordinate Descent Framework

The BCD framework is a widely-applicable strategy in nonlinear optimization problems. It divides variables into several disjoint subgroups and iteratively minimize the objective function with respect to the variables of each subgroup at a time.

Algorithm 1 The BCD framework for solving NMF: $\min_{W,H\geq 0} \|A - WH\|_F^2$

1: Input: Matrix $A \in \mathbb{R}^{m\times n}$, tolerance parameter $0 < \epsilon << 1$, upper limit of the number of iterations T
2: Initialize H
3: **repeat**
4: Obtain the optimal solution of subproblem (7.8a)
5: Obtain the optimal solution of subproblem (7.8b)
6: **until** A particular stopping criterion based on W, H, ϵ is satisfied *or* the number of iterations reaches upper limit T
7: Output: W, H

In the formulation of NMF (7.2), A is given as an input and the entries of W and H are the variables to be solved. A natural partitioning of the variables is the two blocks representing W and H, respectively. That is to say, we take turns solving

$$W \leftarrow \arg\min_{W\geq 0} f(W, H), \tag{7.7a}$$

$$H \leftarrow \arg\min_{H\geq 0} f(W, H). \tag{7.7b}$$

These subproblems can be written as

$$\min_{W\geq 0} \|H^T W^T - A^T\|_F^2, \tag{7.8a}$$

$$\min_{H\geq 0} \|WH - A\|_F^2. \tag{7.8b}$$

The subproblems (7.8) are called nonnegativity constrained least squares (NLS) problems [32], and the BCD framework has been called the alternating nonnegative least squares (ANLS) framework [23, 39]. It is summarized in Algorithm 1.

Note that we need to initialize H and solve (7.8a) and (7.8b) iteratively, as stated in Algorithm 1. Alternatively, we can also initialize W and solve (7.8b) and (7.8a) iteratively. Different initializations may lead to different solutions for NMF. A common strategy is to run an NMF algorithm starting from different random initializations and pick the solution with the smallest objective function value. Other strategies for initializing W and/or H were also proposed in the literature. For example, in the context of clustering, we can run spherical K-means, i.e. K-means with $1 - \mathbf{a}_i^T \mathbf{a}_j$ (one minus cosine similarity) as the distance function, and use the resulting centroids as the initialization of W [50].

Even though the subproblems are convex, they do not have a closed-form solution, and anumerical algorithm for the subproblems has to be provided. Many approaches for solving the NLS subproblems have been proposed in the NMF literature, e.g., an active-set method [23], a block principal pivoting [27, 28], a projected gradient descent [39], a quasi-Newton method [24]. We skip these details in this book chapter, but refer the readers to several software packages that solve NLS efficiently.[2,3]

7.3.1.1 Convergence Property

The objective function of NMF is a fourth-order polynomial with respect to W and H, and thus is nonconvex. For a nonconvex optimization problem, most algorithms only guarantee the stationarity of a limit point [4], not necessarily a local minimum. In practice, we often run an NMF algorithm in the BCD framework for multiple times with different initializations of W and H, and select the output with the smallest objective function value.

We have the following theorem regarding the convergence property of the BCD framework:

Theorem 1. *If a minimum of each subproblem in (7.8) is attained at each step, every limit point of the sequence* $\{(W, H)^{(i)}\}$ *generated by the BCD framework is a stationary point of (7.2).*

The BCD framework requires that the optimal solution be returned for each NLS subproblem. Note that the minimum of each subproblem is not required to be unique for the convergence result to hold because the number of blocks is two, as proved by Grippo and Sciandrone [19].

We remark that at a stationary point solution, the Karush-Kuhn-Tucher (KKT) condition is satisfied:

$$W \geq 0, \quad H \geq 0, \tag{7.9a}$$

$$\nabla f_W = 2WHH^T - 2AH^T \geq 0, \quad \nabla f_H = 2W^T WH - 2W^T A \geq 0, \tag{7.9b}$$

$$W. * \nabla f_W = 0, \quad H. * \nabla f_H = 0. \tag{7.9c}$$

In contrast, the MU algorithm does not have the convergence property stated in Theorem 1. We consider the MU algorithm proposed by Lee and Seung [34]. This algorithm has an advantage of being simple and easy to implement, and it has contributed greatly to the popularity of NMF. Though it also has a form of updating W and H alternately, it is different from the BCD framework in the sense that its solutions for the subproblems (7.8) are not optimal. That is, the update rule of MU is:

$$W \leftarrow W. * \frac{AH^T}{WHH^T}, \quad H \leftarrow H. * \frac{W^T A}{W^T WH}, \tag{7.10}$$

[2]http://www.cc.gatech.edu/~hpark/nmfsoftware.php

[3]http://www.csie.ntu.edu.tw/~cjlin/nmf/index.html

where the division operator indicates entrywise division. This update rule can be seen as a gradient descent algorithm with specifically chosen step lengths. The step lengths are conservative enough so that the result is always nonnegative. However, we cannot achieve the optimal solution of every NLS subproblem using this update rule.

Lee and Seung [34] showed that under the update rule (7.10), the objective function in NMF (7.2) is non-increasing. However, it is unknown whether it converges to a stationary point or a local minimum [18]. In fact, even though the papers using MU algorithms claimed that the solution satisfied the KKT condition, such as in [14], often their proofs did not include all the components of the KKT condition in (7.9), for example, the sign of the gradients (7.9b).

Furthermore, since the values are updated only through multiplications in MU algorithms, the entries of W and H typically remain nonzero. Hence, their solution matrices typically are denser than those from algorithms in the BCD framework empirically, and thus it is harder to interpret the solution matrices as clustering results.

7.3.1.2 Stopping Criterion

Iterative methods have to be equipped with a criterion for stopping iterations. A naive approach is to stop when the decrease of the objective function becomes smaller than a pre-defined threshold:

$$|f(W^{(i-1)}, H^{(i-1)}) - f(W^{(i)}, H^{(i)})| \leq \epsilon. \tag{7.11}$$

Although this method is commonly adopted, it is potentially misleading because the decrease of the objective function may become small before a stationary point is achieved. A more principled criterion was proposed by Lin [39] as follows. Recall the KKT condition (7.9) for the objective function of NMF. Let us define the projected gradient $\nabla^P f_W \in \mathbb{R}^{m\times k}$ as

$$(\nabla^P f_W)_{ij} = \begin{cases} \nabla(f_W)_{ij}, & \text{if } (\nabla f_W)_{ij} < 0 \text{ or } W_{ij} > 0; \\ 0, & \text{otherwise,} \end{cases} \tag{7.12}$$

for $i = 1, \cdots, m$ and $j = 1, \cdots, k$, and $\nabla^P f_H$ similarly. Then, conditions (7.9) can be rephrased as

$$\nabla^P f_W = 0 \text{ and } \nabla^P f_H = 0. \tag{7.13}$$

Let us denote the projected gradient matrices at the i-th iteration by $\nabla^P f_W^{(i)}$ and $\nabla^P f_H^{(i)}$ and define

$$\Delta(i) = \sqrt{\|\nabla^P f_W^{(i)}\|_F^2 + \|\nabla^P f_H^{(i)}\|_F^2}. \tag{7.14}$$

Using this definition, the stopping criterion is written by

$$\frac{\Delta(i)}{\Delta(1)} \leq \epsilon, \tag{7.15}$$

where $\Delta(1)$ is from the first iterate of (W, H). Unlike (7.11), (7.15) guarantees the stationarity of the final solution. For caveats when using (7.15), see [25].

7.3.2 Extension 1: Sparse NMF

With only nonnegativity constraints, the resulting factor matrix H of NMF contains the fractional assignment values corresponding to the k clusters represented by the columns of W. Sparsity constraints on H have been shown to facilitate the interpretation of the result of NMF as a hard clustering result and improve the clustering quality [21, 22, 26]. For example, consider two different scenarios of a column of $H \in \mathbb{R}_+^{3\times n}$: $(0.2, 0.3, 0.5)^T$ and $(0, 0.1, 0.9)^T$. Clearly, the latter is a stronger indicator that the corresponding data point belongs to the third cluster.

To incorporate extra constraints or prior information into the NMF formulation (7.2), various regularization terms can be added. We can consider an objective function

$$\min_{W,H\geq 0} \|A - WH\|_F^2 + \phi(W) + \psi(H), \tag{7.16}$$

where $\phi(\cdot)$ and $\psi(\cdot)$ are regularization terms that often involve matrix or vector norms. The L_1-norm regularization can be adopted to promote sparsity in the factor matrices [22,47]. When sparsity is desired on H, the L_1-norm regularization can be set as

$$\phi(W) = \alpha\|W\|_F^2 \text{ and } \psi(H) = \beta \sum_{i=1}^{n} \|H(:, i)\|_1^2, \tag{7.17}$$

where $H(:, i)$ represents the i-th column of H. The L_1-norm term of $\psi(H)$ in (7.17) promotes sparsity on the columns of H while the Frobenius norm term of $\phi(W)$ is needed to prevent W from growing too large. Scalar parameters α and β are used to control the strength of regularization.

The sparse NMF can be easily computed using the BCD framework. We can reorganize the terms in the sparse NMF formulation (7.16) and (7.17) and the two subproblems in the BCD framework become:

$$\min_{W \geq 0} \left\| \begin{pmatrix} H^T \\ \sqrt{\alpha} I_k \end{pmatrix} W^T - \begin{pmatrix} A^T \\ 0_{k \times m} \end{pmatrix} \right\|_F^2, \tag{7.18a}$$

$$\min_{H \geq 0} \left\| \begin{pmatrix} W \\ \sqrt{\beta} \mathbf{1}_k^T \end{pmatrix} H - \begin{pmatrix} A \\ \mathbf{0}_n^T \end{pmatrix} \right\|_F^2. \tag{7.18b}$$

where $\mathbf{1}_k \in \mathbb{R}^{k \times 1}, \mathbf{0}_n \in \mathbb{R}^{n \times 1}$ are the column vectors whose elements are all 1's and 0's, respectively, and I_k is a $k \times k$ identity matrix. Hence, the two subproblems (7.18a) and (7.18b) for the sparse NMF can be solved as NLS problems, similar to Algorithm 1 for the original NMF.

7.3.3 Extension 2: Weakly-Supervised NMF

The flexible BCD framework allows another important variant called weakly-supervised NMF (WS-NMF).[4] WS-NMF can incorporate a variety of user inputs so that the clustering and topic modeling results of NMF can be improved in a user-driven way. In this section, we describe the formulation and the algorithm of WS-NMF. Later in Sect. 7.6, we further discuss how WS-NMF can be utilized to support various types of user interactions in a visual analytics environment.

In WS-NMF, such user inputs are manifested in the form of reference matrices for W and H. These reference matrices play a role of making W and H become similar to them. That is, given reference matrices $W_r \in \mathbb{R}_+^{m \times k}$ for W and $H_r \in \mathbb{R}_+^{k \times n}$ for H, diagonal mask/weight matrices $M_W \in \mathbb{R}_+^{k \times k}$ and $M_H \in \mathbb{R}_+^{n \times n}$, a data matrix $A \in \mathbb{R}_+^{m \times n}$, and an integer $k \ll \min(m, n)$, WS-NMF has the additional regularization terms that penalize the differences between H_r and H (up to a column-wise scaling via D_H) and those between W_r and W as

$$f(W, H, D_H) = \min_{W, H, D_H} \|A - WH\|_F^2 + \|(W - W_r) M_W\|_F^2 + \|(H - H_r D_H) M_H\|_F^2 \tag{7.19}$$

for $W \in \mathbb{R}_+^{m \times k}$ and $H \in \mathbb{R}_+^{k \times n}$ and a diagonal matrix $D_H \in \mathbb{R}_+^{n \times n}$.

Through these regularization terms, WS-NMF can incorporate various types of users' prior knowledge. Each column of H_r specifies the soft clustering membership of data items. A diagonal matrix D_H accounts for a possible scaling different between H_r and H and is a variable to be computed. For example, two vectors, (0.1, 0.3, 0.6) and (0.2, 0.6, 1.2), are interpreted the same in terms of cluster membership coefficients, and D_H allows them to be treated as same. WS-NMF also supports partial supervision on a subset of column (or data items) in H_r.

[4]The term "weakly-supervised" can be considered similar to semi-supervised clustering settings, rather than supervised learning settings such as classification and regression problems.

The diagonal matrix M_H achieves this by masking/down-weighting the columns or data items in H_r with no prior information.

Next, W_r supervise the basis representations. In document clustering and topic modeling, the columns of W_r specify the keyword-wise topic representations in W. Similar to the role of M_H for the partial supervision on H, the diagonal matrix M_W allows the supervision on a subset of columns in W by masking/down-weighting those columns in W_r with no supervision. However, unlike the supervision on H, the regularization on W via W_r does not involve any diagonal matrix analogous to D_H because scaling on either W or H suffices due to the relationship (7.6), which indicates that if W and H are the solution of a particular NMF problem, then so are WD and $D^{-1}H$ for an arbitrary diagonal matrix D.

Finally, note that (7.19) does not have typical regularization parameters that balance between different terms since M_W and M_H can account for the effects of the parameters. In other words, $\alpha \|(W - W_r) M_W\|_F^2$ is equivalent to $\|(W - W_r) M_W^{new}\|_F^2$ when $M_W^{new} = \alpha M_W$, and the same argument holds for M_H.

The optimization of (7.19) follows the BCD framework by iteratively solving W, H, and D_H. Given initial values for these variables, W is updated as

$$W \leftarrow \arg\min_{W \geq 0} \left\| \begin{bmatrix} H^T \\ M_W \end{bmatrix} W^T - \begin{bmatrix} A^T \\ M_W W_r^T \end{bmatrix} \right\|_F^2. \tag{7.20}$$

Next, each column of H is updated one at a time by solving

$$H(:, i) \leftarrow \arg\min_{H(:,i) \geq 0} \left\| \begin{bmatrix} W \\ M_H(i) I_k \end{bmatrix} H(:, i) - \begin{bmatrix} A(:, i) \\ M_H(i) D_H(i) H_r(:, i) \end{bmatrix} \right\|_F^2, \tag{7.21}$$

where $(:, i)$ indicates the i-th column of a matrix. Finally, the i-th diagonal component $D_H(i)$ of D_H is obtained in a closed form as

$$D_H(i) \leftarrow \begin{cases} \frac{H_r(:,i)^T \cdot H(:,i)}{\|H_r(:,i)\|_2^2} & \text{if } M_H(i) \neq 0 \\ 0 & \text{otherwise} \end{cases}. \tag{7.22}$$

7.4 Choosing the Number of Clusters

The number of clusters k is one of the input parameters that NMF requires. It is often difficult to choose an appropriate k to achieve the best clustering quality. In this section, we introduce our method to choose k based on random sampling and consensus clustering.

Monti et al. [42] proposed a model selection method that used the notion of *stability* of the clusters with respect to random sampling of the data points. Let $\mathscr{A}_1$ and $\mathscr{A}_2$ be two subsets sampled randomly from a data set of n data points. Suppose two data points $\mathbf{a}_i$ and $\mathbf{a}_j$ appear in both subsets generated by random sampling, that is to say, $\mathbf{a}_i, \mathbf{a}_j \in \mathscr{A}_1 \cap \mathscr{A}_2$. Let us run a clustering algorithm on both $\mathscr{A}_1$ and $\mathscr{A}_2$, and the correct number of clusters k is given. Conceptually, we expect that if $\mathbf{a}_i$ and $\mathbf{a}_j$ belong to the same cluster derived from $\mathscr{A}_1$, they also belong to the same cluster derived from $\mathscr{A}_2$. Based on this reasoning, Monti et al. [42] proposed *consensus clustering* to aggregate the results of a clustering method over many runs and achieve a consensus partitioning of data points.

We formulate the idea of a *consensus matrix* in the context of NMF-based document clustering. For a data set with n documents, the (i, j)-th entry of a consensus matrix $\tilde{C} \in \mathbb{R}^{n \times n}$ is the co-clustered frequency of the i-th and j-th documents over multiple runs of NMF. More rigorously, let r be the sampling rate, the fraction of documents selected in each random sample. We generate T subsets $\mathscr{A}_1, \cdots, \mathscr{A}_T$ by random sampling, each with sampling rate r, and run an NMF algorithm on each subset with the same number of clusters k. Define the matrices $C^{(t)}$ and $S^{(t)}$ as the following ($1 \le t \le T$):

$$c_{ij}^{(t)} = \begin{cases} 1, & \text{if the } i\text{-th and the } j\text{-th documents belong to the same cluster using } \mathscr{A}_t; \\ 0, & \text{otherwise,} \end{cases} \tag{7.23}$$

$$s_{ij}^{(t)} = \begin{cases} 1, & \text{if both the } i\text{-th and the } j\text{-th documents appear in } \mathscr{A}_t; \\ 0, & \text{otherwise.} \end{cases} \tag{7.24}$$

Clearly, $c_{ij}^{(t)} = 1$ implies $s_{ij}^{(t)} = 1$. Then we can define the consensus matrix $\tilde{C}$:

$$\tilde{c}_{ij} = \frac{\sum_{t=1}^{T} c_{ij}^{(t)}}{\sum_{t=1}^{T} s_{ij}^{(t)}}. \tag{7.25}$$

The entries of $\tilde{C}$ have values in the interval [0, 1]. In the ideal scenario where no ambiguity exists for the co-membership of any two documents, the entries of $\tilde{C}$ could be 0 or 1 only. To measure the dispersion of a consensus matrix $\tilde{C}$, we define the dispersion coefficient ρ as:

$$\rho = \frac{1}{n^2} \sum_{i=1}^{n} \sum_{j=1}^{n} 4(\tilde{c}_{ij} - 0.5)^2. \tag{7.26}$$

For an ideal consensus matrix where all the entries are 0 or 1, we have $\rho = 1$; for a scattered consensus matrix, $0 \le \rho < 1$. After obtaining ρ_k values for various k's, we can determine the number of clusters as the one with the maximal ρ_k.

Now we illustrate the above method for choosing the number of clusters with a real-world text data set. We extracted the three largest clusters based on ground-truth labels from the entire TDT2 data set. For running NMF, we applied the ANLS

Table 7.1 Dispersion coefficients for $k = 2, 3, 4, 5$ using the three largest clusters based on ground-truth labels from the TDT2 data set

	$k = 2$	$k = 3$	$k = 4$	$k = 5$
$\rho(k)$	0.5642	0.9973	0.8515	0.9411

algorithm with block principal pivoting [27, 28] solving the NLS subproblems. To construct the consensus matrix, we used the parameters $T = 50$ and $r = 0.8$ for $k = 2, 3, 4, 5$. Table 7.1 shows the dispersion coefficients for these k's. We can see that $k = 3$ corresponds to the largest ρ and is thus chosen as the most appropriate number of clusters.

Note that our method for choosing the number of clusters differs from the work of Brunet et al. [6] in two aspects. First, the authors of [6] assessed the stability of clustering results with respect to random initialization of NMF. In contrast, our method reveals the stability of the cluster structure by examining whether the clusters can be reproduced using a random sample of the data points. Second, the rows and the columns of the consensus matrix were reordered in [6], and if the reordered matrix exhibited a block-diagonal structure, the number of clusters was determined to be appropriate. However, the optimal reordering was obtained by a hierarchical clustering of the items using the consensus matrix as similarity values between all the item pairs. Thus, it was very expensive to compute for large-scale data sets. We experienced difficulty in computing the optimal reordering for a few thousand documents. Therefore, we did not adopt the model selection method in [6] but rather used the dispersion coefficient (7.26) to assess the stability of clusters.

7.5 Experimental Results

In this section, we present the empirical evidences that support NMF as a successful document clustering and topic modeling method. We compare the clustering quality between K-means and NMF; Within the NMF algorithms, we compare the multiplicative updating (MU) algorithm and the alternating nonnegative least squares (ANLS) algorithm in terms of their clustering quality and convergence behavior, as well as sparseness and consistency in the solution.

7.5.1 Data Sets and Algorithms

We used the following text corpora in our experiments. All these corpora have ground-truth labels for evaluating clustering quality but not given as an input to the clustering algorithms.

1. **TDT2** contains 10,212 news articles from various sources (e.g., NYT, CNN, and VOA) in 1998.
2. **Reuters**[5] contains 21,578 news articles from the Reuters newswire in 1987.
3. **20 Newsgroups**[6] (20News) contains 19,997 posts from 20 Usenet newsgroups. Unlike previous indexing of these posts, we observed that many posts have duplicated paragraphs due to cross-referencing. We discarded cited paragraphs and signatures in a post by identifying lines starting with ">" or "- -". The resulting data set is less tightly-clustered and much more difficult to apply clustering or classification methods.
4. From the more recent Reuters news collection **RCV1**[7] [35] that contains over 800,000 articles in 1996–1997, we selected a subset of 23,149 articles. Labels are assigned according to a topic hierarchy, and we only considered leaf topics as valid labels.
5. The research paper collection **NIPS14-16**[8] contains NIPS papers published in 2001–2003 [17], which are associated with labels indicating the technical area (algorithms, learning theory, vision science, etc).

For all these data sets, documents with multiple labels are discarded in our experiments. In addition, the ground-truth clusters representing different topics are highly unbalanced in their sizes for TDT2, Reuters, RCV1, and NIPS14-16. We selected the largest 20, 20, 40, and 9 ground-truth clusters from these data sets, respectively. We constructed term-document matrices using tf-idf features [40], where each row corresponds to a term and each column to a document. We removed any term that appears less than three times and any document that contains less than five words. Table 7.2 summarizes the statistics of the five data sets after pre-processing. For each data set, we set the number of clusters to be the same as the number of ground-truth clusters.

We further process each term-document matrix A in two steps. First, we normalize each column of A to have a unit L_2-norm, i.e., $\|\mathbf{a}_i\|_2 = 1$. Conceptually,

Table 7.2 Data sets used in our experiments

Data set	# Terms	# Documents	# Ground-truth clusters
TDT2	26, 618	8, 741	20
Reuters	12, 998	8, 095	20
20 Newsgroups	36, 568	18, 221	20
RCV1	20, 338	15, 168	40
NIPS14-16	17, 583	420	9

[5] http://www.daviddlewis.com/resources/testcollections/reuters21578/

[6] http://qwone.com/jason/20Newsgroups/

[7] http://jmlr.csail.mit.edu/papers/volume5/lewis04a/lyrl2004rcv1v2README.htm

[8] http://robotics.stanford.edu/gal/data.html

this makes all the documents have equal lengths. Next, following [52], we compute the normalized-cut weighted version of A:

$$D = \text{diag}(A^T A\mathbf{1}_n), \quad A \leftarrow AD^{-1/2}, \tag{7.27}$$

where $\mathbf{1}_n \in \mathbb{R}^{n\times 1}$ is the column vector whose elements are all 1's, and $D \in \mathbb{R}_+^{n\times n}$ is a diagonal matrix. This column weighting scheme was reported to enhance the clustering quality of both K-means and NMF [29, 52].

For K-means clustering, we used the standard K-means with Euclidean distances. The Matlab `kmeans` function has a batch-update phase that re-assigns the data points all at once in each iteration, as well as a more time-consuming online-update phase that moves a single data point each time from one cluster to another if such a movement reduces the sum of squared error [16]. We used both phases and rewrote this function using BLAS3 operations and boosted its efficiency substantially.[9]

For the ANLS algorithm for NMF, we used the block principal pivoting algorithm[10] [27, 28] to solve the NLS subproblems (7.8) and the stopping criterion (7.15) with $\epsilon = 10^{-4}$. For the MU algorithm for NMF, we used the update formula in (7.10). The MU algorithm is not guaranteed to converge to a stationary point and thus could not satisfy the stopping criterion in (7.15) after a large number of iterations in our experiments. Therefore, we used another stopping criterion

$$\|H^{(i-1)} - H^{(i)}\|_F / \|H^{(i)}\|_F \leq \epsilon \tag{7.28}$$

with $\epsilon = 10^{-4}$ to terminate the algorithm.

For the sparse NMF, we used the formulations (7.16) and (7.17). The choice of the parameters α, β that control the regularization strength and the sparsity of the solution can be determined by cross validation, for example, by tuning α, β until the desired sparseness is reached. Following [22, 23], we set α to the square of the maximum entry in A and $\beta = 0.01$ since these choices have been shown to work well in practice.

7.5.2 Clustering Quality

We used two measures to evaluate the clustering quality against the ground-truth clusters.

Clustering accuracy is the percentage of correctly clustered items given by the maximum bipartite matching (see more details in [52]). This matching associates each cluster with a ground-truth cluster in an optimal way and can be found by the Kuhn-Munkres algorithm [31].

---

[9] http://www.cc.gatech.edu/grads/d/dkuang3/software/kmeans3.html

[10] https://github.com/kimjingu/nonnegfac-matlab

Normalized mutual information (NMI) is an information-theoretic measure of the similarity between two flat partitionings [40], which, in our case, are the ground-truth clusters and the generated clusters. It is particularly useful when the number of generated clusters is different from that of ground-truth clusters or when the ground-truth clusters have highly unbalanced sizes or a hierarchical labeling scheme. It is calculated by:

$$\mathrm{NMI} = \frac{I(C_{\text{ground-truth}}, C_{\text{computed}})}{\left[H(C_{\text{ground-truth}}) + H(C_{\text{computed}})\right]/2} = \frac{\sum_{h,l} n_{h,l} \log \frac{n \cdot n_{h,l}}{n_h n_l}}{\left(\sum_h n_h \log \frac{n_h}{n} + \sum_l n_l \log \frac{n_l}{n}\right)/2}, \tag{7.29}$$

where $I(\cdot,\cdot)$ denotes mutual information between two partitionings, $H(\cdot)$ denotes the entropy of a partitioning, and $C_{\text{ground-truth}}$ and C_{computed} denote the partitioning corresponding to the ground-truth clusters and the computed clusters, respectively. n_h is the number of documents in the h-th ground-truth cluster, n_l is the number of documents in the l-th computed cluster, and $n_{h,l}$ is the number of documents in both the h-th ground-truth cluster and the l-th computed cluster.

Tables 7.3 and 7.4 show the clustering accuracy and NMI results, respectively, averaged over 20 runs with random initializations. All the NMF algorithms have the same initialization of W and H in each run. We can see that all the NMF algorithms consistently outperform K-means except one case (clustering accuracy evaluated on the Reuters data set). Considering the two algorithms for standard NMF, the clustering quality of NMF/ANLS is either similar to or much better than that of NMF/MU. The clustering quality of the sparse NMF is consistently better than that of NMF/ANLS except on the 20 Newsgroups data set and always better than NMF/MU.

Table 7.3 The average clustering accuracy given by the four clustering algorithms on the five text data sets

	K-means	NMF/MU	NMF/ANLS	Sparse NMF/ANLS
TDT2	0.6711	0.8022	0.8505	0.8644
Reuters	0.4111	0.3686	0.3731	0.3917
20News	0.1719	0.3735	0.4150	0.3970
RCV1	0.3111	0.3756	0.3797	0.3847
NIPS14-16	0.4602	0.4923	0.4918	0.4923

Table 7.4 The average normalized mutual information given by the four clustering algorithms on the five text data sets

	K-means	NMF/MU	NMF/ANLS	Sparse NMF/ANLS
TDT2	0.7644	0.8486	0.8696	0.8786
Reuters	0.5103	0.5308	0.5320	0.5497
20News	0.2822	0.4069	0.4304	0.4283
RCV1	0.4092	0.4427	0.4435	0.4489
NIPS14-16	0.4476	0.4601	0.4652	0.4709

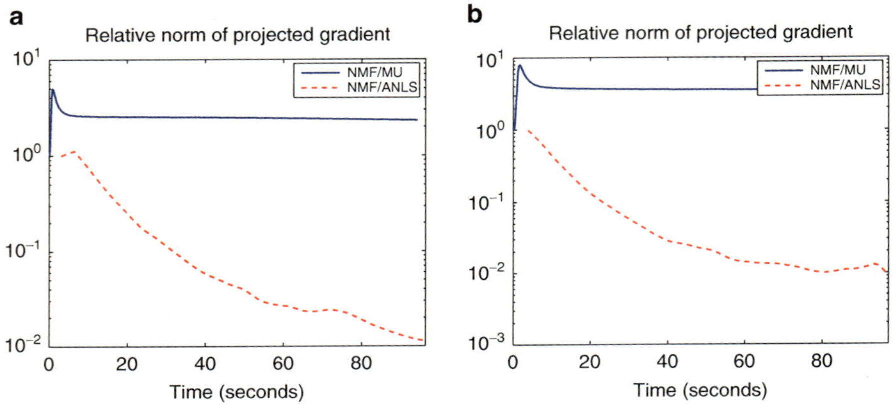

Fig. 7.2 The convergence behavior of NMF/MU and NMF/ANLS on the 20 Newsgroups data set ($k = 20$) and RCV1 data set ($k = 40$). (**a**) 20 Newgroups. (**b**) RCV1

7.5.3 Convergence Behavior

Now we compare the convergence behaviors of NMF/MU and NMF/ANLS. Figure 7.2 shows the relative norm of projected gradient $\Delta/\Delta(1)$ as the algorithms proceed on the 20 Newsgroups and RCV1 data sets. The quantity $\Delta/\Delta(1)$ is not monotonic in general but is used to check stationarity and determine whether to terminate the algorithms. On both data sets, the norm of projected gradient for NMF/ANLS has a decreasing trend and eventually reached the given tolerance ϵ, while NMF/MU did not converge to stationary point solutions. This observation is consistent with the result that NMF/ANLS achieved better clustering quality and sparser low-rank matrices.

7.5.4 Sparseness

We also compare the sparseness in the W and H matrices between the solutions of NMF/MU, NMF/ANLS, and the sparse NMF/ANLS. Table 7.5 shows the percentage of zero entries for the three NMF versions.[11] Compared to NMF/MU, NMF/ANLS does not only lead to better clustering quality and smaller objective values, but also facilitates sparser solutions in terms of both W and H. Recall that each column of W is interpreted as the term distribution for a topic. With a sparser W, the keyword-wise distributions for different topics are more orthogonal,

[11]Results given by the sparseness measure based on L_1 and L_2 norms in [21] are similar in terms of comparison between the three NMF versions.

Table 7.5 The average sparseness of W and H for the three NMF algorithms on the five text data sets. $\%(\cdot)$ indicates the percentage of the matrix entries that satisfy the condition in the parentheses

	NMF/MU		NMF/ANLS		Sparse NMF/ANLS	
	$\%(w_{ij}=0)$	$\%(h_{ij}=0)$	$\%(w_{ij}=0)$	$\%(h_{ij}=0)$	$\%(w_{ij}=0)$	$\%(h_{ij}=0)$
TDT2	21.05	6.08	55.14	50.53	52.81	65.55
Reuters	40.92	12.87	68.14	59.41	66.54	72.84
20News	46.38	15.73	71.87	56.16	71.01	75.22
RCV1	52.22	16.18	77.94	63.97	76.81	76.18
NIPS14-16	32.68	0.05	50.49	48.53	49.90	54.49

and one can select important terms for each topic more easily. A sparser H can be interpreted as clustering indicators more easily. Table 7.5 also validates that the sparse NMF generates an even sparser H in the solutions and often better clustering results.

7.5.5 Consistency from Multiple Runs

We analyze the consistency of the clustering results obtained from multiple runs of a particular method. We have chosen three methods: K-means, LDA,[12] and NMF/ANLS. The detailed procedure is as follows. First, we run each method multiple times, e.g., 30 times in our experiment. Second, for each pair of different runs, e.g., 435 cases, we measure the relative number of documents of which the (hard) clustering membership results differ from each other. To solve the correspondence between the two set of cluster indices generated independently from multiple runs, we apply the Kuhn-Munkres algorithm [31] before comparing the clustering memberships. Finally, we compute the consistency measure as the average value of the relative numbers of documents over all the pairs of runs, e.g., 435 cases.

Table 7.6 shows the results of these consistency measures for the five text data sets. It can be seen that NMF/ANLS generates the most consistent results from multiple runs compared to K-means and LDA. Combined with the accuracy and NMI results shown in Tables 7.3 and 7.4, this indicates that NMF generally produces the best clustering result with the least amount of variance. On the other hand, K-means or LDA may require users to check many results by running them multiple times until finding satisfactory results.

[12]For LDA, we used Mallet [41], a widely-accepted software library based on a Gibbs sampling method.

Table 7.6 The consistency measure of three clustering algorithms on the five text data sets

	K-means	LDA	NMF/ANLS
TDT2	0.6448	0.7321	0.8710
Reuters	0.6776	0.6447	0.8534
20News	0.7640	0.6166	0.7244
RCV1	0.6636	0.5996	0.7950
NIPS14-16	0.6421	0.5352	0.8399

7.6 UTOPIAN: User-driven Topic Modeling via Interactive NMF

In this section, we present a visual analytics system called UTOPIAN (User-driven Topic Modeling Based on Interactive NMF)[13] [9], which utilizes NMF as a main tool to steer topic modeling results in a user-driven manner. As seen in Fig. 7.3, UTOPIAN provides a visual overview of the NMF topic modeling result as a 2D scatter plot where dots represent documents. The color of each dot corresponds to the topic/clustering membership computed by NMF. The position of each dot is determined by running a modified version [9] of t-distributed stochastic neighborhood embedding [48] to the cosine similarity matrix of bag-of-words vectors of documents. Additionally, the topics are summarized as their representative keywords.

Beyond the visual exploration of the topic modeling result in a passive manner, UTOPIAN provides various interaction capabilities that can actively incorporate user inputs to topic modeling processes. The interactions supported by UTOPIAN include topic keyword refinement, topic merging/splitting, and topic creation via seed documents/keywords, all of which are built upon WS-NMF. In the following, we describe each interaction in more detail.

Topic Keyword Refinement. In the topic modeling using NMF, the i-th topic, which corresponds to the i-th column vector $W^{(i)}$ of W is represented as a weighted combination of keywords. This interaction allows users to change the weights corresponding to keywords, corresponding to each component of the vector $W^{(i)}$, so that the meaning of the topic can be refined. For instance, users might want to remove some of the uninformative terms by setting its weight value to zero. In addition, users could increase/decrease the weight of a particular keyword to make the corresponding topic more/less relevant to the keyword. In turn, this refined vector $W^{(i)}$ is placed in the corresponding i-th column vector of W_r in (7.19) as the reference information during the subsequent WS-NMF. We also set a nonzero value of $M_W^{(l)}$ to make this reference vector in effect.

Topic Merging. This interaction merges multiple topics into one. To this end, we utilize the reference information H_r for H as follows. We first interpret the

[13] http://tinyurl.com/2013utopian

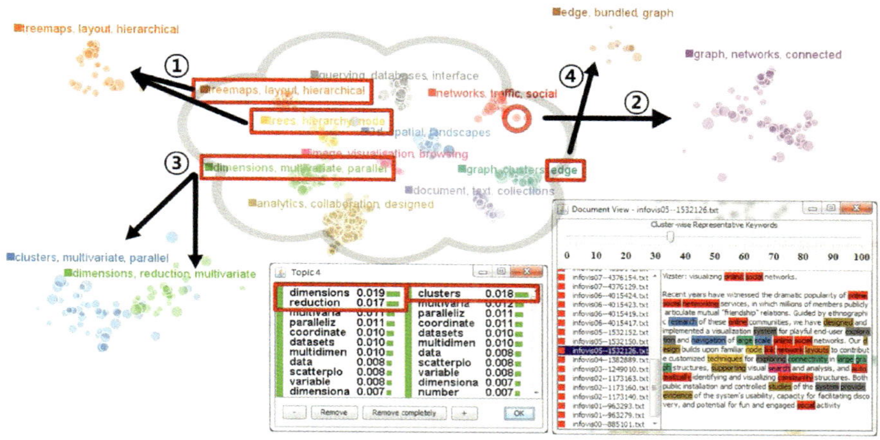

Fig. 7.3 An overview of UTOPIAN. Given a scatter plot visualization generated by a modified t-distributed stochastic neighborhood embedding, UTOPIAN provides various interaction capabilities: (1) topic merging, (2) document-induced topic creation, (3) topic splitting, and (4) keyword-induced topic creation. Additionally, the user can refine topic keyword weights (not shown here). The document viewer highlights the representative keywords from each topic

columns of H as hard clustering results and identify the set of documents clustered to one of the merged topics. For these documents, we obtain their $H^{(i)}$'s and merge the values corresponding to the merged topics by adding them up to a single value, and set the corresponding columns of H_r to the resulting $H^{(i)}$'s. For example, suppose two documents, whose $H^{(i)}$'s are represented as (0.6, 0.3, 0.1) and (0.4, 0.5, 0.1), respectively, corresponding to the three original topics. The corresponding column of H_r will be set to (0.6+0.4, 0.1) and (0.3+0.5, 0.1), respectively, where the first component corresponds to the merged topic. Alternatively, for topic merging, one could use the reference information for W, but we found our approach empirically works better.

Topic Splitting. UTOPIAN also support a topic splitting interaction. It splits a particular topic, e.g., $W^{(i)}$ of W , into the two topics. To guide this splitting process, users can assign the reference information for the two topics as follows. First, both vectors are initialized as the same as $W^{(i)}$. Now users can specify these two topics differently using the topic keyword refinement interaction.. In this manner, the topic splitting process in WS-NMF is guided by users based on the differently weighted keyword-wise representations of the two topics.

Document-Induced Topic Creation. This interaction creates a new topic by using user-selected documents as seed documents. For example, such seed documents can be a person's own research papers and s/he might want to see the topic formed by them as well as other papers relevant to this topic. To achieve this interaction, we utilize the reference information for documents. That is, for those documents specified by the user, their corresponding vectors in H_r in 7.19

are initialized to zero vectors but are set to one for the corresponding component to the newly created topic. This generates the reference information such that these documents are related only to the newly created topic. WS-NMF then creates a new topic based on it, which is represented as a keyword-wise distribution, and as a result, other relevant documents can be included.

Keyword-Induced Topic Creation. It creates a new topic via user-selected keywords. For instance, given the summary of topics as their representative keywords, users might want to explore more detailed (sub-)topics about particular keywords. A new topic created using these keywords would reveal such information. This interaction works similarly to document-induced topic creation except that we now use the reference information for keywords. Given user-selected keywords, the reference information of a new topic, i.e., a newly added column vector of W_r, is set to a zero vector, but the components corresponding to the keywords are set to ones. Accordingly, WS-NMF will include related documents in this topic, which, in turn, reveals the details about this topic.

In all the above-described interaction capabilities, UTOPIAN provides slider user interfaces with which users can interactively control how strongly the supervision is imposed. The values specified via these user interfaces are used as those for the nonzero elements in M_W and M_H.

7.6.1 Usage Scenarios

We show usage scenarios of the above-described interactions using the 20 Newsgroups data set. For efficient interactive visualization in real time, we randomly sampled 50 data items per each of the 20 categories. Figures 7.4, 7.5, 7.6, and 7.7 shows a sequence of interactions with these data in UTOPIAN.

Figure 7.4a shows an initial visualization result generated by UTOPIAN, which gives a nice visual overview about generated topic clusters. One can see that semantically similar topics are placed closely, e.g., the topics 'christian, ...' and 'jesus, god, hell' (top right) and the topics 'monitor, mac, driving' and 'scsi, id, bus' (bottom middle) while unrelated topics far from each other, e.g., the topics 'games, play, scores' (top left) and 'armenian, arabs, israeli' (bottom right) and the topics 'window, file, spss' (lower left) and 'cars, oil, dealers' (upper right).

Now, we perform a topic splitting interaction on the topic 'games, play, scores.' Initially, both the keywords 'baseball' and 'hockey' are shown to be highly ranked in this topic, but we aim at distinguishing the two topics with respect to these keywords. Thus, as shown in Fig. 7.4b, we increase the weight of the former keyword in the left topic and that of the latter in the right topic. This interaction generates the two split topics, as shown in Fig. 7.5, and the documents included in each topic properly reflect such user intention. Next, we merge semantically similar topics. We select the two topics 'jesus, god, hell' and 'belief, religion, god' to merge. The merged topic is generated as 'god, jesus, sin,' as shown in Fig. 7.6.

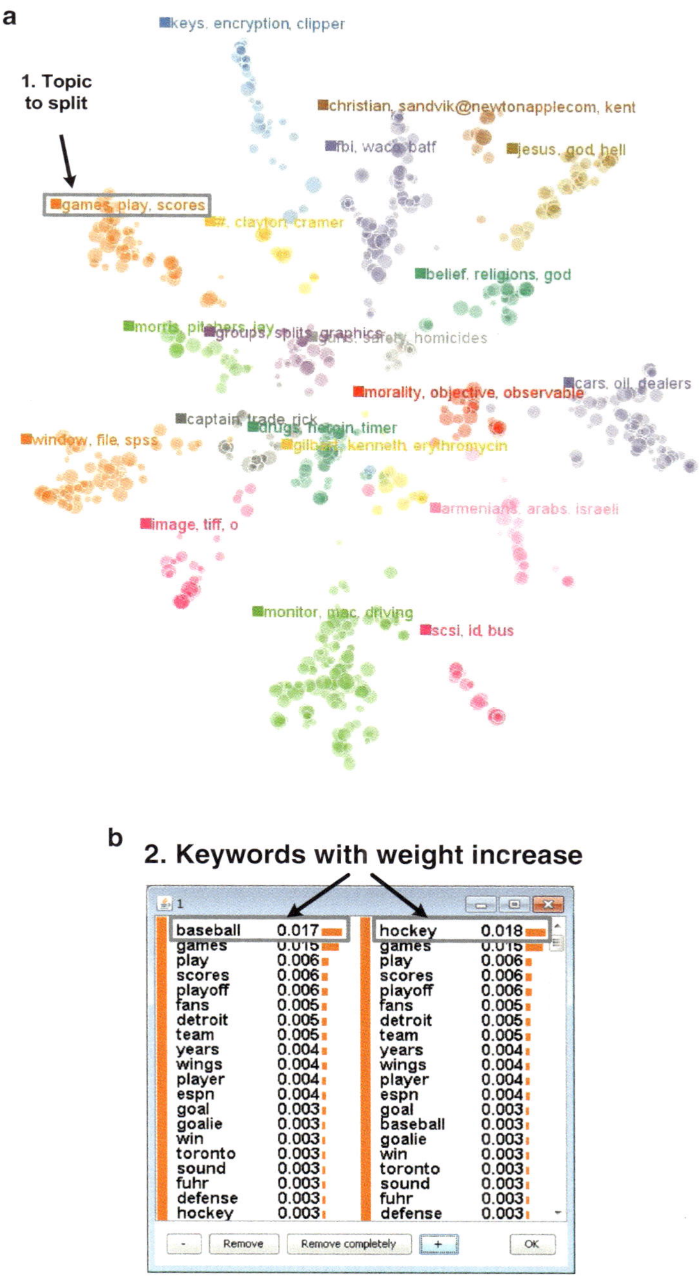

Fig. 7.4 The initial visualization and the topic splitting interaction for the subset of the 20 Newsgroup data set. From the topic 'games, play, scores,' we increase the weight of the keyword 'baseball' in one topic while increasing that of 'hockey' in the other. (**a**) The initial visualization. (**b**) The topic refinement of the two split topics

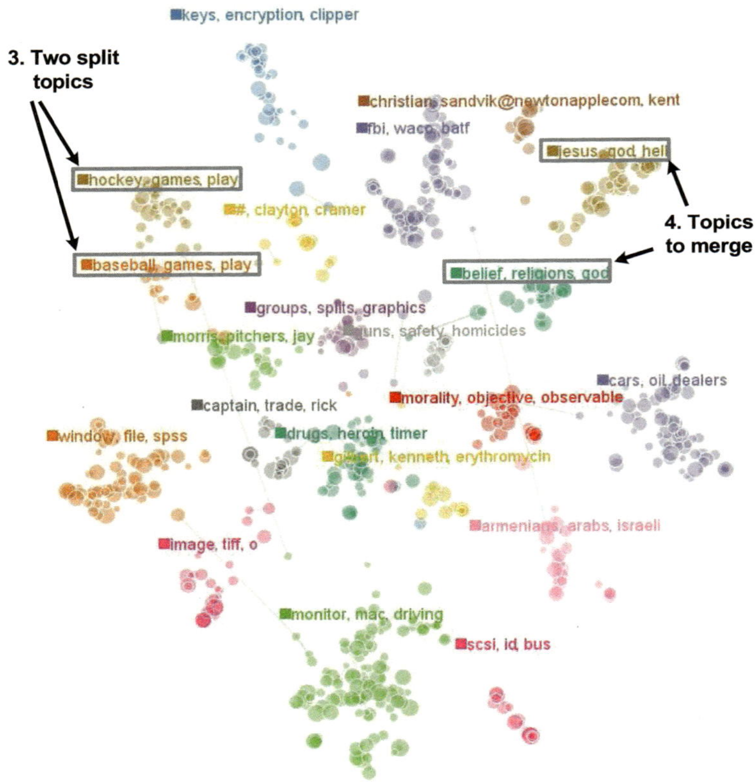

Fig. 7.5 The two splitted topics, one of which is mainly related to baseball and the other to hockey, are visualized. Next, the two topics 'jesus, god, hell' and 'belief, religion, god' are to be merged

Finally, we create a new topic via a keyword-induced topic creation interaction. To this end, we select a keyword 'dealers' from the topic 'cars, oil, dealers.' As shown in Fig. 7.7, the newly created topic 'dealers, invoice, cost' reveals the detailed information about the relevant topic to this keyword.

7.7 Conclusions and Future Directions

In this book chapter, we have presented nonnegative matrix factorization (NMF) for document clustering and topic modeling. We have first introduced the NMF formulation and its applications to clustering. Next, we have presented the flexible algorithmic framework based on block coordinate descent (BCD) as well as its convergence property and stopping criterion. Based on the BCD framework, we discussed two important extensions for clustering, the sparse and the weakly-supervised NMF,

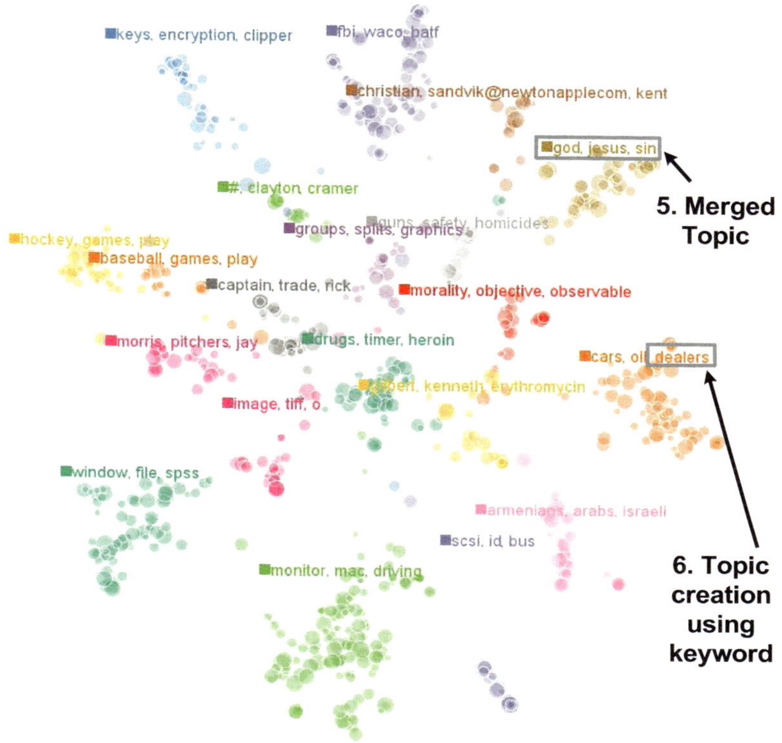

Fig. 7.6 The merged topic 'god, jesus, sin' is generated. Next, a new topic based on the keyword 'dealers' from the topic 'cars, oil, dealers' is to be generated

and our method to determine the number of clusters. Experimental results on various real-world document data sets show the advantage of our NMF algorithm in terms of clustering quality, convergence behavior, sparseness, and consistency. Finally, we presented a visual analytics system called UTOPIAN for interactive visual clustering and topic modeling and demonstrated its interaction capabilities such as topic splitting/merging as well as keyword-/document-induced topic creation.

The excellence of NMF in clustering and topic modeling poses numerous exciting research directions. One important direction is to improve the scalability of NMF. Parallel distributed algorithms are essential for this purpose, but at the same time, the real-time interaction capability can also be considered from the perspective of a human perception [8]. Another direction is to allow users to better understand clustering and topic modeling outputs. In practice, the semantic meaning of document clusters and topics is understood based on several representative

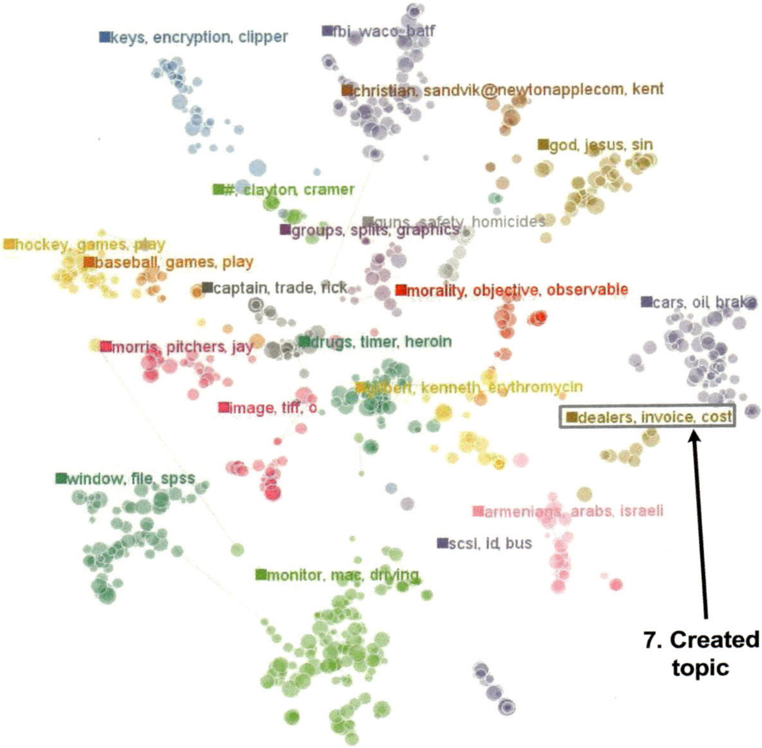

Fig. 7.7 The newly created topic 'dealers, invoice, cost' is shown

keywords and/or documents. However, significant noise in real-world data often makes it difficult to understand the resulting clusters and topics. In this sense, how to provide additional information such as the cluster/topic quality as well as contextual meaning of given keywords has to be addressed.

Acknowledgements The work of the authors was supported in part by the National Science Foundation (NSF) grants CCF-0808863 and the Defense Advanced Research Projects Agency (DARPA) XDATA program grant FA8750-12-2-0309. Any opinions, findings and conclusions or recommendations expressed in this material are those of the authors and do not necessarily reflect the views of the NSF or the DARPA.

References

1. Arora S, Ge R, Kannan R, Moitra A (2012) Computing a nonnegative matrix factorization – provably. In: Proceedings of the 44th symposium on theory of computing (STOC), pp 145–162
2. Arora S, Ge R, Halpern Y, Mimno D, Moitra A, Sontag D, Wu Y, Zhu M (2013). A practical algorithm for topic modeling with provable guarantees. J Mach Learn Res 28(2):280–288

3. Berman A, Plemmons RJ (1994) Nonnegative matrices in the mathematical sciences. SIAM, Philadelphia
4. Bertsekas DP (1999) Nonlinear programming, 2nd edn. Athena Scientific, Belmont
5. Blei DM, Ng AY, Jordan MI (2003) Latent Dirichlet allocation. J Mach Learn Res 3:993–1022
6. Brunet J-P, Tamayo P, Golub TR, Mesirov JP (2004) Metagenes and molecular pattern discovery using matrix factorization. Proc Natl Acad Sci USA 101(12):4164–4169
7. Cai D, He X, Han J, Huang TS (2011) Graph regularized nonnegative matrix factorization for data representation. IEEE Trans Pattern Anal Mach Intell 33(8):1548–1560
8. Choo J, Park H (2013) Customizing computational methods for visual analytics with big data. IEEE Comput Graph Appl 33(4):22–28
9. Choo J, Lee C, Reddy CK, Park H (2013) UTOPIAN: user-driven topic modeling based on interactive nonnegative matrix factorization. IEEE Trans Vis Comput Graph 19(12):1992–2001
10. Cichocki A, Zdunek R, Phan AH, Amari S (2009) Nonnegative matrix and tensor factorizations: applications to exploratory multi-way data analysis and blind source separation. Wiley, London
11. Devarajan K (2008) Nonnegative matrix factorization: an analytical and interpretive tool in computational biology. PLoS Comput Biol 4(7):e1000029
12. Dhillon IS, Sra S (2005) Generalized nonnegative matrix approximations with Bregman divergences. In: Advances in neural information processing systems (NIPS), vol 18, pp 283–290
13. Ding C, He X, Simon HD (2005) On the equivalence of nonnegative matrix factorization and spectral clustering. In: Proceedings of SIAM international conference on data mining (SDM), pp 606–610
14. Ding C, Li T, Jordan M (2008) Nonnegative matrix factorization for combinatorial optimization: spectral clustering, graph matching, and clique finding. In: Proceedings of the 8th IEEE international conference on data mining (ICDM), pp 183–192
15. Ding C, T Li, Jordan MI (2010) Convex and semi-nonnegative matrix factorization. IEEE Trans Pattern Anal Mach Intell 32(1):45–55
16. Duda RO, Hart PE, Stork DG (2000) Pattern classification. Wiley-Interscience, London
17. Globerson A, Chechik G, Pereira F, Tishby N (2007) Euclidean embedding of co-occurrence data. J Mach Learn Res 8:2265–2295
18. Gonzales EF, Zhang Y (2005) Accelerating the Lee-Seung algorithm for non-negative matrix factorization. Technical Report TR05-02, Rice University
19. Grippo L, Sciandrone M (2000) On the convergence of the block nonlinear Gauss-Seidel method under convex constraints. Oper Res Lett 26:127–136
20. Hofmann T (1999) Probabilistic latent semantic indexing. In: Proceedings of the 22nd annual international ACM SIGIR conference on research and development in information retrieval (SIGIR)
21. Hoyer PO (2004) Non-negative matrix factorization with sparseness constraints. J Mach Learn Res 5:1457–1469
22. Kim H, Park H (2007) Sparse non-negative matrix factorizations via alternating non-negativity-constrained least squares for microarray data analysis. Bioinformatics 23(12):1495–1502
23. Kim H, Park H (2008) Nonnegative matrix factorization based on alternating non-negativity-constrained least squares and the active set method. SIAM J Matrix Anal Appl 30(2):713–730
24. Kim D, Sra S, Dhillon I (2007) Fast Newton-type methods for the least squares nonnegative matrix approximation problem. In: Proceedings of SIAM international conference on data mining (SDM), pp 343–354
25. Kim J, He Y, Park H (2014) Algorithms for nonnegative matrix and tensor factorizations: a unified view based on block coordinate descent framework. J Global Optim 58(2):285–319
26. Kim J, Park H (2008) Sparse nonnegative matrix factorization for clustering. Technical Report GT-CSE-08-01, Georgia Institute of Technology
27. Kim J, Park H (2008) Toward faster nonnegative matrix factorization: a new algorithm and comparisons. In: Proceedings of the 8th IEEE international conference on data mining (ICDM), pp 353–362

28. Kim J, Park H (2011) Fast nonnegative matrix factorization: An active-set-like method and comparisons. SIAM J Sci Comput 33(6):3261–3281
29. Kuang D, Park H (2013) Fast rank-2 nonnegative matrix factorization for hierarchical document clustering. In: Proceedings of the 19th ACM international conference on knowledge discovery and data mining (KDD), pp 739–747
30. Kuang D, Ding C, Park H (2012) Symmetric nonnegative matrix factorization for graph clustering. In: Proceedings of SIAM international conference on data mining (SDM), pp 106–117
31. Kuhn HW (1955) The Hungarian method for the assignment problem. Nav Res Logistics Q 2:83–97
32. Lawson CL, Hanson RJ (1974) Solving least squares problems. Prentice Hall, Englewood Cliffs
33. Lee DD, Seung HS (1999) Learning the parts of objects by non-negative matrix factorization. Nature 401:788–791
34. Lee DD, Seung HS (2001) Algorithms for non-negative matrix factorization. In: Advances in neural information processing systems (NIPS), vol 14, pp 556–562
35. Lewis DD, Yang Y, Rose TG, Li F (2004) Rcv1: a new benchmark collection for text categorization research. J Mach Learn Res 5:361–397
36. Li S, Hou XW, Zhang HJ, Cheng QS (2001) Learning spatially localized, parts-based representation. In: Proceedings of the 2001 IEEE conference on computer vision and pattern recognition (CVPR), pp 207–212
37. Li T, Ding C, Jordan MI (2007) Solving consensus and semi-supervised clustering problems using nonnegative matrix factorization. In: Proceedings of the 7th IEEE international conference on data mining (ICDM), pp 577–582
38. Li L, Lebanon G, Park H (2012) Fast Bregman divergence NMF using Taylor expansion and coordinate descent. In: Proceedings of the 18th ACM international conference on knowledge discovery and data mining (KDD), pp 307–315
39. Lin C-J (2007) Projected gradient methods for nonnegative matrix factorization. Neural Comput 19(10):2756–2779
40. Manning CD, Raghavan P, Schütze H (2008) Introduction to information retrieval. Cambridge University Press, Cambridge
41. McCallum AK (2002) MALLET: a machine learning for language toolkit. http://mallet.cs.umass.edu
42. Monti S, Tamayo P, Mesirov J, Golub T (2003) Consensus clustering: a resampling-based method for class discovery and visualization of gene expression microarray data. Mach Learn 52(1–2):91–118
43. Paatero P, Tapper U (1994) Positive matrix factorization: a non-negative factor model with optimal utilization of error estimates of data values. Environmetrics 5:111–126
44. Pauca VP, Shahnaz F, Berry MW, Plemmons RJ (2004) Text mining using non-negative matrix factorizations. In: Proceedings of SIAM international conference on data mining (SDM), pp 452–456
45. Pauca VP, Piper J, Plemmons RJ (2006) Nonnegative matrix factorization for spectral data analysis. Linear Algebra Appl 416(1):29–47
46. Shahnaz F, Berry MW, Pauca VP, Plemmons RJ (2006) Document clustering using nonnegative matrix factorization. Inf Process Manag 42:373–386
47. Tibshirani R (1994) Regression shrinkage and selection via the lasso. J R Stat Soc Ser B Methodol 58:267–288
48. van der Maaten L, Hinton G (2008) Visualizing data using t-SNE. J Mach Learn Res 9:2579–2605
49. Vavasis SA (2009) On the complexity of nonnegative matrix factorization. SIAM J Optim 20(3):1364–1377
50. Wild S, Curry J, Dougherty A (2004) Improving non-negative matrix factorizations through structured initialization. Pattern Recognit 37:2217–2232

51. Xie B, Song L, Park H (2013) Topic modeling via nonnegative matrix factorization on probability simplex. In: NIPS workshop on topic models: computation, application, and evaluation
52. Xu W, Liu X, Gong Y (2003) Document clustering based on non-negative matrix factorization. In: Proceedings of the 26th annual international ACM SIGIR conference on research and development in information retrieval (SIGIR), pp 267–273

Chapter 8
Overview of Overlapping Partitional Clustering Methods

Chiheb-Eddine Ben N'Cir, Guillaume Cleuziou, and Nadia Essoussi

Abstract Identifying non-disjoint clusters is an important issue in clustering referred to as Overlapping Clustering. While traditional clustering methods ignore the possibility that an observation can be assigned to several groups and lead to k exhaustive and exclusive clusters representing the data, Overlapping Clustering methods offer a richer model for fitting existing structures in several applications requiring a non-disjoint partitioning. In fact, the issue of overlapping clustering has been studied since the last four decades leading to several methods in the literature adopting many usual approaches such as hierarchical, generative, graphical and k-means based approach. We review in this paper the fundamental concepts of overlapping clustering while we survey the widely known overlapping partitional clustering algorithms and the existing techniques to evaluate the quality of non-disjoint partitioning. Furthermore, a comparative theoretical and experimental study of used techniques to model overlaps is given over different multi-labeled benchmarks.

Keywords Overlapping clustering • Non-disjoint partitioning • Non-exclusive clusters • Partitional clustering methods • Evaluation of overlapping clustering methods • Small overlaps • Large overlaps

8.1 Introduction

Clustering, also referred to as cluster analysis or learning, has become an important technique in data mining and pattern recognition used either to detect hidden structures or to summarize the observed data. Usually, a clear distinction is made between learning problems that are supervised, also referred to as classification,

C.-E. Ben N'Cir • N. Essoussi
LARODEC, ISG Tunis, University of Tunis, 41 Avenue de la Liberté, Cité Bouchoucha, Le Bardo 2000, Tunisia
e-mail: chiheb.benncir@isg.rnu.tn; nadia.essoussi@isg.rnu.tn

G. Cleuziou (✉)
LIFO, Université d'Orléans, EA 4022, Orléans, France
e-mail: guillaume.cleuziou@univ-orleans.fr

M.E. Celebi (ed.), *Partitional Clustering Algorithms*,
DOI 10.1007/978-3-319-09259-1_8

and those that are unsupervised, referred to as clustering. The first deals with only labeled data while the latter deals with only unlabeled data [22]. In practice, given a description of N data over d variables, clustering aims to find k groups based on a measure of similarity such that similarities between data in the same group are high while similarities between data in different groups are low.

During the last four decades, many researches have been focused in designing clustering methods resulting in many methods that are proposed in the literature which are based on different approaches. Partitional clustering [29], also referred as partitioning relocation clustering [7], constitutes an important approach that several clustering methods are based on. Partitional clustering attempts to directly decompose the data set into a set of disjoint clusters leading to an integer number of clusters that optimizes a given criterion function. The criterion function may emphasize a local or a global structure of the data and its optimization is an *iterative* relocation procedure. Such type of clustering methods considers that clusters are disjoint and does not support intersections. However, for many applications of clustering, it would be recommended to tolerate overlaps between clusters to better fit hidden structures in the observed data. This research's issue is referred to as *overlapping clustering*.

Overlapping clustering has been studied through various approaches during the last half-century. In this paper, we first give a classification of existing methods able to produce a non-disjoint partitioning of data based on their conceptual approach. Then, we review overlapping clustering methods which extends or generalizes k-means and k-medoid for overlapping clustering. We use the concept "overlapping partitional clustering methods" to refer to this kind of methods which aims to build a recovery of a data set containing N objects into a set of k covers or clusters, so to minimize an objective criterion. The sum of the cardinality of the clusters are equal or superior to N leading to k non-exclusive clusters.

The remaining sections are organized as in the following: Section 8.2 gives a classification of existing overlapping methods based on the used approach to build non-disjoint partitioning. Then Sect. 8.3 reviews the existing overlapping partitional clustering methods while Sect. 8.4 reviews the existing techniques to asses the quality of the resulting non-exclusive partitionings. Finally, Sect. 8.5 gives an experimental evaluation of overlapping partitioning clustering methods on different multi-labeled benchmarks.

8.2 Classification of Exiting Approaches for Overlapping Clustering

Traditional learning methods ignore the possibility that an observation can belong to more than one cluster and lead to k exhaustive and exclusive clusters representing the data. Although this approach has been successfully applied in unsupervised learning, there are many situations in which a richer model is needed for representing the data. For example, in social network analysis, community extraction algorithms need to detect overlapping clusters where an actor can belong to multiple communities [23, 43, 46]. In video classification, overlapping clustering is a necessary requirement where videos have potentially multiple genres [42]. In emotion detection, overlapping clustering methods need to detect different emotions for a specific piece of music [47]. In text clustering, learning methods should be able to group document, which discuss more than one topic, into several groups [25, 41], etc. The corresponding research domain has been referred to as *overlapping clustering* and has been studied through various approaches.

Basically, the existing overlapping clustering methods are extensions from usual clustering models such as hierarchical, generative, graph-based or partitional models. Thereby, we propose a classification of existing methods based on the conceptual approach to build non-disjoint partitioning of data. Figure 8.1 shows a classification tree of these methods where the depth of the tree represents the progression in time and the width of the tree represents the different categories and subcategories. We detail in the following the main characteristics of each category.

The overlapping variants of hierarchies aim to reduce the discrepancy between the original dissimilarities over the considered dataset and the ones induced by the hierarchical structure. Although the flexibility in visualization offered by hierarchical methods, they still too restrictive in overlaps while they not study all the possible combinations of clusters for each observation. Examples of these methods are the pyramids [21] and more generally the weak-hierarchies [8]. Concisely, these structures are either too restrictive on the overlaps, as for pyramids, or hard to acquire and visualize as for weak-hierarchies.

Overlapping methods based on graph theory are mostly used in the context of community detection in complex networks [3, 17, 23, 26–28, 36, 45, 51]. In this research area, a network is represented as an undirected or directed graph, depending on the specificity of the problem, where vertices are the studied observations and edges are links between the observations. All these graph-based methods use a greedy heuristic for covering the similarity graph. The difference between them consists on the criterion used for ordering and selecting the sub-graphs. The main shortcoming of these methods is the computational complexity which is usually exponential and could be reduced to $O(N^2)$ as the case for OClustR (Overlapping Clustering based on Relevance) [40].

Overlapping clustering methods using generative mixture models have been proposed [2, 24, 30] as extensions of the EM algorithm [18]. These models are supported by biological processes; they hypothesize that each data is the result of a mixture of distributions : the mixture can be additive [2] or multiplicative [24, 30] and the probabilistic framework makes possible to use not only gaussian components but any exponential family distributions. On the other hand, generative models are not parameterizable and do not allow the user to control the requirements of the overlaps.

Other recent methods for overlapping clustering extend other approaches to address the problem of overlapping clustering. For example, an extension of correlation clustering [10, 11] and topological maps [16] have been recently proposed. Overlapping correlation clustering has been defined as an optimization problem

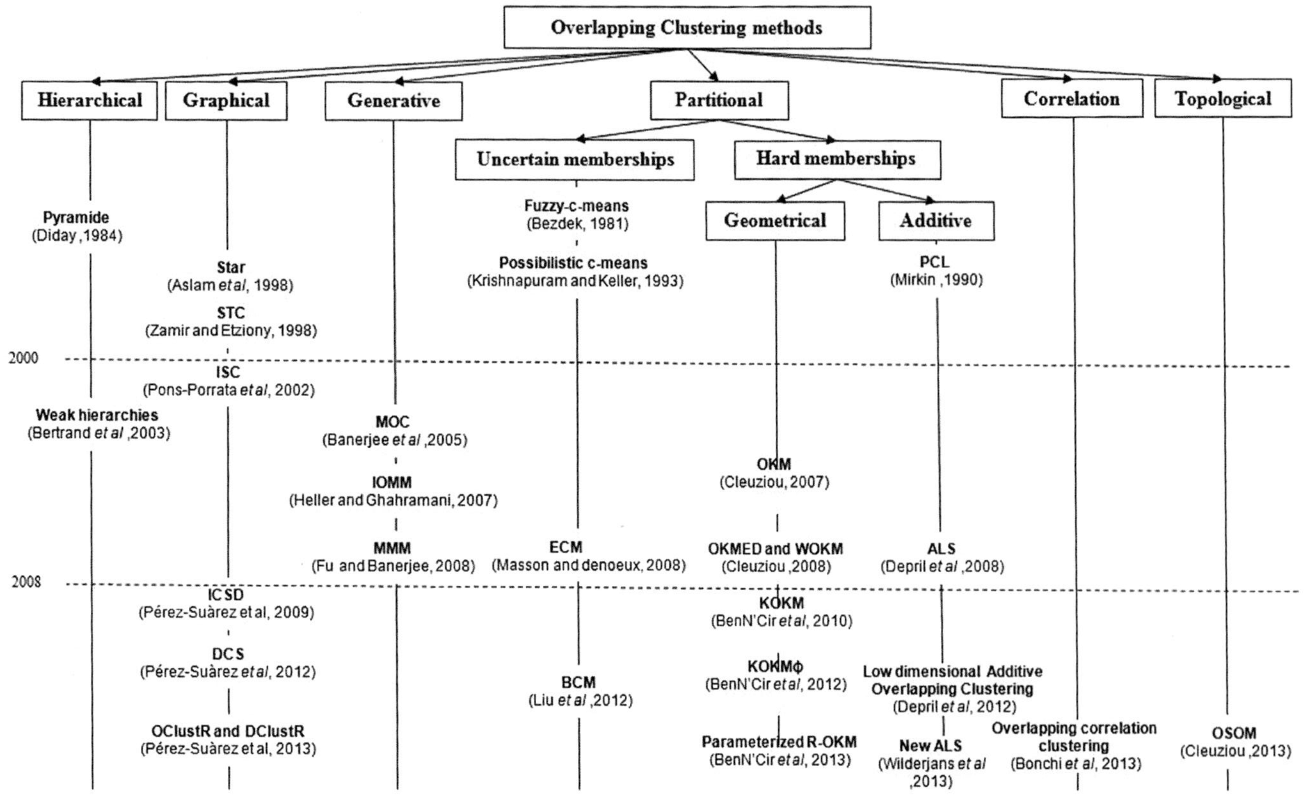

Fig. 8.1 Classification of overlapping clustering methods based on their conceptual approach to build non-disjoint partitioning of data

that extends the framework of correlation clustering to allow overlaps by relaxing the function which measures the similarity of assigned set of labels, instead of one single label, for each data object. For topological maps, an extension of the Self-Organizing-Maps (SOM) has been proposed by allowing for each data to be assigned to one or several neurons on the grid by searching for a subset of neurons winners rather than a single neuron. The main advantage of both correlation and topological methods consists of their ability to learn the right number of overlapping clusters.

Despite the use of all these approaches to build non-disjoint partitioning of data, the Partitional approach remains the most commonly used while several methods are based on. This category of methods consists either in modifying the clusters resulting from a standard method into overlapping clusters or in proposing new objective criteria to model overlaps. This survey's emphasis is on overlapping partitional clustering methods, specifically those extending and generalizing k-means and K-medoid methods [31,35]. In this way, we present in the next section a description of these methods.

8.3 Overlapping Partitional Clustering Methods

Several works have been focused on partitional clustering to build overlapping clusters leading to two main categories of methods: the category of *uncertain memberships* and the category of *hard memberships*. We denote by uncertain memberships the solutions which model clusters' memberships for each data object as uncertainty function using fuzzy, possibilistic or evidential frameworks. The uncertainty function measures the degree of belonging of each data to the underlying group. However, we denote by *hard memberships* the solutions which lead to *hard* and *overlapping* partitioning by considering a binary function to model clusters' memberships.

8.3.1 Uncertain Memberships Based-Methods

Uncertain memberships based-methods consist either in extending results of uncertain methods into overlapping clusters, typically the extension of fuzzy-c-means (FCM) [33,51] and possibilistic c-means (PCM) [32], or in proposing new objective criterion that takes into account the possibility of overlaps between clusters; the Evidential c-means (ECM) introduced by Masson and Denoeux [37] and the Belief c-means (BCM) proposed by Liu et al. [34] are two distinctive examples of such criteria where their optimization processes lead to generate overlapping clusters. All uncertain methods need a post-processing treatment to generate the final overlapping clusters.

We detail in the following the principal uncertain clustering methods which are able to produce non-disjoint partitioning. We consider for all the detailed methods a set of observations $X = \{x_i\}_{i=1}^N$ with $x_i \in \mathbb{R}^d$ and N the number of observations where the aim, of each method, is to find a non-disjoint partitioning matrix $\Pi = \{\pi_c\}_{c=1}^k$ into k clusters and a set $C = \{m_c\}_{c=1}^k$ of k clusters' representatives minimizing an objective criterion.

8.3.1.1 Fuzzy c-Means (FCM) and Possibilistic c-Means (PCM)

The FCM [9] identifies clusters as fuzzy sets where the objective function J_{FCM} allows that an observation belongs to many clusters with a coefficient indicating membership degrees to all clusters in the [0,1] interval (0 stands for no membership and 1 for total membership). FCM is based on the minimization of the following function:

$$J_{FCM}(\Pi, C) = \sum_{c=1}^{k}\sum_{i=1}^{N} \pi_{ic}^{\beta}.||x_i - m_c||^2, \tag{8.1}$$

where Π is the fuzziness membership matrix that indicates the coefficient of closeness of an object to every cluster under the constraints:

$$\begin{gathered} \pi_{ic} \in [0..1] \ \forall i, \quad \forall c \\ \sum_{c=1}^{k}(\pi_{ic}) = 1, \forall i \\ \beta > 1. \end{gathered} \tag{8.2}$$

The parameter β controls the fuzziness of the memberships: for high values of β the algorithm tends to set all the memberships equals while for β tending to one it has the behavior of k-means algorithm with crisp memberships. The minimization of Eq. 8.1 is done iteratively using an alternating least square optimization of the two parameters Π and C. The optimal fuzziness membership matrix Π and the optimal clusters' representatives C are computed in each step as the following.

$$\pi_{ic}^{*} = \frac{1}{\sum_{l=1}^{k}\left(\frac{||x_i - m_c||^2}{||x_i - m_l||^2}\right)^{(\frac{1}{\beta-1})}} \tag{8.3}$$

$$m_c^{*} = \frac{\sum_{i=1}^{N}\pi_{ic}^{\beta}x_i}{\sum_{i=1}^{N}\pi_{ic}^{\beta}} \tag{8.4}$$

	Fuzzy Clustering		
observation	Cluster1	Cluster2	Cluster3
1	0.05	0.90	0.05
2	0.50	0.20	0.30
3	0.30	0.40	0.30
4	0.70	0.20	0.10
5	0.20	0.40	0.40
6	0.00	0.20	0.80

Threshold = 0.3

	Overlapping Clustering		
observation	Cluster1	Cluster2	Cluster3
1	0	1	0
2	1	0	1
3	1	1	1
4	1	0	0
5	0	1	1
6	0	0	1

Fig. 8.2 Extension of the results of fuzzy clustering to obtain overlapping clustering using a threshold value equals to 0.3

The extension of FCM to overlapping clustering can be done by fixing a threshold where all observations having memberships' degrees that exceed this threshold are assigned to the respective clusters. An example of this transformation is shown in Fig. 8.2 where the obtained clusters' memberships are non-disjoints. We note that in some cases, when the fixed threshold is somewhat large, observations having all the memberships lower than this threshold will not be assigned to any cluster. Obtained clusters are usually much sensitive to value of the threshold.

In the same way that FCM, the PCM [32] method is proposed to relax the constrained condition of the fuzzy partition $\left(\sum_{c=1}^{k}(\pi_{ic}) = 1\right)$ in order to obtain a possibilistic type of memberships matrix. The objective function of PCM[1] is described by:

$$J_{PCM}(\Pi, C) = \sum_{c=1}^{k}\sum_{i=1}^{N} \pi_{ic}^{\beta}||x_i - m_c||^2, \tag{8.5}$$

[1]The original objective function of PCM takes into account the identification of outliers, whereas we give in this paper a short structure of the objective function to facilitate the comparison of PCM with the other described methods.

under the constraints:

$$\pi_{ic} \in [0..1]\ \forall i, \quad \forall c$$

$$\sum_{c=1}^{k} (\pi_{ic}) \quad \in [0, k]\ \forall\ i \tag{8.6}$$

$$\beta > 1.$$

Similar to FCM, the objective function J_{PCM} is minimized iteratively where memberships and clusters' representatives are updated as follows:

$$\pi_{ic}^* = \frac{1}{1 + (||x_i - m_c||^2)^{1/(\beta-1)}}, \tag{8.7}$$

$$m_c^* = \frac{\sum_{i=1}^{N} \pi_{ic}^{\beta} x_i}{\sum_{i=1}^{N} \pi_{ic}^{\beta}}. \tag{8.8}$$

Each observation can belong to several clusters with a possibilistic membership. The possibilistic memberships can be extended to overlapping ones by setting memberships with higher values to 1 and the other are replaced with 0.

8.3.1.2 Evidential c-Means (ECM) and Belief c-Means (BCM)

ECM [37] is based on the concept of credal partition which is a general extension of fuzzy and possibilistic partitioning. As opposed to FCM and PCM, the ECM method evaluates all the possible combinations of clusters , denoted A_j, from the set of single clusters $\Omega = \{\omega_1, \ldots, \omega_k\}$ by allocating a mass of belief π_i within each possible combination.

Let the credal partition matrix $\Pi = (\pi_1, \ldots, \pi_N) \in R^{N \times 2^k}$ and the matrix $C = (m_1, \ldots, m_k)$ of clusters' representatives, ECM[2] is based on the minimization of the following objective function:

$$J_{ECM}(\Pi, C) = \sum_{i=1}^{N} \sum_{j/A_j \subseteq \Omega} c_j^{\alpha} \pi_{ij}^{\beta} \left\| x_i - \bar{m_j} \right\|^2 \tag{8.9}$$

[2]The original objective function of ECM takes into account the identification of outliers by considering $\pi_{i\emptyset}$ a mass of belief to belong to any cluster, whereas we consider in this paper that all combinations of clusters are tolerated except the empty set ($A_j \neq \emptyset$) in order to facilitate the comparison of ECM with the other described methods.

under the constraint:

$$\sum_{j/A_j \subseteq \Omega} \pi_{ij} = 1 \; \forall i \in \{1, ..., N\}, \tag{8.10}$$

where π_{ij} denotes the mass of belief for associating the observation x_i to the specific set A_j which can be either a single cluster or a combination of single clusters, c_j denotes the cardinality $|A_j|$ of each set which aims at penalizing the combination of clusters with high cardinality, α and β are two parameters used respectively to control the penalization term c_j and the fuzziness degree π_{ij} and $\|x_i - \bar{m}_j\|^2$ is the distance between x_i and the combination of clusters' representatives $\overline{m_j}$ of the focal set A_j defined by:

$$\overline{m_j} = \frac{\sum_{j/A_j \subseteq \Omega} m_j}{c_j}. \tag{8.11}$$

To minimize the objective function of ECM, an alternating optimization scheme is designed as in FCM where the update of the mass of belief π_{ij} and the clusters' representatives C_k is computed as described in Eqs. 8.12 and 8.13 .

$$\pi_{ij}^* = \frac{c_j^{-\alpha/(\beta-1)} \|x_i - \bar{m}_j\|^{-2/(\beta-1)}}{\sum_{A_k} c_k^{-\alpha/(\beta-1)} \|x_i - \bar{m}_k\|^{-2/(\beta-1)}} \; \forall i \in \{1, .., N\} \; \forall j/A_j \subseteq \Omega, \tag{8.12}$$

$$C_{k\times d}^* = H_{k\times k}^{-1} B_{k\times d}, \tag{8.13}$$

where the elements B_{lq} of the matrix $B_{k\times d}$ for $l \in \{1, ..., k\}$, $q \in \{1, ..., d\}$ and the elements H_{lh} of the matrix $H_{k\times k}$ for $l, h \in \{1, ..., k\}$ are defined respectively by:

$$B_{lq} = \sum_{i=1}^{N} x_{iq} \sum_{j/\omega_l \in A_j} c_j^{\alpha-1} \pi_{ij}^{\beta} \qquad H_{lh} = \sum_{i=1}^{N} \sum_{j/\omega_h, \omega_l \subseteq A_j} c_j^{\alpha-2} \pi_{ij}^{\beta}. \tag{8.14}$$

Similar to ECM, a more recent method, referred to as Belief c-means (BCM) [34], is also developed within the framework of belief function which is an extension of ECM that improves the quality of the credal partitions by assigning only relatively distant observations to the combination of clusters. This method evaluates the distance inter-prototypes before building the mass of believe π_{ij} related to each observation. The objective function optimized by BCM is described by:

$$J_{BCM}(\Pi, C) = \sum_{i=1}^{N} \sum_{j/A_j \subseteq \Omega, |A_j|=1} \pi_{ij}^{\beta} \|x_i - m_j\|^2 + \sum_{i=1}^{N} \sum_{j/A_j \subseteq \Omega, |A_j|>1} c_j^{\alpha} \pi_{ij}^{\beta} \widehat{d_{ij}}^2, \tag{8.15}$$

	Credal clustering						
observation	ω_1	ω_2	ω_3	$\omega_1\ \omega_2$	$\omega_1\ \omega_3$	$\omega_2\ \omega_3$	Ω
1	0.2	0.2	0.4	0.1	0.1	0	0
2	0	1	0	0	0	0	0
3	0.1	0.1	0.5	0.3	0	0	0
4	1	0	0	0	0	0	0
5	0.2	0.1	0.1	0.5	0	0.1	0
6	0	0	0	0	0	0	1

max evidence

	Overlapping Clustering		
observation	Cluster1	Cluster2	Cluster3
1	0	0	1
2	0	1	0
3	0	0	1
4	1	0	0
5	1	1	0
6	1	1	1

Fig. 8.3 Example of the extension of credal partition containing six observations under three clusters: observations are assigned to the combination of clusters having the max mass of evidence

where $\widehat{d_{ij}}^2$ evaluates the distance x_i with respect to inter-distances of clusters' prototypes to which x_i belongs to:

$$\widehat{d_{ij}}^2 = \frac{\sum\limits_{k \in A_j} \|x_i - m_k\|^2 + \sum\limits_{l,p \in A_j} \|m_l - m_p\|^2}{|A_j| + \gamma C^2_{|A_j|}}, \tag{8.16}$$

with γ the weighting factor of the distances among the clusters' prototypes and $C^2_{|A_j|} = \frac{|A_j|!}{2!(|A_j|-2)!}$ the number of combinations of $|A_j|$ taken 2 at a time.

In fact, both ECM and BCM lead to credal partitioning of N data into 2^k possible combinations of k clusters. An example of credal partitioning is reported in Fig. 8.3. We note that both possibilistic and fuzzy partitions can be recovered from the credal partition. A possibilistic partition can be obtained by computing from each m_i the plausibilities (possibilities) of the different clusters and a fuzzy partition can be obtained by calculating the pignistic probabilities of the different clusters from each m_i. The extension of both ECM and BCM to overlapping clustering, called hard credal partition by the authors, can be done by assigning each object x_i to the set of clusters A_j with the highest mass. Overlapping observations are those having highest mass for $|A_j| >= 2$. The process that leads to hard credal partitioning is called "Upper" [37] approximation.

8.3.2 Hard Memberships Based-Methods

The category of hard memberships based methods generalizes the strict k-means to look for optimal overlapping clusters. As opposed to fuzzy, possibilistic, and evidential clustering, these methods produce hard overlapping clusters and do not need any post processing treatment. Two kind of hard memberships based-methods have been proposed: *additive* and *geometrical* based methods. We denote by *additive* the methods which hypothesize that overlaps result in the addition of the representatives of the related clusters. These methods group observations into overlapping clusters while minimizing the sum of distances between each observation and the *sum* of clusters' representatives to which the observation belongs to. In contrast, we denote by *geometrical* based-methods those formalizing overlaps as a barycenter on the related cluster representatives. This category of methods is based on a geometrical reasoning in the data space and groups observations into overlapping clusters while minimizing the sum of distances between each observation and the *average*, instead of the sum, of clusters' representatives to which the observation belongs to.

8.3.2.1 Additive Methods

Examples of additive methods are Principal Cluster Analysis (PCL) [38, 39], the Alternating Least Square algorithms (ALS) [19], the Low dimensional Additive Overlapping Clustering [20] and the Bi-clustering ALS [49].

- **PCL**
  PCL [39] introduces the possibility that an observation belongs to more than one cluster based on the Additive model by considering variable values of an observation equals to the sum of the clusters' representatives to which the observation belongs to. Given a dataset X, a model matrix $M = \Pi C$ is looked for to optimally approximate X. The matrix M can be estimated by minimizing the least squares loss function:

$$J_{PCL}(\Pi, C) = \| X - \Pi C \|_F^2 = \sum_{x_i \in X} \| x_i - \sum_{c \in \Pi_i} m_c \|^2, \tag{8.17}$$

  where $\| \, . \, \|_F^2$ is the Frobenius norm of a matrix and m_c is the representative of cluster c.

  To minimize the objective criterion (8.17), PCL proceeds by building clusters one by one from a data set until achieving the expected number of clusters k. PCL builds the memberships of each cluster c independently from the memberships of the other clusters by minimizing the following criterion for each cluster c:

$$J_{PCL}^c = \sum_{x_i \in X} \pi_{ic} \| x_i - m_c \|^2 . \tag{8.18}$$

The minimization process starts with an empty cluster (i.e., with $\pi_{ic} = 0$, $\forall x_i \in X$) and sequentially add observations to it in a greedy way leading to the smallest value of criterion (8.18). For every observation that is considered for joining the cluster, a new representative m_c and a new value of J^c_{PCL} have to be calculated. The process is continued until there is no further decrease of J^c_{PCL}. The computation of clusters' representatives is also done locally for each cluster independently from the other clusters by:

$$m^*_c = \frac{\sum_{x_i \in X} \pi_{ic} x_i}{\sum_{x_i \in X} \pi_{ic}}. \tag{8.19}$$

The main characteristic of this method consists of its high computational complexity evaluated by $O(2^{NK})$ which makes the method under-used in real life applications of overlapping clustering.

- **ALS**

ALS [19] is based on the same objective criterion of PCL described in Eq. 8.17. However, ALS proposes two other algorithms for minimizing this objective criterion referred to as ALS_{lf1} and ALS_{lf2}. For both algorithms, the minimization of the objective criterion starts from an initial binary membership matrix Π_0. This membership matrix can be initialized using a Bernoulli distribution with parameter $\tau = 0.5$ or by computing the conditionally optimal memberships upon k randomly drawn representatives from the initial data. Then, ALS estimates the conditionally optimal representatives C upon Π; subsequently it estimates the conditionally optimal memberships Π upon C, and this process will be repeated until convergence.

The estimation of optimal memberships are computed separably for each observation x_i by enumerating all possible binary configurations that lead to decrease the objective criterion given the conditionally optimal representative and the conditionally optimal memberships for the other observations. The algorithm repeats this procedure for the next observation and so on. After a pass through all observations, the new value of the objective criterion computed with new memberships Π_1 is compared to the one computed with old memberships Π_0. The process stops when there is no further decrease in the objective criterion. We note that ALS_{lf1} differs from ALS_{lf2} in that for each memberships update Π_i the conditionally optimal representatives C are recalculated immediately, whereas in ALS_{lf2} the representatives are only updated at the end of the membership updating step.

In the other side, the optimal representatives are updated equivalently for ALS_{lf1} and ALS_{lf2} based on the memberships matrix Π as follows:

$$C^* = (\Pi'\Pi)^{-1}\Pi' X. \tag{8.20}$$

In fact, we notice that ALS with its two variants explores all the possible 2^k combinations of clusters and takes the optimal one that leads to decrease the objective criterion. This property makes this methods highly time consuming when the number of clusters becomes large.

- **Low dimensional Additive Overlapping Clustering**
  The Low dimensional Additive Overlapping Clustering [20] extends ALS method by establishing an overlapping clustering of the observations and a dimensional reduction of the variables (or dimensions) simultaneously. This method is designed in order to perform relevant non-disjoint partitioning when data contains a high number of dimensions. Given a set of observations X described over d variables, the aim of this method is to find a recovery Π of k overlapping clusters and a matrix $\tilde{C}$ of clusters' representatives described over $\tilde{d} < d$. The low dimensional additive Overlapping Clustering is based on the same objective criterion used for ALS and PCL. The process of optimizing this objective criterion uses an alternating optimization procedure similar to that of ALS, except that optimal reduced clusters' representatives are computed by:

$$\tilde{C}^* = (\Pi'\Pi)^{-1}\Pi' T T' X, \tag{8.21}$$

  where columns of matrix T represent the $\tilde{d}$ orthogonal eigenvectors of the product of matrixes $ZXX'Z$ with $Z = \Pi(\Pi'\Pi)^{-1}\Pi'$ denotes the orthogonal projector operator.

8.3.2.2 Geometrical Methods

Examples of geometrical methods are the Overlapping k-means (OKM) [14], Overlapping k-Medoid (OKMED) [15], Weighted Overlapping k-means (WOKM) [15], Kernel Overlapping K-means (KOKM) [4, 5] and Parameterized R-OKM [6]. We also note that ECM and BCM, described with uncertain memberships methods, can be categorized with geometrical methods while they are based on a geometrical reasoning to model the different combinations of clusters.

- **OKM**
  The OKM [14] method is an extension of the k-means algorithm which allows observations to belong to one or different clusters. Given a set of N observations $X = \{x_i\}_{i=1}^N$ with $x_i \in \mathbb{R}^d$, OKM aims to find a recovery $\Pi = \{\pi_c\}_{c=1}^k$ of k overlapping clusters such that the following objective function is minimized:

$$J_{OKM}(\Pi, C) = \sum_{i=1}^{N} \|x_i - \overline{(x_i)}\|^2. \tag{8.22}$$

  This objective function minimizes the sum of squared Euclidean distances between each observation x_i and its representatives $\overline{(x_i)}$ for all $x_i \in X$. The representative $\overline{(x_i)}$ is defined as the barycenter of clusters' representatives to which the observation x_i belongs to:

$$\overline{(x_i)} = \sum_{c \in \Pi_i} \frac{m_c}{|\Pi_i|}. \tag{8.23}$$

where Π_i is the set of clusters to which x_i belongs to and m_c is the representative of cluster c. The minimization of the objective function is performed by alternating two principal steps: computation of clusters' representatives (C) and the assignment of observations to one or several clusters (Π). The update of representatives is performed locally for each cluster as described in Eq. 8.24.

$$m_c^* = \frac{\sum_{x_i \in \pi_c} \frac{1}{|\Pi_i|^2} . \widetilde{x_i}^c}{\sum_{x_i \in \pi_c} \frac{1}{|\Pi_i|^2}}, \tag{8.24}$$

where $\widetilde{x_i}^c = |\Pi_i|.x_i - (|\Pi_i| - 1).\overline{(x_i)_{\Pi \setminus c}}$ and π_c denotes the set of observations assigned to cluster c. For the multiple assignment step, OKM uses an heuristic to explore part of the combinatorial set of possible assignments. The heuristic consists, for each observation, in sorting clusters from closest to the farthest, then assigning the observation in the order defined while assignment minimizes the distance between the observation and its representative.

- **Parameterized R-OKM**
  The Parameterized R-OKM offers the possibility to regulate the overlaps using a parameter α. As well as α becomes large ($\alpha \rightarrow +\infty$), Parameterized R-OKM builds clusters with more reduced overlaps. However, overlaps become more large as well as $\alpha \rightarrow 0$. When $\alpha = 0$, Parameterized R-OKM coincides exactly with OKM. In fact, Parameterized R-OKM restricts the assignments of a data point x_i to multiple clusters according to the cardinality of the set of assignments $|\Pi_i|$. The Parameterized R-OKM is based on the minimization of the following objective criterion:

$$J_{R-OKM}(\Pi, C) = \sum_{i=1}^{N} |\Pi_i|^{\alpha}.||x_i - \overline{(x_i)}||^2 \tag{8.25}$$

where $\alpha \geq 0$ is fixed by the user. We note that $|\Pi_i|^{\alpha}$ has the same role of C_j^{α} used within ECM and BCM methods which consists on penalizing or favoring overlaps. For the minimization of the objective criterion, Parameterized R-OKM uses the same minimization steps used for OKM as for the assignment and for the update of clusters' representatives. The later can be updated for each cluster as follows:

$$m_c^* = \frac{\sum_{x_i \in \pi_c} \frac{1}{|\Pi_i|^{2-\alpha}} . \widetilde{x_i}^c}{\sum_{x_i \in \pi_c} \frac{1}{|\Pi_i|^{2-\alpha}}}. \quad (8.26)$$

- **Kernel Overlapping k-means**

  While most of geometrical overlapping methods build non-disjoint partitioning of data with linear separations between clusters, KOKM hypothesizes that data structuring in real life applications is usually complex; thus requiring nonlinear separations to better fit the existing structures in data. In order to perform nonlinear separations between clusters, KOKM introduces the use of kernel methods for overlapping clustering. Two variants are proposed: the first [5] proposes a kernelization of the Euclidean metric used in OKM using the kernel induced distance measure while the second [4], referred as KOKMϕ, proposes to perform all clustering steps in a high dimensional space where data are implicitly mapped.

  Given a set of observations $X = \{x_i\}_{i=1}^N$ and given an implicit nonlinear mapping function $\phi : X \longrightarrow F$ which maps the input space X to a high dimensional feature space F, the objective functions minimized by both variants are respectively described by:

$$J_{KOKM}(\Pi, C) = \sum_{i=1}^{N} \|\phi(x_i) - \phi(\overline{x_i})\|^2, \quad (8.27)$$

$$J_{KOKM\phi}(\Pi, C) = \sum_{i=1}^{N} \|\phi(x_i) - \overline{\phi(x_i)}\|^2. \quad (8.28)$$

The first variant computes representatives $\overline{x_i}$ and clusters' representatives m_c in the original space and only distances between observations are performed in the mapping space. The optimization steps are similar to OKM except that distances are computed in feature space as described in Eq. 8.29.

$$\begin{aligned} \| \phi(x_i) - \phi(x_j) \|^2 &= (\phi(x_i) - \phi(x_j))((\phi(x_i) - \phi(x_j)) \\ &= \phi(x_i)\phi(x_i) - 2\phi(x_i)\phi(x_j) + \phi(x_j)\phi(x_j) \\ &= K_{ii} - 2K_{ij} + K_{jj}. \end{aligned} \quad (8.29)$$

where K_{ij} is the dot product of mapped data in the feature space which can be computed without using ϕ.

Conversely, the second variant performs all the learning process in the feature space F. The representatives $\overline{\phi(x_i)}$ are computed in feature space as follows:

$$\overline{\phi(x_i)} = \frac{\sum_{c=1}^{k} \pi_{ic}.m_c^{\phi}}{\sum_{c=1}^{k} \pi_{ic}}, \tag{8.30}$$

where $\pi_{ic} \in \{0, 1\}$ is a binary variable that indicates membership of observation x_i to cluster c and m_c^{ϕ} is the representative of cluster c in the feature space. The representative of each cluster in the feature space is defined by the medoid that minimizes all distances over all observations included in this cluster:

$$m_c^* = \arg\min_{x_i \in \pi_c}(x_i) \frac{\sum_{x_j \in \pi_c, x_j \neq x_i} |\Pi_j|[K_{ii} - 2K_{ij} + K_{jj}]}{|\pi_c|. \sum_{x_j \in \pi_c, x_j \neq x_i} |\Pi_j|}. \tag{8.31}$$

- **OKMED**
  OKMED extends the method Partitioning Around Medoid (PAM) for overlapping clustering. It consists in aggregating the data around representatives of the clusters denoted as medoids which are chosen among the data themselves. The objective criterion of OKMED is based on the optimization of the following objective criterion:

$$J_{OKMED}(\Pi, C) = \sum_{x_i \in X} \|x_i - \overline{(\chi_i)}\|^2. \tag{8.32}$$

where $\overline{(\chi_i)}$ is defined as the data from X that minimizes the sum of the dissimilarities with all the medoids of the clusters where x_i belongs to:

$$\overline{(\chi_i)} = \arg\min_{x_j \in X} \sum_{m_c \in A_i} \|x_j - m_c\|^2. \tag{8.33}$$

The optimization of the objective function of OKMED is realized using an alternating optimization between two steps: assignment of each data to its nearest medoid and updating of the medoid for each cluster. The update of medoids consist in searching among the set of data belonging to the cluster, the one that minimizes the sum of the distances with any other data into the cluster. Formally, optimal medoid for each cluster is described by:

$$m_c^* = \arg\min_{x_i \in \pi_c}(x_i) \sum_{x_j \in \pi_c} \|x_j - \overline{(\chi_j)_{x_i}}\|^2, \tag{8.34}$$

where $\overline{(\chi_j)_{x_i}}$ denotes the representative of observation x_j computed by considering x_i the medoid of the cluster. The use of medoids as representatives of clusters makes OKMED more robust to outliers and offers the possibility to use any metric since it only requires a proximity matrix over the data.

- **WOKM**
  The WOKM [15] method is a generalization of both OKM and Weighted k-means methods to detect overlapping clusters. The WOKM introduces a vector of local feature weighting λ_c, relative to each cluster c, which allows data to be assigned to a cluster as regards to a subset of attributes that are important for the cluster concerned. WOKM is based on the minimization of the following objective function:

$$J_{WOKM}(\Pi, C) = \sum_{i=1}^{N} \sum_{v=1}^{d} \gamma_{i,v}^{\beta} \|x_{i,v} - \overline{(x_{i,v})}\|^2, \tag{8.35}$$

where d the number of features and γ_i a vector of weights relative to the representative $\overline{(x_i)}$ which is defined for each feature v as the following:

$$\gamma_{i,v} = \frac{\sum_{c \in \Pi_i} \lambda_{c,v}}{|\Pi_i|}. \tag{8.36}$$

The representative $\overline{(x_{i,v})}$ is defined, for each feature v, as the weighted barycenter of clusters' representatives to which the observation x_i belongs to:

$$\overline{(x_{i,v})} = \frac{\sum_{c \in \Pi_i} \lambda_{c,v}^{\beta} . m_{c,v}}{\sum_{c \in \Pi_i} \lambda_{c,v}^{\beta}}. \tag{8.37}$$

The optimization of the objective criterion is performed by iterating three steps. The first step consists in assigning each data object to the nearest cluster while minimizing the local error $\sum_{v=1}^{d} \gamma_{i,v}^{\beta} \|x_{i,v} - \overline{(x_{i,v})}\|^2$. The second step consists in updating clusters' representatives using the following criterion:

$$m_{c,v}^{*} = \frac{\sum_{x_i \in \pi_c} \frac{\lambda_{i,v}^{\beta}}{|\Pi_i|^2} . \widetilde{x_i}^{c}}{\sum_{x_i \in \pi_c} \frac{\lambda_{i,v}^{\beta}}{|\Pi_i|^2}}. \tag{8.38}$$

The third step concerns the update of the set of clusters weights $\{\lambda_c\}_{c=1}^{k}$ by using:

$$\lambda_{c,v} = \frac{\left(\sum_{x_i \in \pi_c} \|x_{i,v} - m_{c,v}\|^2\right)^{1/(1-\beta)}}{\sum_{u=1}^{d}\left(\sum_{x_i \in \pi_c} \|x_{i,u} - m_{c,u}\|^2\right)^{1/(1-\beta)}}. \tag{8.39}$$

The computational complexity of WOKM stills linear, similar to OKM, evaluated by $O(N.k.\lg k)$.

8.3.3 Summary of Overlapping Partitional Methods

This section offers an overview of the main characteristics of overlapping partitional clustering methods presented in a comparative way. Table 8.1 summarizes these main characteristics. Our study is based on the following features of the methods: (1) model of overlaps in the objective criterion, (2) requirement of the method to use a post-assignment step to generate the final overlapping clusters, (3) type of clusters' representatives, (4) type of data supported by each method, (5) type of separations between clusters, (6) computational complexity, (7) ability to handle noise and outliers and (8) ability to regulate the sizes of overlaps.

Methods which integrate overlaps in their optimized criteria lead to two main categories, additive and geometrical, which differers in the assumptions and in the context of use. The adoption of additive or geometrical methods is motivated by the requirement of the application. Additive-based methods have been well applied in grouping patients into diseases. Each patient may suffer from more than one disease and therefore could be assigned to multiple syndrome clusters. Thus, the final symptom profile of a patient is the sum of the symptom profiles of all syndromes he is suffering from. However, this type of methods needs sometimes to prepare data to have zero mean to avoid false analysis. For example, if symptom variable represents the body temperature, then when a patient simultaneously suffers from two diseases, it is not realistic to assume that his body temperature equals to the sum of body temperatures as associated with two diseases.

Conversely, geometrical-based methods have been well applied to group music signals into different emotions and films into several genres. These methods consider that overlapping observations must appear in the extremity surface between overlapping clusters. For example, if a film belongs to action and horror genres, it should have some shared properties with these categories of films but it can neither be a full action film neither a full horror one. So, overleaping films belonging to action and horror categories may appear in the limit surface between full horror and full action films.

Table 8.1 The main characteristics of the overlapping partitional clustering methods

Method	Overlap model	Post-assignment step	Representatives	Type of data	Separation between clusters	Complexity	Outliers identification	overlap regulation
FCM	–	Threshold	Centroid	Numeric	Linear	$O(N.k)$	No	Yes
PCM	–	Threshold	Centroid	Numeric	Linear	$O(N.k)$	Yes	Yes
ECM	Geometrical	Max evidence	Centoid	Numeric	Linear	$O(N.2^k)$	Yes	Yes
BCM	Geometrical	Max evidence	Centoid	Numeric	Linear	$O(N.k.2^k)$	Yes	Yes
OKM	Geometrical	–	Centroid	Numeric	Linear	$O(N.k.\lg k)$	No	No
P. R-OKM	Geometrical	–	Centroid	Numeric	Linear	$O(N.k.\lg k)$	No	Yes
KOKM	Geometrical	–	Centroid	Any type	Nonlinear	$O(N.k.\lg k)$	No	No
KOKMϕ	Geometrical	–	Medoid	Any type	Nonlinear	$O(N^2.k)$	No	No
OKMED	Geometrical	–	Medoid	Any type	Linear	$O(N^3.k)$	No	No
PCL	Additive	–	Centroid	Numeric	Linear	$O(2^{N.k})$	No	No
ALS_{lf1}	Additive	–	Centroid	Numeric	Linear	$O(N^3.2^k)$	Yes	No
ALS_{lf2}	Additive	–	Centroid	Numeric	Linear	$O(N^2.2^k)$	Yes	No

While all described overlapping partitional clustering methods offer a richer model to fit the existing structures in data, some parameters need to be estimated before performing the learning. All the described methods require to configure the number of clusters in prior which is not a trivial task in real life applications where the number of expected clusters is usually unknown. As a solution, one could use different model heuristics for determining the optimal number [20, 49]. For example, the user can test different clusterings with increasingly number of clusters and then, takes the clustering having the best balance between the minimization of the objective function and the number of clusters [48]. Furthermore, all the described overlapping partitional methods need to initialize the clusters' representatives or the primary clusters memberships. Clusters' representatives can be set randomly from data themselves or can be determined using existing initialization methods [12, 13]. For initializing memberships, a Bernoulli distribution of parameter $\tau = 0.5$ can be used to generate random memberships. Using either a representative initialization or memberships initialization, the result of the presented methods may be a local optimum of the objective criterion, rather than the global optimum. To deal with this problem, the user should adopt a multi-start procedure by testing different initializations and keeping only the clustering which has the lowest value of the objective criterion.

8.4 Evaluation of Overlapping Clustering

The evaluation of clustering, also referred to as cluster validity, is a crucial process to assess the performance of the learning method in identifying relevant groups. This process allows the comparison of several clustering methods and allows the analysis of whether one method is superior to another one. Most of the validity measures traditionally used for clustering assessment, including both internal and external evaluations, become obsolete for overlapping clustering because of the multiple assignment of each observation. Despite this, some works propose an extension of well known validation measures to validate overlapping partitioning. In particular, internal evaluation measures, such as purity and entropy-based measures, cannot capture this aspect of the quality of a given clustering solution because they focus on the internal quality of the clusters. However, external validation measures, essentially Precision-Recall measures, were designed for overlapping partitioning. We give in the following three evaluation methods used for computing precision-recall measures.

8.4.1 Label Based Evaluation

Label based evaluation [44] is usually used in the field of Information Retrieval (IR) where each document can discuss several topics. This evaluation method is based on the evaluation of each class separately. Given a set of observations

$X = \{x_1, \ldots, x_N\}$ and two partitions over X to compare, $C = \{c_1, \ldots, c_k\}$ a non-exclusive partitioning of X into k classes representing true labels, and $\Pi = \{\pi_1, \ldots, \pi_k\}$ a non-exclusive partitioning of X into k clusters where Π is defined by the clustering algorithm. Known the following:

- True positive TP_{ij}: the number of observations in π_j that exist in c_i,
- False negative FN_{ij}: the number of observations in c_i that not exist in π_j
- False positive FP_{ij}: the number of observations in π_j that not exist in c_i,

the Precision-Recall validation measures are computed for each class i and cluster j as follows:

$$\begin{aligned} Precision_{ij} &= \frac{TP_{ij}}{TP_{ij} + FP_{ij}} \\ Recall_{ij} &= \frac{TP_{ij}}{TP_{ij} + FN_{ij}} \\ F - measure_{ij} &= \frac{(2 * Recall_{ij} * Precision_{ij})}{(Recall_{ij} + Precision_{ij})}. \end{aligned}$$

The computation of Precision-Recall measures for all labels is archived using macro-averaging technique which is usually used in Information Retrieval tasks to evaluate clustering results when the number of classes is not large [50] as follows:

$$\begin{aligned} Recall &= \frac{\sum_{i=1}^{k} \max_j Recall_{ij}}{k} \\ Precision &= \frac{\sum_{i=1}^{k} \max_j Precision_{ij}}{k} \\ Fmeasure &= \frac{\sum_{i=1}^{k} \max_j Fmeasure_{ij}}{k}. \end{aligned} \tag{8.40}$$

8.4.2 Pair Based Evaluation

The pair based Precision-Recall measures are calculated over pairs of observations [2]. For each pair of observations that share at least one cluster in the overlapping clustering results, Precision-Recall measures evaluate whether the prediction of this pair as being in the same cluster is correct with respect to the underlying true class in the data.

Given a set of observation $X = \{x_1, \ldots, x_N\}$ and two non-exclusive partitionings over X to compare, $C = \{c_1, \ldots, c_k\}$ a partition of X into k classes, and $\Pi = \{\pi_1, \ldots, \pi_{k1}\}$ a partition of X into $k1$ clusters and by Considering the following:

- TP : the number of pairs of observations in X that share at least one class in C and share at least one cluster in Π,
- FN : the number of pairs of observations in X that share at least one class in C and do not share any cluster in Π and
- FP : the number of pairs of observations in X that do not share any class in C and share at least one cluster in Π,

the Precision-Recall measures are computed as follows:

$$\text{Precision} = (TP)/(TP + FP)$$
$$\text{Recall} = (TP)/(TP + FN)$$
$$\text{F-measure} = (2 * \text{Recall} * \text{Precision})/(\text{Recall} + \text{Precision}).$$

8.4.3 BCubed Evaluation

Given the importance of observation occurrences in clusters and classes in overlapping partitioning, the BCubed evaluation [1] takes into account the multiplicity of classes and clusters which considers the fact that two observations sharing n classes should share n clusters.

BCubed Precision-Recall measures are computed independently for each observation in the partitioning. Let the following:

$$\begin{aligned} Multiplicity \quad precision(x_i, x_j) &= \frac{Min(|\Im(x_i) \cap \Im(x_j)|, |L(x_i) \cap L(x_j)|)}{|\Im(x_i) \cap \Im(x_j)|} \\ Multiplicity \quad recall(x_i, x_j) &= \frac{Min(|\Im(x_i) \cap \Im(x_j)|, |L(x_i) \cap L(x_j)|)}{|L(x_i) \cap L(x_j)|} \end{aligned} \tag{8.41}$$

where x_i and x_j are two observations, $L(x_i)$ the set of classes and $\Im(x_i)$ the set of clusters associated to observation x_i. In fact, Multiplicity Precision is defined only when the pair of observations (x_i, x_j) share at least one cluster, and Multiplicity Recall is defined only when (x_i, x_j) share at least one class. Multiplicity Precision is maximal, equal to 1, when the number of shared clusters is lower or equal than the number of shared classes and it is minimal, equal to 0, when the two observations do not share any class. Reversely, Multiplicity Recall is maximal when the number of shared classes is lower or equal than the number of shared clusters, and it is minimal when the two observations do not share any cluster.

The BCubed precision associated to one observation will be its averaged multiplicity precision over other observations sharing some of its classes; and the overall BCubed precision will be the averaged precision of all observations. The overall BCubed recall is obtained using the same procedure. The overall BCubed Precision-Recall measures can be formally described by:

$$Precision = AVG_i\,[AVG_{j.C(x_i)\cap C(x_j)\neq\varnothing} Multiplicity \quad precision(x_i, x_j)]\forall i, j \in \{1, .., N\}$$

$$Recall = AVG_i\,[AVG_{j.L(x_i)\cap L(x_j)\neq\varnothing} Multiplicity \quad recall(x_i, x_j)]\forall i, j \in \{1, .., N\}$$

$$F - measure = (2 * Precision * Recall)/(Precision + Recall)$$

8.4.4 Synthesis of Evaluation Methods for Overlapping Clustering

Three evaluation methods for assessing the quality of overlapping clustering were presented. The first evaluation method, label based evaluation, is based on matching between labels while the two others are based on pairs of observations. In fact, the label based evaluation requires to label the obtained clusters by matching between classes and clusters which is not a trivial task for unsupervised learning, especially for datasets with large overlaps. The labeling of clusters could lead to biased matching, and consequently lead to biased validation measures. Moreover, the matching between classes and clusters requires to configure a number of clusters equal to that of classes which limits the process of evaluation. Although these limitations, label based evaluation is usually used in Information Retrieval tasks to evaluate clustering results when the number of classes and the sizes of overlaps are not large [50].

To address the issue of labeling the obtained clusters and to make possible the comparison of partitionings with different number of clusters, the pair based Precision-Recall measures are calculated over pairs of observations. This evaluation method offers a flexible evaluation of obtained clusters independently from the real number of classes in the labeled dataset. However, the pair based evaluation has the issue that obtained Recall could be biased as the built overlap in the partitioning decreases and the actual overlap in the dataset increases. The biased Recall is induced by considering only a *binary* function for assessing the relation between a pair of observations and ignoring the *multiplicity* of shared clusters between pairs of observations. For instance, if two observations share three classes in the actual dataset and just share two clusters in the partitioning, the obtained Recall is 1 which is not correct. This problem also occurs for the Precision when actual overlap in the dataset is large.

Given the importance of observation occurrences in clusters and classes in overlapping partitioning, the relation between two observations can not be represented as a binary function. If two observations share two classes and share just one cluster, then the clustering is not capturing completely the relation between both observations as presented in case 3 in Fig. 8.4. On the other hand, if two observations share three clusters but just two classes, then the clustering is introducing more information than necessary as shown in case 3 in Fig. 8.4. This relation is considered for BCubed validation measures by extending the pair based evaluation to take into account the multiplicity of clusters and classes.

Figure 8.4 shows an illustrative example comparing Precision and Recall computed for a pair of observations (x_1, x_2) using BCubed and pair based evaluations by considering different case-studies. We notice that Precision and Recall using the label based evaluation are not reported because they can not be computed for a pair of observations. This comparison shows that multiplicity Recall is reduced compared to Recall computed with the pair based evaluation if the partitioning gives less shared clusters than needed as described in case 3. In contrast, if the partitioning gives more shared clusters than actual labels as described in case 4, the multiplicity Precision is reduced compared to the one obtained with pair based evaluation.

As a summary, we can conclude that assessing the quality of overlapping clustering using the BCubed evaluation is more suitable than the pair based evaluation while it takes into account the multiplicity of shared clusters and classes between the pair of observations.

8.5 Empirical Evaluation of Overlapping Partitional Clustering Methods

Experiments are performed on real multi-labeled benchmarks from different domains: video classification based on the users ratings (Eachmovie[3] data set), detection of emotion in music songs (Music emotion[4] data set), clustering natural scene image (Scene[5] data set) and clustering on genes (Yeast[6] data set). Table 8.2 shows the statistics for each data set where "*Labels*" is the number of categories and "*Cardinality*" (natural overlaps) is the average number of categories for each observation. These benchmarks were chosen because of their diversity of application domains, their diversity of sizes (75 $\rightarrow$ 2417), their diversity of dimensions (3 $\rightarrow$ 192) and their diversity of overlap rates (1.07 $\rightarrow$ 4.23).

Results are compared using four validation measures: Precision, Recall, F-measure and Overlap's size. The first three measures are computed using both pair-based and Bcubed-based evaluations as described in Sects. 8.4.2 and 8.4.3.

[3]cf. http://www.grouplens.org/node/76.

[4]cf.http://mlkd.csd.auth.gr/multilabel.html

[5]cf.http://mlkd.csd.auth.gr/multilabel.html

[6]cf.http://mlkd.csd.auth.gr/multilabel.html

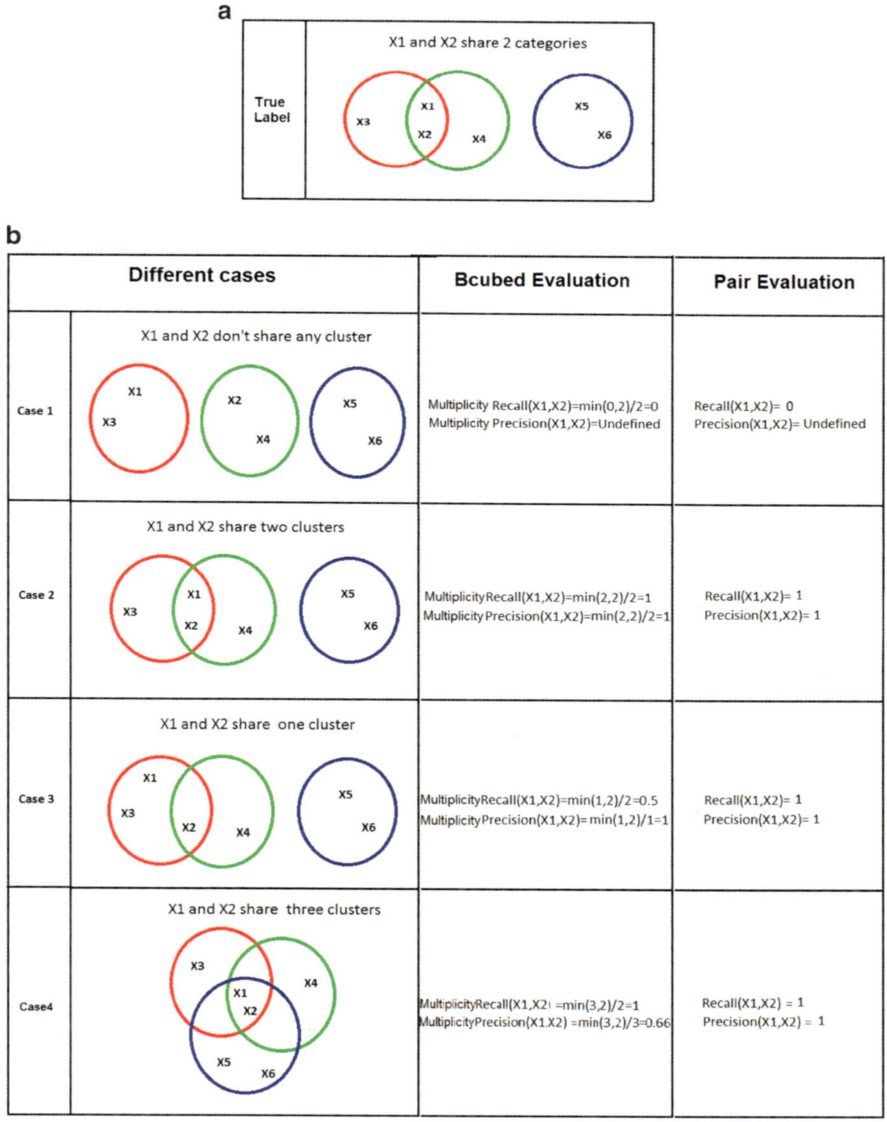

Fig. 8.4 Comparison of Recall and Precision measures computed using the BCubed and the Pair based evaluations: (**a**) true labels and (**b**) different cases of the clustering

The fourth measure, Overlap's size, evaluates the size of overlap built by the learning method. This measure is determined by the average number of clusters to which each observation belongs to:

Table 8.2 Statistics of used data sets

Data set	Domain	N	Dimension	Labels	Cardinality
EachMovie	Video	75	3	3	1.14
Music emotion	Music	593	72	6	1.86
Scene	Images	2,407	192	6	1.07
Yeast	Genes	2,417	103	14	4.23

Table 8.3 Comparison of the performance of overlapping partitioning method in Eachmovie data set

	Pair based evaluation			BCubed Evaluation			
Method	*P.*	*R.*	*F.*	*P.*	*R.*	*F.*	Size of overlap
FCM($\theta = 0.333, \beta = 2$)	0.639	0.675	0.657	0.699	0.683	0.691	**1.12**
ECM ($\alpha = 1, \beta = 2$)	0.566	0.722	0.635	0.575	0.739	0.647	1.32
ECM ($\alpha = 0.5, \beta = 2$)	0.540	0.720	0.617	0.551	0.736	0.631	1.33
OKM	0.465	0.921	0.618	0.399	0.912	0.555	1.70
ALS	0.466	0.855	0.603	0.366	0.819	0.506	1.73
KOKMϕ (RBF)	0.660	0.757	**0.705**	0.685	0.735	**0.709**	1.20
KOKMϕ (Polynomial)	0.459	0.391	0.552	0.416	0.675	0.514	1.48
P. R-OKM ($\alpha = 1$)	0.691	0.731	**0.711**	0.741	0.715	**0.728**	**1.13**
P. R-OKM ($\alpha = 0.1$)	0.505	0.909	0.65	0.440	0.889	0.588	1.57

Table 8.4 Comparison of the performance of overlapping partitioning method in Emotion data set

	Pair based evaluation			BCubed Evaluation			
Method	*P.*	*R.*	*F.*	*P.*	*R.*	*F.*	Size of overlap
FCM($\theta = 0.166, \beta = 2$)	0.492	0.394	0.437	0.480	0.354	0.408	1.43
ECM ($\alpha = 1, \beta = 2$)	0.482	0.675	0.562	0.373	0.648	**0.473**	2.61
ECM ($\alpha = 0.5, \beta = 2$)	0.488	0.703	0.576	0.360	0.711	**0.478**	2.80
OKM	0.483	0.646	0.552	0.353	0.544	0.428	2.35
ALS	0.471	0.999	**0.640**	0.307	0.970	**0.466**	3.46
KOKMϕ (RBF)	0.487	0.548	0.516	0.396	0.462	0.426	2.15
KOKMϕ (Polynomial)	0.481	0.360	0.412	0.446	0.288	0.352	1.52
P. R-OKM ($\alpha = 1$)	0.493	0.351	0.410	0.86	0.278	0.354	1.26
P. R-OKM ($\alpha = 0.1$)	0.487	0.569	0.525	0.392	0.475	**0.430**	**1.88**

$$Overlap \quad size = \frac{\sum_{i=1}^{N} c_i}{N}, \tag{8.42}$$

where c_i is the number of clusters to which observation x_i belongs to. The latter is compared with the true rate of overlaps in the labeled data set.

Table 8.5 Comparison of the performance of overlapping partitioning method in Scene data set

	Pair based evaluation			BCubed Evaluation			
Method	*P.*	*R.*	*F.*	*P.*	*R.*	*F.*	Size of overlap
FCM($\theta = 0.166, \beta = 2$)	0.247	0.946	0.392	0.102	0.946	0.185	3.34
ECM ($\alpha = 1, \beta = 2$)	0.255	0.848	0.393	0.110	0.908	0.171	2.96
ECM ($\alpha = 0.5, \beta = 2$)	0.285	0.813	**0.422**	0.105	0.879	0.188	3.01
OKM	0.233	0.928	0.372	0.192	0.926	0.216	2.85
ALS	–	–	–	–	–	–	–
KOKMϕ (RBF)	0.247	0.813	0.379	0.191	0.809	0.309	2.10
KOKMϕ (Polynomial)	0.236	0.844	0.369	0.156	0.848	0.263	2.36
P. R-OKM ($\alpha = 1$)	0.467	0.426	**0.446**	0.490	0.438	**0.462**	**1.00**
P. R-OKM ($\alpha = 0.1$)	0.30	0.791	**0.435**	0.293	0.797	**0.428**	**1.69**

Table 8.6 Comparison of the performance of overlapping partitioning method in Yeast data set

	Pair based evaluation			BCubed Evaluation			
Method	*P.*	*R.*	*F.*	*P.*	*R.*	*F.*	Size of overlap
FCM($\theta = 0.0714, \beta = 2$)	0.784	1.00	**0.879**	0.148	1.00	0.257	13.60
ECM ($\alpha = 1, \beta = 2$)	–	–	–	–	–	–	–
ECM ($\alpha = 0.5, \beta = 2$)	–	–	–	–	–	–	–
OKM	0.783	0.877	**0.827**	0.587	0.485	**0.531**	**4.80**
ALS	–	–	–	–	–	–	–
KOKMϕ (RBF)	0.785	0.793	**0.789**	0.558	0.467	**0.509**	5.03
KOKMϕ (Polynomial)	0.782	0.755	0.768	0.614	0.398	0.483	**4.59**
P. R-OKM ($\alpha = 1$)	0.801	0.075	0.137	0.806	0.014	0.027	1.00
P. R-OKM ($\alpha = 0.1$)	0.783	0.546	0.643	0.749	0.178	0.287	3.04

Tables 8.3, 8.4, 8.5 and 8.6 report Precision (P.), Recall (R.), F-measure (F.) and Size of overlaps for the best run of each method, which gives the minimal value of the objective criterion, among 20 runs on Eachmovie, Emotion, Scene and Yeast data sets. Values in bold correspond to the best obtained scores. Results of some methods in Yeast and Scene data sets are not reported because of their computational complexity which becomes time consuming and need more than 24 h. A different initialization of prototypes have been used over the 20 runs, whereas within each run the same initialization of prototypes have been used for the different methods. We note that the number of clusters considered within each method was set to the underlying number of labels in each data set. For the other parameters we consider the following:

- for FCM, we set the fuzziness parameter $\beta = 2$ and the threshold value $\theta = 1/k$ with k denotes the number of clusters. However, we note that obtained results of FCM show a high sensitivity to the used threshold as well as the number of clusters increases: for example, in the Yeast data set, using a threshold equal to 0.0714 the obtained F-measure is equal to 0.257 while using the same method with a threshold equal to 0.0715 the F-measure decreases to 0.017

because clusters memberships are almost null. These results show the limit of the extension of FCM to detect overlapping groups.

- for ECM, we fix the fuzziness parameter $\beta = 2$ and we test two values of α equal to 1 and 0.5 for all the data sets.
- for KOKMϕ, we perform different executions using two types of kernels which are RBF (Radial basis Function) and polynomial kernels with parameters $\sigma = 1{,}000$ and $d = 3$ respectively. As for FCM, we show that results are highly sensitive to the type and the considered parameters of the kernel.
- for Parameterized R-OKM, we perform experiments using two values of α equal to 1 and 0.1 for all the data sets. We note that other considered values of α largely improves the performance of Parameterized R-OKM. However, we only report these two values of α to standardize the presented results for the different benchmarks.
- for ALS, reported results are built after reducing data in the interval [-1 1]. We note that ALS fails to build overlapping groups in almost benchmarks (Eachmovie, Emotion and Scene) when data are not centered to have zero mean.

The analysis of the experimental results firstly shows the reliability of the extended Bcubed measures for evaluating overlapping clustering compared to pair based-evaluation: obtained F-measures are higher for clusterings whose overlap rate comes closer to the expected one on the whole. For example, in Scene data set (actual overlap = 1.07) using pair based-evaluation obtained F-measures for ECM and Parameterized R-OKM are nearly equal (0.442 and 0.435 respectively). The F-measure obtained with ECM is induced by high Recall since overlaps rate (overlap = 3) largely exceeds the actual overlap in Scene while for Parameterized R-OKM F-measure is induced by high Precision since built overlap (overlap = 1.68) is near to the actual one. However, using Bcubed evaluation we obtain more reliable evaluation where clustering obtained with Parameterized R-OKM largely exceeds clustering obtained with ECM in terms of F-measure.

Second, empirical results show that Overlap rates have a strong influence on the matching measurement of the clusterings with respect to the expected classes. Methods which allow to control overlap's size, such as Parameterized R-OKM, ECM and FCM can give partitionings which better fit existing structures in the data set, unless good customization of their parameters is required. Whatever the approach used, almost of evaluated methods produce large overlaps exceeding the expected rates, known the actual ones. Producing partitionings which have regulated overlaps improves the F-measure and leads to non-disjoint groups fitting better the data. This fact demonstrates that size of overlaps is an important characteristic that should be controlled while building overlapping groups.

The third conclusion concerns the comparison of uncertain and hard memberships based methods. The results show the limit of using uncertain memberships methods when the number of expected clusters increases. As the case for Yeast data set, uncertain memberships methods require a 10^{-4} precision to fix a threshold value for FCM to be able to produce a non disjoint partitioning of data. However, overlapping groups are easily built using hard memberships methods. We also notice the importance of the computational complexity of overlapping clustering methods when the size of data become large either in terms of number of observations or expected number of clusters.

8.6 Conclusion

We focused in this paper on overlapping clustering, for which we give a classification of existing methods based on the conceptual approach to look for non-disjoint partitioning. Our study is essentially based on overlapping methods which are based on the partitional approach. For that, we survey existing overlapping partitional methods in the literature, classified in two main categories: uncertain memberships and hard memberships. We also gave theoretical and experimental comparisons of existing partitional methods. Furthermore, another important issue that we discussed in this paper is overlapping cluster validity. We presented three external methodologies used to evaluate reliability of non-disjoint partitionings through the evaluation of Precision-recall measures.

At the end of this survey, we claim that overlapping clustering has a growing interest in machine learning research while many real life applications require a non disjoint partitioning of data. Many active challenges in overlapping clustering applications motivate researchers to propose more perfective and efficient learning process. For example, recent works are interested with the identification of non disjoint groups when data contain outliers, or detecting overlapping clusters when data have groups with uneven density. Other works are interested with the identification of overlapping clustering from data having high number of dimensions, or a huge number of observations. All these challenges within overlapping clustering open an exciting directions for future researchers.

References

1. Amigo E, Gonzalo J, Artiles J, Verdejo F (2009) A comparison of extrinsic clustering evaluation metrics based on formal constraints. Inf Retrieval 12(4):461–486
2. Banerjee A, Krumpelman C, Basu S, Mooney RJ, Ghosh J (2005). Model based overlapping clustering. In: International conference on knowledge discovery and data mining, pp 532–537
3. Baumes J, Goldberg M, Magdon-Ismail M (2005) Efficient identification of overlapping communities. In: IEEE international conference on Intelligence and security informatics, pp 27–36
4. BenN'Cir C, Essoussi N (2012) Overlapping patterns recognition with linear and non-linear separations using positive definite kernels. Intern J Comput Appl 56:1–8
5. BenN'Cir C, Essoussi N, Bertrand P (2010) Kernel overlapping k-means for clustering in feature space. In: International conference on knowledge discovery and information retrieval (KDIR), pp 250–256
6. BenN'Cir C, Cleuziou G, Essoussi N (2013) Identification of non-disjoint clusters with small and parameterizable overlaps. In: IEEE international conference on computer applications technology (ICCAT), pp 1–6

7. Berkhin P (2006) A survey of clustering data mining techniques. Grouping Multidimensional Data - Recent Advances in Clustering, Springer pp 28–71
8. Bertrand P, Janowitz M (2003) The k-weak hierarchical representations: an extension of the indexed closed weak hierarchies. Discrete Appl Math 127(2):199–220
9. Bezdek JC (1981) Pattern recognition with fuzzy objective function algorithms. Kluwer Academic Publishers, USA
10. Bonchi F, Gionis A, Ukkonen A (2011) Overlapping correlation clustering. In: 11th IEEE international conference on data mining (ICDM), pp 51–60
11. Bonchi F, Gionis A, Ukkonen A (2013) Overlapping correlation clustering. Knowl Inf Syst 35(1):1–32
12. Celebi ME, Kingravi H (2012) Deterministic initialization of the k-means algorithm using hierarchical clustering Intern J Pattern Recognit Artif Intell 26(7):1250018
13. Celebi ME, Kingravi H, Vela P-A (2013) A comparative study of efficient initialization methods for the k-means clustering algorithm. Expert Syst Appl 40(1):200–210
14. Cleuziou, G. (2008). An extended version of the k-means method for overlapping clustering. In: International conference on pattern recognition (ICPR), pp 1–4
15. Cleuziou G (2009) Two variants of the OKM for overlapping clustering. In: Advances in knowledge discovery and management, Springer pp 149–166
16. Cleuziou G (2013) Osom: a method for building overlapping topological maps. Pattern Recognit Lett 34(3):239–246
17. Davis GB, Carley KM (2008) Clearing the fog: fuzzy, overlapping groups for social networks. Soc Netw 30(3):201–212
18. Dempster AP, Laird NM, Rubin DB (1977) Maximum likelihood from incomplete data via the EM algorithm. J R Stat Soc 39(1):1–38
19. Depril D, Van Mechelen I, Mirkin B (2008) Algorithms for additive clustering of rectangular data tables. Comput Stat Data Anal 52(11):4923–4938
20. Depril D, Mechelen IV, Wilderjans TF (2012) Low dimensional additive overlapping clustering. J Classif 29(3):297–320
21. Diday E (1984) Orders and overlapping clusters by pyramids. Technical Report 730, INRIA
22. Duda RO, Hart PE, Stork DG (2001) Pattern Classification (2nd edition), (John Wiley & Sons, New York, NY)
23. Fellows MR, Guo J, Komusiewicz C, Niedermeier R, Uhlmann J (2011). Graph-based data clustering with overlaps. Discrete Optim 8(1):2–17
24. Fu Q, Banerjee A (2008) Multiplicative mixture models for overlapping clustering. In: 8th IEEE international conference on data mining, pp 791–796
25. Gil-García R, Pons-Porrata A (2010) Dynamic hierarchical algorithms for document clustering. Pattern Recognit Lett 31(6):469–477
26. Goldberg M, Kelley S, Magdon-Ismail M, Mertsalov K, Wallace A (2010). Finding overlapping communities in social networks. In: IEEE second international conference on social computing (SocialCom), pp 104–113
27. Gregory S (2007) An algorithm to find overlapping community structure in networks. In: Knowledge discovery in databases: PKDD 2007, vol 4702, pp 91–102
28. Gregory S (2008) A fast algorithm to find overlapping communities in networks. In: Machine learning and knowledge discovery in databases, vol 5211, pp 408–423
29. Halkidi M, Batistakis Y, Vazirgiannis M (2001) On clustering validation techniques. J Intell Inf Syst 17(2–3):107–145
30. Heller K, Ghahramani Z (2007) A nonparametric Bayesian approach to modeling overlapping clusters. In: 11th International conference on AI and statistics (AISTATS)
31. Jain AK (2010) Data clustering: 50 years beyond k-means. Pattern Recognit Lett 31(8):651–666
32. Krishnapuram R, Keller JM (1993) A possibilistic approach to clustering. IEEE Trans Fuzzy Syst 1(2):98–110
33. Lingras P, West C (2004) Interval set clustering of web users with rough k-means. J Intell Inf Syst 23(1):5–16

34. Liu Z-G, Dezert J, Mercier G, Pan Q (2012) Belief c-means: an extension of fuzzy c-means algorithm in belief functions framework. Pattern Recognit Lett 33(3):291–300
35. MacQueen JB (1967) Some methods for classification and analysis of multivariate observations. In: Fifth Berkeley symposium on mathematical statistics and probability, vol 1, pp 281–297
36. Magdon-Ismail M, Purnell J (2011) Ssde-cluster: fast overlapping clustering of networks using sampled spectral distance embedding and gmms. In: IEEE third international conference on social computing (socialcom), pp 756–759
37. Masson M-H, Denoeux T (2008) Ecm: an evidential version of the fuzzy c-means algorithm. Pattern Recognit 41(4):1384–1397
38. Mirkin BG (1987) Method of principal cluster analysis. Autom Remote Control 48:1379–1386
39. Mirkin BG (1990) A sequential fitting procedure for linear data analysis models. J Classif 7(2):167–195
40. Pérez-Suárez A, Martínez-Trinidad JF, Carrasco-Ochoa JA, Medina-Pagola JE (2013a) Oclustr: a new graph-based algorithm for overlapping clustering. Neurocomputing 109:1–14
41. Pérez-Suárez A, Martínez-Trinidad JF, Carrasco-Ochoa JA, Medina-Pagola JE (2013b) An algorithm based on density and compactness for dynamic overlapping clustering. Pattern Recognit 46(11):3040–3055
42. Snoek CGM, Worring M, van Gemert JC, Geusebroek J-M, Smeulders AWM (2006) The challenge problem for automated detection of 101 semantic concepts in multimedia. In: 14th annual ACM international conference on multimedia, pp 421–430
43. Tang L, Liu H (2009) Scalable learning of collective behavior based on sparse social dimensions. In: ACM conference on information and knowledge management, pp 1107–1116
44. Tsoumakas G, Katakis I, Vlahavas I (2010) Mining multi-label data. In: Data mining and knowledge discovery handbook, Springer pp 667–685
45. Wang Q, Fleury E (2011) Uncovering overlapping community structure. In: Complex networks, vol 116, pp 176–186
46. Wang X, Tang L, Gao H, Liu H (2010) Discovering overlapping groups in social media. In: IEEE international conference on data mining, pp 569–578
47. Wieczorkowska A, Synak P, Ras Z (2006) Multi-label classification of emotions in music. In: Intelligent information processing and web mining. Advances in soft computing, vol 35, pp 307–315
48. Wilderjans T, Ceulemans E, Mechelen I, Depril D (2011) Adproclus: a graphical user interface for fitting additive profile clustering models to object by variable data matrices. Behav Res Methods 43(1):56–65
49. Wilderjans TF, Depril D, Mechelen IV (2013) Additive biclustering: a comparison of one new and two existing als algorithms. J Classif 30(1):56–74
50. Yang Y (1999) An evaluation of statistical approaches to text categorization. J Inf Retrieval 1:67–88
51. Zhang S, Wang R-S, Zhang X-S (2007) Identification of overlapping community structure in complex networks using fuzzy c-means clustering. Phys A Stat Mech Appl 374(1):483–490

Chapter 9
On Semi-Supervised Clustering

Marco Bongini, Friedhelm Schwenker, and Edmondo Trentin

Abstract Due to its capability to exploit training datasets encompassing both labeled and unlabeled patterns, semi-supervised learning (SSL) has been receiving attention from the community throughout the last decade. Several SSL approaches to data clustering have been proposed and investigated, as well. Unlike typical SSL setups, in semi-supervised clustering (SSC) the partial supervision is generally not available in terms of class labels associated with a subset of the training sample. In fact, general SSC algorithms rely rather on additional constraints which bring some kind of a-priori, weak side-knowledge to the clustering process. Significant instances are: COP-COBWEB and COP k-means, HMRF k-means, seeded k-means, constrained k-means, and active fuzzy constrained clustering. This chapter is a survey of major SSC philosophies, setups, and techniques. It provides the reader with an insight into these notions, categorizing and reviewing the major state-of-the-art approaches to SSC.

Keywords Partially supervised learning • Semi-supervised clustering • Active learning • Constrained clustering • Fuzzy clustering

9.1 Introduction[1]

There are at least three major perspectives on data clustering, according to the aim one is specifically interested in. First of all, clustering algorithms are a family of techniques that crunch a bunch of data and come up with a sound partition of the dataset, where "sound" herein means "with a high degree of inner coherence and

[1]Part of this section is based on our paper Schwenker and Trentin [48].

M. Bongini (✉) • E. Trentin
DIISM, University of Siena, Italy
e-mail: bongini@dii.unisi.it; trentin@dii.unisi.it

F. Schwenker
Institute of Neural Information Processing, University of Ulm, Germany
e-mail: friedhelm.schwenker@uni-ulm.de

M.E. Celebi (ed.), *Partitional Clustering Algorithms*,
DOI 10.1007/978-3-319-09259-1_9

outer separation" (*grouping perspective*) [30]. Then, clustering algorithms are a simple and somewhat effective attempt at solving the "data description problem", i.e. the problem of finding adequate and not-too-complex descriptive statistics of a data sample, where "adequate" herein means "as close to sufficient[2] as possible" and "not-too-complex" means "whose size is much smaller than the sample size" (*data description perspective*) [21]. Finally, clustering algorithms may just play the subsidiary role of building blocks (e.g., in a divide and conquer fashion) or initializers (e.g., for the Gaussian kernels of a radial basis functions network [10]) for more sophisticated pattern classifiers (*subservience perspective*) [11]. All in all, the development of robust data clustering from a limited, unlabeled training set of data represented in a proper feature space X (where X is generally a real-valued vector space, i.e., $X \subseteq \mathbb{R}^d$) has long been one of the most relevant tasks in machine learning and statistical pattern recognition [31].

Fully *unsupervised clustering* is the main topic of standard approaches to clustering in traditional unsupervised machine learning. In the unsupervised framework, in fact, a variety of methods and algorithms can be found in the machine learning literature. Aside from clustering, major instances are represented by probability density estimation, self-organizing mapping, and dimensionality reduction (just to mention a few). The goal of the learning process is usually defined through an objective function, where the learning schemes use the data in the training set without any prior knowledge of their labels (e.g. their class labels, if a classification task is faced). In a typical unsupervised learning scenario the training set is defined as

$$\mathcal{U} = \{\mathbf{u}_i \mid \mathbf{u}_i \in \mathbb{R}^d,\ i = 1, \ldots, M\}$$

where the data $\mathbf{u}_i$ are independently drawn from an identical probability density function over $\mathbb{R}^d$ (i.i.d. assumption) [11]. Clearly, the lack of labels (encapsulating any prior expert knowledge) renders unsupervised learning a particularly difficult machine learning problem [29]. In pattern classification tasks, the absence of target class labels during the training phase virtually prevents the machine from resulting in a (more or less reliable) classifier. Although individual clusters yielded by unsupervised clustering are often assumed to form "classes" in an implicit sense, there is no theoretical ground behind such an assumption in the general case. Indeed, from a general standpoint the dataset $\mathcal{U}$ might not even involve any classification task at all. All the learning algorithm can do is analyzing the data, in an attempt to capture either probabilistic (e.g., the probability density function) or geometric/topological information (e.g., some distance/similarity measure, or a partitioning of the data into homogeneous clusters) describing the nature of the data distribution.

---

[2]According to the statistical notion of sufficient statistics.

On the other end, in the fully supervised framework any pattern $\mathbf{x}$ in the training set is uniquely associated with a corresponding (target) label $y \in Y$. It is assumed that $Y = \{y_1, \ldots, y_L\}$ is the set of L (different) class labels, reflecting the ground truth on the classification problem at hand. Intervention from human experts is needed in order to label the training dataset properly. During an initial stage of data collection and annotation, the supervised training set

$$\mathscr{S} = \{(\mathbf{x}_i, y_i) \mid \mathbf{x}_i \in \mathbb{R}^d, y_i \in Y, \ i = 1, \ldots, m\}$$

is therefore prepared. Again, the patterns in $\mathscr{S}$ are i.i.d. according to some unknown probability density function defined on $\mathbb{R}^d \times Y$ [11]. Subsequently, $\mathscr{S}$ is fed into a pre-selected supervised learning algorithm aimed at training a classifier C, that is a mapping $C : \mathbb{R}^d \rightarrow Y$. This algorithm is expected to exploit the information encapsulated within both the feature vectors and the corresponding class labels. Alongside of the algorithm, a hypothesis space has to be fixed, as well. The latter consists of all the potential candidate classifiers C which may become the eventual outcome of the computation on $\mathscr{S}$ of the learning algorithm chosen [1].

It is seen that data annotation is an expensive, time-consuming, and error-prone process. Individual data have to be carefully inspected by one or more domain experts in order to pinpoint the corresponding class labels. Moreover, the exact class labels may not even be explicitly observable. Nonetheless, supervised learning is still far the most prominent branch of machine learning and pattern recognition.

Moving a step beyond the traditional learning frameworks, it is easy to see that a somewhat intermediate scenario occurs when clustering (or, classification) relies on a dataset $\mathscr{T}$ whose data are only partially labeled. Formally, $\mathscr{T} = \mathscr{S} \cup \mathscr{U}$ for a proper, labeled subset $\mathscr{S}$ and its unlabeled counterpart $\mathscr{U}$. While classic clustering techniques do not lead (intrinsically) to any classifier C in this setup, regular supervised classifiers can still be trained over $\mathscr{S}$ all the same. Unfortunately, in so doing all the data in $\mathscr{U}$ would not be exploited, resulting in a waste of potentially useful additional information which could strengthen the very classifier. As a consequence, the framework of *partially supervised learning* (PSL) was introduced [48], having the form of a family of machine learning algorithms lying between supervised and unsupervised learning. Within the realm of PSL, *semi-supervised clustering* (SSC) finds its place. SSC algorithms may offer a precious building block for the development of robust classifiers from partially labeled data. Moreover, they may just as well exploit the presence of the partial supervision (either in the form of $\mathscr{S}$, or in the form of constraints expressing some side-knowledge), and possibly yield a partitioning of the data into clusters that improves over traditional unsupervised clustering techniques. Broadly speaking, in SSC the basic idea may be to exploit some type of prior information on the data at hand, such as: (a) the class labels of some patterns, or (b) certain constraints on pairs of patterns, e.g. "must-link" (ML) or "cannot-link" (CL), as in constrained and seeded k-means clustering [6, 8, 13, 50, 55]. It is a noticeable fact that, to this end, involvement of a human annotator in the pre-processing cycle may not even be required. It will be seen shortly that an interesting relationship holds between the expressiveness of labels and of constraints, respectively, as sources of side-knowledge. Furthermore, some instances of SSC can be seen as "clustering under weak supervision" (for instance, clustering with a fuzzy teacher).

In practical SSC applications, after collecting the raw data, several questions arise concerning the following data processing steps: (1) how can labeled data, unlabeled data, and/or constraints be combined and exploited within a unifying, effective clustering scheme; (2) how do we chose the specific constraints, or patterns in $\mathscr{S}$, such that they are utterly informative; (3) how can the machine deal with soft/fuzzy labels, multiple labels, or soft/fuzzy constraints in a PSL scenario. In an attempt to put forward plausible answers to these (and, further) questions, several prominent directions of research have been developed so far by the community in the SSC area. These directions are the subject of the present chapter.

We made every effort in reviewing and categorizing the different approaches to SSC in a suitable manner. This resulted in the following organization of the chapter. Section 9.2 overviews the whole research area and introduces the reference taxonomy of the different techniques. This section covers also the usual categorization of algorithms into hierarchical vs. partitional procedures, respectively, and elaborates on the nature of the available "supervision" (i.e., labels vs. constraints). The major issues found in SSC and the corresponding keystones are discussed in Sect. 9.3, including the assessment of clustering quality via internal/external measures (Sect. 9.3.1), and a survey of constraints-related questions (Sect. 9.3.2).

The individual techniques and algorithms for SSC are handed out in Sect. 9.4, according to the following arrangement: COP-COBWEB and COP k-means (Sect. 9.4.1), HMRF k-means and the corresponding probabilistic framework (Sect. 9.4.2), SSC based on data-driven learnable similarity measures (Sect. 9.4.3), seeded k-means and constrained k-Means (Sect. 9.4.4), the Zheng-Li algorithm (Sect. 9.4.5), active fuzzy constrained clustering (Sect. 9.4.6), and a kernel approach to semi-supervised graph clustering (Sect. 9.4.7). Finally, Sect. 9.5 pinpoints some directions for future research, while conclusions are drawn in Sect. 9.6.

9.2 Overview and Taxonomy of Algorithms

The goal of this section is to offer a quick overview of the most important SSC algorithms, comparing the most relevant approaches found in the literature. By doing this, we introduce a raw categorization of the methods and highlight their main characteristics. First of all, we discuss the directions along which it is possible to classify SSC approaches. Like in traditional unsupervised clustering, we have the fundamental distinction between hierarchical and partitional clustering algorithms. Due to the easier way in which they accept constraints, partitional SSC approaches have been generally more exploited than hierarchical SSC methods. A second distinction can be made according to the nature of the side-knowledge (supervision) presented to the SSC algorithm. Essentially, amongst the many potential ways to furnish supervision to a SSC procedure, only two of them are actually applied (with few exceptions).

9.2.1 *Types of Supervision (Side-Knowledge)*

In the literature, the most popular and natural way to provide a machine with supervision is through labeled data. Unfortunately, as pointed out in [14], labels may not fit well the general clustering problem. As a matter of fact, when facing a clustering task a labeling might not even be defined at any conceptual level. Moreover, even when a labeling exists (at least in principle), it may be "difficult" for the supervisor (i.e., the human expert) to come up with the correct label in response to a query issued by an active learning SSC machine. On the other hand, assessing whether a specific pair of patterns shall be clustered together or in separate clusters is way easier. This is the rationale behind the introduction of the so-called must-link and cannot-link constraints, defined in [54] and widely applied since. They are straightforwardly defined as follows:

- a must-link (ML) constraint specifies that two patterns should be in the same cluster;
- a cannot-link (CL) constraint specifies that two patterns cannot be in the same cluster.

Therefore, we have the two main types of side-knowledge:

- label-based supervision: widely used in the community, due to the easy access to benchmark datasets, mainly rooted in the pattern classification environment. While it is a valid tool to test an algorithm, in the general case it is not well-suited for real-world clustering applications;
- constraint-based supervision: widely applied in the SSC literature, too. It is better suited for real-world applications, especially for active learning approaches.

Nevertheless, other types of constraints are available in the literature, mainly in the hierarchical SSC framework (as pointed out in Sect. 9.2.3). In the latter case, they are mostly "ordering" constraints that affect the order in which clusters are merged in the framework of agglomerative clustering. A relevant example is represented by the *must link before* (MLB) constraints [4], which are defined as $(\mathbf{x}, (\mathbf{S}_1, \ldots, \mathbf{S}_m))$. The terms S_i are sets of patterns, and the meaning of the rule is that the pattern $\mathbf{x}$ should link with patterns contained in $\mathbf{S}_i$ before than with those in $\mathbf{S}_j$ if $i < j$.

9.2.2 Partitional SSC Algorithms

Traditional unsupervised partitional procedures, starring the k-means algorithm, have possibly been the most popular clustering techniques for decades. Their straightforward contextualization within the semi-supervised framework results in the so-called partitional semi-supervised clustering (PSSC) algorithms. PSSC has been widely investigated in the literature, where the majority of the algorithms proposed so far rely on a k-means-based schema. An utterly relevant instance is represented by the COP-kMeans algorithm [55], where a k-means strategy is integrated with hard (i.e., strictly required) ML and CL constraints. Thus, the algorithm cannot deal efficiently with noise (or, contradictions) in constraints, as we will see in Sects. 9.3.2.3 and 9.4.1. Other relevant approaches can be found in [6], where seeded k-means and constrained k-means algorithms are described. Both of them take advantage from label-based supervision, using the latter to initialize the k-means centroids. In practice, the prototypes (or, centroids) of the clusters are not initialized randomly, but rather relying on the side-knowledge. After the initialization, seeded k-means behaves just like the classical spherical k-means [20]. Therefore, it does not involve the side-knowledge any longer, and inconsistencies w.r.t. the labeling might even emerge at some time during the clustering process (making the approach suitable to cases of noisy side-knowledge). The constrained k-means is based on the spherical k-means, too, but it does not violate the "seeding" (i.e., the labeling specified via $\mathscr{S}$) throughout the following iterations of the algorithm. As a consequence, the technique is not an optimal approach to scenarios involving uncertain supervision. A study of the effects of supervision on PSSC, again carried out using the k-means algorithm, can be found in [56]. Therein, the authors compared the effect of different combinations of distance metric learning and constraints on the results of the consequent clustering process. More recently, an original approach using probabilistic graphical models was described in [8]. It relies on a k-means-based algorithm (HMRF-kMeans) that accounts for the ML/CL constraints, including them in the objective function in the form of a penalty term. This method is presented in detail in Sect. 9.4.2. A different algorithm is described in [36], with a "centroid-less" k-means scheme [58] developed using non-negative matrix factorization. Again, ML and CL constraints are used to express the available supervision. Furthermore, the SS-kernel-kMeans algorithm is introduced in [34]. This approach, which generalizes the HMRF-kMeans, as well as the *Spectral Learning* algorithm [32], is based again on a modified k-means, including a weighted kernel function inspired by the work of Dhillon et al. [19]. Finally, another recent kernel-based algorithm can be found in [22], where sample-to-cluster weights are introduced. The supervision can be given in the form of ML and CL constraints.

9.2.3 Hierarchical SSC Algorithms

Hierarchical clustering is a widely applied class of unsupervised clustering algorithms that can be further divided into agglomerative and divisive clustering approaches [21]. The hierarchical SSC (HSSC) is its natural extension to the semi-supervised scenario. However, to the best of our knowledge, all the HSSC approaches in the literature are agglomerative.

A detailed investigation on the type of constraints and their effects on the HSSC process can be found in [16]. As pointed out in [60] and in [59], the HSSC is not naturally designed to incorporate ML and CL constraints (although this is possible from a theoretical viewpoint [16, 17]), due to the fact that patterns may be linked over possibly different levels of the hierarchy. Therefore, the literature on HSSC investigates other kinds of side-knowledge than the classical ML/CL constraints. A first, relevant example is found in [4], which introduced the aforementioned must-link-before constraint. It is seen that the MLB changes the order in which the clusters are merged during the agglomerative process, insofar that it modifies the distance between nearest clusters whenever their merge results in the violation of a constraint. A different example can be found in [59], where a new kind of ordering constraint is introduced which works on the merge order without affecting the metric. A more recent agglomerative approach was presented by Zheng et al. in [60], relying on *triple-wise relative constraints*, a special case of MLB constraints (a detailed description is given in Sect. 9.4.5). Furthermore, a label-based approach relying on a complete-linkage algorithm was introduced in [15], where the supervision is exploited to empower the dendrogram.

The algorithms listed so far do not use the ML and CL constraints as source of supervision. A relevant exception is the COP-COBWEB[3] [54], based on the COBWEB algorithm [23], and capable of dealing with hard ML and CL constraints. The algorithm requires the satisfaction of all constraints, and it is not an optimal solution in case of uncertainty in constraints. Finally, we should notice a couple of Fuzzy HSSC algorithms presented by Grira et al. [26, 27]. Due to the fuzzy setup, they are able to successfully incorporate the ML/CL constraints, too. The first one is based on competitive agglomeration [26], while the second introduces an active learning strategy [27]. This is described in detail in Sect. 9.4.6.

[3]The authors introduce their algorithm as a partitional method, since it yields a flat partition of the data, corresponding to the top level of the resulting COBWEB hierarchy. Nevertheless, it is our conviction that COP-COBWEB is actually a HSSC approach, due to the hierarchical way in which the process is carried out. The eventual selection of a partition from the dendrogram does not affect this; actually, it is quite a common fact in the hierarchical framework.

9.3 Main Issues and Key Points

In the present section we review some fundamental, general issues of SSC, and some major, related problems. Section 9.3.1 elaborates on the delicate topic of quality assessment of the outcome of clustering procedures. Relevant questions on partial supervision are then faced in Sect. 9.3.2.

9.3.1 *Evaluation of Clustering Quality: Internal/External Measures*

Evaluating the quality of the clustering process (that is, the fitness of the resulting partition of the data to the nature of the problem at hand) is one of the most important and challenging issues in cluster analysis research. In cluster validation a variety of questions arise, including:

- are there clusters in the data?
- How many clusters are in the data?
- What is the correct shape of the clusters?
- Does the clustering result fit the data/agree with prior knowledge?
- Is the result of algorithm A better than the result of algorithm B?

Despite the vast amount of research that has been published on this issues (see for instance [3, 38, 53, 57] and references therein) no conclusive approach to cluster validation has been developed, so far. In the literature, three major types of criteria have been investigated for cluster evaluation [30]:

- *External criteria*: the result of the clustering procedure is evaluated by comparing it to a given external and pre-defined structure, for instance a user-defined labeling of the dataset. Based on this external information the clustering result is tested. A possible test could be to investigate how well a partition computed by a clustering algorithm fits a partition which is pre-defined by a human annotator. For this, in the first set the corresponding contingency matrix must be calculated, and, based on this, an external cluster validation measure is applied (e.g. entropy index, Huberts Γ statistic, Jaccard coefficient, adjusted Rand Index, and many others [44, 57]).
- *Internal criteria*: in this kind of evaluation the results of the clustering algorithm are evaluated through quantities derived from the data themselves, for instance from the distance matrix of the instances, or from distances to the cluster centers. Internal cluster validation is mostly based on compactness and separation measures. Compactness criteria measure how mutually close the instances in a cluster are. For instance, the variance of distances between pairs of objects, or the variance of distances to a cluster center, might be considered as a compactness measure for a single cluster, as well as for a complete partition. Separation is

a measure of how well separated a cluster is from the others. Examples are the distances between cluster centers, or the pairwise distances between instances of clusters [18,38]. The general algorithm for finding the clustering by using a given internal validation measure has the following form:

1. Initialize a set of clustering algorithms.
2. For each clustering algorithm, use different combinations of parameters to get different clustering results.
3. Compute the internal validation measure of each clustering from step 2.
4. Choose the best clustering according to the criteria.

- *Relative criteria*: the idea of relative criteria is to compare different clustering results which stem from running the same algorithm on different parameter values. In so doing, relative cluster criteria are suitable to comparing the results of diverse clusterings, and to pinpoint the best one. External criteria are mostly relative criteria, but also many internal criteria can be seen as relative criteria [53].

For the evaluation of semi-supervised clustering algorithms, typically, external measures are applied to compare clustering results, often together with cross-validation [6]. The problem in SSC settings is to take into account, and to measure, the amount of prior knowledge. For instance, in the scenario of cannot-link and must-link constraints, it is easy to see (for small toy examples) that different sets of CL and ML constraints (with different number of links) can fix the cluster structure. Therefore statistical measures, such as *purity* or *mutual information* as suggest by Strehl et al. [51] or [6] might be useful in SSL, but to the best of our knowledge this topic has not been investigated in depth so far.

9.3.2 Relevant Questions on Supervision

In this section we investigate some of the most relevant supervision-related problems we found in the literature on SSC. For some of them we found, and we report here, a solution (or, several solutions that can be compared to each other), while others are still open.

9.3.2.1 Equivalence Between Constraint-Based and Label-Based Supervision

As stated in Sect. 9.2.1, the supervision can be furnished to the clustering process in several ways, the two most successfully being (a) explicit labels, and (b) ML/CL constraints. A very relevant question arises naturally at this point: are labels and constraints mutually equivalent, at least at some level of abstraction? In the literature the question is not thoroughly investigated, and usually such an investigation is accomplished in an implicit way only. Significant exceptions can be found in [9] and [14]. The answer that emerges is that constraints have a greater modeling capability (i.e. are more general) than labels. Basically, an univocal mapping of labels onto constraints (L2C) is always possible, while in the general case the vice-versa (C2L) is not. In the L2C case, a straightforward approach can be obtained assigning a ML constraint to all pairs of patterns sharing the same label, and a CL constraint to any other pairs. The other way around, a simple case study is enough for demonstrating that the C2L mapping is not always realizable. Let us consider a toy dataset $\mathbf{U} = \{A, B, C, D\}$ embracing 4 patterns to be clustered relying on an active learning scheme. With two queries to the supervisor we can obtain the ML constraints (A, B) and (C, D), respectively. This can be expressed as labeling in two possible forms: $A, B \rightarrow c_1; C, D \rightarrow c_2$ or $A, B, C, D \rightarrow c_1$. Both expressions are not correct: indeed, in the first case we have an implicit CL constraint that was not posed by the user, likewise the second case where another, implicit ML constraint is entailed. Thence, there is no correct label representation for the supervised assignment of constraints at hand. From an applicative point of view, as pointed out in [14] the constraint-based supervision is more natural in those clustering contexts where the user might not know the labeling of the patterns (at times, it is possible that such a labeling does not even exist). On the other hand, the task of assessing whether two patterns shall end up in the same cluster, or not, may be quite easy for the supervisor. Therefore, we can conclude that ML/CL constraints offer a more general tool to furnish supervision to the SSC process in an easier way. In particular, in real-word applications they are easier to use and more realistic for an active learning schema. Nevertheless, labels remain a valid tool, as well. For instance, a label-based SSC approach offers a natural way to testing algorithms applied to benchmark datasets containing labeled data.

9.3.2.2 Selecting the Best Query in Active Learning

Active learning approaches are quite common in SSC environments [7, 14, 27]. It is easy to see that the identity, nature, and number of the specific constraints (or, labels) have a significant influence on the outcome of the active learning SSC algorithm. An intuitive justification of this statement is that obtaining additional information about well-separated clusters is almost useless, while information about overlapping (or, ill-defined) clusters is of the utmost relevance. The literature reports experimental evidence confirming this notion [7, 26, 27]. A similar problem has been investigated also in unsupervised clustering literature [28]. All in all, the goal is to find the best criteria an active SSC learner can apply for selecting the most informative patterns to be submitted to the oracle (i.e., to the human expert) for getting supervision. To this end, basically we have two major approaches that fit two different scenarios, respectively. The first one is the farthest-first traversal method, employed by Basu et al. in [7] with the PCK-Means algorithm, that (ideally) queries the oracle uniformly over all the data space. This is a sound approach whenever all the queries must be asked to the oracle in advance (i.e., before the actual clustering process takes place). With no (even tentative) prior information on position and shape of the clusters, a uniform supervision is an unbiased, likely choice. The idea is to get the best initialization possible for a semi-supervised k-means approach (which is a crucial step, under the circumstances). To achieve this, a farthest first-traversal method is used. First, the algorithm attempts to obtain an example for each cluster. This is consistent with the farthest-first schema that tends to explore all the patterns sets selecting at each step the pattern farthest from the ones already chosen. Then, in the second phase the algorithm tries to consolidate the centroid estimation, by adding new patterns to be queried on the basis of the same approach. For each pattern added to the set we run queries with each other in the set, until we obtain either a ML answer, or CL answers for all the current clusters; finally, the scheme is iterated and used to start a new cluster. Note that for each pattern we have to ask at most $k - 1$ queries.

The second approach was proposed by Grira et al. in [27]. It fits a schema where the queries are submitted to the oracle during the clustering process. In so doing, the algorithm can query for supervision in those regions of the feature space where the separation between clusters does not appear to be neither neat nor trustworthy yet. More specifically, at each new query the algorithm tries to get information in the *weakest* point of the current clustering, that is the worst defined cluster border. To do that, the original algorithm (described in detail in Sect. 9.4.6) uses the *fuzzy hypervolume* to identify the worst-defined cluster. Then, the cluster (fuzzy) membership function is used in order to select the feature vector closest to the frontier to be coupled with the closest pattern in the closest cluster, and jointly queried for supervision. The original setup was defined for a fuzzy clustering technique, but it can be applied (at least, in principle) to a crisp SSC setup, too (just by using different, adequate approaches to identifying the worst-defined cluster and to pinpointing the "borderline" items in need of supervision).

Finally, the optimal query selection strategy depends strongly on the task at hand, as well as on the specific algorithm adopted. If supervision is required for the initialization stage (which is likely to be the case whenever the algorithm is run only once), a method such as the farthest-first traversal is viable. On the other hand, if we have an iterative procedure with several steps of cluster re-assignments, an approach that queries for supervision close to the borders of ill-separated clusters (like the latter described above) may turn out to be best.

9.3.2.3 Uncertain Supervision

In the vast majority of SSC applications the oracle that provides the supervision (either in form of class labels or in form of ML/CL constraints) is assumed to be infallible. Unfortunately, this is not generally the case when we face a real-world problem. Any kind of background knowledge used in SSC, or feedbacks in active learning schemata, are likely to contain noise and/or supervision errors. What sort of problems may arise from such uncertainty? To answer the question we shall severalize the different scenarios of uncertain labels and uncertain constraints. If we have a label-based SSC algorithm, this usually leads to a wrong clustering of the mislabeled data. Thus, the weak/noisy supervision of the clustering process is likely to yield a sub-optimal solution, even worsening the quality offered by a traditional unsupervised clustering approach. Nonetheless, while in the case of supervision-via-labels the process will converge after all (leading to a valid clustering of some sort), constraint-based SSC processes may even get stuck under the circumstances. The common procedure is to perform a transitive closure of the constraints before the actual clustering process takes place. If the noise/errors present in the original constraints lead to an inconsistent closure (i.e., a new set of constraints containing contradictions) then, in general, most constrained-based SSC algorithm will not be able to yield a solution. An example of this situation is the following: let $\mathscr{M} = \{(\mathbf{x}_i, \mathbf{x}_j); (\mathbf{x}_j, \mathbf{x}_k)\}$ be the set of ML constraints and $\mathscr{C} = \{(\mathbf{x}_i, \mathbf{x}_k)\}$ be the set of CL constraints. Clearly, there is no solution if we require the satisfaction of all the constraints. Nevertheless, some algorithms are able to deal with such contradictions, even leading to good solutions [8, 27, 34]. An example of successful strategy is introduced in [9], where the constraints are included in the objective function in the form of a penalty term, such that the algorithm can violate (some of) them if necessary. In practice, both label-based and constraint-based supervision may yield interesting and useful solutions even under noisy/weak supervision. Notwithstanding, when working with constraints we should be aware of the risks, thence opting for clustering algorithms capable of dealing with a faulty supervisor.

9.4 Algorithms for SSC

This section surveys in detail all the SSC algorithms we categorized earlier, in the order, providing the reader with a description of the different techniques, several bibliographic references to the relevant literature, and a concise report on the established performance in the field.

9.4.1 COP-COBWEB and COP-kMeans

COP-COBWEB is a modified version of the COBWEB algorithm [23], which computes a clustering that maximizes the inter-cluster dissimilarity and the intra-cluster similarity [54]. This hierarchical algorithm takes a dataset X, the two sets of constraints $\mathscr{M}$ and $\mathscr{C}$, and it returns a partition of the dataset that fulfilled the specified constraints by utilizing a set of operators Must-link check, Add, New, Merge, Split, Update, and a pre-defined quality measure $Q(\mathscr{P})$ defined for a partition $\mathscr{P}$ of X. The input to the recursive COP-COBWEB algorithm is a dataset $X = \{\mathbf{x}_1, \ldots, \mathbf{x}_N\}$, a set of must-links $\mathscr{M} \subset X \times X$, and a set of cannot-links $\mathscr{C} \subset X \times X$. For an instance $\mathbf{x}_i \in X$, the first step (step 2a) is to check the ML conditions. If there is such a ML constraint that $\mathbf{x}_i$ must be in the same cluster

Algorithm 1 COP-COBWEB

COP-COBWEB ($X, \mathcal{M}, \mathcal{C}$)

1. Initialize clustering $\mathcal{P} := \emptyset$
2. For each instance $\mathbf{x}_i \in X$, try to insert $\mathbf{x}_i$ into the existing clustering
   - 2a. Must-link check: if there exits a must-link $(\mathbf{x}_i, \mathbf{x}_j) \in \mathcal{M}$ and $\mathbf{x}_j \in P$ for some existing cluster $P \in \mathcal{P}$ then include $\mathbf{x}_i$ into cluster P, $\mathcal{P}_{must} = (\mathcal{P} \setminus P) \cup \{P \cup \{\mathbf{x}_i\}\}$ and GoTo Split.
   - 2b. Add: for each cluster $P_j \in \mathcal{P}$, a new clustering is given by $\mathcal{P}_{add-j} = (\mathcal{P} \setminus P_j) \cup \{P_j \cup \{\mathbf{x}_i\}\}$ unless $(\mathbf{x}_i, \mathbf{x}_k) \in \mathcal{C}$ for some instance $\mathbf{x}_k \in P_j$
   - 2c. New: $\mathcal{P}_{new} = \mathcal{P} \cup \{\mathbf{x}_i\}$
   - 2d. Merge: Let P_{max1} and P_{max2} be the two best clusters for $\mathbf{x}_i$ with respect to quality measure Q then merge, $\mathcal{P}_{merge} = ((\mathcal{P} \setminus P_{max1}) \setminus P_{max2}) \cup \{P_{max1} \cup P_{max2} \cup \{\mathbf{x}_i\}\}$ unless $(\mathbf{x}_l, \mathbf{x}_k) \in \mathcal{C}$ for some instances $\mathbf{x}_k \in P_{max1}$ and $\mathbf{x}_l \in P_{max2}$
   - 2e. Split: Let P_{max} the best clusters for $\mathbf{x}_i$ from step (2a.) or (2b.) with respect to quality measure Q, then let $\mathcal{P}_{split} = (\mathcal{P} \setminus P_{max}) \cup$ COP-COBWEB$(P_{max} \cup \{\mathbf{x}_i\}, \mathcal{M}, \mathcal{C})$
   - 2f. Update: Let $m = \text{argmax } C(P_k)$ for $k \in \{must, add-j, new, merge, split\}$ and update $\mathcal{P} := \mathcal{P}_m$.
3. **return** $\mathcal{P}$

P as another instance $\mathbf{x}_j$, then $\mathbf{x}_i$ is added to P. If such a ML constraint is not given, then $\mathbf{x}_i$ is added to any cluster, but taking into account possible cannot-links (in step 2b). In the next step (step 2c), creating a new cluster for an instance $\mathbf{x}_i$ is considered. Merging two suitable clusters (step 2d) needs again a check for CL constraints. A splitting step (2e) is then recursively applied to the cluster that fits best the instance $\mathbf{x}_i$, with respect to quality measure Q (see Algorithm 1 and [54]).

The general concept of CL and ML constraints can be incorporated into existing clustering algorithms, for instance k-means [39, 40] or incremental clustering algorithms, especially neural clustering algorithm (such as Kohonen maps [33], or neural gas [41]). The major modification in such algorithms is to ensure that the given constraints $\mathcal{M}$ and $\mathcal{C}$ are not violated, when adapting the cluster assignments of the instances.

A partitional example is Algorithm 2 [55] where the batch k-means using CLs and ML constraints is given. In k-means the quality measure is usually defined through l_2 empirical discretization:

$$Q(\{P_1, \ldots, P_k\}) = \sum_{1=1}^{N} \|\mathbf{x}_i - \mathbf{y}_{k_i}\|_2^2 \tag{9.1}$$

where $\|\cdot\|$ is the Euclidean norm, $\mathbf{y}_1, \ldots, \mathbf{y}_k$ are the centroids of the clusters $\{P_1, \ldots, P_k\}$ (these are the cluster averages, evaluated over all instances in the corresponding cluster), and $\mathbf{y}_{k_i}$ denotes the cluster center (among all clusters) which is the nearest to pattern $\mathbf{x}_i$. Checking the ML constraints $\mathcal{M}$ means to ensure that there is no must-link into an instance of a cluster P_j with $j \neq k_i$, and checking the cannot-link constraints $\mathscr{C}$ means to avoid a cannot-link $(i, l) \in \mathscr{C}$ with a datapoint $\mathbf{x}_l \in P_{k_i}$. In case of contradictions, the algorithm fails and returns an empty clustering as a result. This behavior is too strict in many applications, thus considering soft, or probabilistic, constraints rather than hard constraints in Algorithms 1 and 2 may lead to more flexible SSC algorithms.

Algorithm 2 COP-kMeans

COP-kMeans ($X, \mathscr{M}, \mathscr{C}$)

1. Initialize clustering $\mathscr{P} := \{P_1, \dots, P_k\}$ and
   compute the corresponding cluster centers $\mathbf{y}_1, \dots, \mathbf{y}_k$
2. For each instance $\mathbf{x}_i \in X$ compute the closest cluster center $\mathbf{y}_{k_i}$ using $Q(\{P_1, \dots, P_k\})$
   and assign $\mathbf{x}_i$ to P_{k_i} if the constraints $\mathscr{M}$ and $\mathscr{C}$ are satisfied
   otherwise Return $\mathscr{P} := \emptyset$
3. Compute cluster centers $\mathbf{y}_1, \dots, \mathbf{y}_k$ by averaging over corresponding clusters $\{P_1, \dots, P_k\}$
4. Until convergence iterate step 2. and 3.
5. Return $\mathscr{P} := \{P_1, \dots, P_k\}$

9.4.2 A Probabilistic Framework for SSC: The HMRF k-Means Method

The method described in this section is a partitional approach to SSC based on probabilistic graphical models, originally proposed in [8]. The approach exploits the benefits of constraints-based supervision in three ways:

- improved initialization;
- assignment of pattern to cluster;
- iterative distance measure learning.

The algorithm takes in input the dataset $X = \{\mathbf{x}_1, \dots, \mathbf{x}_N\}$, the sets $\mathscr{M}$ of ML constraints and $\mathscr{C}$ of CL constraints, and their violation costs W and $\overline{W}$, respectively. Moreover, a distortion measure D should be specified by the user.

The probabilistic graphical model at the root of this approach is a hidden Markov random field (HMRF), which has the following components:

- a hidden field of random variables $L = \{l_i\}_{i=1}^N$ with values in $\{1, \dots, K\}$, corresponding to the K cluster labels;
- an observable set $X = \{x_i\}_{i=1}^N$ of random variables drawn from $P(x_i|l_i)$. An emission Markov assumption holds, such that $P(X|L) = \prod_{x_i \in X} P(x_i|l_i)$.

Each random variable l_i has a neighborhood N_i which is a set including all the other random variables l_j that appear in an ML or CL constraint with l_i. Thus, N_i can be computed from $\mathscr{M}$ and $\mathscr{C}$ in a straightforward manner. Let us assume that a Markov random field is defined over the hidden variables, that is to say that the classic Markov property $P(l_i|L \setminus \{l_i\}) = P(l_i|N_i)$ holds. Therefore, the posterior probability of a certain complete cluster assignment L given the dataset X can be written as:

$$P(L|X) = \frac{1}{CZ} \exp\left(-\sum_i \sum_j V(i,j)\right) \cdot p\left(X, \{\mu_h\}_{h=1}^K\right) \tag{9.2}$$

where μ_h is the hth cluster centroid, Z is the usual normalizing constant for MRFs (i.e., the so-called partition function), and C is another constant that is considered to be "equivalent" to the prior probability $P(X)$ of the dataset. The quantity $V(i,j)$ is the pairwise potential of the MRF, defined as:

$$V(i,j) = \begin{cases} f_{\mathcal{M}}(\mathbf{x}_i, \mathbf{x}_j) \ \text{if}(\mathbf{x}_i, \mathbf{x}_j) \in \mathcal{M} \\ f_{\mathcal{C}}(\mathbf{x}_i, \mathbf{x}_j) \ \ \text{if}(\mathbf{x}_i, \mathbf{x}_j) \in \mathcal{C} \\ 0 \qquad\qquad\qquad \text{otherwise} \end{cases} \tag{9.3}$$

where $f_{\mathcal{M}}(\cdot,\cdot)$ and $f_{\mathcal{C}}(\cdot,\cdot)$ are the non-negative penalty functions for ML and CL violations, respectively. Finally, $p(X, \{\mu_h\}_{h=1}^K)$ is the conditional probability $P(X|L)$ of the dataset given the clustering assignment L. The maximization of Eq. (9.2) is an "incomplete data problem" that can be solved using the Expectation-Maximization (EM) technique. In turn, EM is equivalent to k-means in case of hard (i.e., crisp) clustering assignments. Thus, it is feasible to use k-means in order to maximize Eq. (9.2).

The conditional probability function and the penalty functions are chosen depending on the metric which can be selected to fit a specific problem, thus adapting Eq. (9.2) to different clustering tasks. In [8] the authors restrict their attention to the conditional probability densities of this form:

$$P(X|L) = p(X, \{\mu_h\}_{h=1}^K) = \frac{1}{Z_3} \exp\left(-\sum_{\mathbf{x}_i \in X} D(\mathbf{x}_i . \mu_{l_i})\right)$$

which can include several different cluster models. They propose two suitable penalty functions, weighting the constraint violation proportionally to the pattern distance. In so doing, they penalize the metrics that either (a) set apart patterns that should rather belong to the same cluster, or (b) set close patterns that should lie in different clusters. This is accomplished by defining:

$$f_{\mathcal{M}}(\mathbf{x}_i, \mathbf{x}_j) = w_{ij} \phi_D\left(\mathbf{x}_i, \mathbf{x}_j\right) \mathbb{1}\left[l_i \neq l_j\right] \tag{9.4}$$

and:

$$f_{\mathcal{C}}(\mathbf{x}_i, \mathbf{x}_j) = \overline{w}_{ij}\left(\phi_{D_{MAX}} - \phi_D\left(\mathbf{x}_i, \mathbf{x}_j\right)\right) \mathbb{1}\left[l_i = l_j\right] \tag{9.5}$$

where ϕ_D is a monotonically increasing function of the distance between $\mathbf{x}_i$ and $\mathbf{x}_j$. This choice leads to the following objective function to be minimized:

$$J_{obj} = \sum_{\mathbf{x}_i \in X} D(\mathbf{x}_i, \mu_{l_i}) + \sum_{(x_i, x_j) \in \mathscr{M}, l_i \neq l_j} w_{ij} \phi_D(\mathbf{x}_i, \mathbf{x}_j)$$
$$+ \sum_{(x_i, x_j) \in \mathscr{C}, l_i = l_j} \overline{w}_{ij} \left(\phi_{D_{MAX}} - \phi_D(\mathbf{x}_i, \mathbf{x}_j)\right) + \log(Z) \quad (9.6)$$

The choice of a suitable distortion measure is fundamental in order to obtain a good clustering. The distortion measure should fit the task at hand, i.e. the type of data to be clustered as well as the algorithm used. For the study presented in [8], the authors proposed and investigated the cosine similarity and the Kullback-Leibler divergence. The distortion measure obtained using a parameterized cosine similarity takes the following form:

$$D_{cos_A}(\mathbf{x}_i, \mathbf{x}_j) = 1 - \frac{\mathbf{x}_i^T A \mathbf{x}_j}{\|\mathbf{x}_i\|_A \|\mathbf{x}_j\|_A} \quad (9.7)$$

with norm $\|\mathbf{x}\|_A = \sqrt{\mathbf{x}^T A \mathbf{x}}$. Now, it is trivial to insert $D_{cos_A}(\mathbf{x}_i, \mathbf{x}_j)$ into Eq. (9.6), i.e. $\phi_D(\mathbf{x}_i, \mathbf{x}_j) = D_{cos_A}(\mathbf{x}_i, \mathbf{x}_j)$, to obtain the relative objective function J_{cos_a}.

The second distortion measure proposed is based on I-divergence, strictly related to the Kullback-Leibler divergence. The I-divergence is parameterized trough a vector of non-negative weights **a** as follows:

$$D_{I_a}(\mathbf{x}_i, \mathbf{x}_j) = \sum_{m=1}^{d} a_m x_{im} \log\left(\frac{x_{im}}{x_{jm}}\right) \quad (9.8)$$

where d is the dimensionality of the patterns into the dataset. The selected ϕ_D has the following form (satisfying the symmetry requirement):

$$\phi_D(\mathbf{x}_i, \mathbf{x}_j) = D_{IM_a}(\mathbf{x}_i, \mathbf{x}_j)$$
$$= \sum_{m=1}^{d} a_m \left(x_{im} \log\left(\frac{2x_{im}}{x_{im} + x_{jm}}\right) + x_{jm} \log\left(\frac{2x_{jm}}{x_{im} x_{jm}}\right)\right) \quad (9.9)$$

By putting Eqs. (9.8) and (9.9) in place into Eq. (9.6), the I-parameterized objective function J_{I_a} is readily obtained.

The algorithm for minimizing the selected objective function (either J_{cos_a} or J_{I_a}) is an EM procedure. The initialization of the centroids is performed in two steps:

1. The transitive closure of ML constraints in $\mathscr{M}$ is used to obtain connected components. Let λ be the number of connected components. We can create λ neighborhoods $\{N_p\}_{p=1}^{\lambda}$ from the connected components that correspond to ML neighborhoods on the latent variables of the HMRF.

2. The neighborhoods are then used to initialize the centroids of the clusters in this way: if $\lambda = K$ then the centroids $\{\mu_i\}_{i=1}^K$ are initialized to coincide with the centroids of the neighborhoods. If $\lambda < K$, then we proceed as before for the first λ centroids. The remaining centroids are initialized with random perturbations of the centroid of the whole dataset X. If $\lambda > K$, then K neighborhoods are selected as initial clusters by using a weighted farthest-first traversal on the centroids of the λ neighborhoods. The weights are proportional to the size of the corresponding neighborhoods (in doing so, more relevance is assigned to larger ones).

The expectation-maximization (EM) iterative algorithmic structure of the method thus relies on the following steps:

- *E-step*: in standard k-means this is done by assigning each pattern to the nearest centroid. Things are different in the HMRF model, due to the interaction between neighbor labels. Moreover, computing the assignment of patterns to centroids that minimize the objective function is computationally intractable in any non-trivial HMRF model [49]. Therefore, the ICM approach is adopted to approximate the optimal solution. This algorithm computes the assignment in random order for each point $\mathbf{x}_i$, assigning $\mathbf{x}_i$ to the cluster h that minimizes:

$$J_{obj}(\mathbf{x}_i, \mu_h) = D(\mathbf{x}_i, \mu_h) + \sum_{(\mathbf{x}_i,\mathbf{x}_j)\in\mathcal{M}, h\neq l_j} w_{ij}\phi_D(\mathbf{x}_i, \mathbf{x}_j)$$

$$+ \sum_{(\mathbf{x}_i,\mathbf{x}_j)\in\mathcal{C}, h=l_j} \phi_{D_{MAX}} \overline{w}_{ij}\phi_D(\mathbf{x}_i, \mathbf{x}_j) \tag{9.10}$$

  Once the assignment is completed, the patterns are reordered and reassigned accordingly, until no further label changes are obtained.
- *M-step*: first, the centroids are estimated as follows, for cosine and I-distortion measures respectively:

$$\mu_h = \frac{\sum_{\mathbf{x}_i\in X_h}\mathbf{x}_i}{\left\|\sum_{\mathbf{x}_i\in X_h}\mathbf{x}_i\right\|_A}$$

$$\mu_h = \frac{1}{1+\alpha}\left(\frac{\sum_{\mathbf{x}_i\in X_h}\mathbf{x}_i}{|X_h|} + \alpha\frac{1}{n}\right)$$

  Then, we need to update the parameters of the distortion measure so as to minimize the objective function. This means to adapt the distortion measure so that similar points turn out to be closer to each other, while dissimilar points are set apart. Generally, a closed-form analytical solution does not exist. Resorting to the gradient-descent method yields a viable approximated solution. For both cosine and I-distortion measures, gradient-descent prescribes an update rule having form $a_m = a_m + \eta\frac{\partial J_{obj}}{\partial a_m}$ with:

$$\frac{\partial J_{obj}}{\partial a_m} = \sum_{\mathbf{x}_i \in X} \frac{\partial D(\mathbf{x}_i, \mu_{l_i})}{\partial a_m} + \sum_{(\mathbf{x}_i, \mathbf{x}_j) \in \mathscr{M}, l_i \neq l_j} w_{ij} \frac{\partial D(\mathbf{x}_i, \mathbf{x}_j)}{\partial a_m}$$

$$+ \sum_{(\mathbf{x}_i, \mathbf{x}_j) \in \mathscr{C}, l_i = l_j} \overline{w}_{ij} \left(\frac{\partial D_{MAX}}{\partial a_m} - \frac{\partial D(\mathbf{x}_i, \mathbf{x}_j)}{\partial a_m} \right) \quad (9.11)$$

The gradients $\frac{\partial D(\mathbf{x}_i, \mathbf{x}_j)}{\partial a_m}$ for the proposed distortion measures can be written, respectively, as:

$$\frac{\partial D_{cos_a}(\mathbf{x}_i, \mathbf{x}_j)}{\partial a_m} = \frac{x_{im} x_{jm} \|\mathbf{x}_i\|_A \|\mathbf{x}_j\|_A - \mathbf{x}_i^T A \mathbf{x}_j \frac{x_{im}^2 \|\mathbf{x}_j\|_A^2 + x_{jm}^2 \|\mathbf{x}_i\|_A^2}{2\|\mathbf{x}_i\|_A \|\mathbf{x}_j\|_A}}{\|\mathbf{x_i}\|_A^2 \|\mathbf{x_j}\|_A^2}$$

$$\frac{\partial D_{I_a}(\mathbf{x}_i, \mathbf{x}_j)}{\partial a_m} = x_{im} \log\left(\frac{x_{im}}{x_{jm}}\right) - \left(x_{im} - x_{jm}\right)$$

The HMRF k-means method is tested in [8] on three small datasets of textual documents. This is a challenging task, due to the limited amount of patterns w.r.t the dimensionality of the word space, and many standard clustering and semi-supervised clustering algorithms can reach only local optima. The clustering evaluation measure of choice is the *normalized mutual information* (NMI). The authors compared their method w.r.t. the ablations and the unsupervised k-means. The experiments show a significant improvement in terms of NMI over the unsupervised case, and confirm the complete algorithm yields better results than the ablations does on the three datasets.

9.4.3 Semi-Supervised Learning of the Distance Measure

Both traditional and semi-supervised clustering algorithms rely on some sort of distance measure, between either patterns or clusters [21], Euclidean metric being the most popular choice. A specific algorithm (e.g., agglomerative hierarchical clustering) results in radically different behaviors according to the metric used. Fitness of the resulting algorithm to the task at hand depends on how well the very distance measure fits the underlying nature of the data. Model selection techniques may be applied in order to select the metric a-posteriori from a set of pre-defined measures (such as L_1 or L_2 norms, Mahalanobis distance, etc.). A more general and sound idea is just to let the machine discover the proper distance measure from the available data, whenever possible. An early approach along this line was put forward in [55], where side-knowledge specified by the supervisor tells the clustering algorithm whether certain pairs of patterns shall be put in the same cluster (since they are close to each other) or not (since they are distant). Although this scheme enforces the quality of the resulting clustering, there is no explicit, actual metric learning in it. Furthermore, as observed in [56], since this kind of side-knowledge is defined at a pairwise pattern level, it does not generalize well to unseen regions of the feature space.

The technique proposed in [56] relies on genuine distance measure learning, which may as well fit a SSC scenario with side-information. In this context, the partial supervision is in the form of a set of user-specified pairs of patterns that are known to be similar to (or, possibly, dissimilar from) each other. The metric learning task is formulated in such a way that it turns out to be a convex optimization problem, making it easy to find effective, gradient-based solutions that are free from local optima of the objective function to be extremized.

Formally, distance measures $d_A : \mathbb{R}^d \times \mathbb{R}^d \rightarrow \mathbb{R}$ are considered in [56], having form

$$d_A(\mathbf{x}, \mathbf{y}) = \sqrt{(\mathbf{x} - \mathbf{y})^T A(\mathbf{x} - \mathbf{y})} \tag{9.12}$$

for any pair $(\mathbf{x}, \mathbf{y})$ of patterns in the d-dimensional feature space. Although Eq. (9.12) expresses a linear metric, a family of nonlinear distance measures can be accounted for by applying any nonlinear transformation $\phi(.)$ to the feature vectors and computing $\sqrt{(\phi(\mathbf{x}) - \phi(\mathbf{y}))^T A(\phi(\mathbf{x}) - \phi(\mathbf{y}))}$ instead. The user specifies the side-knowledge by defining two sets, say $\mathscr{S}$ and $\mathscr{D}$, containing examples of "similar" and "dissimilar" pairs of patterns, respectively. Equation (9.12) poses the metric learning task as a parameter learning problem: learn the parameters of the linear transformation A relying on the information contained in $\mathscr{S}$ and $\mathscr{D}$, subject to constraints that ensure that the resulting $d_A(.)$ is a proper metric (i.e., that $d_A(.)$ is non-negative and satisfies the triangular property). To this end, A is required to be positive semi-definite, $A \succeq 0$. In [56] the learning problem is first instantiated as a convex optimization problem: find A that minimizes $\sum_{(\mathbf{x},\mathbf{y})\in\mathscr{S}} d_A^2(\mathbf{x}, \mathbf{y})$ such that $\sum_{(\mathbf{x},\mathbf{y})\in\mathscr{D}} d_A(\mathbf{x}, \mathbf{y}) \geq 1$ and $A \succeq 0$. The constraint $\sum_{(\mathbf{x},\mathbf{y})\in\mathscr{D}} d_A(\mathbf{x}, \mathbf{y}) \geq 1$ prevents degenerate (i.e., null) solutions. If a diagonal A is taken into consideration, Newton's method yields a simple, straightforward, and efficient solution to the problem. Unfortunately, in the general case where A is a full (non-diagonal) matrix, satisfying $A \succeq 0$ is difficult and Newton's method becomes too expensive computationally. For these reasons, in [56] the minimization problem is rewritten in the form of a (theoretically, yet not numerically) equivalent maximization problem as follows: maximize $\sum_{(\mathbf{x},\mathbf{y})\in\mathscr{D}} d_A(\mathbf{x}, \mathbf{y})$ w.r.t. A such that $\sum_{(\mathbf{x},\mathbf{y})\in\mathscr{S}} d_A^2(\mathbf{x}, \mathbf{y}) \leq 1$ and $A \succeq 0$. This problem can be solved via gradient-ascent (to accomplish the maximization) combined with the iterative projection algorithm [45] (to ensure satisfaction of the constraints). In practice, at each iteration of the algorithm the gradient-ascent step is followed by a projection of A onto the sets $\{M \mid \sum_{(\mathbf{x},\mathbf{y})\in\mathscr{S}} d_M^2(\mathbf{x}, \mathbf{y}) \leq 1\}$ and $\{M \mid M \succeq 0\}$ (these projections turn out to be relatively simple and inexpensive). Empirical evidence reported on in [56] shows that these learnable distance measures result in improved clustering techniques.

Algorithm 3 Seeded-kMeans

Seeded-kMeans ($X, \mathscr{S} = \{S_1, \ldots, S_k\}$)

1. Initialize the cluster centers $\mathbf{y}_1, \ldots, \mathbf{y}_k$ by $\mathbf{y}_i = \frac{1}{|S_i|} \sum_{x \in S_i} x$
2. Repeat until convergence
2.a assign each data point $x \in X$ to cluster P_i if $\mathbf{y}_i = \text{argmin}\|x - \mathbf{y}_j\|$
2.b update cluster centers by $\mathbf{y}_i = \frac{1}{|P_i|} \sum_{x \in P_i} x$
3. Return $\mathscr{P} := \{P_1, \ldots, P_k\}$

Algorithm 4 Constrained-kMeans

Constrained-kMeans ($X, \mathscr{S} = \{S_1, \ldots, S_k\}$)

1. Initialize the cluster centers $\mathbf{y}_1, \ldots, \mathbf{y}_k$ by $\mathbf{y}_i = \frac{1}{|S_i|} \sum_{x \in S_i} x$
2. Repeat until convergence
2.a assign each data point $x \in X$ to clusters:
if $x \in S_j$ for some set of seed points, then assign x to P_j
else if $\mathbf{y}_i = \text{argmin}\|x - \mathbf{y}_j\|$ then assign x to P_i
2.c update cluster centers by $\mathbf{y}_i = \frac{1}{|P_i|} \sum_{x \in P_i} x$
3. Return $\mathscr{P} := \{P_1, \ldots, P_k\}$

9.4.4 Seeded k-Means and Constrained k-Means

In this section we describe two simple partitional algorithms based on seeding.

Seeding is possibly one of the basic ideas in semi-supervised clustering approaches. Let us consider a dataset $X = \{\mathbf{x}_1, \ldots, \mathbf{x}_N\}$ and a subset $S \subset X$ of category-labeled instances, the so-called seed points. In Seeded-kMeans (see Algorithm 3) the seed points are assumed to be given in k subsets $\mathscr{S} := \{S_1, \ldots, S_k\}$, used to initialize the cluster centers $\mathbf{y}_1, \ldots, \mathbf{y}_k$ and the corresponding clustering $\mathscr{P} := \{P_1, \ldots, P_k\}$ for a subsequent k-means. Thus, seeded-kMeans is an initialization procedure [6], where the seed points are not used during the clustering process.

In the first step of constrained-kMeans the cluster centers of the k-means are initialized in the same way as in Seeded-kMeans. In the second step (the kMeans-step) of the algorithm (see Algorithm 4) these assignments can not be changed. In the work by Basu et al. [6] a numerical evaluation for Constrained-kMeans, Seeded-kMeans, Random-kMeans and COP-kMeans on two benchmark data sets (*CMU 20 Newsgroup* data and *Yahoo! News* data) is shown. They found that all three SSC algorithms (Constrained-kMeans, Seeded-kMeans, COP-kMeans) outperform Random-kMeans in terms of the mutual information index. In Random-kMeans the K means are initialized by taking the means of K subsets generated at random from the entire dataset (for details refer to [12]) It was shown that Seeded-kMeans is more robust against seeding noise (meaning noise in the external pre-labeled dataset) than Constrained-kMeans and COP-kMeans. In this context it should be mentioned that the transformation of category-based labels into CLs and MLs, as required in Constrained-kMeans and Seeded-kMeans, might be problematic, because a category is typically not given by a single cluster but this is not reflected in the category labels.

9.4.5 The Zheng-Li Algorithm

This approach is a recent SSC hierarchical algorithm proposed by Zheng and Li in [60] and based on an ultra-metric dendrogram distance (the results of hierarchical clustering can be equivalently represented by means of ultra-metric distance matrices [43]). The method incorporates supervision in terms of *triple-wise relative constraints* (TWC). Those constraints are in the form $\left(\mathbf{x}_i, \mathbf{x}_j, \mathbf{x}_k\right)$, meaning that the pattern $\mathbf{x}_i$ shall be closer to $\mathbf{x}_j$ than to $\mathbf{x}_k$, i.e. $d\left(\mathbf{x}_i, \mathbf{x}_j\right) < d\left(\mathbf{x}_i, \mathbf{x}_k\right)$. Note that they can be thought of as a special case of MLB constraints. The rationale behind their use is that the standard ML/CL constraints are not suitable for classical hierarchical clustering approaches, since patterns are linked over different levels of the hierarchy [4, 5].

The Zheng-Li algorithm (ZLA) takes in input the dataset X, the corresponding set of pairwise dissimilarities $D = \{d\left(\mathbf{x}_i, \mathbf{x}_j\right) | \forall \mathbf{x}_i, \mathbf{x}_j \in X\}$ (usually the latter can be easily computed from X) and a set of TWCs $C = \left\{\left(\mathbf{x}_i, \mathbf{x}_j, \mathbf{x}_k\right) | d\left(\mathbf{x}_i, \mathbf{x}_j\right) < d\left(\mathbf{x}_i, \mathbf{x}_k\right)\right\}$. The goal of the algorithm is to return in output a dendrogram $\mathscr{H}$ that

- satisfies as many TWCs as possible;
- keeps the merge order based on pattern dissimilarities as close as possible.

In order to ensure the best results, a two step pre-processing phase takes place before the actual clustering process starts. It consists of the usual transitive closure of constraints, followed by a conflict-removal step. In case of TWCs, the transitive closure can be accomplished adding a new constraint $\left(\mathbf{x}_i, \mathbf{x}_j, \mathbf{x}_l\right)$ for each pair of constraints in the form $\left\{\left(\mathbf{x}_i, \mathbf{x}_j, \mathbf{x}_k\right); \left(\mathbf{x}_i, \mathbf{x}_k, \mathbf{x}_l\right)\right\}$. Then, as long as there are conflicts (e.g. $c_1 = \left(\mathbf{x}_i, \mathbf{x}_j, \mathbf{x}_k\right)$, $c_2 = \left(\mathbf{x}_i, \mathbf{x}_k, \mathbf{x}_l\right)$ and $c_3 = \left(\mathbf{x}_i, \mathbf{x}_l, \mathbf{x}_j\right)$), a conflicting constraint is selected at random and removed, reiterating this step until no further conflicts remain in place.

At each step the (regular) agglomerative hierarchical clustering algorithm takes the pair of nearest clusters formed so far, and merges them into a single, larger cluster. Therefore:

$$
\begin{aligned}
d\left(\mu_i, \mu_j\right) &\leq \min\left(d\left(\mu_i, \mu_k\right), d\left(\mu_j, \mu_k\right)\right) \\
&\Rightarrow \forall i, j, k \min\left(d\left(\mu_i, \mu_k\right), d\left(\mu_j, \mu_k\right)\right) \leq d\left(\mu_{i \cup j}, \mu_k\right)
\end{aligned}
\tag{9.13}
$$

where $\mu_{i \cup j}$ is the centroid of the cluster obtained from the merge of clusters i and j.

The property in Eq. (9.13) is known as the *reducibility property* [46]. Once such condition holds true, the updated dissimiliarities satisfy the ultrametric inequality:

$$d\left(\mathbf{x}_i, \mathbf{x}_j\right) \leq \max\left(d\left(\mathbf{x}_i, \mathbf{x}_k\right), d\left(\mathbf{x}_j, \mathbf{x}_k\right)\right) \forall \mathbf{x}_i, \mathbf{x}_j, \mathbf{x}_k \in X \tag{9.14}$$

Therefore, the hierarchical clustering can be basically understood as an ultra-metric transformation $\hat{D}$ on the initial dissimilarity matrix D which uniquely characterized the ultra-metric tree. That is to say, the hierarchical clustering problem can be seen as a search problem, namely the search for the optimal ultra-metric transformation $\hat{D}$:

$$\arg\min_{\hat{D}} \left(\sum_{\mathbf{x}_i, \mathbf{x}_j \in X} \left(D_{ij} - \hat{D}_{ij}\right)^2 \right) \tag{9.15}$$

that is a NP-hard problem [35], such that an approximation approach is needed.

The ultra-metric distance matrix can also be obtained using transitive dissimilarity. To all intents and purposes, we can see D as the transition matrix on a graph, where each column/row is associated with a node of the graph, and the entries represent the weight associated with the corresponding pair of nodes/patterns (i.e., an edge of the graph). Given any path P_{ij} between $\mathbf{x}_i$ and $\mathbf{x}_j$, its transitive dissimilarity can then be written as

$$T\left(P_{ij}\right) = \max\left(d\left(\mathbf{x}_i, \mathbf{x}_{k_1}\right), d\left(\mathbf{x}_{k_1}, \mathbf{x}_{k_2}\right), \ldots, d\left(\mathbf{x}_{k_n}, \mathbf{x}_j\right)\right) \tag{9.16}$$

Now it is possible to define the *minimal transitive dissimilarity* as

$$m_{ij} = \min_{P_{ij}} \left(T\left(P_{ij}\right)\right) \tag{9.17}$$

and the following theorem holds true:

Theorem 1. *For any weighted dissimilarity graph, the minimal transitive dissimilarity between any pair of vertices satisfies the ultra-metric inequality:*

$$m_{ij} \leq \max\left(m_{ik}, m_{kj}\right) \forall \mathbf{x}_i, \mathbf{x}_j, \mathbf{x}_k$$

In [60] the authors propose two approaches to the problem: the *optimization-based* and the *transitive dissimilarity-based* approaches.

1. Optimization-based: in this approach the dissimilarity matrix is assumed to be non-negative and symmetric. In this case D can be represented with a $m \times 1$ vector $\mathbf{d}$ containing the entries of the upper/lower triangle elements. Then, it is possible to define a $m \times 1$ vector $\mathbf{c}$ such that, for each TWC $\left(\mathbf{x}_i, \mathbf{x}_j, \mathbf{x}_k\right) \in C$,

Algorithm 5 Least-square error minimization with Iterative Projection

LSE Minimization with Iterative Projection ($\mathbf{d}$, C, E)

Init: $\mathbf{a} = \mathbf{d}$ and $\mathbf{u} = \mathbf{0}$
While not converged **do**
  $p = t \bmod r$
  $\mathbf{s} = \mathbf{a}(t-1) + \frac{1}{2} E\mathbf{c}_p \mathbf{u}(t-1)_p$
  for $q = 1$ to r **do**
    if $q = p$
      $\mathbf{u}(t)_q = \max\left(2\mathbf{c}_\mathbf{q}^\mathbf{T}\mathbf{s} / \mathbf{c}_\mathbf{q}^\mathbf{T} E \mathbf{c}_\mathbf{q}, 0\right)$
    else
      $\mathbf{u}(t)_q = \mathbf{u}(t-1)_q$
    end if
  end for
  $\mathbf{a}(t) = \mathbf{s} - \frac{1}{2} E\mathbf{c}_\mathbf{q} \mathbf{u}(t)_q$
  $t = t + 1$
end while
return $\hat{d} = \mathbf{a}$

Algorithm 6 Minimum transitive dissimilarity with modified Floyd-Warshall

Modified Floyd-Warshall (G, C)

Init: $M = G$
For $k = 0$ to N **do**
  For $i = 0$ to N **do**
    For $j = 0$ to N **do**
      For all $c = \left(\mathbf{x}_i, \mathbf{x}_j, \mathbf{x}_l\right)$ **do**
        $minCon = \min\left(minCon, d\left(\mathbf{x}_i, \mathbf{x}_l\right)\right)$
      end for
      $m_{ij} = \min\left(m_{ij}, \max\left(m_{ik}, m_{kj}\right), minCon\right)$
    end for
  end for
end for
return M

the entry corresponding to (i, j) is 1, and the entry corresponding to (i, k) is -1. Thence, we have $\mathbf{d}^T C \geq 0$. Thus, our problem can be written as:

$$\arg_{\hat{d}} \min \left(\left(\mathbf{d} - \hat{\mathbf{d}}\right)^T E \left(\mathbf{d} - \hat{\mathbf{d}}\right) \right)$$

subject to Eq. (9.16) and to $C\mathbf{d} \leq 0$. A possible solution, based on the iterative projections approach, is described in Algorithm 5.

2. Transitive dissimilarity-based: the approach followed in this case is a Floyd-Warshall algorithm [24], modified such that the TWCs are taken into account. The pseudo-code is handed out in Algorithm 6.

The experimental evaluation was accomplished on several datasets from the UCI repository [42] (including the Iris dataset, commonly used in SSC studies), and on benchmark text datasets for document clustering. The experiments show improvement over classical unsupervised approaches, leading to results that compare positively with the performance yielded by HACoc [59], alongside of a reduced computational burden.

9.4.6 Active Fuzzy Constrained Clustering

The Active Fuzzy Constrained Clustering (AFCC) algorithm is described in [27]. It is an interesting fuzzy semi-supervised hierarchical approach that relies on supervision in the form of standard ML/CL constraints. The algorithm takes in input:

1. the set $\mathbf{X}$ of patterns to be clustered;
2. the set $\mathscr{M}$ of ML constraints and the set $\mathscr{C}$ of CL constraints.

At each step it computes:

1. a fuzzy partition matrix U expressing the membership of patterns to clusters;
2. the cluster centroids, $\mu_i\ \forall i = 1 \ldots C$.

This fuzzy approach relies on a cost (i.e., objective) function that is inspired from its crisp counterpart proposed by Basu et al. [6], described in Sect. 9.4.4. The cost function that AFCC aims at minimizing is written as:

$$\gamma(U,V) = \sum_{k=1}^{C}\sum_{i=1}^{N} u_{ik}^2 d^2(\mathbf{x}_i, \mu_k) + \alpha\Bigg(\sum_{(\mathbf{x}_i,\mathbf{x}_j)\in\mathscr{M}}\ \sum_{k=1}^{C}\ \sum_{l=1,l\neq k}^{C} u_{ik}u_{jl} + \sum_{(\mathbf{x}_i,\mathbf{x}_j)\in\mathscr{C}}\ \sum_{k=1}^{C} u_{ik}u_{jk}\Bigg) - \beta\sum_{k=1}^{C}\left[\sum_{i=1}^{N} u_{ik}\right]^2 \tag{9.18}$$

The second term realizes a penalty for the ML and CL constraints that are violated, weighting the penalty via the fuzzy-cluster membership values u_{ik}. The overall penalty is further weighted by the value α, which is a measure of the reliability of the oracle. Finally, the third term comes from the competitive agglomeration method [25], and manages the competition between clusters. Using Lagrange multipliers, the authors proposed the following update equations for the fuzzy partition matrix U:

$$u_{rs} = \frac{\frac{1}{d^2(\mathbf{x}_r,\mu_s)}}{\sum_{k=1}^{C}\frac{1}{d^2(\mathbf{x}_r,\mu_k)}} + \frac{\alpha}{2d^2(\mathbf{x}_r,\mu_s)}\left(\overline{C_{v_r}} - C_{v_{rs}}\right) + \frac{\beta}{d^2(\mathbf{x}_r,\mu_s)}\left(N_s - \overline{N_r}\right) \tag{9.19}$$

where

$$C_{v_{rs}} = \sum_{(\mathbf{x}_r,\mathbf{x}_j)\in\mathcal{M}} \sum_{l=1,l\neq s}^{C} u_{jl} + \sum_{(\mathbf{x}_r,\mathbf{x}_j)\in\mathcal{C}} u_{js} \tag{9.20}$$

$$\overline{C_{v_r}} = \frac{\sum_{k=1}^{C} \frac{\left(\sum_{(\mathbf{x}_r,\mathbf{x}_j)\in\mathcal{M}} \sum_{l=1,l\neq k}^{C} u_{jl} + \sum_{(\mathbf{x}_r,\mathbf{x}_j)\in\mathcal{C}} u_{jk}\right)}{d^2(\mathbf{x}_r,\mu_k)}}{\sum_{k=1}^{C} \frac{1}{d^2(\mathbf{x}_r,\mu_k)}} \tag{9.21}$$

and

$$\overline{N_r} = \frac{\sum_{k=1}^{C} \frac{N_k}{d^2(\mathbf{x}_r,\mu_k)}}{\sum_{k=1}^{C} \frac{1}{d^2(\mathbf{x}_r,\mu_k)}} \tag{9.22}$$

Finally, in order to complete the overview of the AFCC algorithm, we need to review the strategy for constraint selection (thus, the active learning part). As stated in Sect. 9.3.2.2, in an active learning schema the selection of the most informative pair of patterns to be queried for supervision is of the utmost relevance. To put in practice this notion, in [27] the authors focus, at each query, on the frontier of the worst-defined cluster. To select that cluster, they use the fuzzy hypervolume (FHV):

$$FHV = |C_k| \tag{9.23}$$

where $|C_k|$ is the determinant of the covariance matrix of the kth cluster. Once selected the worst-defined cluster accordingly, its frontier is defined as the set of patterns having the lowest membership value. Then, from a theoretical viewpoint, for each element on the frontier we should search for the corresponding nearest cluster. Clearly, this method would entail a high computational burden. Two suitable approximations were proposed instead:

1. pre-define a threshold θ and define the frontier as the set of patterns having membership less than (or, equal to) θ. Usually this frontier turns out to be larger than the true boundary. A pair of these patterns is then selected and presented to the oracle.
2. Non-redundancy: at each selection, choose the pattern (in the frontier) that is the farthest from those already selected.

The pseudo-code for AFCC is summarized in Algorithm 7, which relies on the Mahalanobis distance measure.

Algorithm 7 Active Fuzzy Constrained Clustering

Active Fuzzy Constrained Clustering (X, $\mathcal{M}$, $\mathcal{C}$, number of clusters)
Init: compute randomly the clusters prototypes
initialize memberships with equal probability
repeat
Compute β
Compute α
Compute cluster memberships u_{ik}
Compute cardinalities $N_k \forall 1 \leq k \leq C$
For $k = 1$ to C **do**
if ($N_k \leq$ Threshold)
discard cluster k
end if
end for
Update number of cluster C
Update prototypes
until prototypes stabilize

The two weighting factors α and β can be obtained at each iteration t as follows:

$$\alpha = \frac{N}{M} \frac{\sum_{k=1}^{C} \sum_{i=1}^{N} u_{ik}^2 d^2(\mathbf{x}_i, \mu_k)}{\sum_{k=1}^{C} \sum_{i=1}^{N} u_{ik}^2} \tag{9.24}$$

and

$$\beta(t) = \frac{\eta_0 \exp\left(-\frac{|t-t_0|}{\tau}\right)}{\sum_{j=1}^{C} \left(\sum_{i=1}^{N} u_{ij}\right)^2} \left[\sum_{j=1}^{C} \sum_{i=1}^{N} u_{ij}^2 d^2(\mathbf{x}_i, \mu_j) + \alpha \left(\sum_{(\mathbf{x}_i, \mathbf{x}_j) \in \mathcal{M}} \sum_{k=1}^{C} \sum_{l=1, l \neq k}^{C} u_{ik} u_{jl} + \sum_{(\mathbf{x}_i, \mathbf{x}_j) \in \mathcal{C}} \sum_{k=1}^{C} u_{ik} u_{jk} \right) \right] \tag{9.25}$$

where M is the number of ML constraints.

The experimental results presented in [27] show good performance in comparison with unsupervised clustering approaches, as well as w.r.t. other supervised clustering algorithms like PCCA [26] and constrained k-means on several classification datasets (including the Iris dataset). One of the strongest selling points of the algorithm is its hierarchical schema that discovers the number k of clusters spontaneously, without the need to fix it in advance.

9.4.7 Semi-Supervised Graph Clustering: A Kernel Approach

In a typical clustering algorithm, such as k-means, the input data set $X = \{\mathbf{x}_1, \ldots, \mathbf{x}_N\}$ consists of N feature vectors each with d attribute values, so for instance X can be seen as a matrix $X \in \mathbb{R}^{N \times d}$. Furthermore a proximity measure is assumed to be defined on the entire input space, so that proximities between objects can be computed for pairs of feature vectors. In contrast, the input data in graph clustering is assumed to be a graph $G = (V, E, A)$. Here V is the finite non-empty set of vertices, such that $V = \{1, \ldots, N\}$ coincides with the set of patterns to be clustered; $E \subset V \times V$ is the set of edges, describing a binary relation defined over V that expresses whether pairs of patterns are close together or not; and $A \in \mathbb{R}^{N \times N}$ is the edge adjacency matrix, whose components A_{ij} represent the proximity measure between vertices i and j. Obviously, given a set of feature vectors along with a proximity measure, the problem can be transformed in a graph-based representation easily [2, 47]. Subsequently, we introduce a kernel-based partitional approach for semi-supervised graph clustering. More details, as well as related work, can be found in [34].

In graph clustering the goal is to compute a partition $\mathscr{P} := \{P_1, \ldots, P_k\}$ of V which minimizes a particular objective measure. Let $C, D \subset V$ be two subsets of the vertex set, and define $\text{links}(C, D) := \sum_{i \in C, j \in D} A_{ij}$ and $\deg(C) = \text{links}(C, V)$. Based on these characteristic values several graph clustering objective functions can be defined:

i) Ratio association:

$$\text{maximize}_{\{P_1,\ldots,P_k\}} \sum_{i=1}^{k} \frac{\text{links}(P_i, P_i)}{|P_i|}$$

ii) Ratio cut:

$$\text{minimize}_{\{P_1,\ldots,P_k\}} \sum_{i=1}^{k} \frac{\text{links}(P_i, V \setminus P_i)}{|P_i|}$$

iii) Normalized cut:

$$\text{minimize}_{\{P_1,\ldots,P_k\}} \sum_{i=1}^{k} \frac{\text{links}(P_i, V \setminus P_i)}{\deg(P_i)}$$

The ratio association index tries to maximize the sum of the similarity of instances within a cluster. Normalization by the number of objects is introduced to compensate for differences in the cluster sizes. Ratio cut tries to minimize the similarity between the objects to the other clusters, again each term is normalized by the cluster size. The normalized cut uses in principle the same objective function, but uses $\deg(P_i)$, instead, as the cluster size.

The link from clustering vectors to graph-clustering can be made via weighted kernel k-means. In weighted kernel k-means the goal is to search for a partition $\mathscr{P} := \{P_1, \dots, P_k\}$ of a dataset $X = \{\mathbf{x}_1, \dots, \mathbf{x}_N\} \subset \mathbb{R}^d$ such that the following objective function is minimized:

$$Q(\{P_1, \dots, P_k\}) = \sum_{j=1}^{k} \sum_{\mathbf{x}_i \in P_j} \alpha_i \|\phi(\mathbf{x}_i) - \mathbf{y}_j\|_2^2 \tag{9.26}$$

where the generalized cluster center is given by

$$\mathbf{y}_j = \frac{\sum_{\mathbf{x}_i \in P_j} \alpha_i \phi(\mathbf{x}_i)}{\sum_{\mathbf{x}_i \in P_j} \alpha_i} \tag{9.27}$$

and the function ϕ is a nonlinear transformation from $\mathbb{R}^d$ into some pre-defined Euclidean space, and α_i, $i = 1, \dots, N$ are positive weights. In case of $\phi = id$ (where id is the identity function) and $\alpha_i = 1$ for $i = 1, \dots, N$ we get the standard k-means objective function.

Inserting Eq. (9.27) into the distances of objective function (9.26) yields

$$\begin{aligned}\|\phi(\mathbf{x}_i) - \mathbf{y}_j\|_2^2 = \phi(\mathbf{x}_i)\phi(\mathbf{x}_i) - 2\frac{\sum_{\mathbf{x}_l \in P_j} \alpha_l \phi(\mathbf{x}_i)\phi(\mathbf{x}_l)}{\sum_{\mathbf{x}_l \in P_j} \alpha_l} \\ + \frac{\sum_{\mathbf{x}_l \in P_j, \mathbf{x}_q \in P_j} \alpha_l \alpha_q \phi(\mathbf{x}_l)\phi(\mathbf{x}_q)}{(\sum_{\mathbf{x}_l \in P_j} \alpha_l)^2}\end{aligned} \tag{9.28}$$

By using a positive-semidefinite kernel function k, dot-products can be expressed by $k(\mathbf{x}_i, \mathbf{x}_l) = \phi(\mathbf{x}_i)\phi(\mathbf{x}_l)$, and the kernel matrix $(\mathbf{K}_{i,j})_{i,j=1,\dots,N}$ is given by $\mathbf{K}_{i,j} = k(\mathbf{x}_i, \mathbf{x}_l) = \phi(\mathbf{x}_i)\phi(\mathbf{x}_l)$. Using the kernel matrix the distances can be re-formulated as

$$\|\phi(\mathbf{x}_i) - \mathbf{y}_j\|_2^2 = \mathbf{K}_{i,i} - 2\frac{\sum_{\mathbf{x}_l \in P_j} \alpha_l \mathbf{K}_{i,l}}{\sum_{\mathbf{x}_l \in P_j} \alpha_l} + \frac{\sum_{\mathbf{x}_l \in P_j, \mathbf{x}_q \in P_j} \alpha_l \alpha_q \mathbf{K}_{l,q}}{(\sum_{\mathbf{x}_l \in P_j} \alpha_l)^2} \tag{9.29}$$

For the most popular graph clustering objective functions, i.e. ratio association, ratio cut, and normalized cut, the corresponding weights $\boldsymbol{\alpha}$ and kernel matrix $\mathbf{K}$ are given in Table 9.1. In the table I denotes the identity matrix, D the diagonal matrix whose entry is given by the corresponding row sum of the adjacency matrix A, L the Laplacian matrix $L = D - A$, and σ is a non-negative real value such that the kernel is positive definite.

For a particular set of weights $\boldsymbol{\alpha}$ and kernel matrix $\mathbf{K}$, the generalized Kernel-kMeans-algorithm (see Algorithm 8) is used.

Table 9.1 Weights and kernels for graph clustering objective functions: ratio association, ratio cut, and normalized cut

Objective function	Value of node weights	Kernel Matrix
Ratio association	1 for all objects/nodes	$K = \sigma I + A$
Ratio cut	1 for all objects/nodes	$K = \sigma I - L$
Normalized cut	deg for all objects/nodes	$K = \sigma D^{-1} + D^{-1} A D^{-1}$

Algorithm 8 Kernel-kMeans

Kernel-kMeans ($K, k, \boldsymbol{\alpha}, \mathscr{P}$)

1. Initialize cluster centers $\mathbf{y}_1, \ldots, \mathbf{y}_k$ using $\mathscr{P}$
2. For each instance $\mathbf{x}_i \in X$ and its cluster center $\mathbf{y}_j$ compute $\|\phi(\mathbf{x}_i) - \mathbf{y}_j\|_2^2$ by (9.29)
3. Assign instances $\mathbf{x}_i$ to their closest center P_{j*}
4. Compute and update cluster centers $\mathbf{y}_1, \ldots, \mathbf{y}_k$ by (9.27)
5. Until convergence iterate step 2., 3., and 4.
6. Return $\mathscr{P} := \{P_1, \ldots, P_k\}$

In semi-supervised graph clustering, where the side-knowledge is given through cannot-link constraints $\mathscr{C}$ and must-link constraints $\mathscr{M}$, the corresponding objective functions can be derived in a slightly different way. Let us assume that the constraints define a $N \times N$ matrix W by $W_{ij} = -w_{ij}$ if $(\mathbf{x}_i, \mathbf{x}_j) \in \mathscr{C}$ is a cannot-link and $W_{ij} = w_{ij}$ if $(\mathbf{x}_i, \mathbf{x}_j) \in \mathscr{M}$ is a must-link. Therefore the three semi-supervised objective functions can be derived as follows:

i) Semi-supervised ratio association:

$$\text{maximize}_{\{P_1,\ldots,P_k\}} \sum_{i=1}^{k} \frac{\text{links}(P_i, P_i)}{|P_i|} + \sum_{\substack{(\mathbf{x}_i, \mathbf{x}_j) \in \mathscr{M} \\ k_i = k_j}} \frac{w_{ij}}{|P_{k_i}|} - \sum_{\substack{(\mathbf{x}_i, \mathbf{x}_j) \in \mathscr{C} \\ k_i = k_j}} \frac{w_{ij}}{|P_{k_i}|}$$

ii) Semi-supervised ratio cut:

$$\text{minimize}_{\{P_1,\ldots,P_k\}} \sum_{i=1}^{k} \frac{\text{links}(P_i, V \setminus P_i)}{|P_i|} - \sum_{\substack{(\mathbf{x}_i, \mathbf{x}_j) \in \mathscr{M} \\ k_i = k_j}} \frac{w_{ij}}{|P_{k_i}|} + \sum_{\substack{(\mathbf{x}_i, \mathbf{x}_j) \in \mathscr{C} \\ k_i = k_j}} \frac{w_{ij}}{|P_{k_i}|}$$

iii) Semi-supervised normalized cut:

$$\text{minimize}_{\{P_1,\ldots,P_k\}} \sum_{i=1}^{k} \frac{\text{links}(P_i, V \setminus P_i)}{\deg(P_i)} - \sum_{\substack{(\mathbf{x}_i, \mathbf{x}_j) \in \mathscr{M} \\ k_i = k_j}} \frac{w_{ij}}{\deg(P_{k_i})} + \sum_{\substack{(\mathbf{x}_i, \mathbf{x}_j) \in \mathscr{C} \\ k_i = k_j}} \frac{w_{ij}}{\deg(P_{k_i})}$$

The corresponding transformation as a kernel-based formulation for semi-supervised clustering is given in Table 9.2. Kulis et al. [34] present their

Table 9.2 Weights and kernels for Semi-Supervised graph clustering objective functions: SS ratio association, SS ratio cut, and SS normalized cut

Objective function	Value of node weights	Kernel Matrix
SS ratio association	1 for all objects/nodes	$K = \sigma I + A + W$
SS ratio cut	1 for all objects/nodes	$K = \sigma I - L + W$
SS normalized cut	deg for all objects/nodes	$K = \sigma D^{-1} + D^{-1}(A + W)D^{-1}$

Semi-Supervised Kernel-kMeans algorithm on the basis of the Kernel-kMeans algorithm, where the weights and the kernel are generated through the settings given in Table 9.2. Furthermore, the initial partition for the Kernel-kMeans algorithm is computed according to the cannot-link and must-link constraints. After this initial steps Kernel-kMeans is applied to compute the resulting partition. In [34] a numerical evaluation is performed on six different benchmark datasets. By using the normalized mutual information index [51] as a validity measure, the authors demonstrated that Semi-Supervised Kernel-kMeans is able to outperform several current semi-supervised clustering algorithms.

9.5 A Glimpse of the Future: Some Research Directions

While SSL approaches to classification problems have been widely investigated in the last 15 years or so, research on SSC is mostly in its infancy and several issues are still open (and, far from being solved to date). In the following, we pinpoint some directions (and, challenges) for further developments of the field, aimed at overcoming these issues. The treatment is intentionally and necessarily kept on an abstract, mainly speculative level.

The first direction we recommend is closing the remaining gap between SSC research and the vast knowledge the community has accumulated throughout the decades in the related area of statistical pattern recognition from missing/incomplete data [37]. For instance, the popular mixture-of-mixtures framework [52] used in the estimation of parametric models for the missing data and for the corresponding missingness mechanism may be interpreted in terms of a semi-supervised "clustering of clusters" approach, under Gaussian assumptions, by bearing in mind the usual relationship between maximum-likelihood estimation of mixture of Gaussian components and the k-means algorithm [21]. Moreover, while explicit modeling of the missingness mechanism behind a pattern of missingness M by means of an explicit parametric density function $P(M|., \theta)$ [37] is typical (and fruitful) in statistical SSL, its exploitation in SSC has not been attempted so far, in spite of the fact that accounting for the nature of the mechanism may turn out to be crucial [37]. In SSL the model $P(M|., \theta)$ may be used, along with the data model, for factorizing the joint distribution of M and of the data given M (i.e., the data are explained differently according to the mechanism underlying their missingness). Similarly, in

SSC the clustering itself could be obtained and explained as a corollary of the very missingness mechanism $\psi_M(.) = P(M|., \theta)$.

The presence of an underlying, possibly hidden mechanism $\psi_M(.)$ behind the data distribution, a mechanism which must be accounted for by the SSC algorithm (as if $\psi_M(.)$ were a constraint to satisfy), pinpoints another possible research direction. The general notion is that the side-knowledge (either as labeled subset of the data, or as set of constraints) can be regarded as a highly informative collection of implicit (meta-)constraints $\psi_{\mathscr{M}_1}(.), \ldots, \psi_{\mathscr{M}_m}(.)$ posed on the whole dataset. These meta-constraints, in turn, reverberate on the topological and/or statistical properties of the (unlabeled) data. Hence, any robust SSC algorithm shall comply with $\psi_{\mathscr{M}_1}(.), \ldots, \psi_{\mathscr{M}_m}(.)$, yielding outcomes that satisfy the meta-constraints. In so doing, the meta-constraints are expected to improve effectively the clustering process itself.

Finally, some relevant supervised machine learning paradigms which are not explicitly statistical in nature, such as artificial neural networks (ANN), could benefit from a robust SSC of the data. Robust clustering should be accomplished in compliance with (i.e., constrained by) the underlying laws these (say) ANNs learned to encapsulate. In so doing, improved laws are further learned by ANNs complying, in turn, with the very nature of the consequent SSC of the whole (labeled plus unlabeled) dataset (and so on, in an iterative fashion). An illustrative example may be sketched as follows. Let us consider a mixture of c neural experts $\phi_1(x), \ldots, \phi_c(x)$ aimed at learning a mapping $y = \sum_{i=1}^{c} \alpha_i(x)\phi_i(x)$, where $\alpha_i(x) \in (0, 1)$ is the credit that i-th expert receives over input x. In this example we assume that $\alpha_i(x) = 1$ if $x \in X_i$ and $\alpha_i(x) = 0$ otherwise, where the feature space is partitioned into c non-overlapping regions $X_1, \ldots, X_c$ of competence of the corresponding experts. Also, let us assume that each expert $\phi_i(x)$ is trained on a semi-supervised dataset $\mathscr{T}_i = \{(x_{ij}, y_{ij})|x_{ij} \in X_i, j = 1, \ldots, n_i\}$ (which includes all and only the labeled portion of the whole dataset that lies in X_i) in order to maximize the conditional likelihood $p_i(y|x)$. Starting from a preliminary partition $X_1, \ldots, X_c$ of the feature space, initial maximum-likelihood models $\phi_1(x), \ldots, \phi_c(x)$ are estimated. At this point, SSC is used to obtain a new partition $X'_1, \ldots, X'_c$ by relying on ML-constraints and CL-constraints defined according to a threshold $\xi \in \mathbb{R}^+$ on the conditional-likelihood such that any pair of input patterns (x_1, x_2) (either labeled or unlabeled, and regardless of the preliminary regions X_h, X_k they are originally from) must link if, for a certain i in $1, \ldots, c$, $p_i(\tilde{y}_1|x_1) > \xi$ and $p_i(\tilde{y}_2|x_2) > \xi$, where $\tilde{y}_1 = \phi_i(x_1)$ and $\tilde{y}_2 = \phi_i(x_2)$ (i.e, a ML constraint is introduced over pairs of patterns for which one of the experts expresses high statistical confidence), while they cannot link if $p_i(\tilde{y}_1|x_1) > \xi$ and $p_j(\tilde{y}_2|x_2) > \xi$ with $j \neq i$. Note that normalized probabilistic quantities are assumed, for instance $\sum_{i=1}^{c} p_i(y|x) = 1$ and $\xi \in (0, 1)$. Using the new partition $X'_1, \ldots, X'_c$ yielded by SSC from these constraints, the experts are trained again on their new regions of competence, obtaining maximum-likelihood estimates $\phi'_1(x), \ldots, \phi'_c(x)$ which are used, in turn, to create new ML and CL constraints, and so on.

9.6 Conclusions

Clustering aims at partitioning a dataset into subsets having both high inner coherence and outer separation (grouping perspective), or at describing a data sample in a simple and statistically sound way (data description perspective), or at initializing/boosting sophisticated pattern classifiers (subservience perspective). Traditional clustering algorithms exploit topological or probabilistic properties of the real-valued features of the input patterns, neither requiring nor accounting for the prior knowledge that human experts may associate with (a subset of) the training data. Semi-supervised clustering algorithms try and build on this side-knowledge in order to come up with improved solutions. In so doing, they relate to (and, cross-fertilize with) the general area of partially supervised learning.

In this chapter we have reviewed several algorithms and current concepts of clustering under partial supervision, namely: COP-COBWEB and COP k-means, the HMRF k-means method, SSC with data-driven learnable metric, seeded and constrained k-means, the Zheng-Li algorithm, active fuzzy constrained clustering, and a kernel approach to semi-supervised graph clustering. All in all, although the state-of-the-art of SSC is still in its early developmental stages, the variety of techniques found in the literature to date witnesses the significant potential (both theoretical and practical) behind this branch of science. A potential which, possibly along with some of the directions for further research we pointed out in Sect. 9.5, is likely to unfold to a much greater extent in the years to come.

References

1. Alpaydin E (2010) Introduction to machine learning, 2nd edn. MIT Press, Cambridge
2. Anand R, Reddy CK (2011) Graph-based clustering with constraints. In: Proceedings of the 15th Pacific-Asia conference on advances in knowledge discovery and data mining - volume part II, PAKDD'11, pp 51–62. Springer, New York
3. Arbelaitz O, Gurrutxaga I, Muguerza J, Perez JM, Perona I (2013) An extensive comparative study of cluster validity indices. Pattern Recogn 46(1):243–256
4. Bade K, Nurnberger A (2006) Personalized hierarchical clustering. In: IEEE/WIC/ACM international conference on web intelligence, pp 181–187
5. Bade K, Nurnberger A (2008) Creating a cluster hierarchy under constraints of a partially known hierarchy. In: SDM '08, pp 13–24
6. Basu S, Banerjee A, Mooney R (2002) Semi-supervised clustering by seeding. In: Proceedings of the 19st international conference on machine learning, pp 19–26
7. Basu S, Banerjee A, Mooney R (2004) Active semi-supervision for pairwise constrained clustering. In: Proceedings of the 2004 SIAM international conference on data mining (SDM-04). URL http://www.cs.utexas.edu/users/ai-lab/?basu:sdm04
8. Basu S, Bilenko M, Mooney R (2004) A probabilistic framework for semi-supervised clustering. In: Proc. of the 10th ACM SIGKDD conference on knowledge discovery and data mining (KDD'04), pp 59–68
9. Bilenko M, Basu S, Mooney R (2004) Integrating constraints and metric learning in semi-supervised clustering. In: Proceedings of the 21st international conference on machine learning, Banff, Canada, pp 81–88

10. Bishop CM (1995) Neural networks for pattern recognition. Oxford University Press, New York
11. Bishop CM (2006) Pattern recognition and machine learning. Springer, New York
12. Celebi ME, Kingravi H, Vela PA (2013) A comparative study of efficient initialization methods for the k-means clustering algorithm. Expert Syst Appl 40(1):200–210
13. Chu SM, Tang H, Huang TS (2009) Fishervoice and semi-supervised speaker clustering. In: IEEE international conference on acoustics, speech and signal processing (ICASSP'F09), pp 4089–4092. IEEE, Washington, DC, USA.
14. Cohn D, Caruana R, McCallum A (2003) Semi-supervised clustering with user feedback. Tech. rep.
15. Daniels K, Giraud-Carrier C (2006) Learning the threshold in hierarchical agglomerative clustering. In: Machine learning and applications (ICMLA '06) 5th international conferance, pp 270–278
16. Davidson I, Ravi SS (2005) Agglomerative hierarchical clustering with constraints: Theoretical and empirical results. In: Lecture notes in computer science, pp 59–70. Springer, New York
17. Davidson I, Ravi SS (2007) Intractability and clustering with constraints. In: Proceedings of the 24th international conference on machine learning, ICML '07, pp. 201–208. ACM, New York. DOI 10.1145/1273496.1273522. URL http://doi.acm.org/10.1145/1273496.1273522
18. Deborah L, Baskaran R, Kannan A (2010) A survey on internal validity measure for cluster validation. Int J Comput Sci Eng Survey 1(2):85–102
19. Dhillon I, Guan Y, Kulis B (2004) Kernel k-means: spectral clustering and normalized cuts. In: Proceedings of the tenth ACM SIGKDD international conference on Knowledge discovery and data mining, p 556. ACM, New York
20. Dhillon IS, Fan J, Guan Y (2001) Efficient clustering of very large document collections. In: Grossman RL, Kamath C, Kegelmeyer P, Kumar V, Namburu RR (eds) Data mining for scientific and engineering applications. Springer, New York, pp 357–381
21. Duda RO, Hart PE (1973) Pattern classification and scene analysis. Willey, New York
22. Faußer S, Schwenker F (2012) Semi-supervised kernel clustering with sample-to-cluster weights. In: Schwenker F, Trentin E (eds) Partially supervised learning - First IAPR TC3 workshop, PSL 2011, Ulm, Germany, September 15–16, 2011, Revised Selected Papers, pp 72–81. Springer, New York
23. Fisher DH (1987) Knowledge acquisition via incremental conceptual clustering. Mach Learn 2(2):139–172
24. Floyd RW (1962) Algorithm 97: shortest path. Commun ACM 5(6):345
25. Frigui H, Krishnapuram R (1997) Clustering by competitive agglomeration. Pattern Recogn 7:1109–1119
26. Grira N, Crucianu M, Boujemaa N (2005) Semi-supervised fuzzy clustering with pairwise-constrained competitive agglomeration. In: IEEE international conference on fuzzy systems
27. Grira N, Crucianu M, Boujemaa N (2008) Active semi-supervised fuzzy clustering. Pattern Recogn 41:1834–1844
28. Hofmann T, Buhmann JM (1998) Active data clustering. In: In advances in neural information processing systems 10, pp 528–534
29. Jain AK (2010) Data clustering: 50 years beyond k-means. Pattern Recogn Lett 31(8):651–666
30. Jain AK, Dubes RC (1988) Algorithms for clustering data. Prentice-Hall, Upper Saddle River
31. Jain AK, Duin RPW, Mao J (2000) Statistical pattern recognition: A review. IEEE Trans Pattern Anal Mach Intell 22(1):4–37
32. Kamvar SD, Klein D, Manning CD (2003) Spectral learning. In: IJCAI, pp 561–566
33. Kohonen T (ed) (1997) Self-organizing maps. Springer, New York
34. Kulis B, Basu S, Dhillon I, Mooney R (2009) Semi-supervised graph clustering: A kernel approach. Mach Learn 74(1):1–22
35. Křivánek M, Morávek J (1986) Np-hard problems in hierarchical-tree clustering. Acta Inf 23(3):311–323. DOI 10.1007/BF00289116. URL http://dx.doi.org/10.1007/BF00289116

36. Li T, Ding C, Jordan MI (2007) Solving consensus and semi-supervised clustering problems using nonnegative matrix factorization. In: Proceedings of the 2007 seventh IEEE international conference on data mining, ICDM '07, pp 577–582. IEEE Computer Society, Washington, DC, USA. DOI 10.1109/ICDM.2007.98. URL http://dx.doi.org/10.1109/ICDM.2007.98
37. Little RJA, Rubin DB (2002) Statistical analysis with missing data. Wiley, New York
38. Liu Y, Li Z, Xiong H, Gao X, Wu J (2010) Understanding of internal clustering validation measures. In: Proceedings of the 2010 IEEE international conference on data mining, pp. 911–916. IEEE Computer Society, Washington, DC, USA
39. Lloyd S (2006) Least squares quantization in pcm. IEEE Trans Inform Theory 28(2):129–137
40. MacQueen JB (1967) Some methods for classification and analysis of multivariate observations. In: Cam LML, Neyman J (eds) Proc. of the fifth Berkeley symposium on mathematical statistics and probability, vol 1. University of California Press, California, pp 281–297
41. Martinetz TM, Berkovich SG, Schulten KJ (1993) Neural-gas' network for vector quantization and its application to time-series prediction. IEEE Trans Neural Network 4(4):558–569
42. Newman CBD, Merz C (1998) UCI repository of machine learning databases. URL http://www.ics.uci.edu/~mlearn/MLRepository.html
43. Podani J (2000) Simulation of random dendrograms and comparison tests: Some comments. J Classification 17(1):123–142
44. Rendón E, Abundez IM, Gutierrez C, Zagal SD, Arizmendi A, Quiroz EM, Arzate HE (2011) A comparison of internal and external cluster validation indexes. In: Proceedings of the 2011 american conference on applied mathematics and the 5th WSEAS international conference on computer engineering and applications, pp. 158–163. World Scientific and Engineering Academy and Society (WSEAS)
45. Rockafellar RT (1970) Convex analysis. Princeton Mathematical Series. Princeton University Press, Princeton
46. Sattah S, Tversky A (1977) Additive similarity trees. Psychometrika 3:319–345
47. Schaeffer SE (2007) Survey: graph clustering. Comput Sci Rev 1(1):27–64
48. Schwenker F, Trentin E (2014) Pattern classification and clustering: A review of partially supervised learning approaches. Pattern Recogn Lett 37:4–14
49. Segal E, Wang H, Koller D (2003) Discovering molecular pathways from protein interaction and gene expression data. Bioinformatics 74(19):264–272
50. Soleymani Baghshah M, Bagheri Shouraki S (2010) Kernel-based metric learning for semi-supervised clustering. Neurocomputing 73(7):1352–1361
51. Strehl A, Ghosh J, Mooney R (2000) Impact of similarity measures on web-page clustering. In: Proceedings of the 17th national conference on artificial intelligence: workshop of artificial intelligence for web search (AAAI 2000), 30–31 July 2000. AAAI, Austin, Texas, USA, pp 58–64
52. Streit RL, Luginbuhl TE (1994) Maximum likelihood training of probabilistic neural networks. IEEE Trans Neural Network 5(5):764–783
53. Vendramin L, Campello RJGB, Hruschka ER (2010) Relative clustering validity criteria: A comparative overview. Stat Anal Data Min 3(4):209–235
54. Wagstaff K, Cardie C (2000) Clustering with instance-level constraints. In: Proceeding of the 17th international conference on machine learning, ICML 2000, pp 1103–1110
55. Wagstaff K, Cardie C, Rogers S, Schroedl S (2001) Constrained k-means clustering with background knowledge. In: Proc. of the 18th international conference on machine learning (ICML'01), pp 577–584
56. Xing EP, Ng AY, Jordan MI, Russell S (2003) Distance metric learning, with application to clustering with side-information. In: Advances in neural information processing systems 15, pp 505–512. MIT Press, Cambridge
57. Xiong H, Li Z (2013) Clustering validation measures. In: Data clustering: algorithms and applications, pp 571–606

58. Zha H, He X, Ding C, Simon H, Gu M (2001) Spectral relaxation for k-means clustering. In: NIPS, pp 1057–1064. MIT Press, Cambridge
59. Zhao H, Qi Z (2010) Hierarchical agglomerative clustering with ordering constraints. In: Proceedings of the 2010 third international conference on knowledge discovery and data mining, WKDD '10, pp. 195–199. IEEE Computer Society, Washington, DC, USA. DOI 10.1109/WKDD.2010.123. URL http://dx.doi.org/10.1109/WKDD.2010.123
60. Zheng L, Li T (2011) Semi-supervised hierarchical clustering. In: IEEE international conference on data mining, pp 982–991

Chapter 10
Consensus of Clusterings Based on High-Order Dissimilarities

Helena Aidos and Ana Fred

Abstract Over the years, many clustering algorithms have been developed, handling different issues such as cluster shape, density or noise. Most clustering algorithms require a similarity measure between patterns, either implicitly or explicitly. Although most of them use pairwise distances between patterns, e.g., the Euclidean distance, better results can be achieved using other measures. The dissimilarity increments is a new high-order dissimilarity measure, that uses the information from triplets of nearest neighbor patterns. The distribution of such measure (DID) was recently derived under the hypothesis of local Gaussian generative models, leading to new clustering algorithms. DID-based algorithm builds upon an initial data partition, different initializations producing different data partitions. To overcome this issue, we present an unifying approach based on a combination strategy of all these different initializations. Even though this allows obtaining a robust partition of the data, one must select a clustering algorithm to extract the final partition. We also present a validation criterion based on DID to select the best final partition, consisting in the estimation of graph probabilities for each cluster based on the DID.

Keywords Gaussian mixture decomposition • Dissimilarity increments distribution • Minimum description length • Evidence accumulation • Validity index

10.1 Introduction

Clustering techniques have been used in several areas, like machine learning, artificial intelligence, pattern recognition, web mining, image processing, biology, marketing [13, 48]. The main goal of clustering is to arrange data objects in groups (clusters), such that objects belonging to the same cluster are similar. It is a form of unsupervised learning, since no information about the groups to which the objects belong is known a priori. Several clustering algorithms have been developed

H. Aidos (✉) • A. Fred
Instituto de Telecomunicações, Instituto Superior Técnico,
Universidade de Lisboa, Lisbon, Portugal
e-mail: haidos@lx.it.pt; afred@lx.it.pt

M.E. Celebi (ed.), *Partitional Clustering Algorithms*,
DOI 10.1007/978-3-319-09259-1_10

over the years and, typically, they are suitable for the specialized area they have been designed. This means that usually some assumptions are made in favor of the application of interest and performance may be affected when used in other problems, since the same assumptions may not be verified. Many survey papers on clustering techniques can be found in the literature [26, 27, 56].

Two major clustering strategies have been adopted in published methods: partitional (non-hierarchical) and hierarchical [27, 48, 56]. *Hierarchical clustering techniques* group objects with a sequence of nested partitions, either from singleton clusters to a cluster including all data (agglomerative strategy) or in the opposite way (divisive strategy), while *partitional clustering techniques* divide the data into clusters without the hierarchical structure [13, 27, 48].

There are two main categories of *hierarchical methods*: agglomerative and divisive hierarchical methods. *Agglomerative methods* start by considering each data point as one cluster, and each partition is obtained from the previous one by merging two clusters into a single cluster, according to a proximity criterion. Single-link and complete-link [29] are the most representative algorithms in this class. In the literature can be found various research papers comparing the performance of several agglomerative algorithms in the context of different applications [11, 43]. *Divisive methods* work in the opposite way: one starts with a single cluster with all the objects and a divisive procedure is applied repeatedly until all clusters are singletons. The division of clusters can be made based on all features of the vectors [21, 41] or it is achieved based on a single feature at each step [35, 55]. Most hierarchical clustering methods do not have a probabilistic interpretation, however there are some model-based hierarchical clustering [7, 30, 52]. Traditional hierarchical methods may have problems handling large-scale datasets, therefore more recent hierarchical agglomerative algorithms have been developed [22, 23, 32].

Partitional methods divide data into several small groups, instead of creating a hierarchy of clusters. Traditional hierarchical clustering algorithms, once the clusters are constructed, cannot recover from a bad grouping by revisiting already formed clusters. Partitional techniques can overcome this difficulty imposed by hierarchical approaches by gradually adjusting formed clusters.

One important class of partitional methods is the one of prototype-based methods, such as k-means [24, 25] and k-medoids [33] approaches. Both approaches use a representative for each cluster: k-medoids uses a point of the data as prototype, which turns to be insensitive to outliers, and k-means uses a centroid, which is the mean of points within a cluster, turning this algorithm easier to interpret geometrically (it finds spherical clusters in data) and it is sensitive to outliers. Partitioning around medoids (PAM) and clustering large applications (CLARA) [33] are two versions of k-medoids methods. The k-means method is the simplest and most widespread clustering algorithm, with several extensions. Iterative self-organizing data analysis technique (ISODATA) [5] is a well-known variant of k-means.

Another class of partitional methods is the one of probabilistic approaches, which assume that the data come from a mixture of models whose distributions we want to learn. This category includes methods based on a minimum description length criterion [15], entropy-based expectation-maximization (EM) algorithm (EBEM) [6] and split and merge EM (SMEM) [51].

Density-based partitioning methods try partitioning data by identifying dense areas of data, which leads to the ability of discovering clusters of arbitrary shapes and less sensitivity to outliers. Density based spatial clustering of applications with noise (DBSCAN) [12] is the most representative algorithm of this class and has several variants [45].

Some clustering techniques are *grid-based approaches*, which use a grid-like structure to split the information space, separating the dense grid regions from the less dense ones to form groups. Algorithms using this approach are fast and handle outliers very well. This class of algorithms contains hierarchical techniques, but also partitioning techniques. A very important grid-based algorithm is CLIQUE [1].

Another class of clustering strategies is based on graph theoretical approaches, and consists in partition a graph by simply deleting some edges. Examples of algorithms in this class are spectral clustering methods [31,39].

Clustering ensemble methods is an approach that takes advantage of the diversity of solutions over the same dataset, produced by different algorithms, different initializations or parameters values [4, 16, 34, 46]. They can be generated based on the choice of data representation or on the choice of clustering algorithms or algorithmic parameters. These methods propose a consensus partition, given a set of data partitions, based on a combination strategy. Several authors have shown that these methods tend to reveal more robust and stable cluster structures than the individual clusterings in the clustering ensemble [16,46]. Different paradigms were followed in the literature: (a) similarity between objects, induced by the clustering ensemble [16,17,20,46]; (b) similarity between partitions [4,10,50]; (c) combining similarity between objects and partitions [14]; (d) probabilistic approaches to cluster ensembles [49,53,54].

Most clustering techniques require, either implicitly or explicitly, a similarity measure between patterns. However, choosing such a measure is typically difficult, given no prior knowledge about cluster shapes or structure. Although, most clustering algorithms use pairwise distances between patterns, e.g., the Euclidean distance, better results can be achieved using other measures. Recently, a new high order dissimilarity measure has been proposed, the *dissimilarity increments*, which uses the information from triplets of nearest neighbor patterns. This gives more information about the structure of a cluster, since a smooth evolution of the dissimilarity increments should occur if the patterns are in the same cluster, and high values should occur for patterns lying in different clusters [19].

The dissimilarity increments distribution was recently derived under the hypothesis of local Gaussian generative models for the data in $\mathbb{R}^l$, which lead to the proposal of new clustering algorithms based on this distribution. Here, we give special focus on a partitional clustering algorithm, called GMDID [2]. GMDID builds upon an initial data partition, e.g., a partition given by a Gaussian mixture decomposition. This algorithm consists of a merge strategy, which iteratively accepts or rejects the merging of two clusters based on this new distribution. We have two merging criteria: a likelihood-ratio test, which merges pairs of clusters with a p-value less than a given significance level α, and a parameter-free merge criterion based on the minimum description length principle.

However, GMDID is dependent of a Gaussian mixture decomposition, so different initializations produce different data partitions. This means that if the algorithm used to build the initial partition produces single clusters which in reality should be two separate clusters, the GMDID algorithm does not naturally undo this. To overcome this issue, we present an unifying approach consisting of a consensus function based on a combination strategy of all these different initializations. Even though this allows obtaining a robust partition of the data, one must select a clustering algorithm to extract the final partition. We also present the use of a validation criterion based on dissimilarity increments distribution to select the best final partition. That validation criterion consists in estimating graph probabilities for each cluster based on the dissimilarity increments distribution, and then applying the minimum description length of the graph-based representation of a partition.

Experimental results in both synthetic and real-world datasets show that the GMDID algorithm outperforms other state-of-art clustering algorithms. Furthermore, they show that the presented validation criterion is effective in the choice between results in the presented unified approach, i.e., finding the clustering algorithm leading to the best results.

This chapter is organized as follows: Sect. 10.2 presents the definition, distribution and properties of the dissimilarity increments. Section 10.3 presents a partitional algorithm based on the dissimilarity increments distribution and in Sect. 10.4 a consensus clustering strategy using the partitional clustering method presented in Sect. 10.3. Since a consensus partition is obtained by applying a clustering algorithm, a validation index is required to choose the best final partition. Such criterion is presented in Sect. 10.5. Experimental results for the partitional algorithm and the consensus clustering are presented in Sect. 10.6 and conclusions are drawn in Sect. 10.7.

10.2 High-Order Dissimilarity: The Dissimilarity Increments Principle

In Pattern Recognition, namely in clustering analysis, a pairwise (dis)similarity is used to decide whether two points are similar and should belong to the same group (or cluster) or are very different and should not be in the same group. However, if the data has some sparse groups, a pairwise measure, like Euclidean distance, would have a high value and the points would be interpreted as being very different, even if they share some properties. So, it is important to not only look to pairwise measures, but find a way to understand the data distribution inside each group.

10.2.1 Dissimilarity Increments: Definition and Properties

Definition 1. Consider a set of patterns X. Given $\mathbf{x}_i$, an arbitrary element of X and some dissimilarity measure, $d(\cdot,\cdot)$, between patterns, let $(\mathbf{x}_i, \mathbf{x}_j, \mathbf{x}_k)$ be the triplet of nearest neighbors, obtained as follows:

$$
\begin{aligned}
(\mathbf{x}_i, \mathbf{x}_j, \mathbf{x}_k) - &\ \text{nearest neighbor}\\
\mathbf{x}_j\ : j &= \arg\min_l \{d(\mathbf{x}_l, \mathbf{x}_i), l \neq i\}\\
\mathbf{x}_k\ : k &= \arg\min_l \{d(\mathbf{x}_l, \mathbf{x}_j), l \neq i, l \neq j\}.
\end{aligned}
$$

The **dissimilarity increment** [19] between neighboring patterns is defined as

$$
d_{\text{inc}}(\mathbf{x}_i, \mathbf{x}_j, \mathbf{x}_k) = \left| d(\mathbf{x}_i, \mathbf{x}_j) - d(\mathbf{x}_j, \mathbf{x}_k) \right|. \tag{10.1}
$$

The dissimilarity increment measure gives different information about the structure of a cluster than the pairwise distances, because a cluster is a set of patterns sharing some characteristics. Next, we will explain some useful properties of dissimilarity increments.

Property 1: It Has a Smooth Evolution. Inside a cluster, abrupt changes in the dissimilarity increments should not occur. If abrupt changes takes place, it indicates that we are in the presence of a different cluster.

Property 2: It Works with Different Types of Data Representations. The dissimilarity increment can be computed using a feature representation and some (dis)similarity measure, like the Euclidean distance. However, when no features are available, but a (dis)similarity representation of the objects is available, clustering methods based on this measure can still be applied.

Property 3: It Identifies Sparse Clusters. Most of the distance measures, e.g., Euclidean distance, used in the literature discard samples that are far apart in a sparse cluster. This measure can easily identify those patterns as belonging to the same cluster.

Property 4: It Is Invariant to Shape Features or Orientation. The dissimilarity increment can be applied to identify clusters with odd shapes, since it only takes into account the nearest neighbors.

10.2.2 Dissimilarity Increments Distribution (DID)

The dissimilarity increments distribution (DID) is herein derived, assuming that the Euclidean distance, $d(\cdot,\cdot)$, is the dissimilarity measure used in Definition 1.

10.2.2.1 DID for High-Dimensional Data

The probability density function of the dissimilarity increments measure is derived by assuming that X is a l-dimensional set of patterns (cluster), and that its elements are independent and identically distributed according to a multivariate Gaussian distribution, $\mathbf{x} \sim \mathcal{N}(\boldsymbol{\mu}, \Sigma)$.

In the derivation of DID, firstly, we need to determine the probability density function for the Euclidean distance between two patterns. For that, we transform our data from the multivariate Gaussian distribution to a standard normal distribution through a process called "whitening" or "sphering". The squared Euclidean distance, $(d^*)^2$, in the transformed (normalized) space follows a chi-squared distribution with l degrees of freedom.

After some approximations (see Appendix for details), we obtain the transformation equation from the normalized space (standard normal distribution) to the original space as

$$d^2 = \frac{\pi^{-l/2+1}}{\Gamma(1+l/2)} \operatorname{tr}(\Sigma) \left(d^*\right)^2, \tag{10.2}$$

where d^2 and $(d^*)^2$ are the squared Euclidean distance in the original and normalized spaces, respectively. Therefore, the probability density function of the Euclidean distance, $d \equiv d(\mathbf{x}, \mathbf{y})$, is

$$p(y) = 2G_l(\operatorname{tr}(\Sigma))y^{l-1} \exp\left\{-C_l(\operatorname{tr}(\Sigma))y^2\right\},\ y \in [0, +\infty[, \tag{10.3}$$

where we define

$$G_l(\operatorname{tr}(\Sigma)) \equiv l^{l/2}\Gamma(l/2)^{l/2-1}2^{-l} \operatorname{tr}(\Sigma)^{-l/2}\pi^{l/2(l/2-1)} \tag{10.4}$$

and

$$C_l(\operatorname{tr}(\Sigma)) \equiv l\Gamma(l/2)(4 \operatorname{tr}(\Sigma))^{-1}\pi^{l/2-1}. \tag{10.5}$$

Now, from Definition 1, the dissimilarity increments is defined as the absolute value of the difference of two Euclidean distances, and, in Eq. (10.3) we have the probability density function of the Euclidean distance between two patterns. Therefore, after a convolution (see Appendix for details), the probability density function for the dissimilarity increments is given by

$$p_{d_{\text{inc}}}(w; \operatorname{tr}(\Sigma)) = \frac{G_l(\operatorname{tr}(\Sigma))^2}{2^{l-5/2}C_l(\operatorname{tr}(\Sigma))^{l-1/2}} \exp\left\{-\frac{C_l(\operatorname{tr}(\Sigma))}{2}w^2\right\}$$
$$\left[\sum_{k=0}^{l-1}\sum_{i=0}^{2l-2-k}(-1)^i w^{k+i}\binom{l-1}{k}\binom{2l-2-k}{i}2^{k/2-i/2}C_l(\operatorname{tr}(\Sigma))^{k/2+i/2}\right.$$

$$\Gamma\left(\frac{2l-1-k-i}{2}, \frac{C_l(\mathrm{tr}(\Sigma))}{2} w^2\right)\Bigg], \tag{10.6}$$

with $G_l(\mathrm{tr}(\Sigma))$ and $C_l(\mathrm{tr}(\Sigma))$ defined in Eq. (10.4) and (10.5), respectively, and $\Gamma(a, x)$ is the incomplete gamma function [44].

The DID in Eq. (10.6) requires explicit knowledge of the diagonal covariance matrix, Σ. Therefore, we fit the data model by rewriting the distribution as a function of the mean value of the dissimilarity increment,

$$\lambda \equiv \mathbb{E}[d_{\mathrm{inc}}]. \tag{10.7}$$

Thus, $\mathrm{tr}(\Sigma)$ depends of λ and is given by (see [2] for details)

$$\mathrm{tr}(\Sigma) = \lambda^2 Q_l^2 \left[\sum_{k=0}^{l-1} \sum_{i=0}^{2l-2-k} (-1)^i \binom{l-1}{k} \binom{2l-2-k}{i} 2^k \right.$$
$$\left. B_{1/2}\left(\frac{k+i}{2}+1, l-\frac{k+i+1}{2}\right)\right]^{-2}, \tag{10.8}$$

where

$$Q_l \equiv 2^{l-7/2} l^{1/2} \pi^{l/4-1/2} \Gamma(l/2)^{5/2} \Gamma(l+1/2)^{-1} \tag{10.9}$$

and $B_x(a, b)$ is the incomplete beta function [44]. Plugging Eq. (10.8) into Eq. (10.6) we obtain an approximation for the DID of a cluster, $p_{d_{\mathrm{inc}}}(w; \lambda)$, that only depends on the mean of all increments in that cluster.

10.2.2.2 DID for Two-Dimensional Data

Now, we particularize for a two-dimensional set of patterns in the same conditions as in Sect. 10.2.2.1. So, if we replace l by 2 in Eq. (10.6), we get

$$p_{d_{\mathrm{inc}}}(w; \mathrm{tr}(\Sigma)) = \frac{w}{2\,\mathrm{tr}(\Sigma)} \exp\left\{-\frac{w^2}{2\,\mathrm{tr}(\Sigma)}\right\}$$
$$+ \frac{\sqrt{\pi}}{2\,\mathrm{tr}(\Sigma)^{3/2}} \left(\mathrm{tr}(\Sigma) - \frac{w^2}{2}\right) \exp\left\{-\frac{w^2}{4\,\mathrm{tr}(\Sigma)}\right\} \mathrm{erfc}\left(\frac{w}{2\sqrt{\mathrm{tr}(\Sigma)}}\right), \tag{10.10}$$

where $\mathrm{erfc}(\cdot)$ is the complementary error function.

The empirical estimation based on the expected value of the increments is obtained by replacing l with 2 in Eq. (10.8), which gives

$$\mathrm{tr}(\Sigma) = \frac{4\lambda^2}{2\pi\beta^2}, \tag{10.11}$$

with $\beta \equiv \left(2 - \sqrt{2}\right)$ and λ is defined in Eq. (10.7). Replacing in Eq. (10.10) we get an approximation for the DID of a cluster that only depends of the mean of all the increments in that cluster:

$$p_{d_{\text{inc}}}(w;\lambda) = \frac{\pi\beta^2}{4\lambda^2} w \exp\left\{-\frac{\pi\beta^2}{4\lambda^2}w^2\right\} + \frac{\pi^2\beta^3}{8\sqrt{2}\lambda^3} \exp\left\{-\frac{\pi\beta^2}{8\lambda^2}w^2\right\} \left(\frac{4\lambda^2}{\pi\beta^2} - w^2\right) \operatorname{erfc}\left(\frac{\sqrt{\pi}\beta}{2\sqrt{2}\lambda}w\right). \tag{10.12}$$

The probability density function for the dissimilarity increments in Eq. (10.6) is for data lying in a l-dimensional space. That formula is computationally heavy due to the computation of exponentials of very large numbers, which are troublesome to handle numerically. Therefore, we will mostly work with 2-DID, and later we empirically show that 2-DID is a good fit to data lying in other dimensions different from 2.

10.2.2.3 Characterization and Properties of 2-DID

Here we characterize the 2-DID by presenting its cumulative distribution function and some other properties.

Support of the Function

The dissimilarity increments can only take positive values, because it is an absolute value; then the support of the function given by Eq. (10.12) is $[0, \infty)$.

Probability Density Function

Equation (10.12) is in fact a probability density function, since $\int_0^\infty p_{d_{\text{inc}}}(w;\lambda)\mathrm{d}w = 1$ and it is non-negative. Figure 10.1a shows the influence of λ in the probability density function. The smaller the λ values, the more narrow is the function, which means that the data is more dense. Conversely, the higher the λ values, the wider is the probability density function, indicating that the data is more sparse. Moreover, when $w = 0$, we get $p_{d_{\text{inc}}}(0;\lambda) = \frac{\pi\beta}{2\sqrt{2}\lambda}$, with $\beta = 2 - \sqrt{2}$.

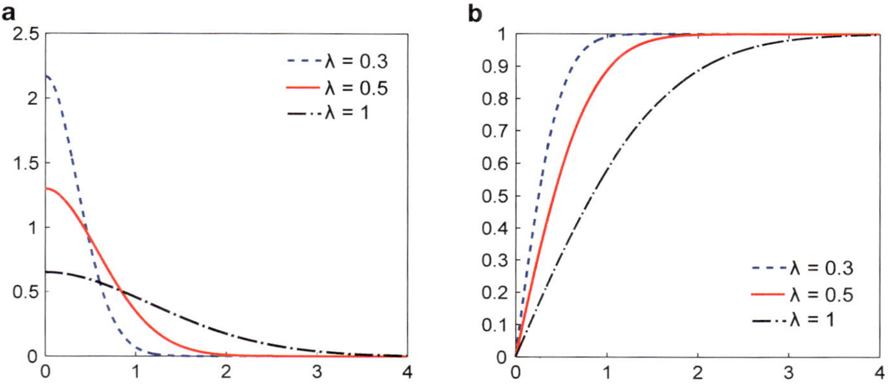

Fig. 10.1 The influence of the parameter λ in the probability density function and cumulative distribution function of 2-DID. (**a**) Probability density function. (**b**) Cumulative distribution function

Cumulative Distribution Function

The cumulative distribution function for 2-DID is defined by

$$F(w) = 1 - \exp\left\{-\frac{\pi\beta^2}{4\lambda^2}w^2\right\} + \frac{\pi\beta}{2\sqrt{2}\lambda}w\exp\left\{-\frac{\pi\beta^2}{8\lambda^2}w^2\right\}\operatorname{erfc}\left(\frac{\sqrt{\pi}\beta}{2\sqrt{2}\lambda}w\right), \tag{10.13}$$

where $\beta = 2 - \sqrt{2}$. Figure 10.1b shows the shape of the cumulative distribution function for different λ values.

Expected Value

The expected value of 2-DID is the parameter λ of the probability density function. Equivalently, $\mathbb{E}[w] = \lambda$.

Variance

The variance of 2-DID is given by

$$\operatorname{var}[w] = \frac{8 - \pi(\beta^2 + 2)}{\pi\beta^2}\lambda^2, \tag{10.14}$$

where $\beta = 2 - \sqrt{2}$.

10.2.3 DID Models and Data Fitting

At the end of Sect. 10.2.2.2 we claimed that Eq. (10.12) is a good fit to data with dimensions different from 2. So, our goal is to empirically show that the DID for any l-dimensional data can be well approximated by a 2-DID. In order to compare two distributions we need some statistical measures, like the Cramér-von-Mises criterion and Jensen-Shannon divergence.

The *Cramér-von-Mises criterion* [3] is a criterion used to determine how well a cumulative distribution function F fits a given empirical cumulative distribution function F_n. It is defined as

$$\omega^2 \equiv \int_{-\infty}^{\infty} [F_n(x) - F(x)]^2 \, \mathrm{d}F(x). \tag{10.15}$$

Let $x_1, x_2, \ldots, x_n$ be the empirically observed values, in nondecreasing order; in [3] it is shown that

$$T \equiv n\omega^2 = \frac{1}{12n} + \sum_{i=1}^{n} \left[\frac{2i-1}{2n} - F(x_i) \right]^2. \tag{10.16}$$

For common distributions, one should use tables of values of T calculated for the distribution F. We do not know the tables for the l-DID or 2-DID distributions, so we will simply compare the values of T directly: smaller values mean that the considered distribution is closer to the empirical distribution.

The *Jensen-Shannon divergence* (D_{JS}) [38] is a measure of the similarity between two probability distributions P and Q. It is similar to the Kullback-Leibler divergence but it is always finite and symmetric. It is defined as

$$D_{JS}(P, Q) \equiv \frac{1}{2} D_{KL}(P, M) + \frac{1}{2} D_{KL}(Q, M), \tag{10.17}$$

where $M = \frac{1}{2}(P + Q)$ and $D_{KL}(P, M) = \int P(x) \log \frac{P(x)}{M(x)} \mathrm{d}x$ is the Kullback-Leibler divergence. Two probability distributions are similar if the D_{JS} has a small value. The higher the value, the more dissimilar the distributions are.

10.2.3.1 Best Approximation to DID for High-Dimensional Data

The two statistical measures defined above are used to empirically show that the DID for two-dimensional data is a good approximation to the DID for l-dimensional data. Moreover, both distributions for dissimilarity increments (2-DID and l-DID) are a much better approximation to the real distribution than the exponential distribution considered by Fred and Leitão [19].

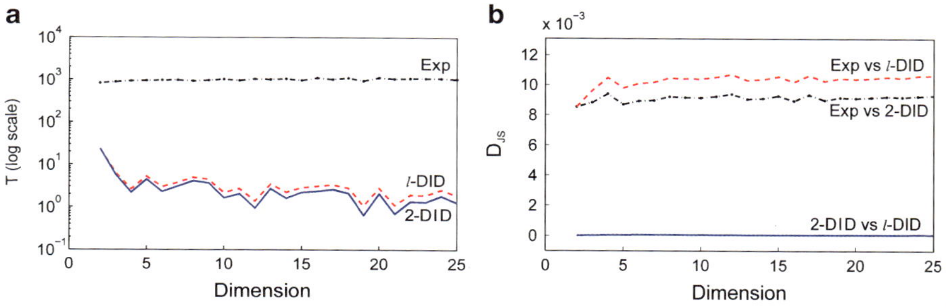

Fig. 10.2 Empirical distribution comparison for Gaussian datasets with 2000 patterns in l dimensions, with l ranging from 2 to 25. Cramér-von-Mises test (T) between each theoretical distribution (Exponential, 2-DID and l-DID) and the empirical distribution, and Jensen-Shannon divergence (D_{JS}) between pairs of theoretical distributions (Exponential, 2-DID and l-DID) [2]. (**a**) Cramér-von-Mises criterion. (**b**) Jensen-Shannon divergence

We generate datasets with 2000 patterns in l dimensions, where l varies from 2 to 25. Without loss of generality, the datasets are normal distributed and the Gaussians are all centered in the origin with diagonal covariance matrices, whose elements are randomly chosen between 0 and 1. The results of the two statistical measures, Cramér-von-Mises and Jensen-Shannon divergence are presented in Fig. 10.2.

Recall that the Cramér-von-Mises criterion is used to compare the fit between a cumulative distribution function and the empirical one. For 2-DID we used the cumulative distribution function defined by Eq. (10.13) and the cumulative distribution function for the exponential distribution is known from the literature. However, it is quite hard to derive the cumulative distribution function for l-DID, and for that reason we computed it numerically using the trapezoidal method.

Figure 10.2a shows that, according to the Cramér-von-Mises criterion (T), both DID distributions are a better fit to the histogram of the real dissimilarity increments distribution than the exponential distribution, i.e., T is better by approximately $10^{2.5}$. Moreover, 2-DID seems slightly better than l-DID, although we suspect, from our experiments, that this is due to approximation errors that occur during the computation of the l-DID and also due to the numerical computation of its cumulative distribution function.

Figure 10.2b compares pairs of distributions (2-DID vs exponential, 2-DID vs l-DID and l-DID vs exponential) through the Jensen-Shannon divergence (D_{JS}). We notice that 2-DID and l-DID are very similar to each other, i.e., D_{JS} is almost zero, while the comparison between exponential distribution and any DID distribution gives higher results. This empirically indicates that 2-DID is a good approximation to l-DID, and since 2-DID is computationally much more manageable, from now on we will always use 2-DID, defined in Eq. (10.12), and will only refer to it as DID.

10.2.3.2 Fitting DID to Non-Gaussian Data

The underlying hypothesis of the DID is that the data comes from a Gaussian distribution with diagonal covariance Σ. However, the distribution in Eq. (10.12) only depends of λ, i.e., the mean of the dissimilarity increments in a cluster. This means that we are able to use DID with non-Gaussian clusters, and it may result in good fits to the histogram of the dissimilarity increments. We now try fitting DID to several continuous and discrete well-known distributions, by generating 2000 patterns in 20 dimensions from Gaussian, Uniform, Exponential, Poisson, and Geometric distributions. The histograms of the dissimilarity increments and corresponding fit of DID is presented in Fig. 10.3.

A visual inspection to Fig. 10.3 indicates that DID has a good fit to the histogram of dissimilarity increments for all the distributions analyzed. We also present the values of the Cramér-von-Mises criterion (T) to demonstrate how good the fit is, and we see that T values are of the same order of magnitude for all these distributions.

10.3 Partitional Clustering

Based on the DID described in Sect. 10.2, a partitional clustering algorithm was derived [2]. The algorithm starts by assuming an initial data partition, denoted by P^{init}, which should have more clusters than the real one. The idea is to use the DID to decide whether two clusters should be merged into one cluster or not, i.e., the decision to merge two clusters will depend on the DID of each separate cluster and the DID of the two clusters combined.

To compare pairs of clusters and decide which pairs of clusters should be tested first, we sort them by ascending order, using the Mahalanobis distance [48]. All pairs of clusters are tested, using the DID, until all the remaining clusters fail the test. The overall procedure of this algorithm is summarized in Algorithm 9.

Algorithm 9 DID-based algorithm [2]

Require: data with N samples
Require: $P^{init} = \{C_1, \ldots, C_K\}$, initial data partition
  Compute D_{ij} as Mahalanobis distance between clusters i and j
  for all pairs (i, j) in ascending order of D_{ij} **do**
    Compute p_i, DID for cluster i, from (Eq. (10.12))
    Compute p_j, DID for cluster j, from (Eq. (10.12))
    Compute p_{ij}, DID for cluster produced merging clusters i and j, from (Eq. (10.12))
    if merge criterion is true **then**
      Merge clusters i and j
    end if
  end for
  return data partition $P = \{C_1, \ldots, C_{K'}\}$, with $K' \leq K$

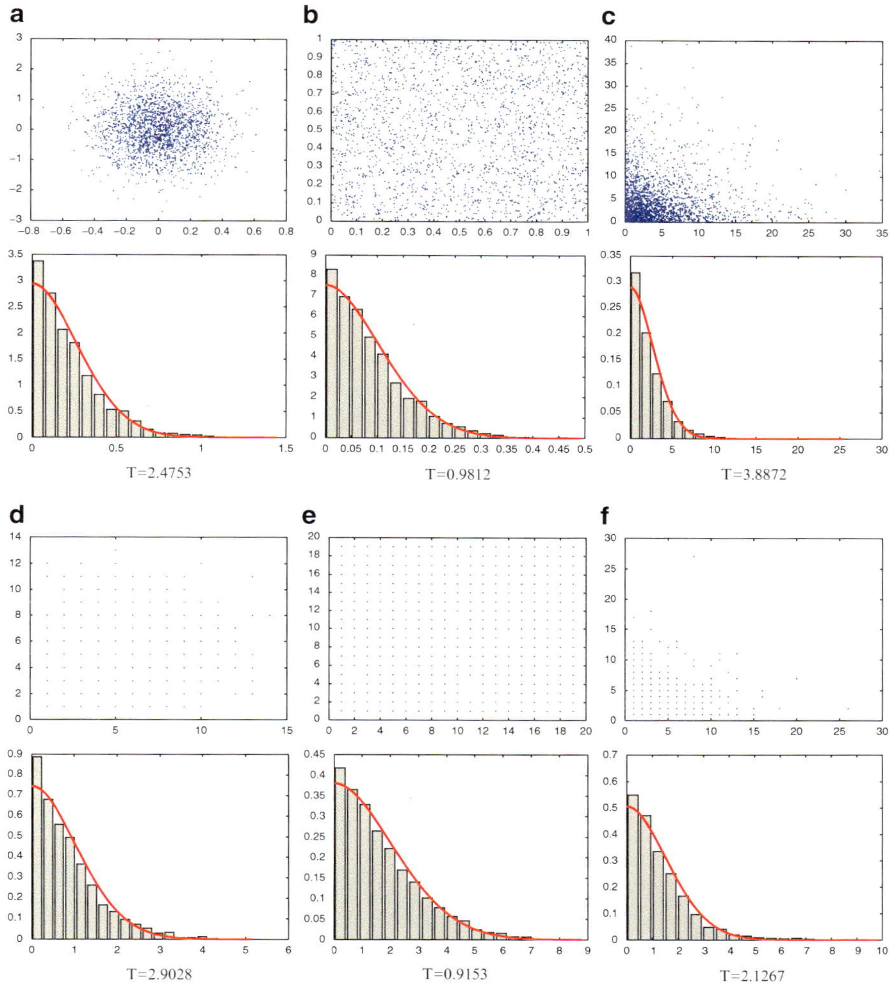

Fig. 10.3 Histograms (*bar plots*) and fitted DID (*solid line curves*) computed on several 20-dimensional datasets. *First row*: scatterplots of the first two dimensions of (**a**) a Gaussian distribution; (**b**) a Uniform distribution; (**c**) an Exponential distribution. *Second row*: corresponding histograms of dissimilarity increments and fit of the DID. *Third row*: scatterplots of the first two dimensions of (**d**) a Poisson distribution; (**e**) a Uniform distribution; (**f**) a Geometric distribution. *Fourth row*: corresponding histograms of dissimilarity increments and fit of the DID. Below each pair of figures, we present the corresponding value of the Cramér-von-Mises criterion (T) [2]

10.3.1 Initial Data Partition

We can obtain an initial data partition using any partitional clustering algorithm, such as Gaussian mixture decomposition (GMD) or k-means. Using GMD to obtain the initial data partition, our clustering algorithm is called GMDID, and using k-means, is called KMDID.

Here, GMDID uses the algorithm proposed in [15], which is an expectation-maximization (EM) algorithm that finds the number of components using the minimum description length criterion. This GMD algorithm produces a partition of the dataset with as many clusters as Gaussian components.

We also run k-means initialized with Variance Partitioning [47], but with several values of k. The choice of the best partition was made using G-DID, a validation criterion presented in Sect. 10.5, and that partition was used as initial partition for the KMDID approach.

10.3.2 Merge Criterion

A likelihood-ratio test or the minimum description length is used as a merge criterion in Algorithm 9. The first one has a parameter, which is a significance level, the other one is parameter-free.

Consider C_i and C_j two clusters candidate for merging, and λ_i and λ_j the parameter of the DID in Eq. (10.12) for clusters C_i and C_j, respectively. We define λ_{ij} as the parameter of the DID of merging clusters C_i and C_j into one cluster, C_{ij}.

Likelihood-Ratio Test. A likelihood-ratio test (LRT) [37] consists of the logarithm of the ratio between the joint likelihood of two models, $\mathscr{L}(W|\lambda_i, \lambda_j)$, and the likelihood of a single model, $\mathscr{L}(W|\lambda_{ij})$. The two models are represented by the DID of clusters C_i and C_j,

$$\mathscr{L}(W|\lambda_i, \lambda_j) = \prod_t p_{d_{\text{inc}}}(W = w_t; \lambda_i) \prod_{t'} p_{d_{\text{inc}}}(W = w_{t'}; \lambda_j) \tag{10.18}$$

and the single model corresponds to the DID of the merged cluster, C_{ij},

$$\mathscr{L}(W|\lambda_{ij}) = \prod_t p_{d_{\text{inc}}}(W = w_t; \lambda_{ij}). \tag{10.19}$$

This LRT is approximated by a chi-square distribution with one degree of freedom (two parameters in the numerator, λ_i and λ_j, corresponding to the expected value of the dissimilarity increments for each of the two clusters C_i and C_j, and one parameter in the denominator, λ_{ij}, corresponding to the expected value of the dissimilarity increments for the two clusters combined). Therefore,

$$-2\log\left(\frac{\mathscr{L}(W|\lambda_i,\lambda_j)}{\mathscr{L}(W|\lambda_{ij})}\right)\sim\chi^2(1). \tag{10.20}$$

Two clusters are merged if the p-value from $\chi^2(1)$ is less than a given significance level α. This significance level is a parameter to be chosen according to some criterion.

Minimum Description Length. The minimum description length (MDL) [40] of the separate clusters, C_i and C_j, and the merged cluster, C_{ij}, is used as another type of merging criterion. The description length of the separate cluster is defined as the sum of the length of the model description, according to the DID hypothesis of each cluster, and the cost of encoding our estimation of the DID. Thus,

$$DL^2_{\text{DID}} = -\left(\sum_{t=1}^{|S_{d_{\text{inc}}}(\lambda_i)|}\log p_{d_{\text{inc}}}(w_t;\lambda_i) + \sum_{t=1}^{|S_{d_{\text{inc}}}(\lambda_j)|}\log p_{d_{\text{inc}}}(w_t;\lambda_j)\right) + \log|S_{d_{\text{inc}}}(\lambda_i)| + \log|S_{d_{\text{inc}}}(\lambda_j)|, \tag{10.21}$$

where $|S_{d_{\text{inc}}}(\lambda_i)|$ and $|S_{d_{\text{inc}}}(\lambda_j)|$ is the total number of increments of clusters C_i and C_j, respectively. Similarly, the description length of the merged cluster is

$$DL^1_{\text{DID}} = -\left(\sum_{t=1}^{|S_{d_{\text{inc}}}(\lambda_{ij})|}\log p_{d_{\text{inc}}}(w_t;\lambda_{ij})\right) + \frac{1}{2}\log|S_{d_{\text{inc}}}(\lambda_{ij})|, \tag{10.22}$$

where $|S_{d_{\text{inc}}}(\lambda_{ij})|$ is the total number of increments of the merged cluster C_{ij}

Consider models M_2 and M_1 corresponding to two separate DID models (two separate clusters, C_i and C_j) and one single DID model (clusters C_i and C_j merged into a single cluster, C_{ij}), respectively. For $s \in \{1,2\}$,

$$\text{choose } M_s : s = \underset{i}{\operatorname{argmin}}\{DL^i_{\text{DID}}\}, \tag{10.23}$$

and consequently, two clusters are merged if DL^1_{DID} is less than DL^2_{DID}.

10.4 Consensus Clustering with DID

Consider $X = \{\mathbf{x}_1,\ldots,\mathbf{x}_N\}$ is a set of N objects and a clustering algorithm that takes X as input and groups the objects into k clusters, forming a partition P. A *clustering ensemble*, $\mathbb{P} = \{P^1, P^2, \ldots, P^M\}$, is a set of M different partitions of the data X:

$$P^1 = \{C_1^1, C_2^1, \dots, C_{k_1}^1\}$$

$$\vdots$$

$$P^M = \{C_1^M, C_2^M, \dots, C_{k_M}^M\},$$

where C_j^i is the j-th cluster in data partition P^i, which has k_i clusters and n_j^i is the cardinality of C_j^i, with $\sum_{j=1}^{k_i} n_j^i = N, i = 1, \dots, M$.

Different clustering algorithms, or one algorithm with different initializations or parameters can be used to obtain the clustering ensemble. Here, we used GMDID described in Sect. 10.3 with different initializations for the Gaussian mixture decomposition. Also, the two merging criteria are used to construct two different clustering ensembles.

Now, the *evidence accumulation* approach [17] takes this ensemble and produces a co-association matrix by taking the co-occurrences of pairs of patterns in the same cluster as votes for their association. The idea is that patterns which should be grouped together are probably going to be assigned to the same cluster in different data partitions. Formally, the M data partitions of N patterns yield a $N \times N$ co-association matrix:

$$\mathscr{C}(i, j) = \frac{n_{ij}}{M}, \tag{10.24}$$

where n_{ij} is the number of times the pattern pair (i, j) is assigned to the same cluster among the M partitions.

Finally, the consensus partition is found by applying a clustering algorithm to the co-association matrix. The overall procedure is summarized in Algorithm 10.

10.5 Validation with DID

In this section we present a validity index for clustering, which can be used to evaluate several data partitions obtained by a single clustering algorithm with different initializations or parameters, or clustering ensemble methods. That cluster validity index, called graph-based dissimilarity increments distribution (G-DID),

Algorithm 10 Evidence accumulation approach

Require: data with N samples
  for $i = 1$ **to** M **do**
    Obtain partition P^i by applying clustering algorithm(s)
  end for
  Compute co-association matrix using Eq. (10.24)
  Apply a clustering algorithm to the co-association matrix
  return Consensus partition

Algorithm 11 Graph generative model [18]

Require: Number of edges, n_{edges}
Require: Initial weight, d_{ini}
Require: Parameter distribution, λ
  Form an initial edge, e_1, with weight d_{ini}
  for $i = 1$ **to** $(n_{edges} - 1)$ **do**
    Randomly select an edge e_j and one of its vertices, v_j
    Let d_j be the weight of e_j
    Draw an increment value, w_i, according to the DID $p_{d_{\text{inc}}}(w; \lambda)$ Eq. (10.12)
    Add a new edge e_i to the graph, connected to v_j and with weight given by $d_i = d_j \pm w_i$
  end for

---

consists in a minimum description length of the graph-based representation of a partition, assuming that each cluster is a subgraph with the topology given by the minimum spanning tree (MST) and each cluster is model by the DID described in Sect. 10.2. This index was presented in [18], but using the exponential distribution to model the dissimilarity increments.

A cluster is modeled by a probabilistic attributed graph (PAG), where nodes represent objects, and edges have two attributes: the dissimilarity between linked objects, and the probability of edge formation. The nearest neighbor relationships presented in the dissimilarity increments definition, lead us to the MST as the graph topology of a cluster. That MST is computed from the completely connected graph with edge weights, which corresponds to dissimilarity between linked objects. Moreover, for each edge e_j (with weight, dissimilarity, d_j), one can connect an edge e_i (with weight, d_i) with probability $p_{d_{\text{inc}}}(w_i; \lambda)$, where $w_i = |d_i - d_j|$. This procedure is called graph generative model between linked objects and the overall procedure is given in Algorithm 11. The probability of edge formation is computed assuming that an edge may have several connections with other edges in the MST, and that all these edges may contribute for the generation of e_i, and it is given by

$$\hat{f}(e_i, \lambda) = \sum_{e_m} p_{d_{\text{inc}}}(w_m; \lambda) P(e_m), \tag{10.25}$$

where e_m is an edge connected to e_i, w_m is the corresponding dissimilarity increment computed from e_i and e_m, and $P(e_m) = 1/|c_{e_i}|$, with $|c_{e_i}|$ the number of edges directly connecting to e_i.

Now, a disconnected graph $G = \{G_1, \ldots, G_{k_P}\}$ represents a partition P with k_P clusters, where each cluster C_i is represented by a subgraph, G_i. The probability of a partition P is given by

$$\hat{f}(P) = \prod_{i=1}^{k_P} \hat{f}(G_i|P) = \prod_{i=1}^{k_P} \prod_{e_j \in G_i} \hat{f}(e_j, \lambda_i). \tag{10.26}$$

Algorithm 12 G-DID index [18]

Require: $P = \{C_1, \ldots, C_{k_P}\}$ a data partition
  for $i = 1$ **to** k_P **do**
    Estimate $\hat{\lambda}_i$ associated with C_i
    Obtain the MST for cluster C_i
    Determine the PAG, G_i:
      (i) G_i topology is given by the MST
      (ii) edge weights are given by the distance between linked objects in G_i
      (iii) estimate edges probabilities $\hat{f}(e_j, \hat{\lambda}_i), \forall e_j \in G_i$ according to Eq. (10.25)
  end for
  Determine partition probability $\hat{f}(P)$ using Eq. (10.26)
  return G-DID index computed using Eq. (10.27)

The *cluster validity index*, G-DID, is a description length of the graph representation of partition P, and is defined as

$$\text{G-DID}(P) = -\log \hat{f}(P) + \frac{k_P}{2} \log |S_{d_{\text{inc}}}|, \tag{10.27}$$

with $|S_{d_{\text{inc}}}|$ the cardinality of the set of increments of the dataset. The first term represents the description length of the partition according to the DID hypothesis, and the second term, the cost of encoding our estimation of the DID. The overall procedure for computing this index is summarized in Algorithm 12. This index can also be used to select among a set of M partitions $\mathbb{P} = \{P^1, P^2, \ldots, P^M\}$, we only need to use a MDL criterion

$$\text{choose } P^i : i = \arg\min_j \{\text{G-DID}(P^j)\}. \tag{10.28}$$

10.6 Experimental Results and Discussion

The performance of the algorithms presented previously are assessed using 10 datasets: four synthetic datasets, four real-world datasets from the UCI Machine Learning Repository[1] and two datasets containing Usenet articles from different discussion groups, which were obtained from the 20-Newsgroups database.[2] The synthetic datasets were chosen to take into account a wide variety of situations: well-separated and touching clusters, Gaussian and non-Gaussian clusters, arbitrary shapes and diverse cluster densities. Figure 10.4 shows these synthetic datasets.

[1] http://archive.ics.uci.edu/ml.

[2] http://www.ai.mit.edu/people/jrennie/20Newsgroups/.

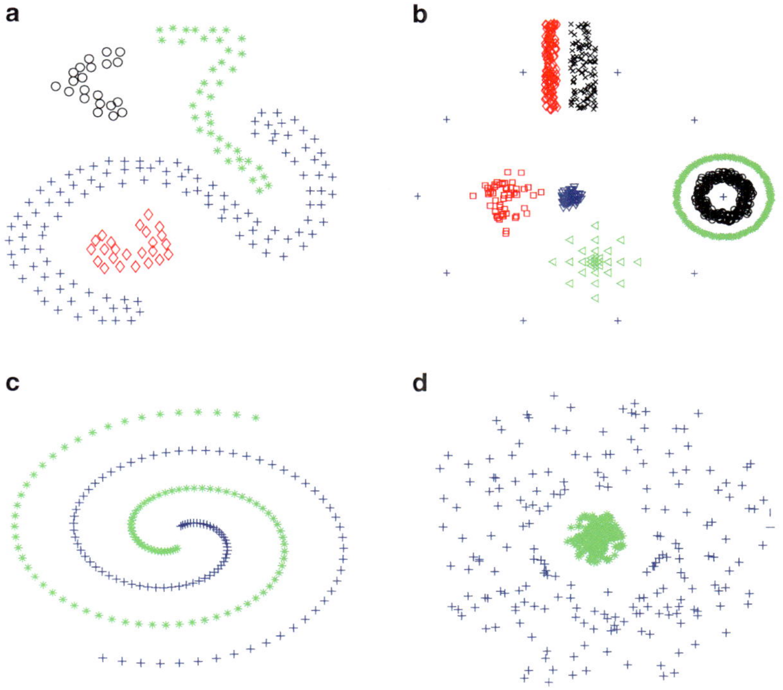

Fig. 10.4 Synthetic datasets. (**a**) d2. (**b**) Mixed Image 2. (**c**) Spiral. (**d**) Circs

The *Breast-cancer* dataset consists of 683 patterns represented by nine features and has two clusters. The *House-votes* dataset consists of votes for each of the U.S. House of Representatives Congressmen on the 16 key votes identified by the Congressional Quarterly Almanac. It is composed by two clusters and only the patterns without missing values were considered, for a total of 232 samples (125 democrats and 107 republicans). The *Iris* dataset consists of three species of Iris plants (Setosa, Versicolor and Virginica). This dataset is characterized by four features and 50 samples in each cluster. The *Crabs* dataset consists of 200 patterns represented by 5 features and has two classes. Crabs and House-votes were normalized to have unit variance. The *Diff-300* consists of 300 documents corresponding to 3 topics in the Usenet discussions, with 10 features per document. There are three topics, each one has exactly 100 documents: alt.atheism, rec.sport.baseball, sci.space from the 20-newsgroups. The *Same-300* consists of 297 documents, with 10 features per document. The documents are very close to each other, even from different classes; the classes are: talk.politics.misc, talk.politics.guns, talk.politics.mideast. A summary of the real-world datasets is given in Table 10.1.

Table 10.1 Real-world datasets

Dataset	Number of samples	Number of features	Number of clusters
Breast-cancer	683	9	2
Crabs	200	5	2
House-votes	232	16	2
Iris	150	4	3
Diff-300	300	10	3
Same-300	297	10	3

We assess the quality of each partition P using the *consistency index* (CI) [16] (also known as accuracy or H index [42]), which is the percentage of agreement between a partition $P = \{C_1, \ldots, C_K\}$ and the ground truth information, $P^{gt} = \{C_1^{gt}, \ldots, C_{K'}^{gt}\}$. Equivalently,

$$CI(P, P^{gt}) = \frac{1}{N} \sum_{k'=match(k)} m_{k,k'}, \tag{10.29}$$

where $m_{k,k'}$ is the contingency table, $m_{k,k'} = |C_k \cap C_{k'}^{gt}|$.

10.6.1 *Partitional Clustering*

This section of experimental results will be split in two case studies: known number of clusters and unknown number of clusters. In the first one, we compare the performances of GMDID for different initializations and merge criteria (LRT and MDL), assuming that the true number of clusters is known. The combination of random initialization and parameter-free merging criterion is then chosen and compared with k-means, and typical hierarchical methods. The second case study assumes that no a priori information about the number of clusters is available, and compare the performances of DID-based algorithms (GMDID and KMDID) with GMD and k-means. Also, we compare GMDID with typical hierarchical methods using the lifetime criterion and G-DID to select the number of clusters.

10.6.1.1 Known Number of Clusters

GMD can be initialized in several ways one possibility consists in choose M random points from the dataset (we denote this as GMD^{rand}). Another possibility is using the final partition given by k-means as part of the initialization of the GMD (denoted by GMD^{init}), adding additional randomly chosen centroids, from the dataset, in a total of 50 initial centroids. In the k-means algorithm different initializations yield different final partitions, due to random initialization. So, we use the Variance

Partitioning method proposed in [47] to initialize k-means, because the results were comparable to the best run out of 20 with random initializations; however, any other kind of initialization could be used [8, 9]. We performed 20 runs of both types of GMD (GMD^{rand} and GMD^{init}) and chose the best run according to the intrinsic criterion of the GMD [15] (which is MDL-based).

DID-based algorithms, described in Sect. 10.3, have a merging criterion that can be MDL-based, which is parameter-free, or a likelihood-ratio test, with a parameter α corresponding to a significance level. We set α to $\{0.001, 0.005, 0.01, 0.05, 0.1, 0.15\}$ and chose the best parameter according to the G-DID index presented in Sect. 10.5.

Table 10.2 presents the results produced by these initializations. Here, we are assuming that the true number of clusters is known, and from the definition of GMDID, this algorithm can use that information only as a lower bound to the number of clusters. This means that, in this experiments (where the true number of clusters is known), we included a stopping criterion for GMDID when the true number of clusters is reached. Therefore, in Table 10.2 this algorithm sometimes overestimates the number of clusters, and underestimates it when GMD does that.

From Table 10.2, we notice that GMDID, using the MDL criterion ($GMDID_M$) and as initial partition the GMD^{rand}, is the best algorithm overall, especially in synthetic datasets. Since k-means is not suitable for situations where clusters have arbitrary shapes and densities, the centroids used to initialize GMD^{init} are wrongly positioned, affecting the results of GMDID.

In Table 10.3 we intend to compare DID-based algorithm with k-means initialized by Variance Partitioning method [47] and some hierarchical methods. To facilitate that comparison, we choose one of the versions of GMDID presented in Table 10.2, because it has a random initialization and is parameter-free in the merging criterion. However, the values in Table 10.3 can be compared to the ones in Table 10.2.

From Table 10.3, we note that $GMDID_M^{rand}$ is overall the best method. Moreover, SL is the best hierarchical clustering algorithm in most synthetic datasets and is a poor choice for the real-world datasets presented. WL is a good choice, for real-world datasets, within the hierarchical clustering algorithms. However, when compared to the partitional methods, WL only achieve the best result for the Breast-cancer dataset. $GMDID_M^{rand}$ is the best method in three real-world datasets and k-means in two datasets.

10.6.1.2 Unknown Number of Clusters

Assuming that no a priori information about the number of clusters is available, we run GMDID using as initial partition the GMD with fully random initialization. As described above, we performed 20 runs of GMD and choose the best run using its intrinsic criterion. Also, we run k-means initialized with Variance Partitioning [47], but with several values of k (k ranges from 2 to 25), and the choice of the best

Table 10.2 Consistency index (in %) for several variants of GMD [15] and GMDID when the true number of clusters (Nc) is known. The values in parentheses correspond to the number of clusters found by each algorithm. GMD^{rand} is the Gaussian mixture decomposition with random initialization and GMD^{init} uses k-means output as initialization; GMDID is the algorithm presented in Sect. 10.3; $(\cdot)_M$ means using MDL criterion and $(\cdot)_L$ means using LRT criterion. The best results for each dataset are shown in bold

Datasets	Nc	GMD^{rand}	GMDID$_M^{rand}$	GMDID$_L^{rand}$	GMD^{init}	GMDID$_M^{init}$	GMDID$_L^{init}$
d2	4	34.00(13)	**100**(4)	90.50(5)	34.50(13)	**100**(4)	90.50(5)
Mixed Image 2	8	34.10(38)	**100**(8)	99.19(8)	41.00(29)	99.86(8)	99.05(8)
Spiral	2	7.00(45)	**100**(2)	**100**(2)	11.00(36)	61.50(5)	58.50(4)
Circs	2	35.75(14)	**99.00**(2)	79.00(3)	49.25(12)	98.25(3)	98.25(2)
Breast-cancer	2	45.10(5)	75.11(3)	75.11(3)	68.23(5)	**79.65**(3)	74.96(4)
Crabs	2	68.50(3)	**74.50**(2)	**74.50**(2)	58.50(2)	58.50(2)	58.50(2)
House-votes	2	87.07(2)	87.07(2)	87.07(2)	**89.22**(2)	**89.22**(2)	**89.22**(2)
Iris	3	71.33(5)	**98.00**(3)	**98.00**(3)	76.00(5)	96.00(3)	96.00(3)
Diff-300	3	**90.67**(3)	**90.67**(3)	**90.67**(3)	73.67(3)	73.67(3)	73.67(3)
Same-300	3	47.47(2)	47.47(2)	47.47(2)	**53.87**(3)	**53.87**(3)	**53.87**(3)

Table 10.3 Consistency index (in %) for clustering algorithms when the true number of clusters (Nc) is known. The values in parentheses correspond to the number of clusters found by each algorithm. $GMDID_M^{rand}$ is GMDID with initial partition given by GMD with random initialization [15] and using MDL as merging criterion. k-means is initialized with Variance Partitioning method [47] and k is equal to Nc. Hierarchical clustering algorithms: single-link (SL), average-link (AL), complete-link (CL) and Ward's link (WL). The best results for each dataset are shown in bold

True number of clusters							
Dataset	Nc	$GMDID_M^{rand}$	k-means	SL	AL	CL	WL
d2	4	**100**(4)	60.50	**100**	61.50	61.50	51.00
Mixed Image 2	8	**100**(8)	41.95	47.50	51.56	51.01	53.99
Spiral	2	**100**(2)	55.00	**100**	52.00	52.00	52.00
Circs	2	99.00(2)	72.75	**100**	61.50	71.00	69.75
Breast-cancer	2	75.11(3)	96.05	65.15	94.29	85.21	**96.63**
Crabs	2	**74.50**(2)	54.00	50.50	51.50	52.00	55.50
House-votes	2	87.07(2)	**89.22**	53.02	88.36	81.03	85.78
Iris	3	**98.00**(3)	89.33	68.00	90.67	84.00	89.33
Diff-300	3	**90.67**(3)	58.67	35.67	36.67	36.67	66.00
Same-300	3	47.47(2)	**54.21**	35.02	33.67	36.70	38.05

partition was made using G-DID. That partition was then used as initial partition for KMDID. The results of these initializations are presented in Table 10.4.

Again, the best DID-based method for the synthetic datasets is when the initial partition is given by GMD. Recall that KM corresponds to k-means with a fixed initialization and k chosen by G-DID. From Table 10.4, we notice that, in most cases, G-DID chooses k less or equal to the true number of clusters. Consequently, KMDID has poor results in almost all datasets, except for Circs dataset. Overall, GMD is the best method for the real-world datasets, but is the worst for datasets with non-Gaussian and/or arbitrary shapes clusters.

Comparing the results of GMDID when the true number of clusters is available ($GMDID_M^{rand}$ from Table 10.2) and when it is not available ($GMDID_M$ from Table 10.4), they are similar except for the Crabs, Iris and Diff-300 datasets. This suggests that the method incorrectly merges clusters due to the fact that those datasets may have the same DID for different clusters. On the other hand, the similarity in the results between $GMDID_M^{rand}$ (Table 10.2) and $GMDID_M$ (Table 10.4) is a strong indication that GMDID can identify correctly the true number of clusters.

Tables 10.5 and 10.6 presents the results for the hierarchical algorithms (similar to Table 10.3), however we are assuming that the number of clusters is unknown. Therefore, the number of clusters is obtained by applying the lifetime criterion [17] (see Table 10.5) and G-DID presented in Sect. 10.5 (see Table 10.6).

Overall, $GMDID_M$ outperforms other clustering algorithms and KM is the best algorithm only for Iris dataset (see Table 10.5). Typical hierarchical clustering algorithms performs poorly when the number of clusters is unknown and the lifetime

Table 10.4 Consistency index (in %) for partitional clustering algorithms when the true number of clusters (Nc) is unknown. The values in parentheses correspond to the number of clusters found by each algorithm. KM is k-means run for several values of k and chosen according to the G-DID, and initialized with Variance Partitioning [47]; GMD is Gaussian mixture decomposition [15] with random initialization; KMDID and GMDID are the DID-based algorithms presented in Sect. 10.3; $(\cdot)_M$ means using MDL criterion and $(\cdot)_L$ means using LRT criterion. The best results for each dataset are shown in bold

Datasets	Nc	KM	$KMDID_M$	$KMDID_L$	GMD	$GMDID_M$	$GMDID_L$
d2	4	44.50(2)	58.00(1)	44.50(2)	34.00(13)	**100**(4)	90.50(5)
Mixed Image 2	8	46.41(7)	45.87(6)	45.87(6)	34.10(38)	**100**(8)	99.19(8)
Spiral	2	55.00(2)	55.00(2)	55.00(2)	7.00(45)	**100**(2)	**100**(2)
Circs	2	57.25(11)	**100**(2)	**100**(2)	35.75(14)	99.00(2)	79.00(3)
Breast-cancer	2	**96.05**(2)	65.01(1)	65.01(1)	45.10(5)	75.11(3)	75.11(3)
Crabs	2	54.00(2)	54.00(2)	54.00(2)	**68.50**(3)	50.00(1)	50.00(1)
House-votes	2	75.43(3)	75.43(3)	75.43(3)	**87.07**(2)	**87.07**(2)	**87.07**(2)
Iris	3	**89.33**(3)	66.67(2)	66.67(2)	71.33(5)	66.67(2)	66.67(2)
Diff-300	3	58.33(2)	33.33(1)	33.33(1)	**90.67**(3)	62.33(2)	62.33(2)
Same-300	3	35.35(2)	35.35(2)	35.35(2)	**47.47**(2)	**47.47**(2)	**47.47**(2)

Table 10.5 Consistency index (in %) for clustering algorithms when the true number of clusters (Nc) is unknown. The values in parentheses correspond to the number of clusters found by each algorithm. $GMDID_M$ with initial partition given by GMD [15] with fully random initialization, and KM is k-means run for several values of k and chosen according to the G-DID, and initialized with Variance Partitioning [47]. Hierarchical clustering algorithms: single-link (SL), average-link (AL), complete-link (CL) and Ward's link (WL). The number of clusters for SL, AL, CL e WL are found by the lifetime criterion [17]. The best results for each dataset are shown in bold

Lifetime criterion							
Datasets	Nc	$GMDID_M$	KM	SL	AL	CL	WL
d2	4	**100**(4)	44.50(2)	68.50(2)	46.00(2)	51.00(3)	46.00(3)
Mixed Image 2	8	**100**(8)	46.41(7)	47.50(9)	47.36(3)	51.56(5)	47.36(3)
Spiral	2	**100**(2)	55.00(2)	**100**(2)	36.00(4)	40.00(4)	28.00(5)
Circs	2	99.00(2)	57.25(11)	**99.75**(3)	62.75(7)	65.00(4)	60.75(7)
Breast-cancer	2	75.11(3)	96.05(2)	65.15(3)	94.00(3)	92.83(2)	**96.63**(2)
Crabs	2	50.00(1)	54.00(2)	50.50(2)	51.50(2)	52.00(2)	**55.50**(2)
House-votes	2	**87.07**(2)	75.43(3)	5.60(160)	5.60(160)	5.6(160)	85.78(2)
Iris	3	66.67(2)	**89.33**(3)	66.67(2)	66.67(2)	66.67(2)	66.67(2)
Diff-300	3	**62.33**(2)	58.33(2)	35.67(3)	34.67(2)	40.67(13)	53.00(11)
Same-300	3	**47.47**(2)	35.35(2)	35.35(4)	34.68(7)	35.69(14)	36.36(2)

Table 10.6 Consistency index (in %) for clustering algorithms when the true number of clusters (Nc) is unknown. The values in parentheses correspond to the number of clusters found by each algorithm. GMDID_M with initial partition given by GMD [15] with fully random initialization, and KM is k-means run for several values of k and chosen according to the G-DID, and initialized with Variance Partitioning [47]. Hierarchical clustering algorithms: single-link (SL), average-link (AL), complete-link (CL) and Ward's link (WL). The number of clusters for SL, AL, CL e WL are found by the G-DID criterion presented in Sect. 10.5. The best results for each dataset are shown in bold

G-DID criterion							
Datasets	Nc	GMDID_M	KM	SL	AL	CL	WL
d2	4	**100**(4)	44.50(2)	**100**(4)	49.00(8)	61.50(6)	61.50(5)
Mixed Image 2	8	**100**(8)	46.41(7)	94.32(45)	22.33(74)	60.49(11)	28.69(42)
Spiral	2	**100**(2)	55.00(2)	**100**(2)	24.00(22)	16.00(36)	16.00(22)
Circs	2	99.00(2)	57.25(11)	**100**(2)	58.75(16)	58.25(16)	69.75(4)
Breast-cancer	2	75.11(3)	**96.05**(2)	65.15(11)	75.26(12)	67.79(16)	82.28(3)
Crabs	2	50.00(1)	**54.00**(2)	50.50(2)	43.50(3)	52.00(2)	33.00(4)
House-votes	2	87.07(2)	75.43(3)	53.02(2)	**87.50**(3)	81.03(2)	76.72(3)
Iris	3	66.67(2)	**89.33**(3)	62.67(15)	79.33(6)	50.00(13)	66.67(2)
Diff-300	3	62.33(2)	58.33(2)	32.67(14)	38.00(19)	41.67(12)	**66.00**(3)
Same-300	3	**47.47**(2)	35.35(2)	35.35(5)	34.68(7)	39.06(10)	41.41(10)

criterion is applied to identify the number of clusters, except WL for Breast-cancer and Crabs datasets. For Breast-cancer dataset, KM, AL, CL and WL are the best algorithms; but for the House-votes dataset, SL, AL and CL are the worst algorithms. Moreover, on average, AL and CL are the worst algorithms.

Again, from Table 10.6, we note that, overall, GMDID_M is the best algorithm, and AL and CL are the worst algorithms, when the true number of clusters is unknown and G-DID is used to obtain the number of clusters. However, AL is the best algorithm for House-votes dataset, although the difference is very small when compared to GMDID_M. Notice that CI has high values even when the number of clusters identified is high, e.g., Mixed Image 2 for SL. For this dataset, G-DID identified 45 clusters for SL with a CI of 94.32 %, this happens because SL identifies some points of this dataset as being singleton clusters.

Comparing the results of the traditional hierarchical algorithms with the number of clusters found by the two criteria mentioned (lifetime and G-DID), there are some improvements in the results when G-DID is used for SL, AL and CL. Only WL had worsen a little compared to the lifetime criterion. Moreover, AL, CL and WL were the best algorithms for the Breast-cancer dataset when the lifetime criterion was used, but with G-DID criterion the results are lower. The highest improvement occurred for SL, in almost all datasets, and AL and CL for the House-votes dataset.

10.6.2 Consensus Clustering

The clustering ensemble, described in Sect. 10.4, was obtained by performing $M = 200$ runs of GMDID algorithm, with different random initializations. The two different merging criteria, described in Sect. 10.3, were used to create different consensus clustering. Since the MDL criterion is parameter-free, only the initial partitions contribute to the diversity of solutions. The LRT criterion has one parameter, so, the clustering ensemble was obtained through the initial partitions and by varying the parameter α, which will take values from $\{0.001, 0.005, 0.01, 0.05, 0.1, 0.15\}$. Finally, the consensus partition was extracted by applying four traditional clustering algorithms: single-link (SL), average-link (AL), complete-link (CL) and Ward's link (WL). The number of clusters were obtained by applying three different strategies: assuming the true number of clusters is known, using the lifetime criterion [17] and the G-DID criterion (see Sect. 10.5). The results for the known true number of clusters, lifetime criterion and G-DID criterion are presented in Tables 10.7, 10.8 and 10.9, respectively. Each table presents the best CI, with the merging criterion for GMDID and the hierarchical algorithm to extract the consensus partition that gives the best CI (columns 2 to 4). Moreover, it also presents the CI values obtained when we selected the best partition according to G-DID index and Dunn's index [48], and the corresponding merging criterion and hierarchical clustering algorithm.

Table 10.7 shows that G-DID index is a good criterion to select the best partition and GMDID merging criterion, it chooses 7 out of 10 datasets, while Dunn's index selects 6 out of 10 datasets. For the Breast-cancer, the partition chosen by G-DID

Table 10.7 Comparison of G-DID and Dunn's index for selecting the best combination of merging criterion and hierarchical algorithm, when the true number of clusters (Nc) is known. *Best final partition(s)* correspond to the best consistency index values (CI, in %), across all combinations of merging criterion and clustering algorithm; column 3 (Crit) and 4 (Alg) indicate the merging criterion and algorithms leading to best CI. *Selection using G-DID and Dunn's index* present the CI values, with corresponding merging criterion and clustering algorithm, when those indexes are used to selected the best final partition (columns 5 to 7 and 8 to 10, respectively). When the CI of the selected partitions by each index (G-DID and Dunn's index) are matched with the best CI, the values are shown in bold

True number of clusters									
	Best final partition(s)			Selection using G-DID			Selection using Dunn's index		
Datasets	CI	Crit	Alg	CI	Crit	Alg	CI	Crit	Alg
d2	**100**	MDL	All	**100**	MDL	All	**100**	MDL	All
Mixed Image 2	**100**	MDL	All	**100**	MDL	All	**100**	MDL	All
Spiral	**100**	LRT	SL,AL,CL	**100**	LRT	SL,AL,CL	**100**	LRT	SL,AL,CL
Circs	**100**	All	All	**100**	All	All	**100**	All	All
Breast-cancer	95.31	MDL	CL	95.17	LRT	AL,CL	64.86	All	SL
Crabs	94.00	All	AL	74.50	All	SL	74.50	All	SL
House-votes	**87.07**	All	All	**87.07**	All	All	**87.07**	All	All
Iris	**96.67**	All	All	**96.67**	All	All	**96.67**	All	All
Diff-300	**76.33**	All	WL	**76.33**	All	WL	33.67	All	SL
Same-30	47.47	All	CL,WL	45.12	All	AL	(*)	All	AL,CL,WL

(*) Dunn's index is the same for all the partitions, but the CI is different across partitions is comparable to the best partition, and the one chosen by Dunn's index is clearly worst. G-DID index only fails the selection of the best partition for Crabs and Same-300 datasets. Moreover, Dunn's index, for Same-300 dataset, does not have CI value, because the index has the same value for all the consensus partitions obtained, but the CI value was different.

Table 10.3 presents results for GMDID when the true number of clusters is known, and we see that consensus clustering improves those results in some datasets, specially Breast-cancer and Crabs dataset. Table 10.3 also presents results for hierarchical clustering algorithms using Euclidean distance, while Table 10.7 presents results for the same algorithms but using the co-association matrix defined in Sect. 10.4, instead of the Euclidean distance. Comparing those results, we notice that hierarchical clustering algorithms using the Euclidean distance have lower results than the ones using the co-association matrix, except for Breast-cancer and House-votes datasets. However, in these two datasets, the difference is not significant: for Breast-cancer, WL has 96.63 % and consensus clustering obtained with CL has 95.31 %, and for House-votes, AL has 88.36 % and consensus clustering obtained with any clustering algorithm has 87.07 %.

Assuming that the true number of clusters is unknown and the lifetime criterion is used to obtain the number of clusters, from Table 10.8, we see that the best solution is chosen in 6 out of 10 datasets using G-DID index and 4 out of 10 datasets using Dunn's index. However, for Breast-cancer dataset, G-DID selection is worsen compared to Table 10.7. Breast-cancer and Same-300 datasets does not have CI values using Dunn's index for the same reason mentioned above.

Comparing the results given in Table 10.5 with the ones in Table 10.8, we note that consensus clustering has better results than GMDID or even than KMDID (see Table 10.4). For synthetic datasets, the results are equally good; for real-world datasets the consensus clustering is better in 4 out of 6 datasets, GMDID is better in 2 out of 6 datasets and WL is better in 1 (Breast-cancer) out of 6 datasets. Also, consensus clustering is better than traditional hierarchical algorithms (algorithms using Euclidean distance) in all datasets, except the Breast-cancer dataset, with 96.63 % for WL and 95.31 % for consensus clustering obtained with CL.

Now, if the true number of clusters is unknown and G-DID criterion is used, from Table 10.9, we note that G-DID chooses the best solutions in 6 out of 10 datasets and Dunn's index 5 out of 10 datasets. Overall, using G-DID to select the number of clusters have lowered the best results in 5 out 10 datasets compared to the lifetime criterion; however, for Diff-300 dataset, the results have improved with G-DID index. Also, note the robustness in the merging criterion when the G-DID index is used to select the number of clusters, i.e., the best solution is achieved, for all datasets, with both merging criteria. Using the lifetime criterion or even considering the true number of clusters, that only happens for some datasets.

Again, comparing the results given in Table 10.9 with the ones without consensus, Table 10.6, GMDID is better than the consensus clustering in 4 out of 10 datasets and has equal results in 3 out of 10 datasets. In terms of hierarchical clustering algorithm with Euclidean distance, SL is better than consensus clustering in the Spiral dataset, WL in the Breast-cancer dataset, and AL in the House-votes and Iris datasets.

Table 10.8 Comparison of G-DID and Dunn's index for selecting the best combination of merging criterion and hierarchical algorithm, when the true number of clusters (Nc) is unknown and the lifetime criterion [17] is used. *Best final partition(s)* correspond to the best consistency index values (CI, in %), across all combinations of merging criterion and clustering algorithm; column 3 (Crit) and 4 (Alg) indicate the merging criterion and algorithms leading to best CI. *Selection using G-DID and Dunn's index* present the CI values, with corresponding merging criterion and clustering algorithm, when those indexes are used to selected the best final partition (columns 5 to 7 and 8 to 10, respectively). When the CI of the selected partitions by each index (G-DID and Dunn's index) are matched with the best CI, the values are shown in bold

Lifetime criterion									
	Best final partition(s)			Selection using G-DID			Selection using Dunn's index		
Datasets	CI	Crit	Alg	CI	Crit	Alg	CI	Crit	Alg
d2	**100**	MDL	CL,WL	**100**	MDL	CL,WL	37.50	LRT	AL
Mixed Image 2	**100**	MDL	AL,CL	**100**	MDL	AL,CL	86.47	MDL	SL
Spiral	**100**	LRT	SL	**100**	LRT	SL	**100**	LRT	SL
Circs	**100**	All	AL,CL,WL	**100**	All	AL,CL,WL	**100**	All	AL,CL,WL
Breast-cancer	95.31	MDL	CL	68.67	MDL	SL	(*)	All	All
Crabs	94.00	All	AL	73.50	All	SL	73.50	All	SL
House-votes	**87.07**	All	All	**87.07**	All	All	**87.07**	All	All
Iris	**96.67**	All	All	**96.67**	All	All	**96.67**	All	All
Diff-300	58.67	All	CL	33.67	All	SL	33.67	All	SL
Same-30	51.52	All	CL	45.12	All	WL	(*)	All	All

(*) Dunn's index is the same for all the partitions, but the CI is different across partitions

Table 10.9 Comparison of G-DID and Dunn's index for selecting the best combination of merging criterion and hierarchical algorithm, when the true number of clusters (Nc) is unknown and the G-DID criterion presented in Sect. 10.5 is used. *Best final partition(s)* correspond to the best consistency index values (CI, in %), across all combinations of merging criterion and clustering algorithm; column 3 (Crit) and 4 (Alg) indicate the merging criterion and algorithms leading to best CI. *Selection using G-DID and Dunn's index* present the CI values, with corresponding merging criterion and clustering algorithm, when those indexes are used to selected the best final partition (columns 5 to 7 and 8 to 10, respectively). When the CI of the selected partitions by each index (G-DID and Dunn's index) are matched with the best CI, the values are shown in bold

G-DID criterion									
	Best final partition(s)			Selection using G-DID			Selection using Dunn's index		
Datasets	CI	Crit	Alg	CI	Crit	Alg	CI	Crit	Alg
d2	**100**	All	All	**100**	All	All	**100**	All	All
Mixed Image 2	**98.64**	All	SL,AL	95.67	All	WL	**98.64**	All	SL,AL
Spiral	**87.50**	All	SL	**87.50**	All	SL	61.5	All	CL
Circs	**100**	All	All	**100**	All	All	**100**	All	All
Breast-cancer	**68.67**	All	All	**68.67**	All	SL	**68.67**	All	All
Crabs	94.00	All	AL	74.50	All	SL	74.50	All	SL
House-votes	87.07	All	WL	81.90	All	SL	86.21	All	AL,CL
Iris	**66.67**	All	All	**66.67**	All	All	**66.67**	All	All
Diff-300	**76.33**	All	WL	**76.33**	All	WL	33.67	All	SL
Same-30	45.12	All	SL	38.72	All	CL	38.72	All	CL

Furthermore, the consensus clustering is a very robust methodology since in most of the datasets we have more than one criterion and/or clustering algorithm producing the same result. For example, Iris dataset has 96.67 % with both merging criteria and all hierarchical clustering methods used to extract the final partition, when the true number of clusters is known and using the lifetime criteria. Also, using G-DID criterion to identify the number of clusters of the final partition, all the hierarchical clustering methods, gave the same result (66.67 %).

Figures 10.5 and 10.6 shows the visualization for the Iris dataset, using the multidimensional scaling [36] to reduce the original space (4-dimensional space) to 2-dimensional space. The multidimensional scaling was applied using the co-association matrix of Fig. 10.5a as similarity measure, and to the Euclidean distance (converted to similarity and shown in Fig. 10.6a).

First of all, a visual inspection to the co-association matrix and to the Euclidean matrix, we notice that the clusters are more distinguishable in the co-association than in the Euclidean matrix, i.e., the co-association matrix presents a more visible separability between clusters. Moreover, using the co-association matrix, multidimensional scaling puts patterns in the same cluster closer, almost cluttered in a single point. Visually it seems that we have three distinct clusters, and those are the ones found by any hierarchical clustering algorithm applied over the co-association matrix (see Fig. 10.5c for an example). Those three distinct clusters

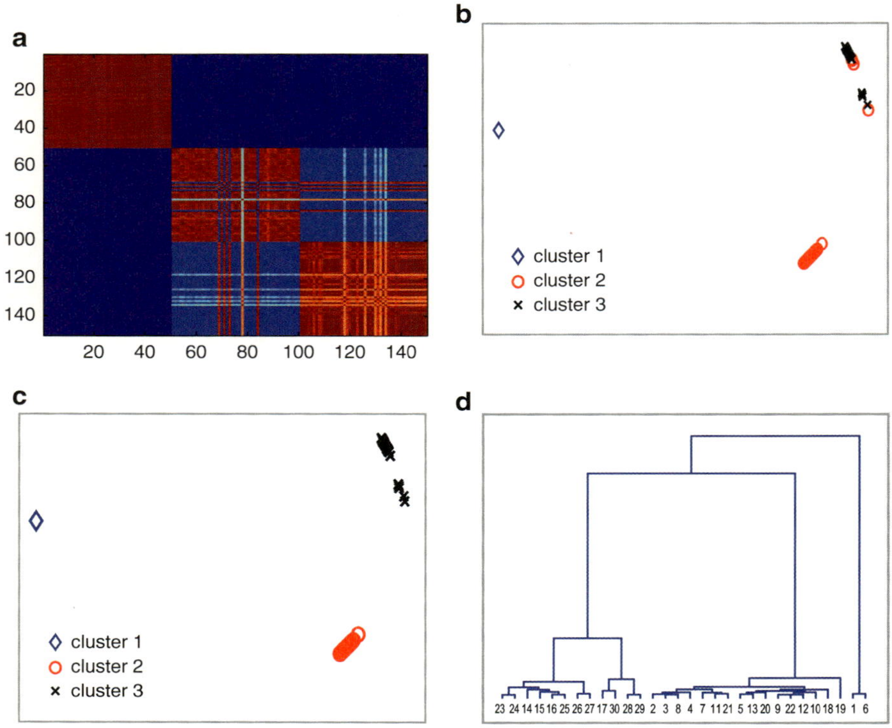

Fig. 10.5 Visualization in a two-dimensional space using multidimensional scaling of the Iris dataset, using the co-association matrix in (**a**) as similarity between patterns. (**b**) and (**c**) are visualizations colored according to the ground truth information and labeling obtained from average-link (AL), respectively. (**d**) presents the dendrogram of the average-link applied to the co-association matrix

are confirmed by the dendrogram in Fig. 10.5d and with a consistency index of 96.67 % (see Table 10.8). While, using the Euclidean distance, multidimensional scaling shows the Gaussian property of each cluster. However, the Euclidean matrix is not so "clean" as the co-association matrix. Visually we can only distinguish two clusters in the visualization, and the dendrogram (Fig. 10.6d) has the highest lifetime for two clusters, with a consistency index of 66.67 % (see Table 10.5).

10.7 Conclusions

In this chapter we presented a statistical model of a high-order dissimilarity measure, called dissimilarity increments. The distribution of the dissimilarity increments (DID) was recently derived under the hypothesis of local Gaussian generative models for the data in $\mathbb{R}^l$, and has been proven that 2-DID is a good approximation

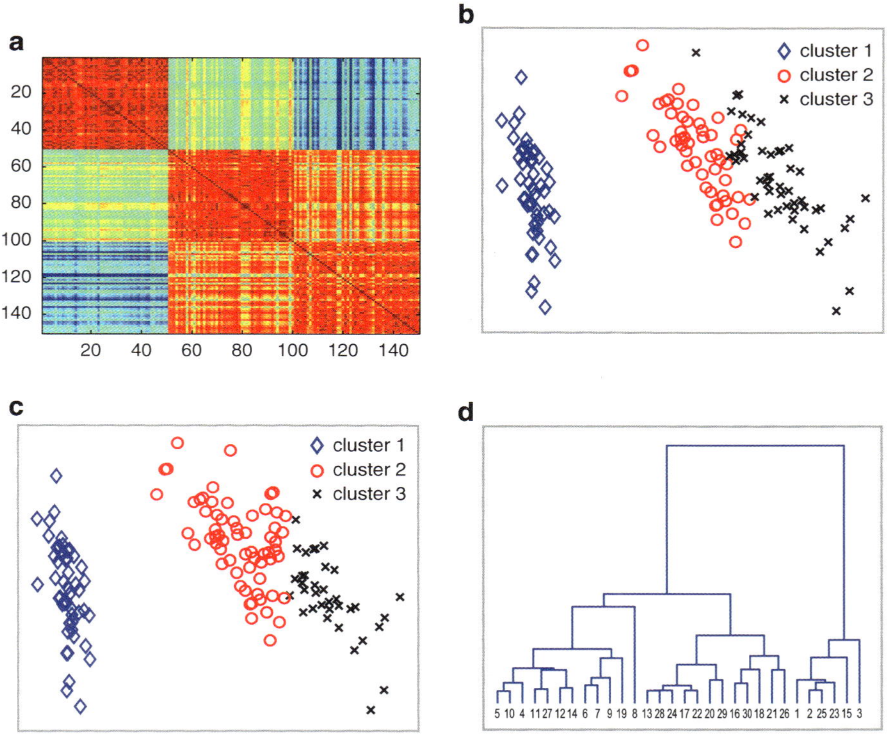

Fig. 10.6 Visualization in a two-dimensional space using multidimensional scaling of the Iris dataset, using the Euclidean distance matrix (which was converted to similarity) in (**a**) as similarity between patterns. (**b**) and (**c**) are visualizations colored according to the ground truth information and labeling obtained from average-link (AL), respectively. (**d**) presents the dendrogram of the traditional average-link, i.e., average-link applied to the Euclidean distance matrix

to DID for high-dimensional data. This means that we only need to estimate 2-DID for clusters, instead of l-DID, which is much simpler.

A partitional clustering algorithm was presented based on DID, with the starting point being a partition given by a Gaussian mixture decomposition or k-means. The decision of merging components is based on a likelihood-ratio test or a minimum description length between the statistical model for the combined components and the statistical model for the separate components. DID-based algorithm builds upon an initial data partition; if the algorithm used to build the initial partition produces single clusters which in reality should be two separate clusters, the presented algorithm does not undo this. Still, DID-based algorithm improves the results, and in most of the datasets it is better than other clustering algorithms, when the true number of clusters is known. If the true number of clusters is unknown a priori, DID-based algorithm can detect the true number, in most cases.

As mentioned, DID-based algorithm is dependent of an initial data partition, and different initializations produce different data partitions. Therefore, an unifying approach consisting of a consensus function based on a combination strategy of all these different initializations was presented. The final partition needs to be extracted using a clustering algorithm and choosing the best strategy automatically is not an easy task. So, we present a validation criterion, consisting in the estimation of graph probabilities for each cluster based on the DID, to select the best final partition. This unifying approach improves results in both synthetic and real-world datasets compared to traditional clustering algorithms, and it improves the results in some datasets of the DID-based algorithm presented here. Moreover, the validation criterion has a good performance in selecting a partition.

Appendix

Assume that $X = \{\mathbf{x}_1, \ldots, \mathbf{x}_N\}$ is a l-dimensional set of patterns, and $\mathbf{x}_i \sim \mathcal{N}(\boldsymbol{\mu}, \Sigma)$. Also, with no loss of generality, assume that $\boldsymbol{\mu} = 0$ and Σ is diagonal; this only involves translation and rotation of the data, which does not affect Euclidean distances. If $\mathbf{x}$ denotes a sample from this Gaussian, we transform $\mathbf{x}$ into $\mathbf{x}^*$ through a process known as "whitening" or "sphering", such that its i-th entry is given by $x_i^* \equiv x_i / \Sigma_{ii}$; x_i^* thus follows the standard normal distribution, $\mathcal{N}(0, 1)$. Now, it is known that the difference between samples, such as $x_i^* - y_i^*$, from two univariate standard normal distributions follows a normal distribution with covariance 2. Therefore,

$$\frac{x_i^* - y_i^*}{\sqrt{2}} \sim \mathcal{N}(0, 1). \tag{10.30}$$

It can be shown that the squared Euclidean distance,

$$(d^*(\mathbf{x}^*/\sqrt{2}, \mathbf{y}^*/\sqrt{2}))^2 = \sum_{i=1}^{l} \left(\frac{x_i^* - y_i^*}{\sqrt{2}} \right)^2 \sim \chi^2(l), \tag{10.31}$$

i.e., follows a chi-square distribution with l degrees of freedom [28]. Thus, the probability density function for $(d^*)^2 \equiv (d^*(\cdot, \cdot))^2$ is given by:

$$p(x) = \frac{2^{-l/2}}{\Gamma(l/2)} x^{l/2-1} \exp\left\{-\frac{x}{2}\right\}, \; x \in [0, +\infty[, \tag{10.32}$$

where $\Gamma(\cdot)$ denotes the gamma function.

After the sphering, the transformed data has circular symmetry in $\mathbb{R}^l$. We define angular coordinates in a $(l-1)$-sphere, with $\theta_i \in [0, \pi[, i = 1, \ldots, l-2$ and $\theta_{l-1} \in [0, 2\pi[$. Define $\mathbf{x} - \mathbf{y} \equiv (b_1, b_2, \ldots, b_l)$, where b_i can be expressed in terms of polar coordinates as

$$b_1 = \sqrt{2\Sigma_{11}}\, d^* \cos\theta_1$$

$$b_i = \sqrt{2\Sigma_{ii}}\, d^* \left[\prod_{k=1}^{i-1} \sin\theta_k\right] \cos\theta_i, i = 2, \ldots, l-1$$

$$b_l = \sqrt{2\Sigma_{ll}}\, d^* \left[\prod_{k=1}^{l-1} \sin\theta_k\right].$$

The squared Euclidean distance in the original space, d^2, is

$$\begin{aligned} d^2 &= \|\mathbf{x}-\mathbf{y}\|^2 \\ &= 2\left[\Sigma_{11}\cos^2\theta_1 + \sum_{i=2}^{l-1} \Sigma_{ii} \left(\prod_{k=1}^{i-1} \sin^2\theta_k\right)\cos^2\theta_i + \Sigma_{ll}\left(\prod_{k=1}^{l-1}\sin^2\theta_k\right)\right]\left(d^*\right)^2 \\ &\equiv 2A(\Theta)\left(d^*\right)^2, \end{aligned} \tag{10.33}$$

where $A(\Theta)$, with $\Theta = (\theta_1, \theta_2, \ldots, \theta_{l-1})$, is called the *expansion factor*. Naturally, this expansion factor depends on the angle vector Θ, and it is hard to properly deal with this dependence. Thus, we will use the approximation that the expansion factor is constant and equal to the expected value of the true expansion factor over all angles Θ. This expected value is given by

$$\mathbb{E}[A(\Theta)] = \int_{S^{l-1}} \left[\prod_{i=1}^{l-2} p_{\theta_i}(\theta_i)\right] p_{\theta_{l-1}}(\theta_{l-1}) A(\Theta)\, \mathrm{d}_{S^{l-1}} V \tag{10.34}$$

where the volume element is $\mathrm{d}_{S^{l-1}} V = \left(\prod_{i=1}^{l-2} \sin^{l-(i+1)}\theta_i\right) \mathrm{d}\theta_1 \ldots \mathrm{d}\theta_{l-1}$. Since we sphered the data, we can assume for simplicity that $\theta_i \sim Unif([0,\pi[)$ for $i = 1, \ldots, l-2$ and that $\theta_{l-1} \sim Unif([0, 2\pi[)$; then $p_{\theta_i}(\theta_i) = {}^1/\pi$ and $p_{\theta_{l-1}}(\theta_{l-1}) = {}^1/2\pi$. Thus, after some computations (see [2] for details), the expected value of the true expansion factor over all angles Θ is given by

$$\mathbb{E}[A(\Theta)] = \frac{\pi^{-l/2+1}}{2\Gamma(1+l/2)} \operatorname{tr}(\Sigma). \tag{10.35}$$

Using Eq. (10.35), the transformation Eq. (10.33) from the normalized space to the original space is given by

$$(d)^2 = \frac{\pi^{-l/2+1}}{\Gamma(1+l/2)} \operatorname{tr}(\Sigma) \left(d^*\right)^2. \tag{10.36}$$

Assume that $Y = aX$, a constant, with $p_X(x)$ the probability density function of X, so $p_Y(y) = p_X(y/a)\mathrm{d}x/\mathrm{d}y = p_X(y/a) \cdot {}^1/a$. From Eq. (10.32), we obtain the probability density function for the squared Euclidean distance in the original space, $(d)^2$. Again, assuming that $Y^2 = X$ and $p_X(x)$ the probability density function of X, we have $p_Y(y) = p_X(y^2)\mathrm{d}x/\mathrm{d}y = p_X(y^2)\cdot 2y$. Therefore, we obtain the probability density function of the Euclidean distance, $d \equiv d(\mathbf{x},\mathbf{y})$, as

$$p(y) = 2G_l(\mathrm{tr}(\Sigma))y^{l-1}\exp\left\{-C_l(\mathrm{tr}(\Sigma))y^2\right\},\ y \in [0,+\infty[, \tag{10.37}$$

where we define

$$G_l(\mathrm{tr}(\Sigma)) \equiv l^{l/2}\Gamma(l/2)^{l/2-1}2^{-l}\,\mathrm{tr}(\Sigma)^{-l/2}\pi^{l/2(l/2-1)} \tag{10.38}$$

and

$$C_l(\mathrm{tr}(\Sigma)) \equiv l\Gamma(l/2)(4\,\mathrm{tr}(\Sigma))^{-1}\pi^{l/2-1}. \tag{10.39}$$

Now, the dissimilarity increment is defined as the absolute value of the difference of two Euclidean distances. Define $d \equiv d(\mathbf{x},\mathbf{y})$ and $d' \equiv d(\mathbf{y},\mathbf{z})$, which follow the distribution in Eq. (10.37). The probability density function of $d_{\mathrm{inc}} = d - d'$ is given by the convolution

$$p_{d_{\mathrm{inc}}}(w;\mathrm{tr}(\Sigma)) = \int_{-\infty}^{\infty}(2G_l(\mathrm{tr}(\Sigma)))^2\,(t(t+w))^{l-1}\exp\left\{-C_l(\mathrm{tr}(\Sigma))\left(t^2+(t+w)^2\right)\right\} \mathbf{1}_{\{t\geq 0\}}\mathbf{1}_{\{t+w\geq 0\}}\mathrm{d}t. \tag{10.40}$$

Therefore, after some calculations (see [2] for details), the probability density function for the dissimilarity increments is given by

$$p_{d_{\mathrm{inc}}}(w;\mathrm{tr}(\Sigma)) = \frac{G_l(\mathrm{tr}(\Sigma))^2}{2^{l-5/2}C_l(\mathrm{tr}(\Sigma))^{l-1/2}}\exp\left\{-\frac{C_l(\mathrm{tr}(\Sigma))}{2}w^2\right\} \left[\sum_{k=0}^{l-1}\sum_{i=0}^{2l-2-k}(-1)^i w^{k+i}\binom{l-1}{k}\binom{2l-2-k}{i}2^{k/2-i/2}C_l(\mathrm{tr}(\Sigma))^{k/2+i/2} \Gamma\left(\frac{2l-1-k-i}{2},\frac{C_l(\mathrm{tr}(\Sigma))}{2}w^2\right)\right], \tag{10.41}$$

where $\Gamma(a,x)$ is the incomplete gamma function [44].

Acknowledgements This work was supported by the Portuguese Foundation for Science and Technology grant PTDC/EEI-SII/2312/2012.

References

1. Agrawal R, Gehrke J, Gunopulos D, Raghavan P (1998) Automatic subspace clustering of high dimensional data for data mining applications. In: Haas LM, Tiwary A (eds) Proceedigns ACM SIGMOD International Conference on Management of Data (SIGMOD 1998), ACM Press, Seattle, WA, USA, pp 94–105
2. Aidos H, Fred A (2012) Statistical modeling of dissimilarity increments for d-dimensional data: Application in partitional clustering. Pattern Recogn 45(9):3061–3071
3. Anderson TW (1962) On the distribution of the two-sample Cramér-von-Mises criterion. Ann Math Stat 33(3):1148–1159
4. Ayad HG, Kamel MS (2008) Cumulative voting consensus method for partitions with variable number of clusters. IEEE Trans Pattern Anal Mach Intell 30(1):160–173
5. Ball GH, Hall DJ (1965) ISODATA, a novel method of data analysis and pattern classification. Tech. rep., Stanford Research Institute
6. Benavent AP, Ruiz FE, Martínez JMS (2006) EBEM: An entropy-based EM algorithm for Gaussian mixture models. In: 18th International Conference on Pattern Recognition (ICPR 2006), IEEE Computer Society, Hong Kong, vol 2, pp 451–455
7. Castro RM, Coates MJ, Nowak RD (2004) Likelihood based hierarchical clustering. IEEE Trans Signal Process 52(8):2308–2321
8. Celebi ME, Kingravi HA (2012) Deterministic initialization of the k-means algorithm using hierarchical clustering. Int J Pattern Recogn Artif Intell 26(7) DOI: 10.1142/S0218001412500188
9. Celebi ME, Kingravi HA, Vela PA (2013) A comparative study of efficient initialization methods for the k-means clustering algorithm. Expert Syst Appl 40(1):200–210
10. Dimitriadou E, Weingessel A, Hornik K (2002) A combination scheme for fuzzy clustering. In: Pal NR, Sugeno M (eds) Advances in soft computing - proceedings international conference on fuzzy systems (AFSS 2002). Lecture Notes in Computer Science, vol 2275. Springer, Calcutta, pp 332–338
11. Duflou H, Maenhaut W (1990) Application of principal component and cluster analysis to the study of the distribution of minor and trace elements in normal human brain. Chemometr Intell Lab Syst 9:273–286
12. Ester M, Kriegel HP, Sander J, Xu X (1996) A density-based algorithm for discovering clusters in large spatial databases with noise. In: Simoudis E, Han J, Fayyad UM (eds) Proceedings of the 2nd international conference on knowledge discovery and data mining (KDD 1996). AAAI Press, Portland, Oregon, USA, pp 226–231
13. Everitt BS, Landau S, Leese M, Stahl D (2011) Cluster analysis, 5th edn. Wiley
14. Fern XZ, Brodley CE (2004) Solving cluster ensemble problems by bipartite graph partitioning. In: Brodley CE (ed) Machine learning - proceedings of the 21st international conference (ICML 2004), ACM, Banff, Alberta, Canada, ACM International Conference Proceeding Series, vol 69
15. Figueiredo MAT, Jain AK (2002) Unsupervised learning of finite mixture models. IEEE Trans Pattern Anal Mach Intell 24(3):381–396
16. Fred A (2001) Finding consistent clusters in data partitions. In: Kittler J, Roli F (eds) Multiple classifier systems - proceedings 2nd international workshop (MCS 2001). Lecture Notes in Computer Science, vol 2096. Springer, Cambridge, pp 309–318
17. Fred A, Jain A (2005) Combining multiple clusterings using evidence accumulation. IEEE Trans Pattern Anal Mach Intell 27(6):835–850
18. Fred A, Jain A (2008) Cluster validation using a probabilistic attributed graph. In: 19th international conference on pattern recognition (ICPR 2008), IEEE, Tampa, Florida, USA, pp 1–4
19. Fred A, Leitão J (2003) A new cluster isolation criterion based on dissimilarity increments. IEEE Trans Pattern Anal Mach Intell 25(8):944–958

20. Fred A, Lourenço A, Aidos H, Bulò SR, Rebagliati N, Figueiredo M, Pelillo M (2013) Similarity-based pattern analysis and recognition, chap Learning similarities from examples under the evidence accumulation clustering paradigm. Springer, New York, pp 85–117
21. Gowda KC, Ravi TV (1995) Divisive clustering of symbolic objects using the concepts of both similarity and dissimilarity. Pattern Recogn 28(8):1277–1282
22. Guha S, Rastogi R, Shim K (1998) CURE: An efficient clustering algorithm for large datasets. In: Haas LM, Tiwary A (eds) Proceedins of the ACM SIGMOD international conference of management of data (SIGMOD 1998). ACM Press, Seattle, Washington, USA, pp 73–84
23. Guha S, Rastogi R, Shim K (1999) ROCK: A robust clustering algorithm for categorical attributes. In: Kitsuregawa M, Papazoglou MP, Pu C (eds) Proceedings of the 15th international conference on data engineering (ICDE 1999). IEEE Computer Society, Sydney, Australia, pp 512–521
24. Hartigan JA, Wong MA (1979) Algorithm AS 136: A k-means clustering algorithm. J Roy Stat Soc C (Appl Stat) 28(1):100–108
25. Jain AK (2010) Data clustering: 50 years beyond k-means. Pattern Recogn Lett 31:651–666
26. Jain AK, Duin RPW, Mao J (2000) Statistical pattern recognition: A review. IEEE Trans Pattern Anal Mach Intell 22(1):4–37
27. Jain AK, Murty MN, Flynn PJ (1999) Data clustering: a review. ACM Comput Surv 31(3):264–323
28. Johnson NL, Kotz S, Balakrishnan N (1994) Continuous univariate distributions. Applied probability and statistics, vol 1, 2nd edn. Wiley, New York
29. Johnson SC (1967) Hierarchical clustering schemes. Psychometrika 32(3):241–254
30. Kamvar S, Klein D, Manning C (2002) Interpreting and extending classical agglomerative clustering algorithms using a model-based approach. In: Sammut C, Hoffmann AG (eds) Machine learning - proceedings of the 19th international conference (ICML 2002). Morgan Kaufmann, Sydney, Australia, pp 283–290
31. Kannan R, Vempala S, Vetta A (2000) On clusterings – good, bad and spectral. In: 41st annual symposium on foundations of computer science (FOCS 2000). IEEE Computer Science, Redondo Beach, CA, USA, pp 367–377
32. Karypis G, Han EH, Kumar V (1999) Chameleon: Hierarchical clustering using dynamic modeling. Computer 32(8):68–75
33. Kaufman L, Rousseeuw PJ (2005) Finding groups in data: an introduction to cluster analysis. Wiley
34. Kuncheva LI, Hadjitodorov ST (2004) Using diversity in cluster ensembles. In: Proceedings of the IEEE international conference on systems, man & cybernetics, vol 2. IEEE, The Hague, Netherlands, pp 1214–1219
35. Lance GN, Williams WT (1968) Note on a new information-statistic classificatory program. Comput J 11:195
36. Lee JA, Verleysen M (2007) Nonlinear dimensionality reduction. Information science and statistics. Springer, New York
37. Lehmann EL, Romano JP (2005) Testing statistical hypotheses, 3rd edn. Springer texts in statistics. Springer, New York
38. Lin J (1991) Divergence measures based on the Shannon entropy. IEEE Trans Inform Theory 37(1):145–151
39. von Luxburg U (2007) A tutorial on spectral clustering. Stat Comput 17(4):395–416
40. MacKay DJ (2006) Information theory, inference, and learning algorithms, 5th edn. Cambridge University Press, Cambridge
41. MacNaughton-Smith P, Williams WT, Dale MB, Mockett LG (1964) Dissimilarity analysis: a new technique of hierarchical sub-division. Nature 202:1034–1035
42. Meila M (2003) Comparing clusterings by the variation of information. In: Schölkopf B, Warmuth MK (eds) Computational learning theory and kernel machines - proceedings 16th annual conference on computational learning theory and 7th kernel workshop (COLT 2003). Lecture Notes in Computer Science, vol 2777. Springer, Washington, USA, pp 173–187

43. Milligan GW, Soon SC, Sokol LM (1983) The effect of cluster size, dimensionality, and the number of clusters on recovery of true cluster structure. IEEE Trans Pattern Anal Mach Intell PAMI-5(1):40–47
44. Olver FWJ, Lozier DW, Boisvert RF, Clark CW (eds) (2010) NIST handbook of mathematical functions. Cambridge University Press, Cambridge
45. Sander J, Ester M, Kriegel HP, Xu X (1998) Density-based clustering in spatial databases: the algorithm GDBSCAN and its applications. Data Min Knowl Discov 2:169–194
46. Strehl A, Ghosh J (2002) Cluster ensembles - a knowledge reuse framework for combining multiple partitions. J Mach Learn Res 3:583–617
47. Su T, Dy JG (2007) In search of deterministic methods for initializing k-means and Gaussian mixture clustering. Intell Data Anal 11(4):319–338
48. Theodoridis S, Koutroumbas K (2009) Pattern recognition, 4th edn. Elsevier Academic, Amsterdam
49. Topchy A, Jain A, Punch W (2003) Combining multiple weak clusterings. In: Proceedings of the 3rd IEEE international conference on data mining (ICDM 2003). IEEE Computer Society, Melbourne, Florida, USA, pp 331–338
50. Topchy A, Jain A, Punch W (2005) Clustering ensemble: Models of consensus and weak partitions. IEEE Trans Pattern Anal Mach Intell 27(12):1866–1881
51. Ueda N, Nakano R, Ghahramani Z, Hinton GE (2000) SMEM algorithm for mixture models. Neural Comput 12(9):2109–2128
52. Vaithyanathan S, Dom B (2000) Model-based hierarchical clustering. In: Boutilier C, Goldszmidt M (eds) Proceedings of the 16th conference in uncertainty in artificial intelligence (UAI 2000). Morgan Kaufmann, Stanford, California, USA, pp 599–608
53. Wang H, Shan H, Banerjee A (2009) Bayesian cluster ensembles. In: Proceedings of the SIAM international conference on data mining (SDM 2009). SIAM, Sparks, Nevada, USA, pp 211–222
54. Wang P, Domeniconi C, Laskey KB (2010) Nonparametric Bayesian clustering ensembles. In: Balcázar JL, Bonchi F, Gionis A, Sebag M (eds) Machine learning and knowledge discovery in databases - proceedings european conference: Part III (ECML PKDD 2010). Lecture notes in computer science, vol 6323. Springer, Barcelona, Spain, pp 435–450
55. Williams WT, Lambert JM (1959) Multivariate methods in plant ecology: 1. association-analysis in plant communities. J Ecol 47(1):83–101
56. Xu R, Wunsch II D (2005) Survey of clustering algorithms. IEEE Trans Neural Network 16(3):645–678

Chapter 11
Hubness-Based Clustering of High-Dimensional Data

Nenad Tomašev, Miloš Radovanović, Dunja Mladenić, and Mirjana Ivanović

Abstract Hubness has recently been established as a significant property of k-nearest neighbor (k-NN) graphs obtained from high-dimensional data using a distance measure, with traits and effects relevant to the cluster structure of data, as well as clustering algorithms. The hubness property is manifested with increasing (intrinsic) data dimensionality. The distribution of data point in-degrees, i.e. the number of times points appear among the k nearest neighbors of other points in the data, becomes highly skewed. This results in hub points that can have in-degrees multiple orders of magnitude higher than expected. In this chapter we review and refine existing work which explains the mechanisms of the phenomenon, establishes the location of hub points near central regions of clusters in the data, and shows how hubness can negatively affect existing clustering algorithms by virtue of hub points lowering between-cluster distance. Next, we review the newly proposed partitional clustering algorithms, based on K-means, which take advantage of hubness by employing hubs in the process of cluster prototype selection. These "soft" K-means extensions avoid premature convergence to suboptimal stable cluster configurations and are able to reach the global optima more often. The algorithms offer significant improvements over the K-means baseline in scenarios involving high-dimensional and noisy data. The improvements stem from a better placement of hub points into clusters, which helps in increasing the between-cluster distance. Finally, we introduce novel clustering algorithms as "kernelized" versions of the most successful hubness-based methods discussed above, that are able to more effectively handle arbitrarily-shaped clusters.

N. Tomašev (✉) • D. Mladenić
Artificial Intelligence Laboratory and Jožef Stefan International, Jožef Stefan Institute, Postgraduate School, Jamova 39, 1000 Ljubljana, Slovenia
e-mail: nenad.tomasev@gmail.com; dunja.mladenic@ijs.si

M. Radovanović • M. Ivanović
Faculty of Sciences, Department of Mathematics and Informatics, University of Novi Sad
Trg Dositeja Obradovića 4, 21000 Novi Sad, Serbia
e-mail: radacha@dmi.uns.ac.rs; mira@dmi.uns.ac.rs

M.E. Celebi (ed.), *Partitional Clustering Algorithms*,
DOI 10.1007/978-3-319-09259-1_11

Keywords Clustering • Hubness • k-nearest neighbor • High-dimensional data • Curse of dimensionality • Kernel methods

11.1 Introduction

High-dimensional data arises naturally in many domains and is known to pose substantial difficulties for traditional clustering approaches, both in terms of efficiency and effectiveness [63]. Sparsity of high-dimensional feature spaces significantly reduces the overall quality of density estimates around data points [61], which are often used for determining the cluster structure. Additionally, many standard distance measures tend to concentrate with increasing dimensionality [2, 26]. Consequently, it becomes more difficult to distinguish between close and distant points and to properly detect cluster boundaries.

Due to the outlined difficulties with applying density-based and distance-based clustering approaches in the high-dimensional case, a different class of techniques is usually used for high-dimensional data clustering. Typically the idea is to observe clusters on a lower dimensional manifold and to automatically detect a proper projection of the original data [39]. Subspace techniques are frequently used for clustering high-dimensional data streams.

We propose to employ a different approach, by exploiting the recently described phenomenon of *hubness* in k-nearest neighbor graphs constructed from intrinsically high-dimensional data using a distance measure.

Before we proceed with describing the main ideas of hubness-based data clustering, we will briefly address other types of high-dimensional clustering techniques in Sect. 11.2 and also give a detailed introduction to the phenomenon of hubness in Sect. 11.3. Section 11.4 examines how hubness affects clustering in general and Sect. 11.5 deals with hubness under kernel mappings some of which are used for kernel-based extensions of the clustering methods that are discussed in this chapter. Finally, Sect. 11.6 describes hubness-based clustering approaches and Sect. 11.7 evaluates hubness-based techniques by comparing them to the baselines on a series of high-dimensional clustering tasks. We conclude the chapter by discussing the application domains for hubness-based methods, related work, as well as possible future development directions in Sects. 11.8 and 11.9.

11.2 High-Dimensional Data Clustering

In many practical applications, for instance document clustering and topic detection [37, 38, 40], meaningful clusters can be found by projecting data onto certain lower-dimensional feature subspaces and manifolds. Sometimes, there might exist multiple subspaces holding different cluster structures that correspond to different contexts. In general, there are two types of subspace clustering approaches—those that try to find an actual formal feature subspace, and those that simulate the process by automatically assigning weights to features in order to increase the influence of certain features on the proximity measure and decrease the influence of others [7,15,44]. One idea is to increase the weights of the low-variance dimensions in the distance function [9]. Such weights help with detecting groups of points that have a low variance in one or more dimensions. Subspace clustering has recently been successfully applied to the problem of clustering high-dimensional data streams [1,49].

It is possible to rely on standard clustering methods to perform the partitioning in the selected lower-dimensional subspace [38]. However, this might be suboptimal in those cases when potential lower-dimensional subspaces are not really low-dimensional and even the simplified clustering task can still be non-trivial. Many subspace clustering methods are density-based [3] or an extension of K-means, but it is possible to use decision trees and other partitioning techniques as well [42].

Subspace clustering is not the only approach to clustering high-dimensional data. Expectation maximization (EM) can be modified to learn a mixture model in many-dimensional spaces, despite the apparent limitations [20]. Instead of using the covariance matrix, the probabilities in the Gaussian mixture model can be estimated from the eigenvectors and eigenvalues of the PCA decomposition of the data.

Another popular approach to high-dimensional clustering is based on shared-neighbor similarity measures, that reduce the negative effects of the dimensionality curse [22, 34, 47, 50, 82, 86]. Cardinality of the intersection between k-nearest neighbor sets can be used for measuring pairwise similarity between points and these secondary distances have been shown to be less susceptible to concentration and usually exhibit more favourable properties than the original metrics [33]. Metric learning can therefore help with improving clustering performance. An additional advantage of using shared-neighbor approaches and relevant set correlations in particular for data clustering is that they are independent of the data representation and the primary similarity measure. They merely assume an existence of an oracle that accepts a query and returns the most relevant data points for that query. The relevant set correlation approach was shown to be very promising for high-dimensional data clustering [32] and it has also been adapted for partitional clustering [79].

In certain cases when dimensionality reduction offers unstable clustering performance, it is advisable to revert to using clustering ensembles that can help with stabilizing the process and can improve the overall clustering quality [23,24].

11.3 The Hubness Phenomenon

Hubness is an aspect of the curse of dimensionality pertaining to nearest neighbors which has only recently come to attention, unlike the much discussed distance concentration phenomenon. Let $D \subset \mathbb{R}^d$ be a set of data points and let $N_k(x)$ denote the number of *k-occurrences* of point $x \in D$, i.e., the number of times x

occurs in k-nearest-neighbor lists of other points from D. As the dimensionality of data increases, the distribution of k-occurrences becomes considerably skewed [54]. As a consequence, some data points, which we will refer to as *hubs*, are included in many more k-nearest-neighbor lists than other points. In the rest of the text we will refer to the number of k-occurrences of point $x \in D$ as its *hubness score*. It has been shown that hubness, as a phenomenon, appears in high-dimensional data as an inherent property of high dimensionality, and is not an artefact of finite samples nor a peculiarity of some specific data sets [54]. Naturally, the exact degree of hubness may still vary and is not uniquely determined by dimensionality.

11.3.1 The Emergence of Hubs

The concentration of distances enables one to view unimodal high-dimensional data as lying approximately on a hypersphere centered at the data distribution mean [54]. However, the variance of distances to the mean remains non-negligible for any finite number of dimensions [26, 27], which implies that some of the points still end up being closer to the data mean than other points. It is well known that points closer to the mean tend to be closer (on average) to all other points, for any observed dimensionality. In high-dimensional data, this tendency is amplified [54]. Such points will have a higher probability of being included in k-nearest neighbor lists of other points in the data set, which increases their influence, and they emerge as neighbor-hubs.

To illustrate the above discussion, Fig. 11.1 depicts, for i.i.d. normally distributed data (for simplicity, we consider single-cluster data), the distribution of Euclidean distances of all points to the true data mean (the origin) for several d values. By definition, the distribution of distances is actually the Chi distribution with d degrees of freedom [54]. In this setting, distance concentration refers to the fact that the standard deviation of distance distributions is asymptotically constant with respect

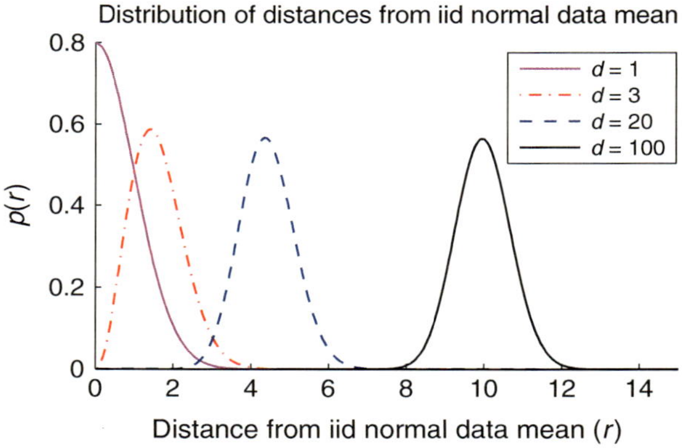

Fig. 11.1 Probability density function of observing a point at distance r from the mean of a multivariate d-dimensional normal distribution, for $d = 1, 3, 20, 100$ [54]

to increasing d, while the means of the distance distributions asymptotically behave like $\sqrt{d}$ [26, 54], making the their ratio tend to 0 as $d \to \infty$, thus enabling the aforementioned hypersphere view of high-dimensional data.

To understand why points closer to the data (cluster) mean become hubs in high dimensions, i.e., to illustrate the mechanism that enables high dimensionality to amplify the tendency of points close to the mean to be closer to other points from the cluster as dimensionality increases, let us consider the following example. We observe, within the i.i.d. normal setting, two points drawn from the data, but at specific positions with respect to the origin: point $\mathbf{b}_d$ which is at the expected distance from the origin, and point $\mathbf{a}_d$ which is two standard deviations closer. In light of the above, the distances of $\mathbf{a}_d$ and $\mathbf{b}_d$ from the origin change with increasing d, and it could be said that different $\mathbf{a}_d$-s (and $\mathbf{b}_d$-s) occupy analogous positions in the data spaces, with respect to changing d. The distances of $\mathbf{a}_d$ (and $\mathbf{b}_d$) to all other points, again following directly from the definition [54], are distributed according to *noncentral* Chi distributions with d degrees of freedom and noncentrality parameter λ equaling the distance of $\mathbf{a}_d$ ($\mathbf{b}_d$) to the origin. Figure 11.2a plots the probability density functions of these distributions for several values of d. It can be seen that, as d increases, the distance distributions for $\mathbf{a}_d$ and $\mathbf{b}_d$ move away from each other. This tendency is depicted more clearly in Fig. 11.2b which plots the difference between the means of the two distributions with respect to d. A general theorem which proves the behavior illustrated in Fig. 11.2b for two more arbitrarily selected points is presented by [54], along with a comprehensive discussion various other properties of data with respect to the hubness phenomenon.

In the preceding example, two simplifying assumptions were adopted: (1) the data is distributed in a single cluster, and (2) features are identically distributed and *independent*. It was established that hubs also exist in clustered (multimodal) data, tending to be situated in the proximity of cluster centers [54]. In addition, the degree of hubness does not depend on the embedding dimensionality, but rather on the *intrinsic* data dimensionality, which is viewed as the minimal number of variables needed to account for all pairwise distances in the data [54].

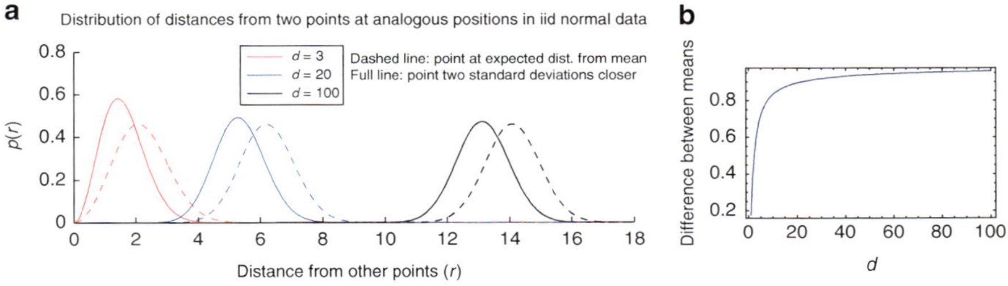

Fig. 11.2 (**a**) Distribution of distances to other points from i.i.d. normal random data for a point at the expected distance from the origin (*dashed line*), and a point two standard deviations closer (*full line*). (**b**) Difference between the means of the two distributions, with respect to increasing d [54]

Generally, the hubness phenomenon is relevant to (intrinsically) high-dimensional data regardless of the distance or similarity measure employed. Its existence was verified for Euclidean (l_2) and Manhattan (l_1) distances, l_p distances with $p > 2$, fractional distances (l_p with rational $p \in (0, 1)$), Bray-Curtis, normalized Euclidean, and Canberra distances, cosine similarity, and the dynamic time warping distance for time series [54–56]. In this paper, unless otherwise stated, we will assume the Euclidean distance. The methods we discuss in Sect. 11.4, however, depend mostly on neighborhood relations that are derived from the distance matrix, and are therefore independent of the particular choice of distance measure.

Before continuing, we should clearly define what constitutes a hub. Similarly to [54], we will say that hubs are points x having $N_k(x)$ more than two standard deviations higher than the expected value k (in other words, significantly above average). However, in the discussion that follows, we will mostly concern ourselves with one major hub in each cluster, i.e. the point with the highest hubness score.

It is possible to take a "soft" approach and introduce fuzziness into the hub concept by saying that each non-orphan point that occurs more frequently than average is a hub to a certain degree and that the degree to which a point can be considered a hub is proportional to its neighbor occurrence frequency and somehow related either to the maximal previously observed neighbor occurrence frequency or the standard deviation of neighbor occurrence frequencies. Different cut-off thresholds are also possible.

11.3.2 Relation of Hubs to Data Clusters

The question of how well high-hubness elements cluster, as well as the general impact of hubness on clustering algorithms, was first investigated by [54]. A correlation between low-hubness elements (i.e., *anti-hubs* or *orphans*) and outliers was also observed. A low hubness score indicates that a point is on average far from the rest of the points and hence probably an outlier. In high-dimensional spaces, however, low-hubness elements are expected to occur by the very nature of these spaces and data distributions. These data points will lead to an average increase of within-cluster distance. It was also shown for several clustering algorithms that hubs do not cluster well compared to the rest of the points. This is due to the fact that some hubs are actually close to points in different clusters. Hence, they lead to a decrease of between-cluster distance. This has been observed on real data sets clustered using state-of-the art prototype-based methods, and was identified as a possible area for performance improvement [54]. We will revisit this point with a more refined study in Sect. 11.4.

Several other natural questions regarding hubs and clustering were raised and answered by [78]. It was shown that hubs are not the same as cluster medoids (although the two notions can overlap for some data points). Also, regarding the relationship between cluster centroids produced by executing an iterative algorithm such as K-means++, it was observed that as iterations progress, centroids become closer and closer to data hubs, hinting at the possibility of developing an iterative approximation procedure which employs hubs as cluster prototypes [78].

Additional simulations with synthetic data by [78] explored the relationships between distance to the true cluster center (as an oracle), hubness and measured density, demonstrating that in low dimensions density pinpoints the location of the cluster center with great accuracy, while in high dimensions, density loses this power, and is no longer a good indicator of the main part of the cluster. Hubness, on the other hand, exhibited the opposite trend to density: it had weak correlation with the distance to the true cluster center in low dimensions, while in the high-dimensional setting of the correlation became much stronger, meaning that the hubness score of a point represents a very good indicator of its proximity to the cluster center. This is shown in Fig. 11.3, for high-dimensional synthetic Gaussian data. Regarding medoids, [78] have shown, based on the ratio between the average distance to the strongest hub and average distance to the medoid, that in high dimensions the hub is equally informative about the location of the cluster center as the medoid, while in low dimensions the hub and medoid are unrelated. At the same time, generally the locally strongest hub and the medoid are in neither case the same point.

The preceding observations offer justification for the idea to use hubs as cluster prototypes. As was shown, it is expected of points with high hubness scores to be closer to centers of clustered subregions of high-dimensional space than other data points, making them viable candidates for representative cluster elements. We are not limited to observing only points with the highest hubness scores, we can also take advantage of hubness information for any given point. More generally, in presence of irregularly shaped clusters, hubs are expected to be found near the centers of compact sub-clusters, which is also beneficial. In addition, hubness of points is straightforward to compute exactly, while the computation of cluster centroids and medoids must involve some iterative inexact procedure intrinsically tied to the process of cluster construction. The remaining question of how to assign individual hubs to particular clusters will be addressed in Sect. 11.6.

11.4 Effects of Hubness on Clustering

In this section we revisit and refine the study by [54, Section 7.3.1], which demonstrated the negative effects of hubness on existing clustering algorithms.

Two of the main objectives of (distance-based) clustering algorithms are to produce such grouping of points which minimizes within-cluster distances, while maximizing between-cluster distances. The hubness phenomenon has an impact on both of these objectives.

As we discussed in Sect. 11.3, anti-hubs act as distance-based outliers which are far from all other points. Thus, they tend to increase within-cluster distance for any clusters they are associated with. Distance-based outliers and their influence

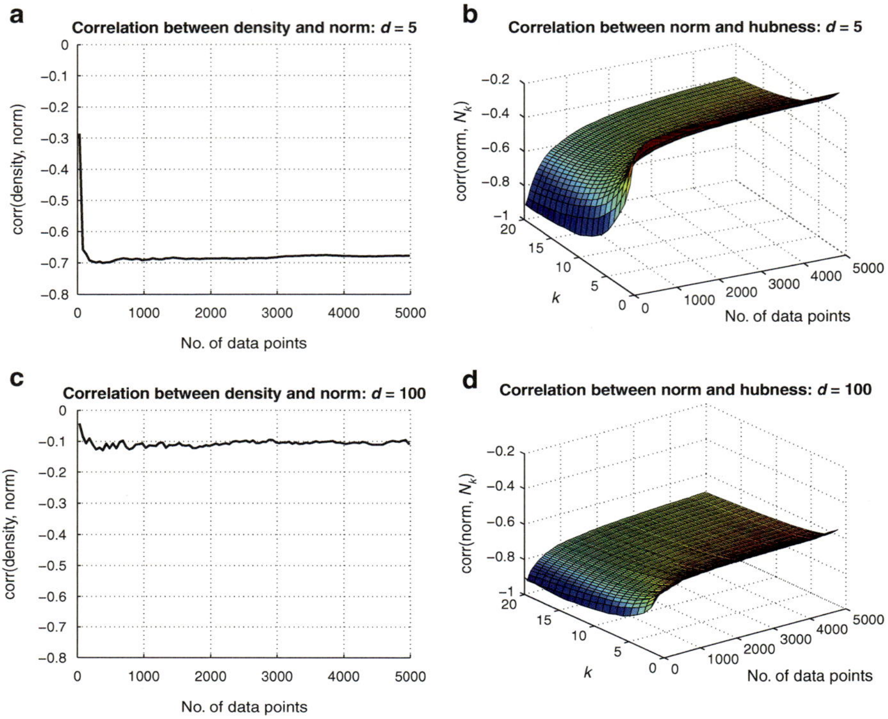

Fig. 11.3 Interaction between norm, hubness and density in low- and high-dimensional synthetic Gaussian data [78]. (**a**) Correlation between density and norm for $d = 5$. (**b**) Correlation between norm and hubness for $d = 5$. (**c**) Correlation between norm and hubness for $d = 100$. (**d**) Correlation between norm and hubness for $d = 100$

on clustering are well-studied subjects [65]: since outliers do not cluster well because of the high within-cluster distance, they are often discovered and eliminated beforehand. The existence of outliers is attributed to various reasons (for example, erroneous measurements). Nevertheless, the skewness of N_k suggests that in high-dimensional data outliers are also expected due to inherent properties of vector space.

On the other hand, between-cluster distance may be reduced for points with high k-occurrences, i.e., hubs, since they tend to be close to many other points. For this reason, hubs also do not cluster well, but unlike anti-hubs (outliers) this is because they have low between-cluster distance. In contrast to outliers, the influence of hubs on clustering traditionally has not attracted significant attention.

To examine the influence of both outliers and hubs, we used the popular silhouette coefficients (SC) [65]. For the ith point, let a_i be the average distance to all points in its cluster (a_i corresponds to within-cluster distance), and b_i the minimum average distance to points from other clusters (b_i corresponds to between-cluster distance). This is shown in Eq. (11.1), where π_c denotes the c-th cluster. The SC of the ith point is $(b_i - a_i)/\max(a_i, b_i)$, ranging between -1 and 1 (higher values are preferred). The SC of a set of points is obtained by averaging the silhouette coefficients of the individual points.

$$
\begin{aligned}
a_i &= \frac{\sum_{x_j \in \pi_{c_i}, x_j \neq x_i | x_i \in \pi_{c_i}} d(x_i, x_j)}{|\pi_{c_i}|} \\
b_i &= \min_{\pi_c, x_i \notin \pi_c} \frac{\sum_{x_j \in \pi_c} d(x_i, x_j)}{|\pi_c|} \\
\mathrm{SC}_i &= \frac{b_i - a_i}{\max(a_i, b_i)}
\end{aligned}
\tag{11.1}
$$

In order to explore the effects on clustering quality of hubs and anti-hubs, we employ the popular K-means++ algorithm, but differently from [54, Section 7.3.1] we vary the number of clusters K. For a given data set, we select as hubs those points $\mathbf{x}$ with $N_k(\mathbf{x})$ more than two standard deviations higher than the mean (we use the number of neighbors $k = 10$). Let n_h be the number of hubs selected. Next, we select as outliers the n_h points with the lowest $N_k(\mathbf{x})$. Finally, we designate all remaining points as regular points. To examine the K-means++ clustering success for hubs and anti-hubs, we measure the *relative SC* of hubs (anti-hubs): the mean SC of hubs (anti-hubs) divided by the mean SC of regular points. For several real-world data sets used in Sect. 11.7, Fig. 11.4 shows with solid lines the relative silhouette coefficients, obtained as averages over 50 runs of K-means++. For lower values of K, outliers have relative SC less than one, meaning that they cluster worse than regular points. As K increases, outliers begin to cluster better, at some instant exceeding the regular points. This can be considered natural, since with the increased number of clusters, outliers will tend to be isolated in their own small clusters, which prevents them from spoiling within-cluster distance of larger clusters they are "forced" to be included in when K is small, at the same time producing increased between-cluster distance for such small "outlying" clusters. In Fig. 11.4, this trend is clearly illustrated with plots of relative a and b components of the SC for the anti-hubs, with the a components starting to drop after initially increasing, and b components continually increasing or staying approximately constant.

On the other hand, for hubs exactly opposite trends emerge. For smaller K values hubs generally cluster well, which is exemplified in Fig. 11.4 with solid black lines indicating their relative SC being below 1, since they "snugly" fit the central regions of the large clusters. However, as K increases the relative SC of hubs begins to drop. In this situation, in whichever cluster a hub is placed, it will tend to decrease between-cluster distance, since hubs are close to many other points, and in the scenario involving a large number of small clusters, many of the

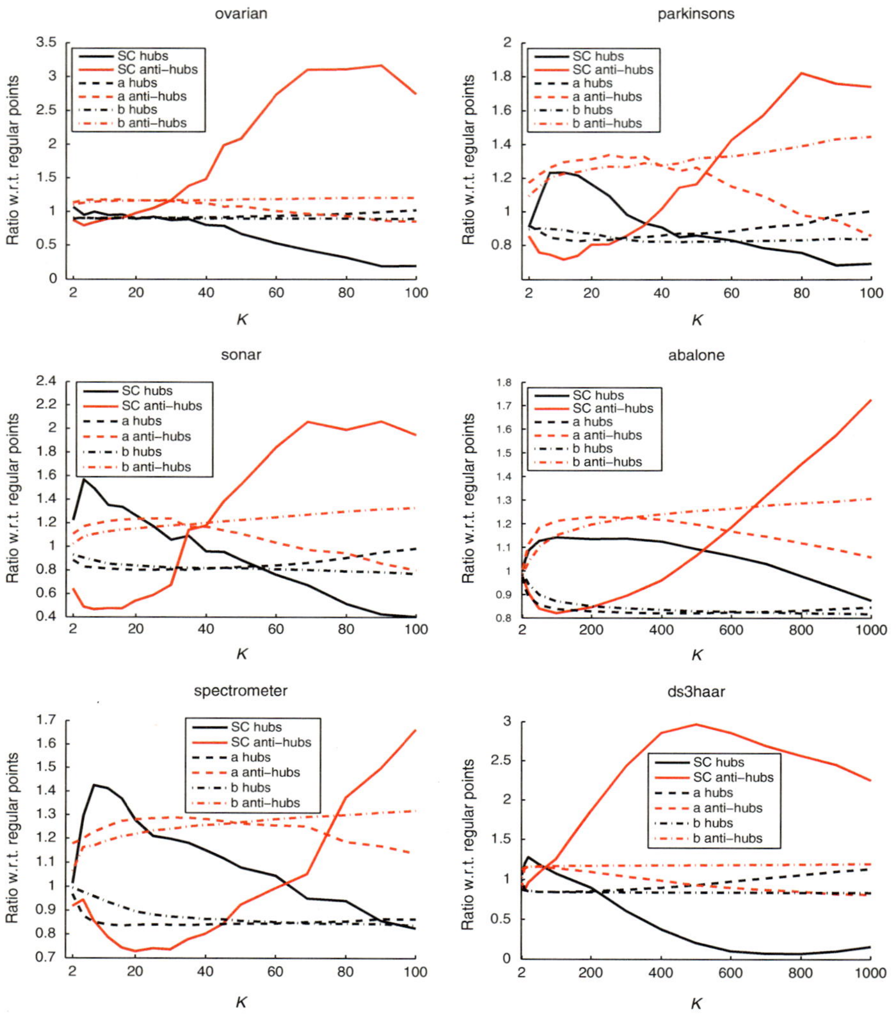

Fig. 11.4 Silhouette coefficients (SC) of hubs (*black solid lines*) and anti-hubs (*red solid lines*), relative to the SC of regular points, for varying numbers of clusters in K-means++. Also plotted are the relative values for *a* (*dashed lines*) and *b* (*dash-dot lines*) components of the SC

points hubs are close to will originate from different clusters. This is illustrated in the figure with plots of relative *b* components for the hubs, which tend to decrease as the number of clusters increases. In addition, somewhat surprisingly (but completing the analogy with the anti-hubs/outliers), the relative *a* component increases for the hubs, signifying that in the many-small-cluster scenario the hubs' advantage of being close to many other points and having small within-cluster distance diminishes compared to the regular points.

Putting the striking analogy between hubs and anti-hubs (in the context of clustering quality) aside, regardless of whether hubs cluster better or worse than regular points according to full silhouette coefficients, the between-cluster b component of the SC is always low. This constitutes a major area for improvement of clustering algorithms and, as will be demonstrated, is the principal reason why the reviewed and proposed hub-based algorithms work well.

11.5 Interaction Between Kernels and Hubness

In Sect. 11.6, we will consider a "kernelized" version of existing hubness-based clustering algorithms. Before that, let us examine the interaction of kernel functions and hubness, i.e., the differences (if any) between hubness in the original space and kernel space. For a given kernel function $K(\mathbf{x}, \mathbf{y})$, norm distance metric $D(\mathbf{x}, \mathbf{y})$ in Hilbert space, such as Euclidean distance on $\mathbb{R}^d$ used throughout this chapter, the square of the distance between $\Psi(\mathbf{x})$ and $\Psi(\mathbf{y})$, the projections of $\mathbf{x}$ and $\mathbf{y}$ to kernel space, can be expressed as $D^2(\Psi(\mathbf{x}), \Psi(\mathbf{y})) = K(\mathbf{x}, \mathbf{x}) - 2K(\mathbf{x}, \mathbf{y}) + K(\mathbf{y}, \mathbf{y})$. Hubness in kernel space is, therefore, straightforward to compute using the distances expressed by K.

We will consider three of the most popular choices of kernel functions—the polynomial, radial basis function (RBF), and sigmoid kernel [41, 83]:

$$\begin{aligned} K_{\text{poly}}(\mathbf{x}, \mathbf{y}) &= (1 + \langle \mathbf{x}, \mathbf{y} \rangle)^p, \\ K_{\text{RBF}}(\mathbf{x}, \mathbf{y}) &= \exp(-\gamma \|\mathbf{x} - \mathbf{y}\|^2), \\ K_{\text{sigm}}(\mathbf{x}, \mathbf{y}) &= \tanh(a \langle \mathbf{x}, \mathbf{y} \rangle + r), \end{aligned}$$

where usually $p \in \{1, 2, \ldots\}$, and $\gamma, a, r \in \mathbb{R}$, $\gamma > 0$.

For the polynomial kernel, with its one parameter p, in the case when $p = 1$ (for which the kernel is also referred to as the linear kernel) it can be shown that distances in the kernel space are exactly the same as in the original space [83], which implies that hubness will be the same as well (and also that kernelized clustering algorithms will produce the same groupings as the original algorithms). For other values of p this is not the case, and to illustrate the difference Fig. 11.5 plots the Spearman correlations between N_{10} of points in the original space and in the kernel space ($C_{N_{10}}$), as well as the skewness of the distribution of N_{10} in kernel space ($S_{N_{10}}$), for all values of $p \in \{1, 2, \ldots, 20\}$, on UCI and image data sets used in Sect. 11.7. Starting with the expected perfect correlation for $p = 1$, the general trend is for correlation to weaken with increasing p, albeit at different rates for different data sets. Skewness of N_{10}, on the other hand, increases for the majority of data sets, while for some it remains approximately constant. Notably, we have not detected significant drops of skewness with increasing p.

The RBF kernel is a smooth monotone function on distances in the original space, and as such produces distances in the kernel space that preserve the ordering of k

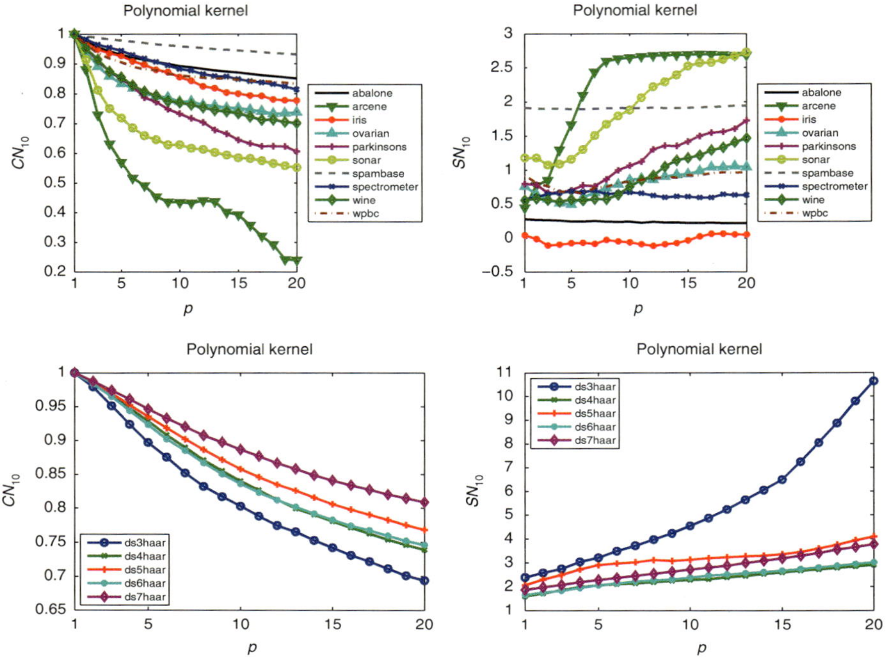

Fig. 11.5 Correlation between N_{10} in original and kernel space (*left*), and skewness of the distribution of N_{10} in kernel space (*right*), for the polynomial kernel with different values of parameter p

nearest neighbors, i.e., preserve kNN graphs from the original space [83]. Therefore, hubness in the kernel space remains the same as in the original space. Unlike the linear kernel, distances are not exactly preserved, which is why it still makes sense to employ the RBF kernel in clustering algorithms which operate with distances (and not only with nearest neighbors). Furthermore, numeric concerns limit the usable values of a to a relatively small range (depending on the data set and numeric precision used).

Finally, sigmoid kernel's two free parameters produce kernel functions with different properties [41], which are more or less meaningful to use in practical situations. We explore the range of parameter values similar to those considered by [41]: $\log(a) \in \{-10, -8, \ldots, 10\}$, and $r \in \{-5, -4, \ldots, 5\}$. Figure 11.6 shows surface plots of $C_{N_{10}}$ and $S_{N_{10}}$ on three data sets representative of the general trend. The influence of the a parameter is similar to the general influence of p in the polynomial kernel: for small values of a, hubness in kernel space is closely correlated with hubness in the original space, and as a increases the correlation significantly weakens, while the r parameter is not particularly influential in this regard except in the transitional regions where it can speed up the drop in correlation with increasing a. As for skewness, it generally increases substantially with

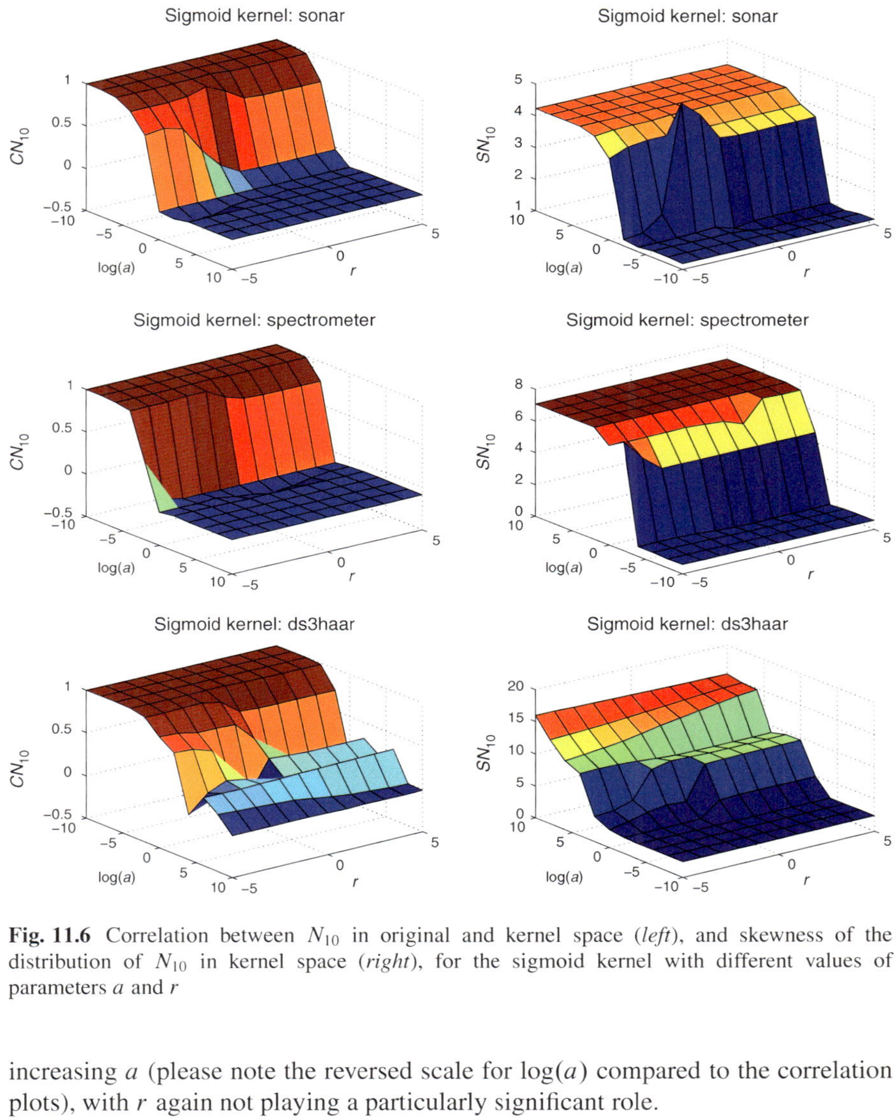

Fig. 11.6 Correlation between N_{10} in original and kernel space (*left*), and skewness of the distribution of N_{10} in kernel space (*right*), for the sigmoid kernel with different values of parameters a and r

increasing a (please note the reversed scale for $\log(a)$ compared to the correlation plots), with r again not playing a particularly significant role.

To summarize, the parameters of the considered kernels, except for the RBF kernel, can produce mappings of distances which may or may not preserve hubness from the original space, and in the case where hubness is not preserved it is usually amplified, meaning that in the kernel space there is a tendency to have strong hub points which are not hub points at all in the original space. Thus, we can conclude that "kernelizing" hubness-based clustering algorithms can have a significant impact on the groupings that are produced.

11.6 Clustering Algorithms Based on Hubness

It was shown that pointwise hubness scores correlate well with local cluster centrality in intrinsically high-dimensional Gaussian mixture data [78] and neighbor occurrence frequency ($N_k(\mathbf{x})$) can be used as an estimate of centrality in many dimensions.

This information can be integrated with different types of clustering approaches. Several hubness-based extensions of K-means were proposed by [78], as a way of exploiting $N_k(\mathbf{x})$ information in partitional clustering. In global K-hubs (GKH), hubs are used as cluster prototypes instead of cluster centroids in K-means iterations. Initial experiments have shown that this approach is somewhat rigid, as it is prone to premature convergence to decent but suboptimal cluster configurations. GKH was "softened" by introducing stochasticity, so that hubs are used as prototypes in iterations with a certain probability and that other points might become prototypes as well, with a probability proportional to $N_k(\mathbf{x})^2$. The algorithm was named global hubness-proportional clustering (GHPC). The alternating probabilistic framework was controlled by a temperature parameter, as in simulated annealing. The motivation behind using hubs as cluster prototypes is outlined in Fig. 11.7.

Essentially, when a cluster in the current iteration is composed of multiple compact components, centroids and medoids do not necessarily represent meaningful prototypes. Ideally, we might like to assign these separate components to different clusters in the following iterations, while searching for the best split and the optimal cluster configuration. However, centroid and medoid updates might either lead to minor differences between iterations, inflating the iteration count or even converge into a suboptimal configuration. Furthermore, centroids and medoids of such multi-component clusters do not really correspond to local component centers. Therefore, using hubs as prototypes and hubness while guiding the prototype search might help with overcoming these issues in certain cases in intrinsically high-dimensional data.

It should be noted that there are more benefits to using hubness instead of density for guiding the prototype search in clustering, other than the mere fact that it correlates better with local cluster centrality. Namely, hubness does not depend

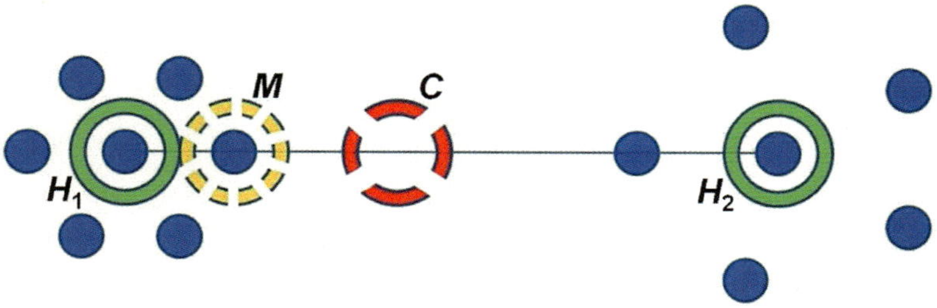

Fig. 11.7 An illustrative example of the difference between using hubs and centroids/medoids as cluster prototypes in partitional clustering iterations: The *red dashed circle* marks the centroid (C), *yellow dotted circle* the medoid (M), and *green circles* denote two elements of highest hubness (H_1, H_2), for neighborhood size 3 [78]

on scale. This plays an important role in data sets that incorporate several levels of granularity by containing clusters of highly varying densities. Scaling a cluster up or down does not change its neighbor structure and does not, in general, affect the hubness scores of data points.

Naturally, hub prototype cluster configurations are not necessarily optimal on any given data. Sometimes, there are still benefits when using centroids as cluster prototypes instead. Therefore, an extension of GHPC was examined, where the deterministic iteration step was to take centroids as cluster prototypes and the probabilistic step was still used to choose prototype points proportional to their $N_k(\mathbf{x})^2$ scores. This hybrid approach was named global hubness-proportional K-means (GHPKM) and was shown to perform somewhat better than GHPC in several real-world clustering tasks.

As GKH, GHPC and GHPKM are only capable of capturing hyperspherical clusters, it was mentioned by [78] that it might be promising to explore alternative ways of clustering with hubs that would be able to handle clusters of different shapes as well. It was argued that kernels might be used to introduce additional non-linearity to the approaches, similarly to what was done with kernel K-means [18].

We will first formally present and examine in more detail the previously proposed hubness-based partitional approaches that were mentioned above. We will proceed by proposing and describing a novel hubness-based extension of the kernel K-means algorithm [18].

11.6.1 Deterministic Approach

A simple way to employ hubs for clustering is to use them as one would normally use centroids. This is a simple extension of the K-means method. The algorithm, referred to as *K-hubs*, is given in Algorithm 13.

This initial approach converges to final cluster configurations very quickly. In earlier experimentation, it took no more than four iterations on all the data sets used for testing, most of which contained around 10,000 data instances. Nevertheless, its instability and sensitivity to initialization [11, 12] render it ineffective in most practical cases.

Algorithm 13 K-hubs

```
initializeClusterCenters();
Cluster[] clusters = formClusters();
repeat
   for all Cluster c ∈ clusters do
      DataPoint h = findClusterHub(c);
      setClusterCenter(c, h);
   end for
   clusters = formClusters();
until noReassignments
return clusters
```

11.6.2 Probabilistic Approach

Local hubs, that is, points with the highest hubness scores among their respective temporary clusters during iterations, are certainly good candidates for cluster prototypes. K-hubs illustrates that by being able to reach meaningful cluster configurations quickly. However, hubness of other points in the data set contains potentially useful information related to the cluster structure. Global hubness-proportional clustering algorithm (GHPC), described in Algorithm 14, implements a squared hubness-proportional stochastic scheme based on the widely used simulated annealing approach to optimization [17]. The temperature factor was introduced to the algorithm, so that it may start as being entirely probabilistic and eventually end by executing deterministic K-hubs iterations.

The skewness of the k-occurrence distribution is why this approach becomes meaningful when clustering in many dimensions. A vast majority of points have low $N_k(\mathbf{x})$ scores, which means that they will be disregarded by GHPC, as they will be considered bad candidates for cluster prototypes and assigned a very low probability of being selected. The square of the actual neighbor occurrence frequency is used ($N_k(\mathbf{x})^2$) in order to emphasize this feature of the algorithm. This is conceptually similar to the use of squared point-center distances for determining initial cluster seeds in K-means++ [4], the important difference being that the squared neighbor occurrence frequencies are used for determining local point centrality, while the D^2 initialization scheme is used for achieving a good spread of the initial points.

Algorithm 14 GHPC

```
initializeClusterCenters();
Cluster[] clusters = formClusters();
float t = t0; {initialize temperature}
repeat
   float θ = getProbFromSchedule(t);
   for all Cluster c ∈ clusters do
      if randomFloat(0,1) < θ then
         DataPoint h = findClusterHub(c);
         setClusterCenter(c, h);
      else
         for all DataPoint x ∈ c do
            setChoosingProbability(x, N_k^2(x));
         end for
         normalizeProbabilities();
         DataPoint h = chooseHubProbabilistically(c);
         setClusterCenter(c, h);
      end if
   end for
   clusters = formClusters();
   t = updateTemperature(t);
until noReassignments
return clusters
```

The actual exponent can be varied and more experiments would be needed to determine its optimal value in general. A high exponent would increase the skewness in selection probabilities of different points, thereby decreasing the exploratory potential of the algorithm and increasing its greedy exploitation. The exploration vs exploitation dilemma is well-known in various optimization problems and the optimal trade-off between the two depends on particular data characteristics. In terms of data hubness, it might be advisable to use higher exponents for low- and medium-dimensional data and lower exponents for very high-dimensional data.

GHPC uses a rather trivial temperature schedule in the getProbFromSchedule(t) function. The number of probabilistic iterations N_{Prob} is passed as an argument to the algorithm and the probability $\theta = \min(1, t/N_{Prob})$. It is possible to use more complex cooling schemes and this might lead to slightly better results in practice. The basic temperature schedule implemented in GHPC, however, has proven sufficient in many cases.

GHPC executes a search through the data space based on hubness as an estimate of local cluster centrality. To justify the use of the proposed stochastic scheme, a series of tests was performed on Gaussian mixtures [78], for dimensionality $d = 50$, $n = 10{,}000$ instances, and $K = 25$ clusters in the data. Neighborhood size was set to $k = 10$ and for each preset number of probabilistic iterations in the annealing schedule, the clustering was run 50 times, each time re-initializing the seeds. The results are displayed in Fig. 11.8. The silhouette index [65] was used to estimate the clustering quality. Due to a significant skewness of the squared hubness scores, adding more probabilistic iterations helps in achieving better clustering, up to a certain plateau that is eventually reached.

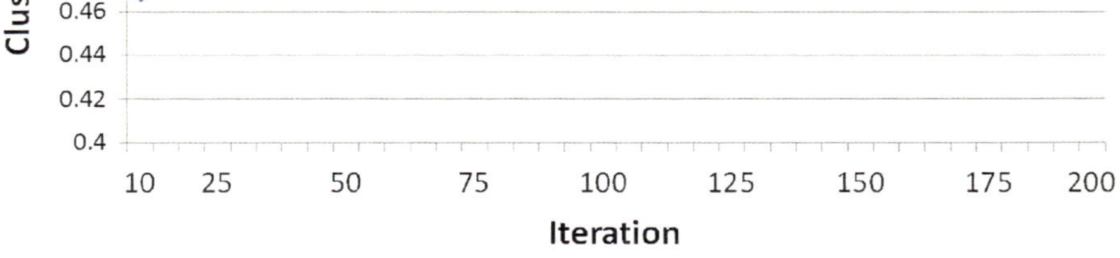

Fig. 11.8 Estimated quality of clustering for various durations of probabilistic search in GHPC, on synthetic Gaussian mixtures [78]

11.6.3 A Hybrid Approach: Extending K-Means

K-hubs and GHPC share a property that they do not assume any underlying data representation and can handle arbitrary object types, as long as some measure of similarity can be established between objects. In cases when features are available as vectors and centroids are computable, it is possible to use hubness information for clustering in conjunction with the basic principles of K-means. In this case, hubness scores can be used to guide the prototype search and help K-means to avoid premature convergence to local cluster configuration optima. This simple extension of K-means has been shown to perform well in practice [78]. It is formally described in Algorithm 15 and will be referred to as global hubness-proportional K-means (GHPKM). The only difference between GHPKM and GHPC is in the deterministic part of each iteration, where the deterministic option executes standard K-means updates by selecting centroids for cluster prototypes. In the probabilistic branch, points are selected with a probability proportional to $N_k(\mathbf{x})^2$, as in GHPC.

11.6.4 Using Kernels in Hubness-Based Clustering

A major problem with K-hubs, GHPC and GHPKM is that they are limited to discovering hyperspherical clusters. In many applications this is not sufficient. It is therefore natural to explore other options, including the use of kernels as means of

Algorithm 15 GHPKM

```
initializeClusterCenters();
Cluster[] clusters = formClusters();
float t = t0; {initialize temperature}
repeat
   float θ = getProbFromSchedule(t);
   for all Cluster c ∈ clusters do
      if randomFloat(0,1) < θ then
         DataPoint h = findClusterCentroid(c);
         setClusterCenter(c, h);
      else
         for all DataPoint x ∈ c do
            setChoosingProbability(x, N_k^2(x));
         end for
         normalizeProbabilities();
         DataPoint h = chooseHubProbabilistically(c);
         setClusterCenter(c, h);
      end if
   end for
   clusters = formClusters();
   t = updateTemperature(t);
until noReassignments
return clusters
```

inducing additional non-linearity to the cluster discovery process. Kernel K-means is a natural extension of K-means that supports the use of kernels [18]. As K-hubs, GHPC and GHPKM are all essentially extensions and variations of K-means as well, it is possible to combine these approaches in order to enable kernel clustering via the hubness-based methodology.

Let π_c be the data clusters, where $c \in \{1..K\}$. Using the non-linear function ϕ, the objective function of kernel GHPKM is, similarly to the objective function of kernel K-means, as follows:

$$F(\{\pi_c\}_{c=1}^{K}) = \sum_{c=1}^{K} \sum_{\mathbf{x} \in \pi_c} w(\mathbf{x}) \cdot \|\phi(\mathbf{x}) - p_c\|^2 \tag{11.2}$$

In Eq. (11.2), $w(\mathbf{x})$ are weights that may be assigned to different points, if needed. In our experiments, we have used $w(\mathbf{x}) = 1$ for all x. Cluster prototypes are denoted by p_c. This is where the first difference between kernel K-means and kernel GHPKM appears, as p_c points in kernel GHPKM are not necessarily centroids, but can also be hubs, or rather—points that are taken as prototypes with a probability proportional to $N_k(\mathbf{x})^2$, as in GHPC and GHPKM. As the iterations progress, this difference disappears, as they become more and more deterministic, due to the cool-down in the simulated annealing procedure that governs the optimization. Therefore, kernel GHPKM will eventually end up minimizing the same function as kernel K-means, but it will also perform a slightly more randomized initial search that might help with avoiding premature convergence to local optima, as in GHPKM.

Denote the centroid of cluster c by m_c. The centroid under the mapping ϕ is then calculated according to Eq. (11.3). We will similarly denote the currently selected prototype hub of cluster c as $p_c = h_c$, when there is a need to clarify the difference.

$$m_c = \frac{\sum_{\mathbf{x} \in \pi_c} w(\mathbf{x}) \cdot \phi(\mathbf{x})}{\sum_{\mathbf{x} \in \pi_c} w(\mathbf{x})} \tag{11.3}$$

When defined as in Eq. (11.3), centroids minimize the weighted squared distance to other mapped points in the cluster, as suggested by Eq. (11.4). Using hubs as prototypes does not minimize this squared distance sum, but might be beneficial for different reasons.

$$m_c = \text{argmin}_{\mathbf{z} \sum_{\mathbf{x} \in \pi_c} w(\mathbf{x}) \cdot \|\phi(\mathbf{x}) - \mathbf{z}\|^2} \tag{11.4}$$

The squared distance to the prototype can be easily calculated in the case when the prototype is selected from the cluster point pool. This actually speeds up the distance calculations somewhat. This dichotomy is given in Eq. (11.5), and the formula for calculating the distance to the cluster centroid according to the kernel trick is given in Eq. (11.6). This is necessary, as we are not explicitly mapping the points via ϕ, so there is no way to explicitly calculate the centroid itself in an effective manner.

$$\|\phi(\mathbf{x}) - p_c\|^2 = \begin{cases} \|\phi(\mathbf{x}) - m_c\|^2 & \text{if } p_c = m_c \\ K(\mathbf{x},\mathbf{x}) + K(h_c,h_c) - 2 \cdot K(\mathbf{x},h_c), & \text{if } p_c = h_c \end{cases} \tag{11.5}$$

$$\|\phi(\mathbf{x}) - m_c\|^2 = \|\phi(\mathbf{x}) - \frac{\sum_{\mathbf{z}\in\pi_c} w(\mathbf{z}) \cdot \phi(\mathbf{z})}{\sum_{\mathbf{z}\in\pi_c} w(\mathbf{z})}\|^2 = K(\mathbf{x},\mathbf{x}) - 2 \cdot$$

$$\frac{\sum_{\mathbf{z}\in\pi_c} w(\mathbf{z}) \cdot K(\mathbf{x},\mathbf{z})}{\sum_{\mathbf{z}\in\pi_c} w(\mathbf{z})} + \frac{\sum_{\mathbf{z}_1,\mathbf{z}_2\in\pi_c} w(\mathbf{z}_1) \cdot w(\mathbf{z}_2) \cdot K(\mathbf{z}_1,\mathbf{z}_2)}{(\sum_{\mathbf{z}\in\pi_c} w(\mathbf{z}))^2} \tag{11.6}$$

The details of Kernel GHPKM are given in pseudo-code in Algorithm 16. For each cluster we keep track of whether we have chosen the prototype point according to the squared hubness-proportional procedure or have chosen to use the centroid implicitly for determining future point assignments. While looping through the points, Eqs. (11.5) and (11.6) are used interchangeably, depending on the situation.

11.6.5 Scalability of Hubness-Based Approaches

Calculating all the $N_k(\mathbf{x})$ factors for guiding the prototype search in clustering iterations requires computing the entire kNN graph. Using a naive approach that does not take advantage of any spatial data structures or approximate calculations can be costly and prohibitive for very large data sets that are frequently encountered in modern practical applications. In such large data sets, there might be millions of examples that need to be clustered in a certain way.

There exist, however, fast approximate methods which can be used to construct fairly accurate approximations in reasonable time. It is possible to use either a generic approach [14] or some metric-specific approximation method based on locality-sensitive hashing [52,57,80,81]. The performance of different approximate kNN search methods depends on data characteristics of the data set in question and the intrinsic difficulty of the k-nearest neighbor search problem in the particular context [30].

As the neighbor occurrence frequency of a point is inferred from its reverse k-nearest neighbor sets, fast approximate reverse k-nearest neighbor queries are also potentially relevant for on-line, incremental hubness-based approaches, where new observations come in streams and the training set needs to be frequently updated. Early approaches were based on running approximate kNN queries and then analyzing and modifying the approximate kNN sets [62]. Many approaches use tree-based indexing techniques or influence zones [13] and some focus on multi-type queries [46]. Some attention has been given recently to probabilistic reverse k-nearest neighbor queries on fuzzy and uncertain data [8, 84].

The data which we had available for our experiments was not prohibitively big, so we only report the results on the accurate, complete, k-nearest neighbor graphs. It is worth noting, however, that it was already shown for some other hubness-aware

Algorithm 16 Kernel GHPKM

```
initializeClusterCenters();
float t = t_0; {initialize temperature}
repeat
  float θ = getProbFromSchedule(t);
  for all Point x ∈ dataset do
    closestCluster = NULL;
    minimalDistance = MAX_VALUE;
    for all Cluster c ∈ clusters do
      if getClusterCenter(c) NOT NULL then
        distance = getDistanceToHub(c);{Equation (11.5)}
        if distance ≤ minimalDistance then
          minimalDistance = distance;
          closestCluster = c;
        end if
      else
        distance = getDistanceToCentroid(c);{Equation (11.6)}
        if distance ≤ minimalDistance then
          minimalDistance = distance;
          closestCluster = c;
        end if
      end if
    end for
    assignPointToFutureCluster(x, closestCluster)
  end for
  updateClusterAssignments();
  for all Cluster c ∈ clusters do
    if randomFloat(0,1) < θ then
      setClusterCenter(c, NULL);
    else
      for all DataPoint x ∈ c do
        setChoosingProbability(x, N_k^2(x));
      end for
      normalizeProbabilities();
      DataPoint h = chooseHubProbabilistically(c);
      setClusterCenter(c, h);
    end if
  end for
  t = updateTemperature(t);
  calculateErrorFunction(){Equation (11.2)};
until convergenceCriterion
clusters = formClusters();
return clusters
```

methods to be quite robust with regards to the approximate neighbor sets [67] and that there is usually no significant decrease in accuracy even for nearly linear graph construction time complexities.

However, not all complexity is due to the kNN set calculations. In kernel GHPKM, there is also the known complexity of the kernel K-means base algorithm, that equals $O(n^2(\tau + d))$, where τ is the total number of iterations and d is data dimensionality. It is also worth noting that kernel GHPKM tends to have more iterations than kernel K-means itself, due to its stochastic nature and the fact that it uses some more time for the hubness-guided exploration of the cluster configuration space before it switches to the exploitation phase. Therefore, this exact implementation is not well suited for large-scale clustering but rather when trying to achieve a higher quality of clustering on small-to-medium size intrinsically high-dimensional data. Nevertheless, it should be noted that there exist approximate kernel K-means implementations as well [16] that are able to handle very large data sets, so the approximate implementations of kernel GHPKM will also be evaluated in the future, in order to enable hubness-based clustering of very large data sets.

11.7 Experimental Comparisons

Experimental comparisons performed by [78] have shown that hubness-based clustering methods offer promising performance. Extensive evaluation was done both on high-dimensional synthetic Gaussian mixture data as well as a series of real world data sets, some of which exhibited substantial hubness. Promising performance was also observed in presence of high noise levels, where GHPKM was the best among the basic hubness-based approaches in detecting the underlying Gaussian clusters that were placed in volumes of uniform noise. These improvements were more pronounced in cases of higher dimensionality. A quick comparison of the evolution of silhouette index in those experiments is shown in Fig. 11.9.

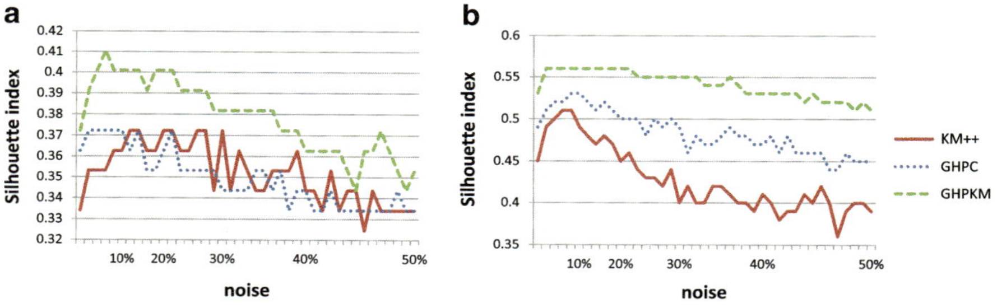

Fig. 11.9 Gradual change in cluster quality measures with rising noise levels. Differences between algorithm performance are more pronounced in the high-dimensional case [78]. (**a**) Silhouette index values for $d = 10, k = 50$. (**b**) Silhouette index values for $d = 50, k = 50$

We will not review the experiments on synthetic data in much detail here. Interested readers can look up the original results in the paper [78]. Instead, we will focus on presenting comparisons on real-world data, extended by including the evaluation of kernel GHPKM. The downside is, of course, that real-world test cases are less controlled as optimal configurations are usually not available a priori in the high-dimensional case. Nevertheless, these tests are more challenging and more practically relevant.

There were two different experimental setups on real-world data sets. In the first setup, a single data set was clustered for many different K-s, in order to compare algorithm performance over a range of cluster numbers. In the second setup, 20 different data sets were all clustered by the number of classes known to be present in the data. For all centroid-based algorithms, including KM, we used the D^2 (K-means++) initialization procedure [4]. As for choice of k, there is no consensus on what might be the best or easiest way for choosing proper neighborhood size in algorithms that use k-nearest neighbor sets. One option would be to do initial cross-validation in order to detect potential parameter values. However, the initial experiments have shown that hubness-based clustering algorithms are not very sensitive to the choice of k, as long as it remains in a reasonably low range. The performance of GHPC was rather constant for neighborhood sizes $k \in [1, 20]$ [78].

The clustering quality in these experiments was measured by two quality indices, the silhouette index and the isolation index [28], which measures a percentage of k-neighbor points that are clustered together.

In the first experimental setup, the algorithms were compared on the two-part Miss-America data set (cs.joensuu.fi/sipu/datasets/). Each part consists of 6,480 instances having 16 dimensions. Results were compared for various predefined numbers of clusters in algorithm calls. Each algorithm was tested 50 times for each number of clusters. Neighborhood size was 5. Histogram intersection kernel was used in the kernel methods and Euclidean distance in other cases. Histogram intersection kernel is non-parametric and is believed to perform well on image data, which is why it was employed in these experiments.

The results for both parts of the data set are given in Table 11.1. Kernel GHPKM clearly outperformed all remaining tested approaches on this data set, both in terms of silhouette and isolation index. If we compare the value of the silhouette index for kernel GHPKM of 0.75 for $K = 2$ on part I to the second best value achieved by kernel K-means of 0.33, we see that the difference is quite substantial. Similar and even more pronounced differences can be seen for other values of K. On the other hand, GHPC clearly outperformed the other considered non-kernel clustering algorithms on this data set. This shows that hubs can serve as good cluster prototypes.

As intrinsically high-dimensional, high-hubness data, we have taken several subsets of the ImageNet public repository (www.image-net.org). These data sets are described in detail by [72, 73]. Two separate cases are examined: Haar wavelet representation and SIFT codebook + color histogram representation [43, 85]. This totals to 10 different clustering problems. The neighborhood size k was set to 20. Silhouette values are given in Table 11.2 and the average values of the isolation

Table 11.1 Clustering quality on the Miss-America data set Parts I and II, measured by silhouette index and isolation index, for various cluster numbers. Kernel-GHPKM clearly outperforms all other approaches on this data set. Also, GHPC clearly outperforms all non-kernel tested approaches here. Highest scores are given in bold

(a) Silhouette index

	K	2	4	6	8	10	12	14	16
Part I	GKH	0.28	0.14	0.12	0.08	0.07	0.05	0.06	0.05
	GHPC	0.38	0.29	0.25	0.21	0.15	0.10	0.10	0.09
	KM++	0.14	0.12	0.09	0.08	0.07	0.07	0.07	0.07
	GHPKM	0.28	0.18	0.17	0.14	0.13	0.11	0.10	0.08
	ker-KM++	0.33	0.36	0.36	0.34	0.35	0.22	0.28	0.14
	ker-GHPKM	**0.75**	**0.75**	**0.77**	**0.76**	**0.78**	**0.75**	**0.76**	**0.73**
Part II	GKH	0.33	0.21	0.13	0.08	0.08	0.07	0.06	0.06
	GHPC	0.33	0.27	0.22	0.26	0.18	0.19	0.12	0.11
	KM++	0.18	0.12	0.10	0.08	0.07	0.08	0.07	0.07
	GHPKM	0.33	0.22	0.18	0.14	0.12	0.11	0.10	0.08
	ker-KM++	0.46	0.30	0.41	0.46	0.29	0.28	0.24	0.23
	ker-GHPKM	**0.54**	**0.48**	**0.50**	**0.54**	**0.51**	**0.50**	**0.51**	**0.52**

(b) Isolation index

	K	2	4	6	8	10	12	14	16
Part I	GKH	0.83	0.58	0.53	0.38	0.27	0.22	0.21	0.15
	GHPC	**0.91**	0.89	0.71	0.53	0.42	0.33	0.30	0.26
	KM++	0.62	0.46	0.34	0.23	0.19	0.16	0.13	0.12
	GHPKM	0.85	0.54	0.45	0.38	0.29	0.26	0.24	0.23
	ker-KM++	0.77	0.92	0.93	0.92	0.95	0.91	0.91	0.80
	ker-GHPKM	0.88	**0.94**	**0.94**	**0.94**	**0.95**	**0.93**	**0.94**	**0.91**
Part II	GKH	0.82	0.56	0.35	0.26	0.21	0.17	0.15	0.14
	GHPC	0.80	0.64	0.45	0.48	0.37	0.35	0.26	0.23
	KM++	0.62	0.35	0.28	0.20	0.16	0.14	0.11	0.09
	GHPKM	0.77	0.50	0.36	0.29	0.26	0.24	0.22	0.19
	ker-KM++	**0.88**	0.78	0.90	**0.94**	**0.91**	0.89	**0.90**	0.91
	ker-GHPKM	**0.88**	**0.88**	**0.91**	**0.94**	0.90	**0.90**	**0.90**	**0.92**

index in Table 11.3. Histogram intersection kernel was again used in kernel-based approaches and Manhattan distance in other clustering methods. Unlike in the original paper [78], feature values were not previously normalized to the [0, 1] range. This leads to an improved clustering performance on SIFT representations and comparable performance on the Haar ImageNet representations. However, it seems that kernel K-means performs much worse when excluding normalization, so the comparisons of kernel methods in both cases are shown in Fig. 11.10.

As Table 11.2 shows, kernel GHPKM clearly outperforms kernel K-means on Haar ImageNet representations. Both methods perform poorly with the histogram intersection kernel on the non-normalized SIFT representations. Other kernels might

Table 11.2 Clustering quality expressed as silhouette index on high-hubness data sets from the ImageNet repository. We examine both SIFT and Haar image feature representations. Histogram intersection kernel was used in the kernel-based algorithms and Manhattan distance in other cases. Features were not normalized. The skewness of N_k for $k = 1$ is shown as S_{N_1}. Highest scores are given in bold

Data set	Size	d	K	S_{N_1}	GKH	GHPC	KM++	GHPKM	ker-KM	ker-GHPKM
ds3haar	2731	100	3	2.27	0.55	0.57	0.68	0.70	−0.02	0.04
ds4haar	6054	100	4	2.44	0.41	0.49	0.60	0.61	−0.08	0.83
ds5haar	6555	100	5	2.43	0.49	0.52	0.63	0.64	−0.11	0.87
ds6haar	6010	100	6	2.13	0.42	0.39	0.57	0.57	−0.29	0.78
ds7haar	10544	100	7	4.60	0.25	0.44	0.64	0.66	−0.25	0.78
AVG-Haar					0.42	0.44	0.62	0.64	−0.15	0.66
ds3sift	2731	416	3	15.85	0.13	0.16	0.23	0.24	0.03	−0.09
ds4sift	6054	416	4	8.88	0.07	0.08	0.19	0.20	−0.06	0.10
ds5sift	6555	416	5	26.08	0.07	0.08	0.23	0.24	−0.03	−0.07
ds6sift	6010	416	6	13.19	0.06	0.05	0.14	0.15	−0.12	−0.06
ds7sift	10544	416	7	9.15	0.03	0.02	0.13	0.16	−0.13	0.01
AVG-Sift					0.07	0.08	0.19	0.20	−0.06	−0.02
AVG-Total					0.25	0.26	0.40	0.42	−0.10	0.32

Table 11.3 Average isolation index values achieved by the compared algorithms on UCI and ImageNet data sets. Highest scores are given in bold

Data sets	GKH	GHPC	KM++	GHPKM	ker-KM	ker-GHPKM
AVG-UCI	0.48	0.47	0.44	0.47	**0.77**	0.60
AVG-Haar	0.65	0.68	0.81	0.82	0.38	**0.98**
AVG-Sift	0.29	0.29	0.48	0.51	0.50	**0.97**
AVG-Img	0.47	0.49	0.65	0.66	0.44	**0.97**
AVG-Total	0.47	0.48	0.55	0.57	0.61	**0.78**

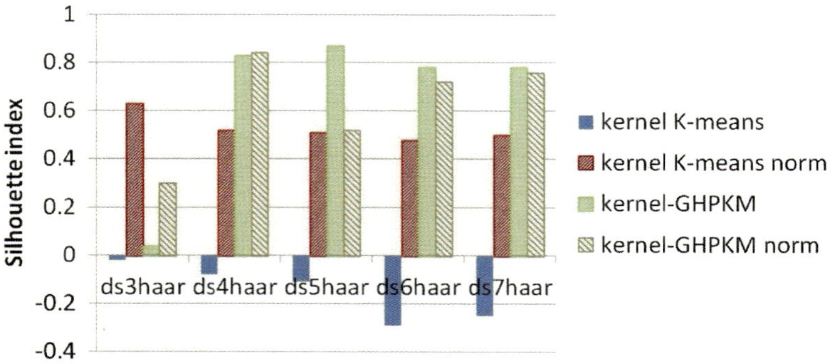

Fig. 11.10 The influence of feature normalization to the [0, 1] range on the performance of kernel-based methods on Haar representations of ImageNet data. Kernel K-means and kernel GHPKM are compared in both cases

be more appropriate in that case. On SIFT representations, GHPKM was slightly better than K-means++, though the difference is not significant.

As mostly low-to-medium hubness data (with the exception of spambase), several UCI data sets (archive.ics.uci.edu/ml/datasets.html) were randomly selected for comparisons. Values of all the individual features in the data sets were normalized prior to testing, as such preprocessing seems to slightly increase the hubness of the data in this case. The data sets were mostly simple, composed only of a few clusters. The value of k was set to 20. The results are shown in Table 11.4. RBF kernel was used in the kernel methods and the γ parameter was set to $\gamma = \frac{1}{d}$ by default, which is a common practice. In these experiments we did not try to optimize parameter values or kernel choices, due to the time complexity of the computations in the current implementations.

In the absence of hubness, purely hubness-based methods usually do not perform well. Note, however, that they score comparably to KM++ and kernel K-means on several UCI data sets and sometimes even outperform these baselines. Nevertheless, these examples show that in order to expect significant performance gains from using hubness-based clustering methods, a significant neighbor k-occurrence skewness is required.

Table 11.4 Clustering quality expressed as silhouette index on mostly low to medium-hubness data sets from the UCI repository. Hubness-based methods are much less effective when data hubness is not pronounced. The skewness of N_k for $k = 1$ is shown as S_{N_1}. Highest scores are given in bold

Data set	Size	d	K	S_{N_1}	GKH	GHPC	KM++	GHPKM	ker-KM	ker-GHPKM
wpbc	198	33	2	0.64	0.16	0.16	0.16	0.16	**0.32**	0.18
spambase	4601	57	2	21.46	0.29	**0.38**	0.31	**0.50**	0.48	0.37
arcene	100	1000	2	1.08	0.21	0.22	0.20	0.23	0.36	**0.37**
ovarian	253	15154	2	1.20	0.17	0.17	0.18	0.19	**0.25**	0.18
iris	158	4	3	0.46	0.48	0.47	0.49	0.49	**0.67**	0.61
parkinsons	195	22	2	0.39	0.25	0.30	0.37	0.37	**0.64**	0.28
sonar	208	60	2	1.35	0.11	0.11	0.19	0.15	**0.26**	0.17
wine	178	13	3	0.76	0.27	0.33	0.34	**0.35**	0.25	**0.35**
abalone	4177	8	29	0.92	0.22	0.20	0.26	0.27	−0.20	**0.40**
spectrometer	531	100	10	1.20	0.16	0.16	0.23	**0.25**	0.23	0.24
AVG-UCI					0.23	0.25	0.27	0.30	**0.33**	0.31

The improvements in silhouette index that were observed on high-dimensional data were explained by separately observing the index values and index component values for different types of points—hubs, outliers and regular points [78]. It was shown that hubness-based clustering methods mostly help with improving the b-component in the silhouette index for hub points. This result suggests that hubness-based methods achieve most of their improvements by ensuring that hubs remain distributed among clusters in such a way as to increase the overall between-cluster distances.

11.8 Application Domains and Other Types of Hubness-Aware Methods

Hubness is a phenomenon that was first observed in music recommendation and retrieval [5, 6, 25, 29, 45]. Hub songs were occurring very frequently in top-k result sets even in those cases when there was little or no observable semantic correlation with the queries, according to expert analysis. While the existence of hubs is quite natural in many types of complex real-world networks where power laws frequently arise, such as protein-protein interaction networks [21, 31, 51], this finding was considered surprising at the time, as kNN graphs were not closely examined in this way before. Similar problems have been noted in a related task of speaker recognition [19, 59]. The phenomenon was initially attributed to the particular representations and metrics that were standard at the time. However, as subsequent analysis has shown, hubness is ubiquitous in intrinsically high-dimensional data and has since been confirmed in images [53, 71, 72], text [55, 75, 77], time series [56], collaborative filtering data [48], etc. In general, most data that is encountered today in practical machine learning and data mining applications is highly complex and intrinsically high-dimensional. Therefore, it is expected to exhibit certain hubness. The exact skewness of the occurrence distribution may vary, but hubs are still expected to emerge. Hubness removal and/or hubness-aware learning have recently been advocated for high-hubness data [60].

In many tasks, data hubness is detrimental to analysis, as hubs often tend to act as semantic singularities when labels are present. This is both a geometric consequence of their emergence, as well as a consequence of frequent and severe cluster assumption violation in many real-world data sets. Therefore, hubness-aware methods have been proposed for instance selection [10, 66], metric learning [35, 36, 58, 64, 70] and classification [67, 69, 74, 76]—in order to reduce the negative impact of detrimental hub points in supervised learning. Hubness has also been shown to affect the performance of learning methods under class imbalance in high-dimensional data [68].

As for clustering, we have shown that hubness can actually be exploited for improving clustering performance and that it is possible to base clustering approaches for intrinsically high-dimensional data on the assumption that neighbor occurrence frequencies correlate well with local cluster centrality. Hubness-based clustering algorithms excel precisely in those cases where most standard clustering methods start facing problems—in cases of high dimensionality and high sparsity. In fact, they are not really applicable to the low-dimensional case and this should be kept in mind. As our experimental comparisons suggest, biggest improvements can be seen in cases where data hubness is most pronounced, as the skewness of the neighbor k-occurrence distribution is their working assumption.

11.9 Possible Future Directions

Our experiments support hubness-based clustering as a promising approach for handling intrinsically high-dimensional data and here we have presented several simple partitional clustering methods based on exploiting hubness in the search for cluster prototypes. Neighbor occurrence frequency acts as a good local centrality measure in high-dimensional data, so it easily conceivable that it can be integrated with ease into other types of existing clustering approaches as well, not only into extensions of K-means. It is especially tempting to consider hubs in those algorithms that use the k-nearest neighbor graph for clustering—and there is a considerable number of algorithms that fall into that category. Also, it would be interesting to see how hubness can be exploited to possibly improve or enrich other types of clustering methods that are tailored specifically for intrinsically high-dimensional data, like shared-neighbor clustering methods or subspace clustering methods. Another tempting idea would be to extend the existing hubness-based approaches to include label information when available, in order to adapt them to semi-supervised clustering tasks. These are some general directions that we intend to pursue in the future.

In kernel GHPKM, a more thorough investigation into the effects of parameter values of different kernels and the consequences of preserving/changing the original kNN structure in the maps is required. The initial results presented in this chapter need to be expanded in order to obtain a fuller understanding of the underlying processes and to consequently be able to use the algorithms to their fullest potential.

Additionally, as scalability is becoming crucial in many practical applications, it is necessary to develop efficient hubness-based approximations that would be applicable to large data sets.

Acknowledgements Miloš Radovanović and Mirjana Ivanović gratefully acknowledge the support of this work by the Serbian Ministry of Education, Science and Technological Development through project *Intelligent Techniques and Their Integration into Wide-Spectrum Decision Support*, no. OI174023.

References

1. Aggarwal C (2009) On high dimensional projected clustering of uncertain data streams. In: Proceedings of the 25th IEEE international conference on data engineering (ICDE), pp 1152–1154
2. Aggarwal CC, Hinneburg A, Keim DA (2001) On the surprising behavior of distance metrics in high dimensional spaces. In: Proceedings of the 8th international conference on database theory (ICDT), pp 420–434
3. Agrawal R, Gehrke J, Gunopulos D, Raghavan P (2005) Automatic subspace clustering of high dimensional data. Data Min Knowl Discov 11(1):5–33
4. Arthur D, Vassilvitskii S (2007) k-means++: The advantages of careful seeding. In: Proceedings of the 18th Annual ACM-SIAM symposium on discrete algorithms (SODA), pp 1027–1035
5. Aucouturier JJ (2006) Ten experiments on the modelling of polyphonic timbre. PhD thesis, University of Paris 6, Paris, France
6. Aucouturier JJ, Pachet F (2004) Improving timbre similarity: How high is the sky? J Negative Results Speech Audio Sci 1(1):1–13
7. Bai L, Liang J, Dang C, Cao F (2011) A novel attribute weighting algorithm for clustering high-dimensional categorical data. Pattern Recogn 44(12):2843–2861
8. Bernecker T, Emrich T, Kriegel HP, Renz M, Zankl S, Züfle A (2011) Efficient probabilistic reverse nearest neighbor query processing on uncertain data. Proc VLDB Endowment 4(10):669–680
9. Bohm C, Kailing K, Kriegel HP, Kroger P (2004) Density connected clustering with local subspace preferences. In: Proceedings of the Fourth IEEE international conference on data mining (ICDM), pp 27–34
10. Buza K, Nanopoulos A, Schmidt-Thieme L (2011) INSIGHT: Efficient and effective instance selection for time-series classification. In: Proceedings of the 15th Pacific-Asia conference on knowledge discovery and data mining (PAKDD), Part II, pp 149–160
11. Celebi ME, Kingravi HA (2012) Deterministic initialization of the k-means algorithm using hierarchical clustering. Int J Pattern Recogn Artif Intell 26(7):1250,018
12. Celebi ME, Kingravi HA, Vela PA (2013) A comparative study of efficient initialization methods for the k-means clustering algorithm. Expert Syst Appl 40(1):200–210
13. Cheema M, Lin X, Zhang W, Zhang Y (2011) Influence zone: Efficiently processing reverse k nearest neighbors queries. In: Proceedings of the 27th IEEE international conference on data engineering (ICDE), pp 577–588
14. Chen J, ren Fang H, Saad Y (2009) Fast approximate kNN graph construction for high dimensional data via recursive Lanczos bisection. J Mach Learn Res 10:1989–2012
15. Chen X, Ye Y, Xu X, Huang JZ (2012) A feature group weighting method for subspace clustering of high-dimensional data. Pattern Recogn 45(1):434–446
16. Chitta R, Jin R, Havens TC, Jain AK (2011) Approximate kernel k-means: Solution to large scale kernel clustering. In: Proceedings of the 17th ACM SIGKDD international conference on knowledge discovery and data mining, pp 895–903
17. Corne D, Dorigo M, Glover F (1999) New ideas in optimization. McGraw-Hill, London
18. Dhillon IS, Guan Y, Kulis B (2004) Kernel k-means: spectral clustering and normalized cuts. In: Proceedings of the 10th ACM SIGKDD international conference on knowledge discovery and data mining, pp 551–556
19. Doddington G, Liggett W, Martin A, Przybocki M, Reynolds D (1998) SHEEP, GOATS, LAMBS and WOLVES: A statistical analysis of speaker performance in the NIST 1998 speaker recognition evaluation. In: Proceedings of the 5th international conference on spoken language processing (ICSLP), paper 0608
20. Draper B, Elliott D, Hayes J, Baek K (2005) Em in high-dimensional spaces. IEEE Trans Syst Man Cybern 35(3):571–577

21. Ekman D, Light S, Björklund A, Elofsson A (2006) What properties characterize the hub proteins of the protein-protein interaction network of saccharomyces cerevisiae? Genome Biol 7:1–13
22. Ertz L, Steinbach M, Kumar V (2003) Finding topics in collections of documents: A shared nearest neighbor approach. In: Wu W, Xiong H, Shekhar S (eds) Clustering and information retrieval. Kluwer, Dordrecht
23. Fern XZ, Brodley CE (2003) Random projection for high dimensional data clustering: A cluster ensemble approach. In: Proceedings of 20th international conference on machine learning (ICML), pp 186–193
24. Fern XZ, Brodley CE (2004) Cluster ensembles for high dimensional clustering: An empirical study. Tech. Rep. CS06-30-02, Oregon State University
25. Flexer A, Schlueter J, Schnitzer D (2012) Putting the user in the center of music information retrieval. In: Proceedings of the 13th international society for music information retrieval conference (ISMIR), pp 385–390
26. François D, Wertz V, Verleysen M (2007) The concentration of fractional distances. IEEE Trans Knowl Data Eng 19(7):873–886
27. France S, Carroll D (2009) Is the distance compression effect overstated? Some theory and experimentation. In: Proceedings of the 6th international conference on machine learning and data mining in pattern recognition (MLDM), pp 280–294
28. Frederix G, Pauwels EJ (2004) Shape-invariant cluster validity indices. In: Proceedings of the 4th industrial conference on data mining (ICDM), pp 96–105
29. Gasser M, Schnitzer D, Flexer A (2010) Hubs and orphans – an explorative approach. In: Proceedings of the 7th sound and music computing conference (SMC)
30. He J, Kumar S, Chang SF (2012) On the difficulty of nearest neighbor search. In: Proceedings of the 29th international conference on machine learning (ICML), pp 1127–1134
31. He X, Zhang J (2006) Why do hubs tend to be essential in protein networks? PLoS Genetics 2(6):826–834
32. Houle ME (2008) The relevant-set correlation model for data clustering. J Stat Anal Data Min 1(3):157–176
33. Houle ME, Kriegel HP, Kröger P, Schubert E, Zimek A (2010) Can shared-neighbor distances defeat the curse of dimensionality? In: Proceedings of the 22nd international conference on scientific and statistical database management (SSDBM), pp 482–500
34. Jarvis RA, Patrick EA (1973) Clustering using a similarity measure based on shared near neighbors. IEEE Trans Comput 22:1025–1034
35. Jégou H, Harzallah H, Schmid C (2007) A contextual dissimilarity measure for accurate and efficient image search. In: Proceedings of the IEEE conference on computer vision and pattern recognition (CVPR), pp 1–8
36. Jégou H, Schmid C, Harzallah H, Verbeek J (2010) Accurate image search using the contextual dissimilarity measure. IEEE Trans Pattern Anal Mach Intell 32(1):2–11
37. Jing L, Ng M, Xu J, Huang J (2005) Subspace clustering of text documents with feature weighting k-means algorithm. In: Ho T, Cheung D, Liu H (eds) Advances in knowledge discovery and data mining, lecture notes in computer science, vol 3518. Springer, New York, pp 802–812
38. Jing L, Ng M, Huang J (2007) An entropy weighting k-means algorithm for subspace clustering of high-dimensional sparse data. IEEE Trans Knowl Data Eng 19(8):1026–1041
39. Kriegel HP, Kröger P, Zimek A (2009) Clustering high-dimensional data: A survey on subspace clustering, pattern-based clustering, and correlation clustering. ACM Trans Knowl Discov Data 3(1):1:1–1:58
40. Li T, Ma S, Ogihara M (2004) Document clustering via adaptive subspace iteration. In: Proceedings of the 27th annual international ACM SIGIR conference on research and development in information retrieval, pp 218–225
41. Lin HT, Lin CJ (2003) A study on sigmoid kernels for SVM and the training of non-PSD kernels by SMO-type methods. Tech. rep., Department of Computer Science, National Taiwan University

42. Liu B, Xia Y, Yu PS (2000) Clustering through decision tree construction. In: Proceedings of the 26th ACM SIGMOD international conference on management of data, pp 20–29
43. Lowe D (1999) Object recognition from local scale-invariant features. In: Proceedings of the 7th IEEE international conference on computer vision (ICCV), vol 2, pp 1150–1157
44. Lu Y, Wang S, Li S, Zhou C (2011) Particle swarm optimizer for variable weighting in clustering high-dimensional data. Mach Learn 82(1):43–70
45. M S, A F (2012) A mirex meta-analysis of hubness in audio music similarity. In: Proceedings of the 13th international society for music information retrieval conference (ISMIR), pp 175–180
46. Ma X, Zhang C, Shekhar S, Huang Y, Xiong H (2011) On multi-type reverse nearest neighbor search. Data Knowl Eng 70(11):955–983
47. Moëllic PA, Haugeard JE, Pitel G (2008) Image clustering based on a shared nearest neighbors approach for tagged collections. In: Proceedings of the international conference on content-based image and video retrieval (CIVR), pp 269–278
48. Nanopoulos A, Radovanović M, Ivanović M (2009) How does high dimensionality affect collaborative filtering? In: Proceedings of the 3rd ACM conference on recommender systems (RecSys), pp 293–296
49. Ntoutsi I, Zimek A, Palpanas T, Kröger P, Kriegel HP (2012) Density-based projected clustering over high dimensional data streams. In: Proceedings of the 12th SIAM international conference on data mining (SDM), pp 987–998
50. Patidar A, Agrawal J, Mishra N (2012) Analysis of different similarity measure functions and their impacts on shared nearest neighbor clustering approach. Int J Comput Appl 40:1–5
51. Patil A, Kinoshita K, Nakamura H (2010) Hub promiscuity in protein-protein interaction networks. Int J Mol Sci 11(4):1930–1943
52. Paulevé L, Jégou H, Amsaleg L (2010) Locality sensitive hashing: A comparison of hash function types and querying mechanisms. Pattern Recogn Lett 31(11):1348–1358
53. Pracner D, Tomašev N, Radovanović M, Mladenić D, Ivanović M (2011) WIKImage: Correlated image and text datasets. In: Proceedings of the 14th international multiconference on information society (IS), Jožef Stefan Institute, Ljubljana, Slovenia, vol A, pp 141–144
54. Radovanović M, Nanopoulos A, Ivanović M (2010) Hubs in space: Popular nearest neighbors in high-dimensional data. J Mach Learn Res 11:2487–2531
55. Radovanović M, Nanopoulos A, Ivanović M (2010) On the existence of obstinate results in vector space models. In: Proceedings of the 33rd annual international ACM SIGIR conference on research and development in information retrieval, pp 186–193
56. Radovanović M, Nanopoulos A, Ivanović M (2010) Time-series classification in many intrinsic dimensions. In: Proceedings of the 10th SIAM international conference on data mining (SDM), pp 677–688
57. Satuluri V, Parthasarathy S (2012) Bayesian locality sensitive hashing for fast similarity search. Proc VLDB Endowment 5(5):430–441
58. Schnitzer D, Flexer A, Schedl M, Widmer G (2011) Using mutual proximity to improve content-based audio similarity. In: Proceedings of the 12th international society for music information retrieval conference (ISMIR), pp 79–84
59. Schnitzer D, Schlüter J, Flexer A (2012) The relation of hubs to the Doddington zoo in speaker verification. In: Proceedings of the 21st european signal processing conference (EUSIPCO)
60. Schnitzer D, Flexer A, Tomašev N (2014) A case for hubness removal in high-dimensional multimedia retrieval. In: Advances in information retrieval, lecture notes in computer science, vol 8416. Springer, New York, pp 687–692
61. Scott D, Thompson J (1983) Probability density estimation in higher dimensions. In: Proceedings of the 15th symposium on the interface, pp 173–179
62. Singh A, Ferhatosmanoglu H, Tosun AŞ (2003) High dimensional reverse nearest neighbor queries. In: Proceedings of the 12th international conference on information and knowledge management (CIKM), pp 91–98
63. Steinbach M, Ertöz L, Kumar V (2004) The challenges of clustering high dimensional data. In: Wille LT (ed) New directions in statistical physics. Springer, New York, pp 273–309

64. Suzuki I, Hara K, Shimbo M, Saerens M, Fukumizu K (2013) Centering similarity measures to reduce hubs. In: Proceedings of the conference on empirical methods in natural language processing (EMNLP), pp 613–623
65. Tan PN, Steinbach M, Kumar V (2005) Introduction to data mining. Addison Wesley, Reading
66. Ting-ting Z, Zhen-feng H (2012) Instance selection algorithms of balanced class distribution based on hubness for time series. J Comput Appl cations 32:3034–3037
67. Tomašev N, Mladenić D (2012) Nearest neighbor voting in high dimensional data: Learning from past occurrences. Comput Sci Inform Syst 9(2):691–712
68. Tomašev N, Mladenić D (2013) Class imbalance and the curse of minority hubs. Knowl Based Syst 53:157–172
69. Tomašev N, Mladenić D (2013) Hub co-occurrence modeling for robust high-dimensional knn classification. In: Machine learning and knowledge discovery in databases, lecture notes in computer science, vol 8189. Springer, New York, pp 643–659
70. Tomašev N, Mladenić D (2013) Hubness-aware shared neighbor distances for high-dimensional k-nearest neighbor classification. Knowl Inform Syst 39(1):89–122
71. Tomašev N, Mladenić D (2013) Image hub explorer: Evaluating representations and metrics for content-based image retrieval and object recognition. In: Machine learning and knowledge discovery in databases. Springer, Berlin, pp 637–640
72. Tomašev N, Brehar R, Mladenić D, Nedevschi S (2011) The influence of hubness on nearest-neighbor methods in object recognition. In: Proceedings of the 7th IEEE international conference on intelligent computer communication and processing (ICCP), pp 367–374
73. Tomašev N, Radovanović M, Mladenić D, Ivanović M (2011) Hubness-based fuzzy measures for high-dimensional k-nearest neighbor classification. In: Proceedings of the 7th international conference on machine learning and data mining (MLDM), pp 16–30
74. Tomašev N, Radovanović M, Mladenić D, Ivanović M (2011) A probabilistic approach to nearest-neighbor classification: Naive hubness bayesian kNN. In: Proceedings of the 20th ACM international conference on information and knowledge management (CIKM), pp 2173–2176
75. Tomašev N, Leban G, Mladenić D (2013) Exploiting hubs for self-adaptive secondary re-ranking in bug report duplicate detection. In: Proceedings of the conference on information technology interfaces (ITI)
76. Tomašev N, Radovanović M, Mladenić D, Ivanović M (2013) Hubness-based fuzzy measures for high-dimensional k-nearest neighbor classification. Int J Mach Learn Cybern. DOI 10.1007/s13042-012-0137-1
77. Tomašev N, Rupnik J, Mladenić D (2013) The role of hubs in cross-lingual supervised document retrieval. In: Proceedings of the Pacific-Asia conference on knowledge discovery and data mining (PAKDD). Springer, New York, pp 185–196
78. Tomašev N, Radovanović M, Mladenić D, Ivanović M (2014) The role of hubness in clustering high-dimensional data. IEEE Trans Knowl Data Eng 26(3):739–751
79. Vinh NX, Houle ME (2010) A set correlation model for partitional clustering. In: Zaki M, Yu J, Ravindran B, Pudi V (eds) Advances in knowledge discovery and data mining, lecture notes in computer science, vol 6118. Springer, New York, pp 4–15
80. Wang J, Kumar S, Chang SF (2012) Semi-supervised hashing for large-scale search. IEEE Trans Pattern Anal Mach Intell 34(12):2393–2406
81. Xia H, Wu P, Hoi SC, Jin R (2012) Boosting multi-kernel locality-sensitive hashing for scalable image retrieval. In: Proceedings of the 35th international ACM SIGIR conference on research and development in information retrieval, pp 55–64
82. Yin J, Fan X, Chen Y, Ren J (2005) High-dimensional shared nearest neighbor clustering algorithm. In: Fuzzy systems and knowledge discovery, lecture notes in computer science, vol 3614. Springer, New York, pp 484–484
83. Yu K, Ji L, Zhang X (2002) Kernel nearest-neighbor algorithm. Neural Process Lett 15(2):147–156

84. Zhang P, Cheng R, Mamoulis N, Renz M, Zufle A, Tang Y, Emrich T (2013) Voronoi-based nearest neighbor search for multi-dimensional uncertain databases. In: Proceedings of the 29th IEEE international conference on data engineering (ICDE), pp 158–169
85. Zhang Z, Zhang R (2009) Multimedia data mining. Chapman and Hall, Boka Raton
86. Zheng L, Huang D (2012) Outlier detection and semi-supervised clustering algorithm based on shared nearest neighbors. Comput Syst Appl 29:117–121

Chapter 12
Clustering for Monitoring Distributed Data Streams

Maria Barouti, Daniel Keren, Jacob Kogan, and Yaakov Malinovsky

Abstract Monitoring data streams in a distributed system is a challenging problem with profound applications. The task of feature selection (e.g., by monitoring the information gain of various features) is an example of an application that requires special techniques to avoid a very high communication overhead when performed using straightforward centralized algorithms.

Motivated by recent contributions based on geometric ideas, we present an alternative approach that combines system theory techniques and clustering. The proposed approach enables monitoring values of an arbitrary threshold function over distributed data streams through a set of constraints applied independently on each stream and/or clusters of streams. The clusters are designed to evolve in time and to adapt themselves to the data stream. A correct choice of clusters yields a reduction in communication load. Unlike many clustering algorithms that attempt to collect together similar data items, monitoring requires clusters with *dissimilar* vectors canceling each other as much as possible. In particular, sub-clusters of a good cluster do not have to be good. This novel type of clustering dictated by the problem at hand requires development of new algorithms and/or modification of the existing ones, and the chapter is a step in this direction.

We report experiments on real-world data with a newly devised clustering algorithm. The experiments detect instances where communication between nodes is required, and show that the clustering approach reduces communication load. We then propose an application of the well known clustering algorithms to the monitoring problem.

Keywords Data streams • Distributed system • Clustering • Adaptive stream mining

M. Barouti (✉) • J. Kogan • Y. Malinovsky
Department of Mathematics and Statistics, UMBC, Baltimore, MD 21250, USA
e-mail: bmaria2@umbc.edu; kogan@umbc.edu; yaakovm@umbc.edu

D. Keren
Department of Computer Science, Haifa University, Haifa 31905, Israel
e-mail: dkeren@cs.haifa.ac.il

M.E. Celebi (ed.), *Partitional Clustering Algorithms*,
DOI 10.1007/978-3-319-09259-1_12

12.1 Introduction

In many emerging applications one needs to process a continuous stream of data in real time. Sensor networks [21], network monitoring [11], and real-time analysis of financial data [30, 31] are examples of such applications. Monitoring queries are a particular class of queries in the context of data streams. Previous work in this area deals with monitoring simple aggregates [11], or term frequency occurrence in a set of distributed streams [22]. The current contribution is motivated by results recently reported in [24, 25] where a more general type of monitoring query is described as follows:

Let $\mathbf{S} = \{\mathbf{s}_1, \ldots, \mathbf{s}_n\}$ be a set of data streams collected at n nodes $\mathbf{N} = \{\mathbf{n}_1, \ldots, \mathbf{n}_n\}$. Let $\mathbf{v}_1(t), \ldots, \mathbf{v}_n(t)$ be d-dimensional, real-valued, time varying vectors derived from the streams. For a function $f : \mathbf{R}^d \to \mathbf{R}$ we would like to monitor the inequality

$$f\left(\frac{\mathbf{v}_1(t) + \cdots + \mathbf{v}_n(t)}{n}\right) > 0 \tag{12.1}$$

while minimizing communication between the nodes. Often the threshold might be a constant r other than 0. In what follows, for notational convenience, we shall always consider the inequality $f > 0$, and when one is interested in monitoring the inequality $f > r$ we will modify the threshold function and consider $g = f - r$, so that the inequality $g > 0$ yields $f > r$. In e.g. [5, 14, 15, 24] a few real-life applications of this monitoring problem are described; see also Sect. 12.2 here.

The difference between monitoring problems involving linear and non-linear functions f is discussed and illustrated by a simple example involving a quadratic function f in [24]. The example demonstrates that, for a non-linear f, it is often very difficult to determine from the values of f at the nodes whether its value evaluated at the average vector is above the threshold or not. The present chapter deals with the information gain function (see Sect. 12.2 for details), and rather than focus on the values of f we consider the location of the vectors $\mathbf{v}_i(t)$ relative to the boundary of the subset of $\mathbf{R}^d$ where f is positive. We denote this set by $\mathbf{Z}_+(f) = \{\mathbf{v} : f(\mathbf{v}) > 0\}$, and state (12.1) as

$$\mathbf{v}(t) = \frac{\mathbf{v}_1(t) + \cdots + \mathbf{v}_n(t)}{n} \in \mathbf{Z}_+(f). \tag{12.2}$$

Thus, the functional monitoring problem is transformed to the monitoring of a *geometric* condition. As a simple illustration, consider the case of three scalar functions $v_1(t)$, $v_2(t)$ and $v_3(t)$, and the identity function f (i.e. $f(x) = x$). We would like to monitor the inequality

$$v(t) = \frac{v_1(t) + v_2(t) + v_3(t)}{3} > 0$$

while keeping the nodes silent as long as possible. One strategy is to verify the initial inequality $v(t_0) = \frac{v_1(t_0) + v_2(t_0) + v_3(t_0)}{3} > 0$ and to keep the nodes silent while

$$|v_i(t) - v_i(t_0)| < \delta = v(t_0), \ t \geq t_0, \ i = 1, 2, 3.$$

The first time t when one of the functions, say $v_1(t)$, crosses the boundary of the local constraint, i.e. $|v_1(t) - v_1(t_0)| \geq \delta$ the nodes communicate, t_1 is set to be t, the mean $v(t_1)$ is computed, the local constraint δ is updated and made available to the nodes. The nodes are kept silent as long as the inequalities

$$|v_i(t) - v_i(t_1)| < \delta, \ t \geq t_1, \ i = 1, 2, 3$$

hold. This type of monitoring was suggested in [19] for a variety of vector norms. The numerical experiments conducted in [19] with the dataset described in Sect. 12.5.1 show that:

1. The number of time instances the mean violates (12.1) is a small fraction ($< 1\%$) of the number of time instances when the local constraint is violated at the nodes.
2. The lion's share of communications (about 75 %) is required because of a single node violation of the local constraint δ.
3. The smallest number of communications is required when one uses the l_1 norm.

We note that if, for example, the local constraint is violated at $\mathbf{n}_1$, i.e. $|v_1(t) - v_1(t_0)| \geq \delta$, and at the same time

$$v_1(t) - v_1(t_0) = -[v_2(t) - v_2(t_0)],$$

while $|v_3(t) - v_3(t_0)| < \delta$ then $|v(t) - v(t_0)| < \delta$, $f(v(t)) > 0$, and update of the mean can be avoided. Separate monitoring of the two node cluster $\{\mathbf{n}_1, \mathbf{n}_2\}$ would require communication involving two nodes only, and could reduce communication load. We aim to extend this idea to the general case—involving many nodes, arbitrary functions, and high-dimensional data.

Clustering in general is a difficult problem, and many clustering problems are known to be NP-complete [4]. Unlike standard clustering that attempts to collect together similar data items [23], we are seeking clusters with *dissimilar data items*, which cancel out each other as much as possible. While sub-clusters of a "classical" good cluster are usually good, this may not be the case when a cluster contains dissimilar objects. These observations indicate that a straightforward application of common clustering methods to our problem is not possible.

A basic attempt to cluster nodes was suggested in [20] with results reported for the dataset presented in Sect. 12.5.1. Clustering together just two nodes reported in [20] reduces communication by about 10 %.

In this chapter we advance clustering approach to monitoring. The main contribution of this work is twofold:

1. We suggest a specific clustering strategy applicable to a variety of vector norms, and report the communication reduction achieved when the proposed strategy is applied with l_1, l_2, and l_∞ norms.
2. We suggest an application of well known clustering algorithms to monitoring.

The chapter is organized as follows. In Sect. 12.2 we present a relevant Text Mining application. Section 12.3 provides motivation for node clustering. A specific implementation of node clustering is presented in Sect. 12.4. Experimental results are reported in Sect. 12.5. In Sect. 12.6 we discuss a possible application of classical clustering algorithms to monitoring. Section 12.7 concludes the chapter and indicates new research directions. Appendix 1 summarizes some useful properties of the first and second moments. Appendix 2 details the accounting of message transmission.

In the next section we provide a Text Mining related example that leads to a non linear threshold function f.

12.2 Text Mining Application

Let $\mathbf{T}$ be a textual database (for example a collection of mail or news items). We denote the size of the set $\mathbf{T}$ by $|\mathbf{T}|$. We will be concerned with two subsets of $\mathbf{T}$:

1. $\mathbf{R}$-the set of "relevant" texts (e.g. texts not labeled as "spam"),
2. $\mathbf{F}$-the set of texts that contain a "feature" (word or term for example).

We denote complements of the sets by $\overline{\mathbf{R}}$, $\overline{\mathbf{F}}$ respectively (i.e. $\mathbf{R} \cup \overline{\mathbf{R}} = \mathbf{F} \cup \overline{\mathbf{F}} = \mathbf{T}$), and consider the relative size of the four sets $\mathbf{F} \cap \overline{\mathbf{R}}$, $\mathbf{F} \cap \mathbf{R}$, $\overline{\mathbf{F}} \cap \overline{\mathbf{R}}$, and $\overline{\mathbf{F}} \cap \mathbf{R}$ as follows:

$$x_{11}(\mathbf{T}) = \frac{|\mathbf{F} \cap \overline{\mathbf{R}}|}{|\mathbf{T}|}, \quad x_{12}(\mathbf{T}) = \frac{|\mathbf{F} \cap \mathbf{R}|}{|\mathbf{T}|},$$

$$x_{21}(\mathbf{T}) = \frac{|\overline{\mathbf{F}} \cap \overline{\mathbf{R}}|}{|\mathbf{T}|}, \quad x_{22}(\mathbf{T}) = \frac{|\overline{\mathbf{F}} \cap \mathbf{R}|}{|\mathbf{T}|}. \tag{12.3}$$

Note that

$$0 \le x_{ij} \le 1, \text{ and } x_{11} + x_{12} + x_{21} + x_{22} = 1. \tag{12.4}$$

The function f is defined on the simplex (i.e. $x_{ij} \ge 0$, $\sum x_{ij} = 1$), and given by

$$\sum_{i,j} x_{ij} \log \left(\frac{x_{ij}}{(x_{i1} + x_{i2})(x_{1j} + x_{2j})} \right), \tag{12.5}$$

where $\log x = \log_2 x$ throughout the chapter. It is well-known that (12.5) provides the *information gain* for the "feature" (see e.g. [16]).

As an example, we consider n agents installed on n different servers, and a stream of texts arriving at the servers. Let $\mathbf{T}_h = \{\mathbf{t}_{h1}, \ldots, \mathbf{t}_{hw}\}$ be the last w texts received at the hth server, with $\mathbf{T} = \bigcup_{h=1}^{n} \mathbf{T}_h$. Note that

$$x_{ij}(\mathbf{T}) = \sum_{h=1}^{n} \frac{|\mathbf{T}_h|}{|\mathbf{T}|} x_{ij}(\mathbf{T}_h),$$

i.e., entries of the global contingency table $\{x_{ij}(\mathbf{T})\}$ are the weighted average of the local contingency tables $\{x_{ij}(\mathbf{T}_h)\}$, $h = 1, \ldots, n$.

To check that the given "feature" is sufficiently informative with respect to the target relevance label r, one may want to monitor the inequality

$$f\left(x_{11}(\mathbf{T}), x_{12}(\mathbf{T}), x_{21}(\mathbf{T}), x_{22}(\mathbf{T})\right) - r > 0 \tag{12.6}$$

with f given by (12.5) while minimizing communication between the servers.

We next provide arguments in support of clustering for monitoring distributed data streams.

12.3 Monitoring Threshold Functions Through Clustering: Motivation

In what follows we denote a norm of a vector $\mathbf{v}$ by $\|\mathbf{v}\|$. While the experiments reported in this chapter have been conducted with l_1, l_2, and l_∞ norms, the proposed monitoring and node clustering procedures can be applied with any norm. We shall identify a specific norm used when needed. For a vector set $\mathbf{X} = \{\mathbf{x}_1, \ldots, \mathbf{x}_m\} \subset \mathbf{R}^d$ we denote the arithmetic mean $\frac{\mathbf{x}_1 + \cdots + \mathbf{x}_m}{m}$ of the set by $\mu(\mathbf{X})$. With slight abuse of notations the central second moment $\sum_{i=1}^{m} (\mathbf{x}_i - \mu(\mathbf{X}))^T (\mathbf{x}_i - \mu(\mathbf{X}))$ is denoted by $\sigma^2(\mathbf{X})$. The zero set $\{\mathbf{v} : \mathbf{v} \in \mathbf{R}^d, f(\mathbf{v}) = 0\}$ of a function f is denoted by $\mathbf{Z}_f$.

Our aim is to monitor data streams with as little communication as possible over a sequence of discrete time instances that we shall denote by t. The time instances that require communication between nodes are denoted by t_i, $i = 0, 1, 2, \ldots$. The approach suggested in this chapter builds on the monitoring strategy involving no clustering proposed in [19] and briefly described as follows:

Algorithm 12.3.1 *Monitoring Threshold Function*

- *A node is designated as a root* **r**.
- *The root sets* $i = 0$.
- *Until end of stream*

  1. *The root sends a request to each node* **n** *for the vectors* $\mathbf{v_n}(t_i)$. *The nodes respond to the root. The root computes the distance* δ *between the mean* $\frac{1}{n}\sum_{\mathbf{n}\in\mathbf{N}} \mathbf{v}_n(t_i)$ *and the zero set* $\mathbf{Z}_f$ *of the function* f. *The root transmits* δ *to each node.*
  2. *do for each* $\mathbf{n} \in \mathbf{N}$
     If $||\mathbf{v_n}(t) - \mathbf{v_n}(t_i)|| < \delta$
     the node **n** *is silent*
     else
     n *notifies the root about the violation of its local constraint* δ
     the root sets $i = i + 1$
     go to Step 1.
     endif

- *Stop*

An application of the above procedure to data streams generated from the Reuters Corpus RCV1-V2 (see Sect. 12.5 for detailed description of the data and experiments) leads to 4006 time instances in which the local constraints are violated, and the root is updated. Results presented in Table 12.1 show that in 3034 out of 4006 time instances, communications with the root are triggered by constraint violations at exactly one node.

The results immediately suggest to cluster nodes to further reduce communication load. Each cluster will be equipped with a "coordinator" **c** (one of the cluster's nodes). If a cluster node **n** violates its local constraint at time t, then the coordinator collects vectors $\mathbf{v_n}(t) - \mathbf{v_n}(t_i)$ from all the nodes in the cluster, computes the mean of the vectors, and checks whether the mean violates the coordinator constraint δ (at this point, node and coordinator constraints are identical, see Sect. 12.4 for discussion pertaining to constraints). We shall follow [24] and refer to this step as "the balancing process." If the coordinator constraint is violated, the coordinator alerts the root, and the mean of the entire dataset is recomputed by the root (for detailed description of the procedure see Sect. 12.4).

For the problem at hand we would like to partition the set of nodes **N** into k clusters $\Pi = \{\pi_1, \ldots, \pi_k\}$ so that

Table 12.1 number of local constraint violations simultaneously by k nodes, $r = 0.0025$, l_2 norm, the feature is "bosnia"

# of nodes violators	1	2	3	4	5	6	7	8	9	10
# of violation instances	3034	620	162	70	38	26	34	17	5	0

$$\mathbf{N} = \bigcup_{i=1}^{k} \pi_i, \text{ and } \pi_i \bigcap \pi_j = \emptyset \text{ if } i \neq j.$$

We denote the size of π_i by $|\pi_i|$. If for each cluster π_i one has

$$\frac{1}{|\pi_i|} \left\| \sum_{\mathbf{n} \in \pi_i} [\mathbf{v_n}(t) - \mathbf{v_n}(t_j)] \right\| < \delta,$$

then, due to convexity of any norm, one has

$$\left\| \frac{1}{n} \sum_{\mathbf{n} \in \mathbf{N}} \mathbf{v_n}(t) - \frac{1}{n} \sum_{\mathbf{n} \in \mathbf{N}} \mathbf{v_n}(t_j) \right\| \leq \sum_{i=1}^{k} \frac{|\pi_i|}{n} \left[\frac{1}{|\pi_i|} \left\| \sum_{\mathbf{n} \in \pi_i} [\mathbf{v_n}(t) - \mathbf{v_n}(t_j)] \right\| \right] < \delta.$$

Hence the "new" mean $\frac{1}{n} \sum_{\mathbf{n} \in \mathbf{N}} \mathbf{v_n}(t)$ belongs to $\mathbf{Z}_+(f)$ if the "old" mean $\frac{1}{n} \sum_{\mathbf{n} \in \mathbf{N}} \mathbf{v_n}(t_j)$ belongs to this set. We therefore may attempt to define the quality of a k cluster partition Π as

$$Q(\Pi) = \max_i \left\{ \frac{1}{|\pi_i|} \left\| \sum_{\mathbf{n} \in \pi_i} [\mathbf{v_n}(t) - \mathbf{v_n}(t_j)] \right\| \right\}, i = 1, \ldots, k. \tag{12.7}$$

Our aim is to identify k and a k cluster partition Π^o that **minimizes** (12.7). The monitoring problem requires to assign nodes $\{\mathbf{n}_{i_1}, \ldots, \mathbf{n}_{i_k}\}$ to the same cluster π so that the total average change within cluster π

$$\left\| \frac{1}{|\pi|} \sum_{\mathbf{n} \in \pi} [\mathbf{v_n}(t) - \mathbf{v_n}(t_j)] \right\| \text{ for } t > t_j$$

is minimized, i.e., nodes with **different** variations $\mathbf{v_n}(t) - \mathbf{v_n}(t_j)$ that cancel out each other as much as possible are assigned to the same cluster.

A standard clustering problem is often described as "...finding and describing cohesive or homogeneous chunks in data, the clusters" (see e.g. [23]). Unlike classical clustering procedures, this one needs to combine "dissimilar" nodes together.

The proposed partition quality $Q(\Pi)$ (see (12.7)) generates three immediate problems:

1. Since the arithmetic mean $\overline{a}$ of a finite set of real numbers $\{a_1, \ldots, a_k\}$ satisfies

$$\min\{a_1, \ldots, a_k\} \leq \overline{a} \leq \max\{a_1, \ldots, a_k\}$$

the single cluster partition always minimizes $Q(\Pi)$. Considering the entire set of nodes as a single cluster with its own coordinator that communicates with the root introduces an additional unnecessary "bureaucracy" layer that only increases communications. We seek a trade-off which yields clusters with "good" sizes (this is rigorously defined in Sect. 12.4).
2. Computation of $Q(\Pi)$ involves future values $\mathbf{v_n}(t)$, which are not available at time t_j when the clustering is performed.
3. Since the communication overhead of the balancing process is proportional to the size of a cluster, the individual clusters' sizes should affect the clustering quality $q(\pi)$.

In the next section we address these problems.

12.4 Monitoring Threshold Functions Through Clustering: Implementation

We argue that in addition to the average magnitude of the variations $\mathbf{v_n}(t) - \mathbf{v_n}(t_j)$ inside the cluster π, the cluster's size also affects the frequency of updates, and, as a result, the communication load. We therefore define the quality of the cluster π by

$$q(\pi) = \frac{1}{|\pi|} \left\| \sum_{\mathbf{n} \in \pi} [\mathbf{v_n}(t) - \mathbf{v_n}(t_j)] \right\| + \alpha |\pi|, \tag{12.8}$$

where α is a nonnegative scalar parameter. The quality of the partition $\Pi = \{\pi_1, \dots, \pi_k\}$ is defined by

$$Q(\Pi) = \max_{i \in \{1,\dots,k\}} q(\pi_i), \tag{12.9}$$

When $\alpha = 0$ the partition that minimizes $Q(\Pi)$ is a single cluster partition (that we would like to avoid). When $\max_{\mathbf{n}} \left\| \mathbf{v_n}(t) - \mathbf{v_n}(t_j) \right\| \leq \alpha$ the optimal partition is made up of n singleton clusters. In this chapter we focus on

$$0 < \alpha < \max_{\mathbf{n} \in \mathbf{N}} \left\| \mathbf{v_n}(t) - \mathbf{v_n}(t_j) \right\|. \tag{12.10}$$

The constant α depends on t and t_j, and below we show how to avoid this dependence.

Computation of $Q(\Pi)$ required for the clustering procedure is described below. In order to compute $Q(\Pi)$ at time t_j one needs to know $\mathbf{v_n}(t)$ at a future time $t > t_j$ which is not available. While the future behavior is not known, we shall use past values of $\mathbf{v_n}(t)$ for prediction. For each node $\mathbf{n}$ we build "history" vectors $\mathbf{h_n}(t_j)$ defined as follows:

1. $\mathbf{h_n}(t_0) = 0$
2. if ($\mathbf{h_n}(t_j)$ is already available)
   $\mathbf{h_n}(t_{j+1}) = \mathbf{h_n}(t_j)$
   for t increasing from t_j to t_{j+1} do
   $$\mathbf{h_n}(t_{j+1}) = \frac{1}{2}\mathbf{h_n}(t_{j+1}) + [\mathbf{v_n}(t) - \mathbf{v_n}(t_j)]$$

The vectors $\mathbf{h_n}(t_j)$ accumulate the history of changes, with older changes assigned smaller weights. We shall use the vectors $\{\mathbf{h_n}(t_j)\}$ to generate a node partition at time t_j. We note that normalization of the vector set that should be clustered does not change the induced optimal partitioning of the nodes. When the vector set is normalized by the magnitude of the longest vector in the set, the range for α conveniently shrinks to $[0, 1]$. In what follows we set $h = \max_{\mathbf{n}\in\mathbf{N}} ||\mathbf{h_n}(t_j)||$, assume that $h > 0$, and describe a "greedy" clustering procedure for the normalized vector set

$$\{\mathbf{a}_1, \dots, \mathbf{a}_n\},\ \mathbf{a}_i = \frac{1}{h}\mathbf{h}_{\mathbf{n}_i}(t_j),\ i = 1, \dots, n.$$

We start with the n cluster partition Π^n (each cluster is a singleton). We set $k = n$ and loop the following procedure until the number of clusters reduces to $k = 2$.

Algorithm 12.4.1 (Incremental Clustering)

- *Set* $k = n$.
- *do until* $k > 2$:

  1. *in partition* Π^k *identify cluster* π_j *of maximal quality, i.e.,*
  $$q(\pi_j) \geq q(\pi_i)\ i \neq j.$$
  2. *identify cluster* π_i *so that the merger of* π_i *with* π_j *produces a cluster of smallest possible quality, i.e.,*
  $$q\left(\pi_j \bigcup \pi_i\right) \leq q\left(\pi_j \bigcup \pi_l\right),\ l \neq j,$$
  where cluster's quality is defined by (12.8).
  3. *Build partition* Π^{k-1} *by merging clusters* π_j *and* π_i.
  4. *Set* $k = k - 1$.
  5. *go to Step 1.*

- *Stop.*

The final partition is selected from the $n - 1$ *partitions* $\{\Pi^2, \dots, \Pi^n\}$ *as the one that minimizes* Q.

Note that node constraints δ do not have to be equal (see Algorithm 12.4.2, Step 2). Taking into account the distribution of the data streams at each node can further reduce communication. We illustrate this statement by a simple example involving two nodes. If, for example, there is reason to believe that the inequality

$$2\|\mathbf{v}_1(t) - \mathbf{v}_1(t_i)\| \le \|\mathbf{v}_2(t) - \mathbf{v}_2(t_i)\| \tag{12.11}$$

always holds, then the number of node violations may be reduced by imposing node dependent constraints

$$\|\mathbf{v}_1(t) - \mathbf{v}_1(t_i)\| < \delta_1 = \frac{2}{3}\delta, \text{ and } \|\mathbf{v}_2(t) - \mathbf{v}_2(t_i)\| < \delta_2 = \frac{4}{3}\delta$$

so that the wider varying stream at the second node enjoys larger "freedom" of change, while the inequality

$$\left\|\frac{\mathbf{v}_1(t) + \mathbf{v}_2(t)}{2} - \frac{\mathbf{v}_1(t_i) + \mathbf{v}_2(t_i)}{2}\right\| < \frac{\delta_1 + \delta_2}{2} = \delta$$

holds true. Assigning "weighted" local constraints requires information provided by (12.11). With no additional assumptions about the stream data distribution this information is not available. Unlike [17] we refrain from making assumptions regarding the underlying data distributions; instead, we estimate the weights through past values $\mathbf{v}_\mathbf{n}(t)$, $\mathbf{n} \in \mathbf{N}$.

At the initial time t_0 all nodes report their vectors $\mathbf{v}_\mathbf{n}(t_0)$ to the root, the root computes the average, and the distance $\delta(\mathbf{r})$ from the average to the boundary of $\mathbf{Z}_+(f)$. At this point we define $\delta(\mathbf{n}) = \delta(\mathbf{r})$, for each $\mathbf{n} \in \mathbf{N}$.

We now focus on a particular node $\mathbf{n}$. Consider first m time instances $t^1, t^2, \dots, t^m$ and the vector set

$$\mathbf{V}'_\mathbf{n} = \{\mathbf{v}'_\mathbf{n}(t^1), \dots, \mathbf{v}'_\mathbf{n}(t^m)\},$$

where

$$\mathbf{v}'_\mathbf{n}(t^m) = \mathbf{v}_\mathbf{n}(t^m),\ \mathbf{v}'_\mathbf{n}(t^{m-1}) = \frac{1}{2}\mathbf{v}_\mathbf{n}(t^{m-1}),\ \dots,\ \mathbf{v}'_\mathbf{n}(t^1) = \frac{1}{2^{m-1}}\mathbf{v}_\mathbf{n}(t^1).$$

The node constraint $\delta(\mathbf{n})$ introduced below depends on the arithmetic mean $\mu(\mathbf{V}'_\mathbf{n})$ and the central second moment

$$\sigma^2(\mathbf{V}'_\mathbf{n}) = \sum_{i=0}^{m} \left(\mathbf{v}'_\mathbf{n}(t_i) - \mu(\mathbf{V}'_\mathbf{n})\right)^T \left(\mathbf{v}'_\mathbf{n}(t_i) - \mu(\mathbf{V}'_\mathbf{n})\right)$$

of the node $\mathbf{n}$. We denote $\boldsymbol{\mu}(\mathbf{V}'_{\mathbf{n}})$ by $\boldsymbol{\mu}_{\mathbf{n}}$, and $\sigma^2(\mathbf{V}'_{\mathbf{n}})$ by $\sigma^2_{\mathbf{n}}$. Since $\left\|\mathbf{v}'_{\mathbf{n}}(t^i) - \boldsymbol{\mu}_{\mathbf{n}}\right\| \leq \sqrt{\frac{m-1}{m}\sigma^2_{\mathbf{n}}}$, $i = 1, \ldots, m$ (see Appendix 1) we define

$$W_{\mathbf{n}}(t^m) = W_{\mathbf{n}} = ||\boldsymbol{\mu}_{\mathbf{n}}|| + \sqrt{\frac{m-1}{m}\sigma^2_{\mathbf{n}}}.$$

We note that although the bound $\sqrt{\frac{m-1}{m}\sigma^2_{\mathbf{n}}}$ may be very conservative, the same conservative criterion is applied uniformly to every node.

If at time t^m the root constraint $\delta(\mathbf{r})$ is updated, each node $\mathbf{n}$ broadcasts $W_{\mathbf{n}}(t^m) = W_{\mathbf{n}}$ to the root, the root computes $W = \sum_{\mathbf{n}\in\mathbf{N}} W_{\mathbf{n}}$, and transmits the updated $\delta(\mathbf{n}) = w_{\mathbf{n}}\delta(\mathbf{r})$ where $w_{\mathbf{n}} = n \times \frac{W_{\mathbf{n}}}{W}$ (so that $\sum_{\mathbf{n}\in\mathbf{N}} w_{\mathbf{n}} = n$) back to node $\mathbf{n}$. For a coordinator $\mathbf{c}$ of a node cluster π the constraint $\delta(\mathbf{c}) = \frac{1}{|\pi|}\sum_{\mathbf{n}\in\pi}\delta(\mathbf{n})$.

Algorithm 12.4.2 *Monitoring Threshold Function with Clustering*

- *A node is designated as a root* $\mathbf{r}$.
- *The root sets* $i = 0$.
- *Until end of stream*

1. *The root sends a request to each node* $\mathbf{n}$ *for the vectors* $\mathbf{v}_{\mathbf{n}}(t_i)$. *The nodes respond to the root. The root computes the distance* δ *between the mean* $\frac{1}{n}\sum_{\mathbf{n}\in\mathbf{N}} \mathbf{v}_n(t_i)$ *and the zero set* $\mathbf{Z}_f$ *of the function* f. *The root transmits* δ *to each node.*
2. *set* $t = t_i$
   do
   - *set violation*$= 0$, $t = t + 1$
   - *for each* $\mathbf{n} \in \mathbf{N}$
     - *If* $||\mathbf{v}_{\mathbf{n}}(t) - \mathbf{v}_{\mathbf{n}}(t_i)|| \geq \delta$
       - *violation++*
     - *endif*
   - *while (violation=0)*
3. *set* $i = i + 1$, *and* $t_i = t$
4. *violator node* $\mathbf{n}$ *notifies the root about the violation of its local constraint* δ
5. *The root requests vectors* $\mathbf{v}_{\mathbf{n}}(t_i)$ *and weights* $W_{\mathbf{n}}(t_i)$.
   The root forms a partition $\Pi = \{\pi_1, \ldots, \pi_k\}$ *and sends node and coordinator constraints* $\delta(\mathbf{n})$ *and* $\delta(\mathbf{c})$ *to nodes and coordinators.*
6. *do for each* $\pi \in \Pi$
   - *do for each* $\mathbf{n} \in \pi$
     - *If* $\delta(\mathbf{n}) \leq ||\mathbf{v}_{\mathbf{n}}(t) - \mathbf{v}_{\mathbf{n}}(t_i)||$

$$\textit{If}\ \delta(\mathbf{c}) \le \frac{1}{|\pi|}\left\|\sum_{\mathbf{n}\in\pi}\mathbf{v}_{\mathbf{n}}(t) - \sum_{\mathbf{n}\in\pi}\mathbf{v}_{\mathbf{n}}(t_i)\right\|$$

notify root about coordinator violation

goto Step 3

endif

endif

- *Stop*

Node constraints $\delta(\mathbf{n})$ based on the first moment only are introduced in [2]. In the next section we provide monitoring results for node constraints based on the first and second moments, and compare the results with those reported in [2] as well as monitoring with no clustering reported in [19].

12.5 Experimental Results

This section presents few experimental results of monitoring with clustering. Monitoring algorithms are always applied to the data considered in [24] which is described next.

12.5.1 Data

The data streams analyzed in this section are generated from the Reuters Corpus RCV1-V2.[1] The data consists of $781,265$ tokenized documents with document ID ranging from 2651 to 810596. We simulate n streams by arranging the feature vectors in ascending order with respect to document ID, and selecting feature vectors for the stream in the round-robin fashion.

Each document in the Reuters Corpus RCV1-V2 is labeled as belonging to one or more categories. We label a vector as "relevant" if it belongs to the "CORPORATE/INDUSTRIAL" ("CCAT") category, and "spam" otherwise. Following [24] we focus on three features: "bosnia," "ipo," and "febru." Each experiment was performed with 10 nodes, where each node holds a sliding window containing the last 6,700 documents it received.

First we use 67,000 documents to generate initial sliding windows. The remaining 714,265 documents are used to generate datastreams, hence the selected feature information gain is computed 714,265 times. Based on all the documents contained in the sliding window at each one of the 714,266 time instances we compute and graph 714,266 information gain values for the feature "bosnia" (see Fig. 12.1). For

[1] http://leon.bottou.org/projects/sgd.

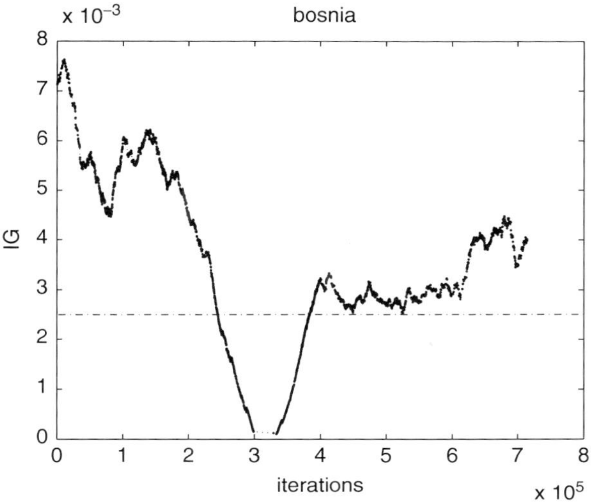

Fig. 12.1 information gain values for the feature "bosnia"

the experiments described below, the threshold value r is predefined, and the goal is to monitor the inequality $f(\mathbf{v}) - r > 0$ while minimizing communication between the nodes. We also assume that new texts arrive simultaneously at each node. We define a broadcast as one time transmission of information between different nodes.

12.5.2 Monitoring with Incremental Clustering

The previous work [2] reported monitoring results obtained with the incremental clustering algorithm (Algorithm 12.4.1) with weights $W_{\mathbf{n}} = \|\boldsymbol{\mu}_{\mathbf{n}}\|$ only. We shall call this implementation of the algorithm "first moment incremental clustering" (FMIC). The clustering algorithm with weights $W_{\mathbf{n}} = ||\boldsymbol{\mu}_{\mathbf{n}}|| + \sqrt{\frac{m-1}{m}\sigma_{\mathbf{n}}^2}$ introduced in this chapter will be referred to as the "second moment incremental clustering" (SMIC). In this section we report and compare results generated by the algorithms for the threshold $r = 0.0025$ and $\alpha = 0.05, 0.10, \ldots, 0.95$.

The best result with respect to α obtained by an application of FMIC to the feature "febru," is presented in Table 12.2. The clustering approach in this case is particularly successful—coordinators' constraints are not violated, and the root mean updates are decreased significantly as compares, for example, to results reported in [19].

The corresponding results generated by SMIC are provided in Table 12.3. Results in Table 12.3 show a significant decrease in the number of broadcasts as compared to results in Table 12.2.

Next we turn to the features "ipo" and "bosnia." In both cases we run monitoring with FMIC and SMIC, allowing $\alpha = 0.05, 0.10, \ldots, 0.95$, and report results with

Table 12.2 Number of root and coordinator mean computations, and total broadcasts for feature "febru" with threshold $r = 0.0025$ and the "first moment clustering"

Norm	Best α	Root mean update	Coordinator mean update	Total broadcasts
l_1	0.70	1431	0	38665
l_2	0.80	1317	0	35597
l_∞	0.65	1409	0	38093

Table 12.3 Number of root and coordinator mean computations, and total broadcasts for feature "febru" with threshold $r = 0.0025$ and the "second moment clustering"

Norm	Best α	Root mean update	Coordinator mean update	Total broadcasts
l_1	0.85	883	0	23859
l_2	0.75	833	0	22509
l_∞	0.75	854	0	23076

Table 12.4 Number of root and coordinator mean computations, and total broadcasts for feature "ipo" with threshold $r = 0.0025$ and the "first moment clustering"

Norm	Best α	Root mean update	Coordinator mean update	Total broadcasts
l_1	0.15	5455	829	217925
l_2	0.10	7414	1782	296276
l_∞	0.10	9768	2346	366300

Table 12.5 Number of mean computations, and broadcasts, for feature "bosnia" with threshold $r = 0.0025$, no clustering

Norm	Mean updates	Broadcasts
l_1	3053	79378
l_2	4006	104156
l_∞	3801	98826

the lowest number of broadcasts. The results obtained for "ipo" with FMIC are presented in Table 12.4. Application of SMIC leads to results provided in Table 12.4. The tables demonstrates significant inside cluster activity, and a significant decrease in broadcasts due to the second moment.

Finally we turn to the feature "bosnia." Monitoring procedure presented in [19] and involving no clustering produced results collected in Table 12.5. Application of FMIC to monitoring this feature information gain reported in [2] was described as the "less successful." FMIC leads to a slight decrease in the number of broadcasts

Table 12.6 Number of root and coordinator mean computations, and total broadcasts for feature "bosnia" with threshold $r = 0.0025$ and the "first moment clustering"

Norm	Best α	Root mean update	Coordinator mean update	Total broadcasts
l_1	0.65	3290	2	89128
l_2	0.55	3502	7	97602
l_∞	0.60	3338	2	91306

Table 12.7 Number of root and coordinator mean computations, and total broadcasts for feature "bosnia" with threshold $r = 0.0025$ and the "second moment clustering"

Norm	Best α	Root mean update	Coordinator mean update	Total broadcasts
l_1	0.65	1749	8	47717
l_2	0.75	1940	4	52510
l_∞	0.65	1756	8	47958

Table 12.8 Number of root and coordinator mean computations, and total broadcasts for feature "ipo" with threshold $r = 0.0025$ and the "second moment clustering"

Norm	Best α	Root mean update	Coordinator mean update	Total broadcasts
l_1	0.50	4585	121	127345
l_2	0.35	6304	421	180536
l_∞	0.30	8405	842	240455

in case of the l_2 and l_∞ norms (see Table 12.6). In case of the l_1 norm, the number of broadcasts increases. The second moment clustering further significantly reduces the number of broadcasts, see Table 12.7.

In the next section we briefly recall a number of well known clustering algorithms, and indicate how those can be used for monitoring data streams.

12.6 Conventional Clustering Algorithms

This section briefly reviews a number of classical clustering algorithms we propose to use for node clustering. The algorithms are:

1. Principal Direction Divisive Partitioning (PDDP), [3];
2. batch k-means (see e.g. [13, 28]),
3. incremental k-means (e.g. [12, 26]),
4. the combination of batch and incremental k-means [26], and [27].

An application of k-means requires an initial partition of the data. While a number of initialization methods for the k-means clustering algorithm are available in the literature (see e.g. recent survey [6]) we focus on PDDP which is briefly described next.

12.6.1 PDDP

A good partitioning of a vector set into a number of subsets is a difficult problem even in the case when the required number of subsets is only two. There is, however, an exception. When the dimension of the vector space is one, i.e. one has to deal with a scalar set, the problem is relatively easy. While real-life data is rarely one dimensional, a least squares one dimensional approximation can be constructed and used to cluster a multidimensional vector set. The Principal Direction Divisive Partitioning algorithm briefly recalled below does just that. In the reminder of this section we demote by $\|\mathbf{a}\|$ the l_2 norm of a vector $\mathbf{a}$.

For a vector $\mathbf{a}$ and a line $\mathbf{l}$ in $\mathbf{R}^d$ denote by $\mathrm{P}_\mathbf{l}(\mathbf{a})$ the orthogonal projection of $\mathbf{a}$ on $\mathbf{l}$. For a set of vectors $\mathcal{A} = \{\mathbf{a}_1, \ldots, \mathbf{a}_m\}$ and a line $\mathbf{l}$ in $\mathbf{R}^d$ denote by $\mathrm{P}_\mathbf{l}(\mathcal{A})$ the set of projections $\{\mathrm{P}_\mathbf{l}(\mathbf{a}_1), \ldots, \mathrm{P}_\mathbf{l}(\mathbf{a}_m)\}$. For a fixed vector set $\mathcal{A}$ the quantity

$$\sum_{i=1}^{m} ||\mathbf{a}_i - \mathrm{P}_\mathbf{l}(\mathbf{a}_i)||^2$$

depends on the line $\mathbf{l}$. A line that minimizes this quantity (and provides the best least squares fit for the set $\mathcal{A}$) defines a principal direction. This line passes through the arithmetic mean $\boldsymbol{\mu} = \boldsymbol{\mu}(\mathcal{A})$ of the vector set $\mathcal{A}$, and its direction vector is an eigenvector of the matrix BB^T that corresponds to the maximal eigenvalue. Here $B = [\mathbf{a}_1, \ldots, \mathbf{a}_m] - \boldsymbol{\mu}\mathbf{e}^T$, and $\mathbf{e}$ is a vector of ones (for details see e.g. [18]).

A basic step of the Principal Direction Divisive Partitioning algorithm (PDDP) is the following:

1. Given a set of vectors $\mathcal{A}$ in $\mathbf{R}^d$ determine the one dimensional line $\mathbf{l}$ that provides the "best" approximation to $\mathcal{A}$.
2. Project $\mathcal{A}$ onto $\mathbf{l}$, and denote the projection of the set $\mathcal{A}$ by $\mathcal{P}$ (note that $\mathcal{P}$ is just a set of scalars). Denote the projection of a vector $\mathbf{a}$ by p.
3. Partition $\mathcal{P}$ into two subsets $\mathcal{P}_1$ and $\mathcal{P}_2$.
4. Generate the induced partition $\{\mathcal{A}_1, \mathcal{A}_2\}$ of $\mathcal{A}$ as follows:

$$\mathcal{A}_1 = \{\mathbf{a} \,:\, p \in \mathcal{P}_1\}, \text{ and } \mathcal{A}_2 = \{\mathbf{a} \,:\, p \in \mathcal{P}_2\} \tag{12.12}$$

The algorithm divides the entire collection into two clusters by using the principal direction. Each of these two clusters will be divided into two sub-clusters using the same process recursively. The subdivision of a cluster is stopped when the cluster satisfies a certain "quality" criterion (such as, for example, cluster size, number of clusters, or cluster quality).

Implementation of the algorithm requires computation of the largest eigenvalue of the symmetric matrix BB^T. In many cases this task may not be performed analytically. While in the text mining application described in Sect. 12.2 the space dimension $d = 4$ one of the coordinates is an affine function of three others (see (12.4)). We now consider the case when each d dimensional data vector $\mathbf{a}$ can be written as

$$\mathbf{a} = \begin{bmatrix} \mathbf{b} \\ C\mathbf{b} + \mathbf{d} \end{bmatrix},$$

where $\mathbf{b} \in \mathbf{R}^{d_1}$, $\mathbf{d} \in \mathbf{R}^{d_2}$, $d_1 + d_2 = d$, and C is an $d_2 \times d_1$ matrix. For a vector set $\mathcal{B} = \{\mathbf{b}_1, \ldots, \mathbf{b}_m\}$ one has $\boldsymbol{\mu}(\mathcal{A}) = \begin{bmatrix} \boldsymbol{\mu}(\mathcal{B}) \\ C\boldsymbol{\mu}(\mathcal{B}) + \mathbf{d} \end{bmatrix}$. We now turn to the matrix BB^T. Denoting $\mathbf{b}_i - \boldsymbol{\mu}(\mathcal{B})$ by $\mathbf{v}_i$ we obtain

$$B = \begin{bmatrix} \mathbf{v}_1 \ldots \mathbf{v}_m \\ C\mathbf{v}_1 \ldots C\mathbf{v}_m \end{bmatrix} = \begin{bmatrix} I \\ C \end{bmatrix} \begin{bmatrix} \mathbf{v}_1 \ldots \mathbf{v}_m \end{bmatrix},$$ where I is the $d_1 \times d_1$ identity matrix.

Since rank $B \leq d_1$ one has rank $BB^T \leq d_1$, and the number of nonzero eigenvalues of BB^T does not exceed d_1. For our text mining application $d_1 = 3$, and the nonzero eigenvalues can be obtained by solving a cubic equation, i.e., the eigenvalues for BB^T corresponding to the largest eigenvalue can be obtained just by solving a system of linear equations.

PDDP by itself generates good clustering results. Those could be further improved by applying k-means clustering to partitions generated by PDDP (see e.g. [18]). Next we briefly recall a number of versions of k-means.

12.6.2 Batch k-Means

A work horse of clustering mentioned already in [28] and most often attributed to [13] batch k-means is by far most popular clustering algorithm. The algorithm is scalable, and easy to implement. The algorithm is centered around the concept of "centroid"-the best vector representative for a vector set introduced below (see [9, 10]).

For a set of vectors $\mathcal{A} = \{\mathbf{a}_1, \ldots, \mathbf{a}_m\} \subset \mathbf{R}^n$, and a "distance" function $d(\mathbf{x}, \mathbf{a})$ define a centroid $\mathbf{c} = \mathbf{c}(\mathcal{A})$ of the set $\mathcal{A}$ as a solution of the minimization problem

$$\mathbf{c} = \arg\min \left\{ \sum_{\mathbf{a} \in \mathcal{A}} d(\mathbf{x}, \mathbf{a}), \ \mathbf{x} \in \mathcal{C} \right\}, \tag{12.13}$$

where $\mathcal{C}$ is a predefined subset of the space.

We call d a "distance" function because even in the classical implementation of k-means $d(\mathbf{x}, \mathbf{a}) = \|\mathbf{x} - \mathbf{a}\|_2^2$, the square of the l_2 norm, that fails to be a distance function (the triangle inequality does not hold). Further, k-means works with a wide class of functions called Bregman divergences and failing to be distances (Kullback–Leibler divergence is one of them, see [1]).

The quality of the set $\mathscr{A}$ is denoted by $q(\mathscr{A})$ and is defined by

$$q(\mathscr{A}) = \sum_{\mathbf{a} \in \mathscr{A}} d(\mathbf{c}, \mathbf{a}), \quad \text{where } \mathbf{c} = \mathbf{c}(\mathscr{A}) \tag{12.14}$$

(we set $q(\emptyset) = 0$ for convenience). Let $\Pi = \{\pi_1, \ldots, \pi_k\}$ be a partition of $\mathscr{A}$, i.e.

$$\bigcup_i \pi_i = \mathscr{A}, \text{ and } \pi_i \cap \pi_j = \emptyset \text{ if } i \neq j.$$

We define the quality of the partition Π by

$$Q(\Pi) = q(\pi_1) + \cdots + q(\pi_k). \tag{12.15}$$

We aim to find a partition $\Pi^{\min} = \{\pi_1^{\min}, \ldots, \pi_k^{\min}\}$ that *minimizes* the value of the objective function Q. The problem is known to be NP-hard, and we are looking for algorithms that generate "reasonable" solutions. It is easy to see that centroids and partitions are associated as follows:

1. Given a partition $\Pi = \{\pi_1, \ldots, \pi_k\}$ of the set $\mathscr{A}$ one can define the corresponding centroids $\{\mathbf{c}(\pi_1), \ldots, \mathbf{c}(\pi_k)\}$ by:

$$\mathbf{c}(\pi_i) = \arg\min \left\{ \sum_{\mathbf{a} \in \pi_i} d(\mathbf{x}, \mathbf{a}), \ \mathbf{x} \in \mathscr{C} \right\}. \tag{12.16}$$

2. For a set of k "centroids" $\{\mathbf{c}_1, \ldots, \mathbf{c}_k\}$ one can define a partition $\Pi = \{\pi_1, \ldots, \pi_k\}$ of the set $\mathscr{A}$ by:

$$\pi_i = \{\mathbf{a} : \mathbf{a} \in \mathscr{A}, \ d(\mathbf{c}_i, \mathbf{a}) \leq d(\mathbf{c}_l, \mathbf{a}) \text{ for each } l = 1, \ldots, k\} \tag{12.17}$$

(we break ties arbitrarily). Note that, in general, $\mathbf{c}(\pi_i) \neq \mathbf{c}_i$.

The classical batch k-means algorithm is a procedure that iterates between the two steps described above to generate a partition Π' from a partition Π.

While $0 \leq Q(\Pi') \leq Q(\Pi)$ and the process described above converges, it rarely converges to the global minimum. Even in a simple scalar case $\mathscr{A} = \{0, 2, 3\}$, and the initial partition $\Pi^{(0)} = \left\{\pi_1^{(0)}, \pi_2^{(0)}\right\}$ where $\pi_1^{(0)} = \{0, 2\}$, and $\pi_2^{(0)} = \{3\}$ an application of batch k-means to $\Pi^{(0)}$ does not change the partition, and misses a better partition $\Pi^{(1)} = \left\{\pi_1^{(1)}, \pi_2^{(1)}\right\}$ with $\pi_1^{(1)} = \{0\}$, and $\pi_2^{(1)} = \{2, 3\}$. The reason for this phenomenon along with a possible remedy is suggested in [7]. Before the relevant material is briefly recalled in the next section we remark that an application of PDDP to the scalar dataset $\mathscr{A}$ generates partition $\Pi^{(1)}$. This observation suggests to use PDDP to generate initial partitions to k-means like algorithms.

12.6.3 Incremental k-Means

The failure of batch k-means to discover a better partition $\Pi^{(1)}$ stems from a simple fact that Step 2 of the procedure ignores change of centroids due to data-vectors' movement governed by (12.17). A way to accurately account for the centroid change is to allow a single data-vector movement during one iteration of the algorithm. This version of k-means is described, for example, in the classical manuscript [12].

While more accurate, incremental k-means changes cluster affiliation of only one vector per iteration. As compared to batch k-means the algorithm requires many more iterations to converge, hence is time consuming. We next discuss a "merger" of two algorithms.

12.6.4 Batch k-Means Followed by Incremental k-Means

While more accurate incremental k-means is not as fast as the batch algorithm. To benefit from speed of the batch algorithm and accuracy of the incremental k-means a number of contributions suggested to "merge" both algorithms as follows:

1. run batch k-means until it stops.
2. run one iteration of incremental k-means
3. if the iteration incremental k-means changed the partition
        goto Step 1
    else
        Stop.

All numerical computations associated with Step 2 of the algorithm have been already performed in Step 1. The improvement over batch k-means comes, therefore, at virtually no additional computational expense [8]. The possibility of the "merger" was first indicated, perhaps, in [26], and formally introduced in [27]. Later the "merger" was independently rediscovered by many other authors.[2] Sequential application

$$\text{PDDP}\longrightarrow\text{batch } k\text{-means}\longrightarrow\text{incremental } k\text{-means}$$

generates good tight clusters. Next we describe how these tight clusters can be used for node clustering.

[2]Confirming the old adage that "success has many parents while failure is an orphan."

12.6.5 Node Clustering with Classical Clustering Algorithms

To simplify the exposition we first consider a two cluster $\{\pi_1, \pi_2\}$ partition problem for a given set of n vectors $\mathscr{A} = \{\mathbf{a}_1, \ldots, \mathbf{a}_n\} \subset \mathbf{R}^d$. We are seeking a partition $\Pi = \{\pi_1, \pi_2\}$ so that

$$\mathscr{A} = \pi_1 \bigcup \pi_2, \ \pi_1 \bigcap \pi_2 = \emptyset$$

and the partition Π quality $Q(\Pi)$ given by

$$Q(\Pi) = \max\left\{\left\|\frac{1}{|\pi_1|}\sum_{\mathbf{a}\in\pi_1}\mathbf{a}\right\|, \left\|\frac{1}{|\pi_2|}\sum_{\mathbf{a}\in\pi_2}\mathbf{a}\right\|\right\}$$

is minimized. Due to convexity of any norm one has

$$\begin{aligned}\left\|\frac{1}{|\mathscr{A}|}\sum_{\mathbf{a}\in\mathscr{A}}\mathbf{a}\right\| &= \left\|\frac{|\pi_1|}{|\mathscr{A}|}\frac{1}{|\pi_1|}\sum_{\mathbf{a}\in\pi_1}\mathbf{a} + \frac{|\pi_2|}{|\mathscr{A}|}\frac{1}{|\pi_2|}\sum_{\mathbf{a}\in\pi_2}\mathbf{a}\right\| \\ &\leq \frac{|\pi_1|}{|\mathscr{A}|}\left\|\frac{1}{|\pi_1|}\sum_{\mathbf{a}\in\pi_1}\mathbf{a}\right\| + \frac{|\pi_2|}{|\mathscr{A}|}\left\|\frac{1}{|\pi_2|}\sum_{\mathbf{a}\in\pi_2}\mathbf{a}\right\| \\ &\leq \frac{|\pi_1|}{|\mathscr{A}|}Q(\Pi) + \frac{|\pi_2|}{|\mathscr{A}|}Q(\Pi) = Q(\Pi).\end{aligned}$$

This inequality shows that the norm of the mean is a lower bound for $Q(\Pi)$. We next show how to build an optimal partition for a special particular case of the data set.

Assume that $n = 2m$, and the vector set $\mathscr{A}$ consists of two identical copies of m vectors, i.e.

$$\mathscr{A} = \{\mathbf{a}_1, \ldots, \mathbf{a}_m, \mathbf{a}_1, \ldots, \mathbf{a}_m\}.$$

If $\pi_1^o = \pi_2^o = \{\mathbf{a}_1, \ldots, \mathbf{a}_m\}$, then $|\pi_1^o| = |\pi_2^o| = \frac{1}{2}|\mathscr{A}|$, and

$$\max\left\{\frac{1}{|\pi_1^o|}\left\|\sum_{\mathbf{a}\in\pi_1^o}\mathbf{a}\right\|, \frac{1}{|\pi_2^o|}\left\|\sum_{\mathbf{a}\in\pi_2^o}\mathbf{a}\right\|\right\} = \left\|\frac{1}{|\mathscr{A}|}\sum_{\mathbf{a}\in\mathscr{A}}\mathbf{a}\right\|,$$

i.e., $\{\pi_1^o, \pi_2^o\}$ is an optimal partition.

This observation motivates the following two cluster $\Pi = \{\pi_1, \pi_2\}$ partition strategy:

1. Apply any clustering algorithm to the dataset $\mathscr{A}$ to generate clusters of size 2.
2. Select one vector from each cluster generated and assign selected vectors to cluster π_1.
3. Assign remaining vectors to cluster π_2.

Generalization of this strategy to a k cluster partition and full description of the algorithm is beyond the scope of this chapter and will be provided elsewhere.

12.7 Conclusions

In this chapter we consider application of clustering to monitoring data streams in a distributed system. Unlike standard clustering algorithms that aiming at collections of similar data items into same clusters, monitoring requires clusters with *dissimilar* vectors canceling each other as much as possible. A straightforward application of a standard clustering algorithm is, therefore, not possible.

We devise a specific clustering strategy that yields a reduction in communication load. The proposed clustering depends on a scalar parameter α, and may be too slow for applications involving systems with large number of nodes. Dependence of the number of broadcasts on α is not understood at this point and should be further investigated. Figure 12.2 shows the number of broadcasts for 18 values $\alpha = 0.05, 0.10, \dots, 0.95$. The smallest number of broadcasts corresponds to $\alpha = 0.30$ (as reported in Table 12.8). Next we run monitoring for 98 values of $\alpha = 0.01, 0.02, \dots, 0.99$ (see Fig. 12.3). This time the smallest number of broadcasts corresponds to $\alpha = 0.16$. Zooming in does not indicate any particularly useful property of the function.

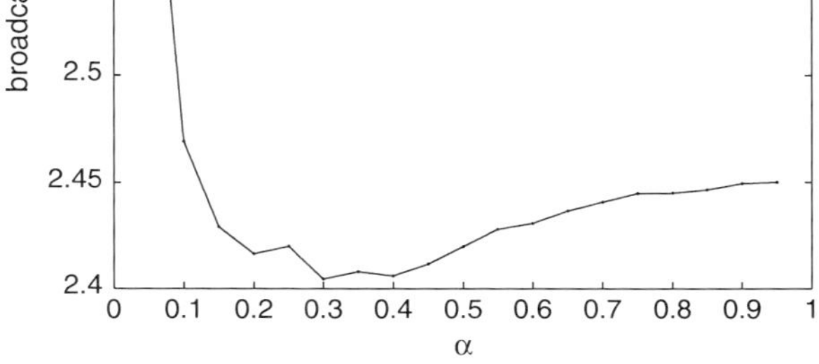

Fig. 12.2 l_∞ norm, $\alpha = 0.05, 0.10, \dots, 0.95$ vs. broadcasts, for feature "ipo"

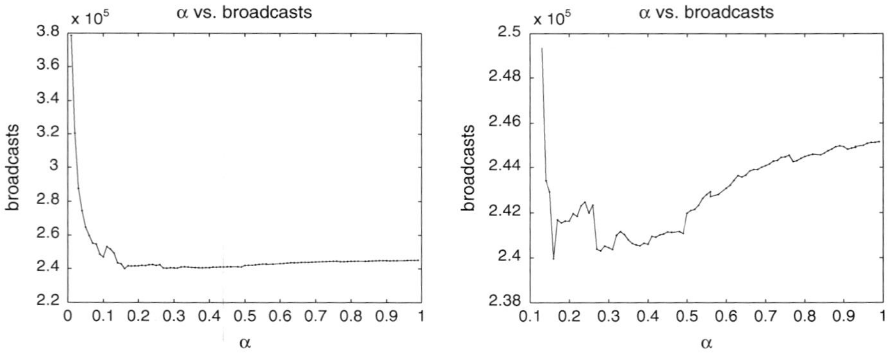

Fig. 12.3 feature "ipo", l_∞ norm, $\alpha = 0.01, 0.02, \ldots, 0.99$ vs. broadcasts, (*left*) and zoom in for $\alpha = 0.14, 0.15, \ldots, 0.99$ (*right*)

Each recomputation of δ (the distance from the mean $\frac{1}{n}\sum_{\mathbf{n}\in\mathbf{N}} \mathbf{v}_n(t_i)$ and the zero set $\mathbf{Z}_f$ of the function f) triggers recomputations of node constraints $\delta(\mathbf{n})$. This chapter uses the first and second moments to recompute node constraints. We plan to consider additional statistical metrics such as, for example, median for node constraints computations.

A possible applications of classical clustering algorithms is an additional research direction that may lead to scalable clustering procedures. While the experimental results demonstrate than communication savings may depend on the choice of a norm or a "distance" function many of the proposed clustering algorithms can be applied with Bregman divergences [29].

Appendix 1: First and Second Moments

In what follows we consider the auxiliary problem: "Let $\mathbf{X}$ be a vector set of size m with mean $\boldsymbol{\mu}$ and variance $\gamma > 0$. How far away from $\boldsymbol{\mu}$ a vector $\mathbf{x} \in \mathbf{X}$ can get?" The answer to this question is provided below. To simplify the exposition we first assume $\boldsymbol{\mu} = 0$.

Let $\mathscr{X}(m, \gamma)$ be a family of sets $\mathbf{X} = \{\mathbf{x}_1, \ldots, \mathbf{x}_m\} \subset \mathbf{R}^d$ with $\boldsymbol{\mu}(\mathbf{X}) = 0$, and $\sigma^2(\mathbf{X}) = \sum_{i=1}^{m} (\mathbf{x}_i - \boldsymbol{\mu}(\mathbf{X}))^T (\mathbf{x}_i - \boldsymbol{\mu}(\mathbf{X})) = \gamma \geq 0$. In this section $||\mathbf{x}||$ stands for $||\mathbf{x}||_2$. For each $\mathbf{X} \in \mathscr{X}(m, \gamma)$ define $r(\mathbf{X})$ and $R(\gamma)$ as follows:

$$r(\mathbf{X}) = \max_{\mathbf{x}\in\mathbf{X}} \|\mathbf{x}\|, \text{ and } R(\gamma) = \sup_{\mathbf{X}\in\mathscr{X}(m,\gamma)} r(\mathbf{X}).$$

In what follows we describe sets $\overline{\mathbf{X}}_\gamma \in \mathscr{X}(m,\gamma)$ that maximize r, and the function $R(\gamma)$.

Lemma 12.1. *The function $R(\gamma)$ is a homogeneous function of degree $\frac{1}{2}$. For each positive scalar c one has $R(c\gamma) = c^{\frac{1}{2}} R(\gamma)$.*

Proof. Note that for a positive scalars t and s one has

$$tR(\gamma) = tr\left(\overline{\mathbf{X}}_\gamma\right) = r\left(t\overline{\mathbf{X}}_\gamma\right) \le r\left(\overline{\mathbf{X}}_{\gamma t^2}\right) = R(\gamma t^2),$$

and

$$sR(\gamma t^2) = sr(\overline{\mathbf{X}}_{\gamma t^2}) = r(s\overline{\mathbf{X}}_{\gamma t^2}) \le r(\overline{\mathbf{X}}_{\gamma t^2 s^2}) = R(\gamma t^2 s^2).$$

In particular when $ts = 1$ one has

$$R(\gamma) \le t^{-1} R(\gamma t^2), \text{ and } R(\gamma t^2) \le s^{-1} R(\gamma).$$

This shows that for positive t one has $tR(\gamma) = R(\gamma t^2)$ and completes the proof. □

Lemma 12.2. *Let $\mathbf{u} \in \overline{\mathbf{X}}_\gamma$ be such that $\|\mathbf{u}\| = r(\overline{\mathbf{X}}_\gamma) = R(\gamma)$. For each $\mathbf{x} \in \overline{\mathbf{X}}_\gamma$ there is a scalar c such that $\mathbf{x} = c\mathbf{u}$.*

Proof. We assume now that the claim is false. Without any loss of generality we assume that $\|\mathbf{u}\| = 1$. Let $\{\mathbf{x}_1, \dots, \mathbf{x}_k\}$ be all nonzero vectors in $\overline{\mathbf{X}}_\gamma$ so that $\mathbf{u}(\mathbf{u}^T\mathbf{x}_i) \ne \mathbf{x}_i, i = 1, \dots, k$. The vectors $\mathbf{x}_i - \mathbf{u}(\mathbf{u}^T\mathbf{x}_i) \ne 0$, $\sum_{i=1}^{k} \left[\mathbf{x}_i - \mathbf{u}(\mathbf{u}^T\mathbf{x}_i)\right] = 0$, and $\sum_{i=1}^{m} \mathbf{u}(\mathbf{u}^T\mathbf{x}_i) = 0$. Consider now the vector set $\mathbf{X}' = \{\mathbf{x}'_1, \dots, \mathbf{x}'_m\}$ where

$$\mathbf{x}'_i = \frac{1}{2}[\mathbf{x}_i - \mathbf{u}(\mathbf{u}^T\mathbf{x}_i)] + \mathbf{u}(\mathbf{u}^T\mathbf{x}_i), \ i = 1, \dots, k, \ \text{ and } \mathbf{x}'_i = \mathbf{x}_i, \ i = k+1, \dots, m.$$

We note that $\boldsymbol{\mu}(\mathbf{X}') = 0$, and $\sigma^2(\mathbf{X}') = \gamma' < \gamma$. Due to Lemma 12.1 one has

$$1 = r(\mathbf{X}') \le R(\gamma') < R(\gamma) = 1.$$

This contradiction completes the proof.

Lemma 12.3. *Let $\mathbf{u} \in \overline{\mathbf{X}}_\gamma$ be such that $\|\mathbf{u}\| = r(\overline{\mathbf{X}}_\gamma) = R(\gamma)$. For each $\mathbf{x} \in \overline{\mathbf{X}}_\gamma$, $\mathbf{x} \ne \mathbf{u}$ there is a scalar $c \le 0$ such that $\mathbf{x} = c\mathbf{u}$.*

Proof. First note that there is at least one $\mathbf{x} \in \overline{\mathbf{X}}_\gamma$ such that $\mathbf{x} = c\mathbf{u}$ with $c < 0$. We denote this c by c_-. Assume that the statement of the lemma is false. Then there is $0 < c_+ \le 1$ such that $c_+\mathbf{u} \in \overline{\mathbf{X}}_\gamma$. Let $\epsilon > 0$ be so small that $c_+ - \epsilon > 0$, and

$c_- + \epsilon > 0$. Define $\mathbf{X}'$ by substituting the vectors $c_+\mathbf{u}$ and $c_-\mathbf{u}$ by $(c_+ - \epsilon)\mathbf{u}$ and $(c_- + \epsilon)\mathbf{u}$ correspondingly, and keeping the other $m-2$ vectors unchanged. We note that $\boldsymbol{\mu}(\mathbf{X}') = 0$, and $\sigma^2(\mathbf{X}') = \gamma' < \gamma$. Due to Lemma 12.1 one has

$$\|\mathbf{u}\| = r(\mathbf{X}') \leq R(\gamma') < R(\gamma) = \|\mathbf{u}\|.$$

This contradiction completes the proof.

Lemma 12.4. *Let* $\mathbf{u} \in \overline{\mathbf{X}}_\gamma$ *be such that* $\|\mathbf{u}\| = r(\overline{\mathbf{X}}_\gamma) = R(\gamma)$. *If* $\mathbf{x} \in \overline{\mathbf{X}}_\gamma$ *and* $\mathbf{x} \neq \mathbf{u}$, *then* $\mathbf{x} = -\frac{1}{m-1}\mathbf{u}$.

Proof. Assume the opposite, i.e., there are $\mathbf{x}_1 = c_1\mathbf{u}$, and $\mathbf{x}_2 = c_2\mathbf{u}$ such that $c_1 < c_2 \leq 0$. Let $\mathbf{X}'$ be a vector set obtained from $\overline{\mathbf{X}}_\gamma$ by substituting $c_1\mathbf{u}$ by $(c_1 - \epsilon)\mathbf{u}$, $c_2\mathbf{u}$ by $(c_2 + \epsilon)\mathbf{u}$, and keeping the other vectors unchanged. Note that $\boldsymbol{\mu}(\mathbf{X}') = 0$, and

$$\sigma^2(\overline{\mathbf{X}}_\gamma) - \sigma^2(\mathbf{X}') = 2\epsilon(c_2 - c_1 - \epsilon).$$

We note that for a small positive ϵ one has $\sigma^2(\overline{\mathbf{X}}_\gamma) > \sigma^2(\mathbf{X}')$. Due to Lemma 12.1 one has

$$\|\mathbf{u}\| = r(\mathbf{X}') \leq R(\gamma') < R(\gamma) = \|\mathbf{u}\|.$$

This contradiction completes the proof.

The next statement summarizes the above results.

Theorem 12.1. *If* $\mathbf{X} = \{\mathbf{x}_1, \ldots, \mathbf{x}_m\} \in \mathbf{R}^d$ *with* $\boldsymbol{\mu}(\mathbf{X}) = \boldsymbol{\mu}$, *and* $\sigma^2(\mathbf{X}) = \gamma \geq 0$, *then*

$$\|\mathbf{x}_i - \boldsymbol{\mu}\|^2 \leq \frac{m-1}{m}\gamma.$$

Further, $\|\mathbf{x}_m - \boldsymbol{\mu}\|^2 = \frac{m-1}{m}\gamma$ *if and only if*

$$\mathbf{x}_1 = \cdots = \mathbf{x}_{m-1} = \boldsymbol{\mu} - \frac{1}{m-1}[\mathbf{x}_m - \boldsymbol{\mu}].$$

We now establish connection between first and second moments for two sets of d dimensional vectors $\mathbf{X} = \{\mathbf{x}_1, \ldots, \mathbf{x}_m\}$ and $\mathbf{X}' = \{\mathbf{x}'_1, \ldots, \mathbf{x}'_m, \mathbf{x}'_{m+1}\}$ where $2\mathbf{x}'_i = \mathbf{x}_i$, $i = 1, \ldots, m$.

$$\boldsymbol{\mu}(\mathbf{X}') = \frac{m}{m+1}\left(\frac{1}{2}\boldsymbol{\mu}(\mathbf{X})\right) + \frac{1}{m+1}\mathbf{x}'_{m+1}$$

$$\sigma^2(\mathbf{X}') = \frac{1}{4}\sigma^2(\mathbf{X}) + \frac{m}{m+1}\left(\mathbf{x}'_{m+1} - \frac{1}{2}\boldsymbol{\mu}(\mathbf{X})\right)^T\left(\mathbf{x}'_{m+1} - \frac{1}{2}\boldsymbol{\mu}(\mathbf{X})\right).$$

Appendix 2: Broadcast Count

Transmission of a double precision real number is defined as a message in [19]. In this chapter, in addition to real numbers typically representing vector coordinates, integer values such as node ID and node "reporting order" should also be transmitted. Transmission of node IDs is needed, for example, to allow the root to cluster nodes. To minimize communication load nodes in smaller clusters report violations of node constraints first, and the reporting order is assigned and communicated to nodes by the root that knows all cluster sizes.

Since every vector **v** associated with a node belongs to a simplex, it is represented by a real number not exceeding 1. We may use the integer part of these real numbers for transmission of integers. There is a variety of coding and compression techniques that can be used to transmit a set of real numbers as a single real. The discussion of these methods is beyond the scope of this chapter. In order to be able to compare different monitoring techniques we shall count a number of broadcasts, where by a broadcast we mean a single communication between two nodes. As an illustration, below we compute the number of broadcasts needed for one iteration of Algorithm 12.3.1 triggered by violation of a node constraint (note that the communication scheme below is different from the one suggested in Algorithm 12.3.1). We first assume that the violator node **n** is different from the root.

1. The violator node **n** notifies all other nodes (except the root) about the violation ($n - 2$ broadcasts).
2. Each node **n** broadcasts its vector $\mathbf{v}_\mathbf{n}$ to the root ($n - 1$ broadcasts).
3. The root recomputes $\delta(\mathbf{r})$ and sends it to each node ($n - 1$ broadcasts).

This leads to $3(n - 1) - 1$ broadcasts. If the violator node **n** is the root itself, the number of broadcasts becomes $3(n - 1)$ (at step 1 above the root has to make $n - 1$ broadcasts).

Next we turn to monitoring with clustering. The monitoring procedure starts with each node **n** sending its initial vector $\mathbf{v}_\mathbf{n}(t_0)$ to the root **r** (that requires $n - 1$ broadcasts). The root computes the mean $\frac{1}{n}\sum_{\mathbf{n}} \mathbf{v}_\mathbf{n}(t_0)$ of the initial vectors, computes $\delta(\mathbf{r})$, and broadcasts $\delta(\mathbf{r})$ to each node ($n - 1$ broadcasts). After exchanging

$$2(n - 1) \tag{12.18}$$

broadcasts the monitoring proceeds with each node being a singleton cluster.

1. As long as the inequality

$$|\mathbf{v}_\mathbf{n}(t) - \mathbf{v}_\mathbf{n}(t_0)| < \delta(\mathbf{r}) \text{ holds true for each node } \mathbf{n}$$

the nodes are silent. At the first time instance t when the inequality is violated for at least one node $\mathbf{n}$, the following actions are triggered:

a. the node $\mathbf{n}$ (if the node itself is not the root) broadcasts its ID and vector $\mathbf{v_n}(t)$ to the root (1 broadcast),
b. the root issues $n-2$ requests for ID and $\mathbf{v_n}(t)$ to the other nodes ($n-2$ broadcasts),
c. $n-2$ nodes report their IDs and $\mathbf{v_n}(t)$ vectors to the root ($n-2$) broadcasts),

This brings the number of broadcasts to $2n-3$. If the violating node is the root, then this number is $2n-2$. To simplify the computations we select the largest number $2n-2$.

At this step, and keeping in mind (12.18), the total number of broadcasts needed to be exchanged is

$$2(n-1)+2n-2=4(n-1). \tag{12.19}$$

2. Next the root recomputes $\delta(\mathbf{r})$, clusters nodes, and broadcasts to each node ($n-1$ broadcasts) its updated local constraint $\delta(\mathbf{n})$, the ID of its coordinator, and the reporting order. If a node is also a coordinator, then IDs of its nodes, and coordinator reporting order are provided to the coordinator by the root. Keeping in mind (12.19), the total number of broadcasts right after the first root mean update and first clustering is

$$5(n-1). \tag{12.20}$$

Clusters are now formed, and we shall count the number of broadcasts needed to be exchanged for each of the three types of possible violations at time t assuming that clusters are formed last time at time t_k, $k=0,1,\ldots$.

1. *A node constraint is violated in a singleton cluster.*

a. the violating node $\mathbf{n}$ reports its ID, $\mathbf{v_n}(t)$, $W_\mathbf{n}$, and the history vector $\mathbf{h_n}$ to the root (1 broadcasts),
b. the root requests all other $n-2$ nodes to provide their input (ID's, $\mathbf{v_n}(t)$ vectors, $W_\mathbf{n}$ weights, and history vectors $\mathbf{h}$, total of $n-2$ broadcasts),
c. the $n-2$ nodes report ID's, $\mathbf{v_n}(t)$ vectors, $W_\mathbf{n}$ weights, and history vectors $\mathbf{h}$ to the root (($n-2$) broadcasts),
d. the root recomputes the constraint $\delta(\mathbf{r})$, node constraints $\delta(\mathbf{n})$, and reports to each node its coordinator ID, $\delta(\mathbf{n})$, and the node "reporting order." Cluster coordinators also receive IDs of the nodes in their respective clusters ($n-1$ broadcasts).

This leads to $3(n-1)-1$ broadcasts if the violating node is not the root, and $3(n-1)$ broadcasts if the violation is at the root. To compute the broadcasts we use the larger number

$$3(n-1). \tag{12.21}$$

2. *A node constraint is violated in a non singleton cluster π with coordinator* $\mathbf{c}$.
   a. the violator $\mathbf{n}$ reports its ID, $\Delta_{\mathbf{n}} = \mathbf{v}_{\mathbf{n}}(t) - \mathbf{v}_{\mathbf{n}}(t_k)$, and $\delta_{\mathbf{n}}$ to the coordinator $\mathbf{c}$ (1 broadcast),
   b. the coordinator $\mathbf{c}$ sends request for $\Delta_{\mathbf{n}}$ vectors and node constraints $\delta_{\mathbf{n}}$ for all nodes in its cluster π other then $\mathbf{n}$ and itself ($|\pi| - 2$ broadcasts)
   c. the nodes broadcast their vectors Δ and constraints δ to the coordinator (total of ($|\pi| - 2$) broadcasts). The total comes to $2|\pi| - 3$, and this number is $2|\pi| - 2$ when the violating node is the coordinator.

   The total of broadcasts needed is:

$$2|\pi| - 2. \tag{12.22}$$

3. *A coordinator constraint is violated.* First we assume the coordinator $\mathbf{c}$ is not the root:
   a. the coordinator $\mathbf{c}$ of cluster π broadcasts requests to all nodes (except itself and the root) to provide the root with their IDs, vectors $\mathbf{v}_{\mathbf{n}}(t)$, weights W, and history vectors $\mathbf{h}$ ($n - 2$ broadcasts).
   b. $n - 1$ nodes ($n - 2$ nodes requested by the coordinator and the coordinator itself) send the requested information to the root ($n - 1$ broadcasts).
   c. the root recomputes $\delta(\mathbf{r})$, clusters nodes and provides each node with updated local constraint $\delta(\mathbf{n})$, the new cluster affiliation (i.e. ID of a new coordinator), and the node "reporting order." Coordinators are also provided with the IDs of their nodes (total of $n - 1$ broadcasts).

   This brings the number of broadcasts to $3(n - 1) - 1$. If $\mathbf{c}$ is the root, then this number is $3(n - 1)$, and this is the number we use to compute broadcasts

$$3(n - 1). \tag{12.23}$$

Acknowledgements The authors thank the editor for bringing a number of significant relevant references to their attention. The research of the second author was supported by Grant No. 2008405 from the United States-Israel Binational Science Foundation (BSF). The work of the fourth author was partially supported by a 2013 UMBC Summer Faculty Fellowship grant.

References

1. Banerjee A, Merugu S, Dhillon IS, Ghosh J (2004) Clustering with Bregman divergences. In: Proceedings of the 2004 SIAM international conference on data mining, pp 234–245
2. Barouti M, Keren D, Kogan J, Malinovsky Y (2014) Monitoring distributed data streams through node clustering. In: Proceedings of the international conference on machine learning (MLDM'2014), July 21–24. St. Petersburg, Russia, Springer-Verlag Lecture Notes in Computer Science (LNAI), pp. 149–162
3. Boley DL (1998) Principal direction divisive partitioning. Data Min Knowl Discov 2:325–344

4. Brucker P (1978) On the complexity of clustering problems. In: Lecture notes in economics and mathematical systems, vol 157. Springer, Berlin, pp 45–54
5. Burdakis S, Deligiannakis A (2012) Detecting outliers in sensor networks using the geometric approach. In: ICDE, pp 1108–1119
6. Celebi ME, Kingravi H, Vela PA (2013) A comparative study of efficient initialization methods for the k-means clustering algorithm. Expert Syst Appl 40(1):200–210
7. Dhillon IS, Guan Y, Kogan J (2002) Iterative clustering of high dimensional text data augmented by local search. In: Proceedings of the 2002 IEEE international conference on data mining, pp 131–138
8. Dhillon IS, Kogan J, Nicholas C (2003) Feature selection and document clustering. In: Berry MW (ed) Survey of text mining. Springer, New York, pp 73–100
9. Diday E (1973) The dynamic cluster method in non–hierarhical clustering. J Comput Inf Sci 2:62–88
10. Diday E (1987) Some recent advances in clustering. Recent developments in clustering and data analysis. In: Proceedings of the Japanise–French scientific seminar, pp 119–136
11. Dilman M, Raz D (2001) Efficient reactive monitoring. In: Proceedings of the twentieth annual joint conference of the IEEE computer and communication societies, pp 1012–1019
12. Duda RO, Hart PE, Stork DG (2000) Pattern classification. Wiley, New York
13. Forgy E (1965) Cluster analysis of multivariate data: efficiency vs. interpretability of classifications. Biometrics 21:768
14. Gabel M, Schuster A, Keren D (2013) Communication-efficient outlier detection for scale-out systems. In: BD3@VLDB, pp 19–24
15. Garofalakis M, Keren D, Samoladas V (2013) Sketch-based geometric monitoring of distributed stream queries. In: PVLDB
16. Gray RM (1990) Entropy and information theory. Springer, New York
17. Keren D, Sharfman I, Schuster A, Livne A (2012) Shape sensitive geometric monitoring. IEEE Trans Knowl Data Eng 24:1520–1535
18. Kogan J (2007) Introduction to clustering large and high-dimensional data. Cambridge University Press, New York
19. Kogan J (2012) Feature selection over distributed data streams through convex optimization. In: Proceedings of the twelfth SIAM international conference on data mining (SDM 2012). SIAM, Anaheim, pp 475–484
20. Kogan J, Malinovsky Y (2013) Monitoring threshold functions over distributed data streams with clustering. In: Proceedings of the workshop on data mining for service and maintenance (held in conjunction with the 2013 SIAM international conference on data mining), pp 5–13
21. Madden S, Franklin MJ (2002) An architecture for queries over streaming sensor data. In: ICDE 02, p 555
22. Manjhi A, Shkapenyuk V, Dhamdhere K, Olston C (2005) Finding (recently) frequent items in distributed data streams. In: ICDE 05, pp 767–778
23. Mirkin B (2005) Clustering for data mining: a data recovery approach. Chapman & Hall/CRC, Boca Raton
24. Sharfman I, Schuster A, Keren D (2007) A geometric approach to monitoring threshold functions over distributed data streams, ACM Trans Database Syst 32:23:1–23:29
25. Sharfman I, Schuster A, Keren D (2010) A geometric approach to monitoring threshold functions over distributed data streams. In: May M, Saitta L (eds) Ubiquitous knowledge discovery. Springer, New York, pp 163–186
26. Späth H (1980) Cluster analysis algorithms for data reduction and classification of objects. Ellis Horwood, Chichester
27. Späth H (1985) Cluster dissection and analysis: theory, FORTRAN programs, examples. Ellis Horwood, Chichester
28. Steinhaus H (1956) Sur la division des corps matèriels en parties. Bull De L'Acadēmie Polonaise Des Sci Classe III Math Astronomie Physique Chimie Geologie et Geographie 4:801–804

29. Teboulle M, Berkhin P, Dhillon I, Guan Y, Kogan J (2006) Clustering with entropy-like k-means algorithms. In: Kogan J, Nicholas C, Teboulle M (eds) Grouping multidimensional data: recent advances in clustering. Springer, New York, pp 127–160
30. Yi B-K, Sidiropoulos N, Johnson T, Jagadish HV, Faloutsos C, Biliris A (2000) Online datamining for co-evolving time sequences. In: ICDE 00, p 13
31. Zhu Y, Shasha D (2002) Statestream: atatistical monitoring of thousands of data streams in real time. In: VLDB, pp 358–369

Printed by Printforce, the Netherlands